PHYSICS UP TO 200 TeV

THE SUBNUCLEAR SERIES

Series Editor: ANTONINO ZICHICHI, *European Physical Society, Geneva, Switzerland*

1. 1963 STRONG, ELECTROMAGNETIC, AND WEAK INTERACTIONS
2. 1964 SYMMETRIES IN ELEMENTARY PARTICLE PHYSICS
3. 1965 RECENT DEVELOPMENTS IN PARTICLE SYMMETRIES
4. 1966 STRONG AND WEAK INTERACTIONS
5. 1967 HADRONS AND THEIR INTERACTIONS
6. 1968 THEORY AND PHENOMENOLOGY IN PARTICLE PHYSICS
7. 1969 SUBNUCLEAR PHENOMENA
8. 1970 ELEMENTARY PROCESSES AT HIGH ENERGY
9. 1971 PROPERTIES OF THE FUNDAMENTAL INTERACTIONS
10. 1972 HIGHLIGHTS IN PARTICLE PHYSICS
11. 1973 LAWS OF HADRONIC MATTER
12. 1974 LEPTON AND HADRON STRUCTURE
13. 1975 NEW PHENOMENA IN SUBNUCLEAR PHYSICS
14. 1976 UNDERSTANDING THE FUNDAMENTAL CONSTITUENTS OF MATTER
15. 1977 THE WHYS OF SUBNUCLEAR PHYSICS
16. 1978 THE NEW ASPECTS OF SUBNUCLEAR PHYSICS
17. 1979 POINTLIKE STRUCTURES INSIDE AND OUTSIDE HADRONS
18. 1980 THE HIGH-ENERGY LIMIT
19. 1981 THE UNITY OF THE FUNDAMENTAL INTERACTIONS
20. 1982 GAUGE INTERACTIONS: Theory and Experiment
21. 1983 HOW FAR ARE WE FROM THE GAUGE FORCES
22. 1984 QUARKS, LEPTONS, AND THEIR CONSTITUENTS
23. 1985 OLD AND NEW FORCES OF NATURE
24. 1986 THE SUPERWORLD I
25. 1987 THE SUPERWORLD II
26. 1988 THE SUPERWORLD III
27. 1989 THE CHALLENGING QUESTIONS
28. 1990 PHYSICS UP TO 200 TeV

Volume 1 was published by W. A. Benjamin, Inc., New York; 2-8 and 11-12 by Academic Press, New York and London; 9-10 by Editrice Compositori, Bologna; 13-28 by Plenum Press, New York and London.

PHYSICS UP TO 200 TeV

Edited by

ANTONINO ZICHICHI

European Physical Society
Geneva, Switzerland

PLENUM PRESS • NEW YORK AND LONDON

Library of Congress Cataloging-in-Publication Data

International School of Subnuclear Physics (28th : 1990 : Erice, Italy)
Physics up to 200 TeV / edited by Antonino Zichichi.
p. cm. -- (The Subnuclear series ; v. 28)
"Proceedings of the twenty-eight course of the International School of Subnuclear Physics ... held July 16-24, 1990, in Erice, Sicily, Italy"--T.p. verso.
Includes bibliographical references and index.
ISBN 0-306-43935-2
1. Quantum chromodynamics--Congresses. 2. particles (Nuclear physics)--Congresses. I. Zichichi, Antonino. II. Title. III. Series.
QC793.3.Q35I54 1990
539.7'2167--dc20 91-20396
CIP

Proceedings of the Twenty-Eighth Course of the International School of Subnuclear Physics on Physics up to 200 TeV, held July 16-24, 1990, in Erice, Sicily, Italy

ISBN 0-306-43935-2

A Division of Plenum Publishing Corporation
233 Spring Street, New York, N.Y. 10013

Printed in the United States of America

PREFACE

During July 1990, a group of 83 physicists from 43 laboratories in 21 countries met in Erice for the 28th Course of the International School of Subnuclear Physics. The countries represented were: Algeria, Canada, Chile, China, Czechoslovakia, Denmark, France, the Federal Republic of Germany, Greece, Holland, India, Italy, Pakistan, Peru, Poland, Sweden, Switzerland, Turkey, the Union of Soviet Socialist Republics, the United Kingdom, and the United States of America. The School was sponsored by the European Physical Society (EPS), the Italian Ministry of Education (MPI), the Italian Ministry of University and Scientific Research, the Sicilian Regional Government (ERS), and the Weizmann Institute of Science.

The new topic discussed in some detail at the School was QCD phenomenology at 200 TeV. This energy frontier represents the goal for subnuclear physics after LHC and SSC. The main lecturers were T.D. Lee, L. Cifarelli, Y. Dokshitzer and A. Ringwald. There are two possibilities, one optimistic, the other pessimistic. LHC and SSC can produce remarkable discoveries. In this case, the need to go to higher energies will be obvious. In the pessimistic case, LHC and SSC will produce the strongest evidence for the validity of the Standard Model. The need for higher energies will be impelled. In neither case can we wait. If the 200 TeV frontier is to be reached within our lifetime we need to start now: thinking and preparing the tools. Theory and R&D for future detectors are the basic points of the 200 TeV frontier.

The first results from LEP were the other news of the School, presented by J. Steinberger. The group of theoretical lectures included some hot topics like "The Derivation of the Minimal Standard Model Lagrangian" (by R. Kleiss), "The Heterotic and Type II Superstrings" (S. Ferrara) and "The Symmetries of Extended Objects" (M. Duff). Stan Brodsky and Rinaldo Baldini discussed two interesting and highly specialized topics, i.e., "The Intrinsic Short-Distance Structure of Hadrons in QCD" and "The Time-Like Form Factor of the Neutron", a crucial test for QCD.

The School has in its tradition the memory of great events. This year "The Glorious Days of Physics" were devoted to the Higgs mechanism. And Peter Higgs was the speaker, to celebrate the quarter of a century that has passed since the publication of his famous paper.

The School closed with Victor Weisskopf's lecture on the "New Role of Science".

A unique feature of the School is the discussion sessions. They are reproduced as faithfully as possible thanks to the work of the Scientific Secretaries. I hope the reader will enjoy these discussions, as did all the participants in Erice. This volume is the effort of lecturers, Scientific Secretaries and collaborators of mine in Erice and Geneva. To all of them my deepest gratitude.

Antonino Zichichi

CONTENTS

THE FUTURE OF HIGH ENERGY PHYSICS

THE GLORIOUS DAYS OF PHYSICS

CLOSING LECTURE

CLOSING CEREMONY

PARTICLE PHYSICS IN THE NINETIES

S.L. GLASHOW

Harvard University
Cambridge, MA
USA

Whereas allegedly elementary particles were known to show a great variety of rather complicated properties, the properties of quarks and leptons are extremely simple; in fact they can be compared in their simplicity to geometrical points. Also, instead of a large number of seemingly elementary hadrons, we are left today with only six essentially different quarks. And in spite of the greatest desire and effort to reduce everything to its simplest form, one cannot possibly reduce something to nothing. Thus, it seems that we have actually hit the bottom in our search for the basic elements from which matter is formed.

The above paragraph is a free paraphrase from George Gamow's book *One, Two, Three, Infinity* which was published in 1947. It was the book that got me hooked on fundamental physics while I was a high-school student. Of course, Gamow was writing about how the chemical elements and their behavior could be reduced to the study of the 'elementary particles' of half a century past — nucleons, electrons, neutrinos, pions and photons. However, in the very year that Gamow's book was published, things began to become very much more complicated. Yukawa's strongly interacting particle was found to decay into a muon: an entirely unanticipated new particle. Rochester and Butler, by studying emulsions exposed on mountain tops, discovered several new particles with curious properties: the strange particles. Hundreds of hadrons would be discovered when large accelerators and their bubble chambers were deployed. Only recently did simplicity emerge once again, so that the preceding paragraph, suitably adapted, could be recycled. Perhaps there is a lesson to be learned beyond *plus ça change, plus c'est la meme chose.*

Today is a time of immense promise, great accomplishment, and growing frustration. The Fermilab collider is busily exploring the two TeV energy frontier. LEP is accumulating and studying millions of Z decays. The first electron- proton collider is about to be launched. Three giant hadron colliders are contemplated:

Physics Up to 200 TeV
Edited by A. Zichichi, Plenum Press, New York, 1991

the LHC, the SSC and the Eloisatron. And Japan's stated highest priority in high-energy physics is the creation of an immense electron-positron linear collider. Exciting new developments are sure to emerge. Within a decade or two, the greatest mystery of the standard model will be solved: the nature of electroweak symmetry breaking will be revealed. This is the time to study particle physics.

The successful unification of the weak and electromagnetic forces pales in comparison with today's greater unification: that of the immeasurably small with the inconceivably large. Cosmology and particle physics are becoming one. What astrophysicsts told us about supernovas has been proven to be true by particle physicists. Cosmological deductions about the number of neutrino species have been verified by experiments performed at LEP and at SLC: only last year, we found (to our dismay) that there are only three conventional families of quarks and leptons. Neonatal cosmologists, using what we have learned about particles, are well on their way to explicating the birth and development of the Universe.

Today, we have an apparently complete, correct and consistent theory of elementary particle physics the so-called minimal standard model. It consists of Quantum-ChromoDynamics and the Electroweak Theory. It is a theoretical system based upon the gauge group $SU(3)\times SU(2)\times U(1)$. The minimal version of the theory involves the existence of precisely three families of fundamental fermions, each containing two quarks and two leptons. Each of the three neutrinos is massless, and the breakdown of electroweak symmetry results in the appearance of a single Higgs boson. All told, there are 17 specimens in today's particle bestiary: three quarks with electric charge $\frac{2}{3}$, three with charge $\frac{-1}{3}$, three charged leptons, three neutrinos, the W, the Z, the photon, the gluon and the elusive Higgs particle. Only two of these particles remain to be seen and cataloged by experimenters: the top quark (known to be heavier than $\sim$ 90 GeV), and the Higgs (known to be heavier than $\sim$ 25 GeV). In the most unlikely case that our list of particles is complete, we are 88.2% to the end of particle physics! Not to worry: Nature surely has more surprises in store for us.

All observed particle phenomena seem to be explicable in terms of the minimal standard model. There are no confirmed flaws or anomalies in sight. Many searches have been dedicated to the search for 'new physics' lying beyond the scope of the standard model: but they have all failed:

1) Neutrinoless double beta decay has not shown up.
2) No fractionally charged particles are seen.
3) No magnetic monopoles are found.
4) There is no evidence for neutrino masses. Both the tritium endpoint experiment and the supernova neutrino data constrain ν_e to weigh less than $\sim 10eV/c^2$.
5) There is no laboratory evidence for neutrino oscillations. (There is however, a 'solar-neutrino problem.' More on this anon.

6) No one has seen wimps, axions, cosmions, champs, tachyons, quixes, queights, charged scalars, axigluons, *et nauseum.*
7) There are no convincing cosmic ray anomalies.(What ever happened to the 'Centauro events'? What of the muon-rich events from Cygnus?)
8) There are not any longer any inexplicable 'monojets.'
9) Right-handed currents produce no measureable effects.
10) There is no indication of a second W or Z.
11) Forbidden decays seem to be truly forbidden, as experimental limits grow ever more impressive.
12) The 'fifth force' (along with cold fusion) was a mirage.
13) are no confirmed departures from the predictions of the minimal standard model.

Furthermore, the theory seems to be mathematically sound (at least for conceivably accessible particle energies), with one important exception — the so-called strong CP problem. The tininess of CP violating effects (especially the immeasurably small neutron electric dipole moment) is unnatural from the viewpoint of the minimal standard model. This could be rectified if the up-quark mass happens to vanish: however, this is a phenomenologically unsatisfactory alternative. Axions could have solved the problem, but there are no axions that are not 'invisiblized'. The invisible axion is a viable (but contrived) retrofit. Other theoretical alternatives have been suggested. (*E.g.*, I received the latest suggestion by Frampton and Kephart just last week.) It may be that CP violation is the clue that will lead us to a new and better theory. More experimental work is required. The B-meson may display CP violating effects which cannot be understood from our present framework. This curiously low-energy arena may turn out to be our most important new frontier. We must learn to do precision B physics at colliders, or we must build dedicated B factories. (And, we must resolve the present contradictory experimental results on the ϵ'/ϵ ratio in K decay, and determine or better constrain the electric dipole moments of electrons and neutrons.)

The 'solar neutrino question' offers another potential challenge to the standard model. Over the past two decades, Ray Davis has measured the flux of high-energy neutrinos emitted by the sun. His result is significantly smaller than the prediction of stellar-structure calculations. (Furthermore, the result to the chlorine experiment has been confirmed recently at Kamiokande.) The disparity has two possible explanations: It may have to do with the nature of the sun, or with the nature of neutrinos. The highest energy solar neutrinos are produced by a secondary reaction which is exceedingly sensitive to temperature. The flux would be halved if the central solar temperature were only 3.7% smaller than it is thought to be. The most promising neutrino-related explanations are (1) matter-enhanced (so-called MSW) neutrino oscillations, or (2) an unexpectedly (and inexplicably) large neutrino magnetic moment. Two large Gallium experiments have been deployed, and their results will soon clarify the situation. Other experiments (especially at the proposed Canadian Sudbury Solar Neutrino facility) are planned. We shall soon get to the bottom of this mystery. For the record, I predict that no new physics beyond the standard model is required: the problem is likely to be one of stellar structure, not of neutrino physics. We shall soon find out.

Departures from the minimal standard model may show up at existing high-energy accelerators. The Tevatron Collider has just begun its exploration of physics at two TeV. New phenomena may well be discovered. Some possibilities include: (1) long-lived heavy hadrons made up of exotic colored fermions such as quixes or quarks with unusual electroweak assignments; (2) exotic vector bosons such as axigluons (reflecting chiral color), massive 'gluons' associated with other extensions of the color group, or additional W's or Z's associated with an extended electroweak group; or, (3) new particles associated with technicolor or low-energy supersymmetry. Of course, the search for the top quark will be extended, and possibly the Higgs boson can show up.

The possibilities for LEP are even more exciting. While data taken at the Z pole are roughly compatible with standard model predictions, there are still several exotic Z decay channels that are not absolutely ruled out: A new $Q = \frac{2}{3}$ 'up' quark which is an electroweak singlet. Such a particle would contribute $\sim$100 MeV to the Z width. A singlet 'down' quark would contribute a mere 25 MeV, and a singlet $Q = \frac{-1}{3}$ quix $\sim$50 MeV. In these cases, the associated exotic hadrons may be expected to be *stable* on the time scale of the detector. The events will be of low multiplicity, of balanced transverse momentum, but of considerable missing energy. Another possibility, recently stressed by Anne Nelson, is the existence of a few-GeV neutral component of a second scalar doublet. Its decay width should be $\sim$ 85 MeV, and these events will consist of exceedingly narrow low-multiplicity hadronic jets. They might be confused with Z decay into tau pairs. To eliminate the above-mentioned exotic possibilities, the Z total width must be measured with 95% confidence to an accuracy of one percent. This point should be reached in mere months.

Everyone is looking forward to LEP running at its design energy. We shall learn of the existence of any new charged particle with a mass up to 100 GeV. Furthermore, we may put to rest speculations about a second low-lying Z. (While a second Z which is coupled to $B{-}L$ is very highly constrained by low-energy neutrino-scattering data, there are other interesting possibilities that are not yet ruled out: Two examples are the Foot-Hernandez scenario, in which color $SU(3)$ is placed within a spontaneously broken low-energy $SU(5)$ group, and the Frampton-Kephart aspon. Neither of these models is seriously constrained by low-energy data.)

In summary, here is a list of the central problems facing the discipline of elementary particle physics:

The Fermi Problem is the gap between potential and realization, between what is known in principle and what is knowable in practice. With an allegedly correct and complete theory, we ought to be able to predict the results to experiments. All hadronic properties (like the mass spectrum of baryons and of mesons) should be computible in terms of α_s and the light quark masses. We must develop superior computational techniques. The most promising possibility would seem to be computer-assisted lattice field theory, although improved analytic procedures may also be developed. The present situation, in which only a few generic properties of hadrons are determinable, is intolerable for a theory that purports to be the last word. Physicists must create the successors to Fermi's Golden Rules.

The Rabi Problem, on the other hand, relates to questions that the standard model does not even confront. In a story that may be apocryphal, I.I. Rabi was lunching a long, long time ago, with his colleagues at a Chinese restaurant. Upon being told that the muon had been shown to be simply an obese variety of electron, the first particle with no apparent relevance to atomic structure and behavior, Rabi is said to have asked: *Who ordered that?* Half a century later, we still have no idea. Why are there three families of fermions, not two or four? Why are there families at all? Why are the gauge groups what they are? Why are the masses of the particles what they are? The minimal standard model involves seventeen arbitrary numerical parameters, while our faith in simplicity demands that the theory of physics be unique, or at least, more unique. Should we not be embarrassed that we cannot compute the masses of the top quark and the Higgs boson (or, the K.-M. matrix) from all that is known so far?

The Landau Problem is more of a philosophical than practical difficulty. Is it sufficient to have a theory which answers questions at energies that are accessible to us now (or in the future), but a theory that fails at inaccessibly high energies? Surely, we should be able to answer even those questions we cannot ask. This problem is especially pertinent as cosmology and particle physics grow closer together. If we are to understand the birth and development of our universe, surely we must be able to comprehend the earliest stages of the Big Bang, when temperatures and densities were so great that quantum gravity played a role.

The Einstein Problem, or, the need for a quantum theory of gravity. Our standard model deals with only three of the four known forces. However, quantum mechanics is contagious, and gravity cannot avoid quantization, even though the particulate nature of gravity - the graviton - can never be detected. (A galactic mass of neutral pions, in decaying, would be unlikely to result in the production of a single hard graviton. If it were somehow produced, such a particle would pass through a galactic mass of one-mil wire without a single scattering. Nonetheless, a quantum theory of gravity must be found to satisfy the purists among us. Here is where string theory has been most successful, for it seems to present us with a consistent quantum gravity. Unfortunately, string theory has not yet descended to low energy to recapture any of the successes of the standard model.

The Zwicky Problem: What is the nature of the dark mass of the universe? The ancient Greeks were right once again. They believed that the heavens were made up of a fifth element: the quintessence. Modern astronomy confirms the fact that most of the matter of the universe is not in the form of stars, or gas, or dust. It is something else, perhaps something not in our present list of nature's building blocks. The range of uncertainty in the mass of dark matter candidates is mind-boggling: from axions of mass 10^{-8} eV/c^2 to black-holes with mega-solar masses. Cosmions and most wimp models seem to have been excluded by recent experiments. Neutrinos (muon or tau) remain a viable explanation.

The 80's have been a remarkably dull decade. Finding the weak intermediaries (particles whose existence the *avant garde* never doubted) was hardly exciting. Learning to our dismay that there are only three families of fundamental fermions was no better. Another decade or two like that and our discipline will achieve the maturity of atomic physics, nuclear physics and chemistry. Our experimental colleagues had better hurry up and find something that doesn't fit into our pat minimal standard model. Wish them luck...

Chairman: S.L. Glashow

Scientific Secretaries: P.V.K.S. Baba, S. Hsu, J.F. Li, H.A. Rizvi

Discussion

– *Brahm:*

Measurements of ρ put the mass of the top quark $\leq$ 200 GeV. Tevatron expects to probe m_t up to 130 GeV in the next few years. Assuming minimal standard model, isn't 2076 a bit pessimistic for the discovery of the top quark?

– *Glashow:*

True, the minimum standard model predicts that 100 GeV $\leq m_t \leq$ 200 GeV. However, the next Tevatron run would bring the limit to 130 GeV in 1992. Perhaps I have been pessimistic. Maybe we will get to 200 GeV before the bimillenium.

– *Brahm:*

I am making an assumption concerning the continuing progress in technology so that the Tevatron would be probing 130 GeV in few years and even higher energies afterwards.

– *Glashow:*

I do not know if there is much chance in pushing beyond 130 or 150 GeV with a sufficiently high luminosity. It looks highly uncertain that they will get the necessary funding to increase the luminosity by a factor of 5. But if it is so, you may be right and we may have the top quark in the near future. It is also not inconceivable to me that the top quark has a mass lighter than what people suppose it to be (- 90 GeV) or even less. But this is a contentious subject in which I am not involved. According to Frank Wilczek, if the top quark is strongly coupled to a very light scalar particle (not the Higgs), then there are infra-red divergences when the top quark decays and one or more scalar mesons can share the energy. This scenario lies beyond the standard model because it involves another light scalar. Present top searches fail if no hard leptons are produced in top decay. These calculations are in progress. There is an amusing possibility that the mass of the top quark is 50 or 60 GeV rather then the much higher value most of us anticipate. So that is another way in which the problem can be solved.

– *Brahm:*

In the minimal standard model, is it not true that the LHC or SSC would find the top quark if its mass is around 200 GeV?

– *Glashow:*

The big machines may be just too big to find the top. Anyway, neither of them yet exists. A machine too powerful might just be unable to extracate the top quark. The optimal machine for such a search would be a 2 or 3 TeV machine working at high luminosity. An upgraded Tevatron is the machine of choice.

– *Grossmann:*

You mentioned that it might be interesting to have answers to questions which <u>in principle</u> cannot be answered by experiments ("the Landau Problem"). Don't you think that these would be questions of mathemathics rather than physics?

– *Glashow:*

You are saying that it would be mere mathemathics to answer questions about experiments that we cannot do. On the other hand, the universe was once a very very hot place, and to be able to understand its earliest history, one has to know what happens at very high temperatures and energies. So it is not just an epistemological question to have a theory that can describe experiments at arbitrarily high energies. It is also philosophically nice to be able to answer such questions: the scattering of one black hole upon another is certainly not pure mathematics.

– *Grossmann:*

If it has consequences for cosmology there would be an experiment to check those questions. I earlier misunderstood them to be questions that are in principle unanswerable and that for me is not natural science.

– *Glashow:*

Let us put aside the universe for a while. If we have a theory that says there would be a violation of causality and divergences at energies beyond our capacity to reach, it would be a very unsatisfactory situation and I can perfectly see devoting ones scientific endeavor remedying the situation. But I agree with you that there are lots of down-to-earth problems concerning the particles we see that should come first.

– *Wexler:*

Will the future successful theory of particle physics that addresses some of the questions you posed be a quantum field theory?

– *Glashow:*

Almost certainly not. I do not see in the cards answers to certain questions that I would like to ask; like why the masses of the particles are what they are, and why do they fit into the families such as they do? Why are the groups nature has chosen, SU(3), SU(2) and U(1)? I cannot imagine answers will come to those questions from a pure quantum field theory. On the other hand, string theorists were under the belief (and some still are) that they can answer questions of that sort. I have confidence that string theory is not the right direction to answer these questions, but there may be another direction. After all, we came upon string theory in an incidental and accidental fashion, and there is almost certainly another, better theory that will have the answers to all these questions.

– *Wexler:*

Looking at this, do you think this other theory will have fewer or more degrees of freedom?

– *Glashow:*

What do you mean by a degree of freedom?

– *Wexler:*

I mean a field at each point in space and time.

– *Glashow:*

Whatever the system, it will have infinite degrees of freedom. About thirty years ago, people like Bohm and Finkelstein thought that the error we were making that led to infinities and other puzzles was the continuum; that we were using real numbers which were the invention of human mind and not the proper way to describe physics. There were attempts to formulate a purely algebraic view of physics that would involve much less complex and less sophisticated mathematical structures. Those ideas were very fascinating but they have not gone anywhere. I do not know who, if anyone is trying to pursue these ideas today except a few computer scientists who regard the universe as a universal computer. But they are also not getting anywhere. They may be other attempts but I do not know about them. The designer of a correct theory of everything has almost certainly not yet been born.

– *Wexler:*

Do you think it might involve not string theory but some other theory of extended objects?

– *Glashow:*

The algebraists go precisely in the opposite direction. They are not dealing with real numbers, coordinates or continuous coordinates. They are working in a discrete algebraic system much more premature than a metric space. Their aim is (or, was) completely different from people who describe objects in terms of strings or membranes. They are aiming towards more primitive mathematical systems which are not compact bodies in metric space.

– *Wexler:*

Do you think that is the way to go to fewer degrees of freedom?

– *Glashow:*

I have no idea. I do not know the right direction. I think that theorists of the future will have to be familiar with lots of different branches of mathematics.

– *Monch:*

What do you think about the ideas of grand unification as suggested by yourself in the past?

– *Glashow:*

According to an Indian theorist I recently encountered, everyone has overestimated the matrix element for proton decay. This fellow did his calculations in 1983 and showed (he says) that the proton lifetime comes out a thousand times longer. This publication exists though I have not checked it myself. It would be nice, however, if everybody had made this error and SU(5) turns out to be right. There is something very elegant and pretty about SU(5); it is a minimal unification. I was very disappointed when people at Kamiokande and the Morton-Thiokol Salt mines did not see it. I like what Dr. Mitter said: that we had never calculated a process in which the proton disappears and that there were fundamental problems in computing the normalization of the matrix elements which he claimed to have done correctly himself. I hope he is right.

– *Zichichi:*

What was the lifetime of the proton in SU(5)?

– *Glashow:*

It was 3×10^{29} years and Mitter brings it up to 10^{32} or 10^{33} years which is acceptable. The experiments are running and even though they are primitive, they will be able to detect the lifetime of this scale in a few years. I should mention here that the Japanese are upgrading the Kamiokande experiment (what is called the super Kamiokande) to 20 times its present volume. I do not know if this upgrade is funded.

– *Brahm:*

Is it true that SU(5) is not ruled out by proton decay, only the non-supersymmetric SU(5) is?

– *Glashow:*

Supersymmetric SU(5) is not SU(5), but your statement is probably correct. Supersymmetric SU(5) is a different theory for which I take neither pride or responsibility.

– *Zhang:*

Some people believe that in the energy range up to 10^{15} GeV, no new physics exists. Do you think we can get some exciting experimental results from the next generation machines?

– *Glashow:*

No new particles have shown up for quite a long time, and the most minimal model suggests no new particles except for Higgs boson and top quark. But Nature may not be a minimalist. Who knows that there is a desert and that there are no new particles? Of course, there are lots of new particles beyond our ken. I would look forward to somebody finding some, since the minimal standard model must be taken as a rock that is our duty to break. Progress is the search for things that do not fit into our picture.

– *Brown:*

You hinted in your lecture that you thought it unlikely that superstrings will provide the correct theory of quantum gravity. Would you expand why you think that is.

– *Glashow:*

I am no superstring expert. What I said was that though the superstrings apparently do offer a consistent quantum gravity, they do not answer any of the questions that they originally set out to answer. They do not make contact with the low energy world of electroweak forces and of quantum chromodynamics; they do not explain why the particles that exist do exist; they do not explain why there are electroweak and color forces. They do not even tell us the dimensionality of the space and time. People who have played lightly with the string theories and have not invested much of their professional experience in them, have generally given up: Weinberg has, Ed Witten has become an excellent mathematician. The field is turning away from particle physics to the realms of early cosmology , abstract mathematics, and other domains where it cannot be readily tested, but it is not providing the theory of everything that in the early halcyon days it promised. The recent literature on string theory is totally removed from particle physics. Who cares about quantum gravity and gravitons we may never see.

– *Hsu:*

You referred to the strong CP problem as strong evidence of physics beyond the standard model. Since you seem to be a fan of the standard model perhaps you could comment on a possible solution on a strong CP problem in the standard model (ie) M_d or $M_u = 0$.

– *Glashow:*

Of course, the simplest solution to the strong CP problem would not be to put $M_d = 0$ but to a phenomenologically less dangerous solution $M_u = 0$. Howard Georgi wrote of and rejected such a possibility long ago. Experiment does not support the hypothesis of a massless up quark. Almost nobody believes in a massless up quark resolution to the strong CP problem. The "invisible" axion scenario is far more popular. I do not think that the problem of CP violation is finished. Experimental surprises may emerge from precision experiments on B-physics. The amount of $\mathrm{B} - \overline{\mathrm{B}}$ mixing is alarmingly large, and so may be the extent of CP violation in the B-meson sector. B-physics may force us to reconsider the minimal standard theory.

– *Zichichi:*

An excellent B-factory is LEP at Z^0 peak.

– *Ypsilantis:*

Not enough B's.

– *Zichichi:*

10^{32} luminosity gives millions of B's.

– *Ypsilantis:*

It has been established that in order for a B-factory to be competitive to be a dedicated B-factory, it has to have the same luminosity , so LEP has to go to $10^{33} - 10^{34}$ before it does the kind of the physics that the B-factories are supposed to do.

– *Glashow:*

How many B's are produced at Cornell?

– *Ypsilantis:*

At the moment it is very low because their luminosity is 10^{32}, it is about 10^5/year.

– *Glashow:*

LEP may be useful, but only if it can reconstruct the B-decay, an unlikely possibility.

– *Zichichi:*

But in the future when LEP is upgraded 10^8/year. How can you compete?

– *Ypsilantis:*

Hadron colliders will do it, as you will demonstrate later.

– *Glashow:*

I am encouraged to see the response for the B-physics is positive, because that may be the domain in which the hints are hiding, from which the next standard model may emerge.

– *Borden:*

Is it possible that the fact that string theory has not answered many questions is due to the deficiency of string theorists and not the string theory?

– *Glashow:*

It is an interesting question which I never heard before. I have the highest respect for the prople who are doing string theory since they are very brilliant and they know physics and they know what they are doing. My feeling is that the fault lies with the theory, not the theorists.

– *Borden:*

They may be very smart, but is it possible that there do not exist mathematical toos, or the previous experience to solve some of the problems. So sometime in the future when these tools are developed string theory will be able to answer the questions that you are asking for.

– *Glashow:*

What is going to happen to the string theory is that it (like classical general relativity) is going to survive for a long time in various contained pockets rather like tuberculosis nodules here and there in lungs throughout the world.

– *Lu:*

In light of its explanation of anomaly cancellation and charge quantization, shouldn't we pay more attention to the search for right-handed neutrinos (as predicted in models such as SO(10)) rather than exotic partciles like axions, wimps, superpartners etc.?

– *Glashow:*

HERA will study electron-proton collisions. It will search for weak right-handed currents. This is the one thing it can do better than any other experiment. Our experimental friends are quite smart. They will find right-handed currents if they exist.

– *Ciuchini:*

In your list of open problems in he standard model, you did not mention the KM matrix. Do you not think that it is an open question to predict these matrix elements, possibly in connection to the mass problem?

– *Glashow:*

I'm sorry. I usually point out that the standard model has 17 dimensionless parameters, among them the the elements of the KM matrix. We are not happy with a theory with 17 parameters. In analogy with a television set, with our universe as "the greatest show on earth", why should it have 17 controls? It should come factory perfect. The four parameters in the KM matrix play the same role as the various particle masses. or other masses. They have to be measured. However, in the standard model, they cannot be calculated. Thus, the standard model is not good enough.

– *Kastening:*

How well do you think we understand the standard model? I'm not only thinking in terms of the strong CP problem or problems in QCD, but rather about non-perturbative effects in the electroweak interactions, resulting in strong interactions and a breakdown in perturbation theory at not so high energies.

– *Glashow:*

I do not think that there are so many things we do not understand about the standard model. Our Russian friends seem to think that the electroweak force becomes strong and non-perturbative at Eloisatron energies. I have no idea whether they are talking sense or nonsense.

– *Larsen:*

This morning you stated that it might be that nothing is found at SSC or even in Eloisatron. Nevertheless, wouldn't one expect that something must happen below, say 1 Tev. Could you review the reasons for believing something to happen soon?

– *Glashow:*

Soon is a relative term. The LHC cannot operate before 1995, the SSC before 2000, and the Eloisatron before 2005. These are optimistic estimates. Detectors will have to be developed that are superior to any in existence. The secret of electroweak symmetry breaking will surely be revealed, but it will take a long time. Meanwhile, what should experimenters do in the next few decades? It is still my hope that exciting discoveries will be made at the Tevatron, at LEP, at B-factories (existing and proposed), and perhaps even at HERA.

– *Brahm:*

Can string theory tell us anything about classical gravity? In particular, anything that has astrophysical or cosmological consequences?

– *Glashow:*

So they say.

– *Hsu (comment):*

There are 'stringy' corrections to classical general relativity which would yield Brans-Dicke type theories at low energies. Modifications such as this have implications for inflation as well as producing long range forces which may be detectable.

– *Glashow:*

Hsu brings us back to the fifth force. We should quit now before the discussion turns to cold fusion.

LIGHT AND HEAVY QUARK JETS IN PERTURBATIVE QCD

Yu. L. Dokshitzer

Leningrad Nuclear Physics Institute
188350 Gatchina, Leningrad, USSR

1. Introduction

The aim of this lecture is to continue the discussion of the Perturbative approach to QCD jet physics[1)] which has been started by the lecture at the Erice School 1989[2)] to which an interested reader is referred for Basics.

Experimental studies of multihadron production at both e^+e^- and hadronic colliders strongly support the hypothesis of the Local Parton-Hadron Duality (LPHD)[3),4)]. According to the LPHD which is naturally connected to the "preconfinement" properties of QCD[5)], one can expect similarity between observable characteristics of hadron spectra and those calculated at the parton level by means of perturbative (PT) QCD.

PT approach to jet physics is based on the so called Modified Leading Logarithmic Approximation (MLLA) which takes care of both double logarithmic and single logarithmic effects in the development of parton cascades in a systematic way[3),6)]. It attempts to describe the gross features of the hadronic systems, such as multiplicity distributions and fluctuations, angular pattern of particle flows, inclusive energy spectra and correlations of particles, etc. without invoking phenomenological fragmentation schemes. In this approach *nonperturbative effects* are reduced to normalizing coefficients relating hadronic characteristics to partonic ones, which, according to the LPHD, must be independent of both the hardness of a jet generating process and the energy range of particles under study.

Here we shall consider some new developments connected with QCD features of jets generated by **heavy quarks** and discuss recent experimental LEP results revealing one of the brightest QCD predictions, namely, the coherent **hump-backed plateau** in inclusive particle spectra, together with some new problems arising from the new knowledge.

Physics Up to 200 TeV
Edited by A. Zichichi, Plenum Press, New York, 1991

2. MLLA Hump-backed Plateau

The MLLA Evolution Equation for particle spectra reads:

$$\frac{d}{d\ln\Theta}\, x\overline{D}_A^B(x,\ln E\Theta) \;=\; \sum_{C=q,\bar{q},g}\int_0^1 \frac{dz}{z}\,\frac{\alpha_s(k_\perp^2)}{2\pi}\,\Phi_A^C(z)\left[\frac{x}{z}\overline{D}_C^B\left(\frac{x}{z},\ln zE\Theta\right)\right]\,, \tag{1}$$

where E, Θ are energy and opening angle of a jet A and the restriction $k_\perp \approx z(1-z)E\Theta \geq Q_0$ terminates PT cascades at the transverse momentum scale Q_0 which regularizes collinear and soft singularities and sets a formal boundary between the perturbative and nonperturbative phases of jet evolution.
Functions $\Phi_A^C(z)$ stand for the **regularized** Gribov-Lipatov-Altarelli-Parisi (GLAP) parton splitting functions[7] which play the role of *kernels* of the Evolution Equation (1). Introducing notations: $\ell = \ln(E/k) = \ln(1/x)$, $Y = \ln(E\Theta/Q_0)$ one arrives at a compact integro–differential equation:

$$\frac{d}{dY}\,D(\omega,Y) = \int_0^\infty d\bar{\ell}e^{-\omega\bar{\ell}}\Phi(\bar{\ell})\,\frac{\alpha_s(Y-\bar{\ell})}{2\pi}\,D(\omega,Y-\bar{\ell}) \tag{2}$$

for the Mellin–transformed distributions:

$$D(\omega,Y) = \int_0^1 \frac{dx}{x}x^\omega\left[x\overline{D}(x,Y)\right] \equiv \int_0^\infty d\ell\,\overline{D}(\ell,Y)\,e^{-\omega\ell}\,. \tag{3}$$

Diagonalization of the matrix of GLAP kernels $\Phi(\omega)$ in the ω-representation results in the two "trajectories"[7]:

$$\nu_\pm(\omega) = \frac{1}{2}\left(\Phi_G + \Phi_F \pm \sqrt{(\Phi_G-\Phi_F)^2 + 8n_f\Phi_F^G\Phi_G^F}\right) \tag{4}$$

that determine the anomalous dimensions of two operators arising from mixing of g and q states in the cascade. At $x \ll 1$ the trajectory $\nu_+(\omega)$, singular at $\omega = 0$,

$$\nu_+(\omega) = \frac{4N_c}{\omega} - a + O(\omega), \quad a = \frac{11}{3}\,N_c\,+\,\frac{2}{3}\,\frac{n_f}{N_c^2}\,, \tag{5}$$

gives the main contribution to $\overline{D}(x,Y)$.
The following chain of transformations expresses Eq. (2) in terms of the known LLA trajectories with the *differential operator* $j = \omega + d/dY$ as an argument of $\Phi(\omega)$:

$$\begin{aligned}\frac{d}{dY}\,D(\omega,Y) &= \left(\int_0^\infty d\bar{\ell}e^{-\bar{\ell}(\omega+d/dY)}\Phi(\bar{\ell})\right)\frac{\alpha_s(Y)}{2\pi}\,D(\omega,Y)\\ &= \Phi(\omega+d/dY)\,\frac{\alpha_s(Y)}{2\pi}\,D(\omega,Y)\,,\end{aligned} \tag{6}$$

or after the diagonalization:

$$\frac{d}{dY}\,D^\pm(\omega,Y) = \nu_\pm(\omega+d/dY)\,\frac{\alpha_s(Y)}{2\pi}\,D^\pm(\omega,Y)\quad. \tag{7}$$

Equation (7) can be used to derive the *anomalous dimension* which we may define as:

$$\gamma_\omega\left(\alpha_s(Y)\right) \equiv \frac{d}{dY}\ln \overline{D}(\omega, Y)\ . \tag{8}$$

Within the DLA accuracy one substitutes $\nu_+(j) = 4N_c/j$ in Eq. (7) and easily gets the known anomalous dimension:

$$(\omega+\gamma)\gamma - \gamma_0^2 = 0, \quad \gamma^{DLA}(\omega, \alpha_s) = \frac{1}{2}\left(-\omega + \sqrt{\omega^2 + 4\gamma_0^2}\right)\ ; \tag{9a}$$

where:

$$\gamma_0 = \gamma_\omega(\alpha_s)|_{\omega=0} \equiv \sqrt{4N_c\,\frac{\alpha_s}{2\pi}}\ . \tag{9b}$$

Notice, that it is γ_0 proportional to the square root of running coupling, and not to α_s itself, which plays the role of the true PT expansion parameter for small-x problems. Approximation (5) generates MLLA anomalous dimension accounting for α_s corrections to Eq. (9b).

Derivation of the analytical solution of Eq. (7) has been presented in Ref. [2]. This solution, with account of initial conditions, is given in the physical region of particle energies $0 \le \ell \le Y$ by:

$$x\overline{D}_i^f(x, Y, \lambda) = \frac{4N_c(Y+\lambda)}{bB(B+1)}\int_{\epsilon-i\infty}^{\epsilon+i\infty}\frac{d\omega}{2\pi i}x^{-\omega}\Phi(-A+B+1, B+2, -\omega(Y+\lambda))K(\omega,\lambda) \tag{10a}$$

where:

$$K(\omega, \lambda) = \frac{\Gamma(A)}{\Gamma(B)}\,(\omega\lambda)^B\,\Psi(A, B+1, \omega\lambda)\cdot C_i^f \tag{10b}$$

where Φ and Ψ stand for *confluent hypergeometric functions* and the following notations have been used:

$$A = \frac{4N_c}{b\omega}\ , \qquad B = \frac{a}{b}\ , \qquad \lambda = \ln\frac{Q_0}{\Lambda}\ . \tag{11}$$

Indices i and f in Eq. (10) stand for the initial parton generating the jet ($i = q, g$) and final one spectrum of which is studied ($f = q, g$). In the leading approximation coefficient functions C_i^f are simply $C_g^g = 1$ and $C_q^g = C_f/N_c = 4/9$.

2.1. Developed QCD cascade and LPHD hypothesis

The parton spectrum Eq. (10) has an interesting property supporting the LPHD hypothesis. Namely, at very high energies when the typical value of the product $\omega\lambda \sim \lambda/\sqrt{Y}$ in Eq. (10) becomes small, the *shape* of the spectrum gets insensitive to the value of the transverse momentum cutoff parameter Q_0:

$$K(\omega, \lambda) \approx \frac{2}{\Gamma(B)}\left(\frac{z_0}{2}\right)^B K_B(z_0) = \text{const at}\quad \omega\lambda \ll 1; \qquad z_0 \equiv \sqrt{\frac{16N_c\lambda}{b}}\ . \tag{12}$$

Therefore when the bremsstrahlung cascade is developed enough one can study the shape of particle spectra with use of simplified version of the main formula (10) where one takes $\lambda = 0$. In this limit factor (12) totally disappears: $K(\omega, 0) = 1$ and, omitting C_i^f factor, we arrive at the expression which we shall refer to hereafter as the **limiting spectrum**:

$$\overline{D}^{\lim}(\ell, Y) \equiv x\overline{D}^{\lim}(x, Y) = \frac{4N_c Y}{bB(B+1)} \int\limits_{\epsilon - i\infty}^{\epsilon + i\infty} \frac{d\omega}{2\pi i} x^{-\omega} \, \Phi(-A+B+1, B+2, -\omega Y) \,. \tag{13}$$

2.2. Evaluation of the Limiting Spectrum

Equation (13) is badly suited for numerical evaluation. To get practically useful formulae let us invoke the integral representation for the confluent hypergeometric function Φ:

$$\frac{1}{\Gamma(c)} \Phi(a, c, x) = x^{1-c} \int_{\gamma - i\infty}^{\gamma + i\infty} \frac{dt}{2\pi i} \, e^{xt} t^{-c} \left(\frac{t}{t-1}\right)^a \quad , \qquad \gamma_0 > 1 \,. \tag{14}$$

Path parallel to the imaginary axis can be replaced by the integration around the cut $0 \le t \le 1$. Substituting this representation into Eq. (13) let us change now the order of ω and t integrations. Integral over ω results then in the Modified Bessel function I_B for $t > \zeta \equiv 1 - \ell/Y$. For smaller values of t, as can be easily seen, $\omega-$integral vanishes. Finally, introducing complex variable α according to:

$$\alpha = \frac{1}{2} \ln \frac{t}{t-1} \quad ; \qquad t = \frac{e^\alpha}{2\sinh\alpha} \quad , \qquad dt = \frac{d\alpha}{2\sinh^2\alpha} \quad , \tag{15}$$

we arrive at the following convenient formula giving D^{lim} by means of finite real integral instead of integration over complex ω-plane in Eq. (13):

$$\overline{D}^{lim} = \frac{4N_c}{b} \Gamma(B) \int\limits_{-\frac{\pi}{2}}^{\frac{\pi}{2}} \frac{d\tau}{\pi} e^{-B\alpha} \left[\frac{\cosh\alpha + (1-2\zeta)\sinh\alpha}{\frac{4N_c}{b} Y \, \frac{\alpha}{\sinh\alpha}} \right]^{B/2} \quad , \tag{16}$$
$$\cdot I_B \left(\sqrt{\frac{16N_c}{b} Y \frac{\alpha}{\sinh\alpha} [\cosh\alpha + (1-2\zeta)\sinh\alpha]} \right)$$

where $\alpha = \alpha_0 + i\tau$ and α_0 is determined by $\tanh\alpha_0 = 1 - 2\ell/Y \equiv 2\zeta - 1$. This spectrum exhibits the abovementioned "hump-backed" structure with a maximum at particle energies $E_{max} = x_{max} E$ asymptotically described by:

$$\zeta_{max} = \frac{1}{2} - \frac{a}{8N_c} \sqrt{\frac{4N_c \alpha_s(Y)}{2\pi}} + \left(\frac{a}{8N_c}\right)^2 \frac{4N_c \alpha_s(Y)}{2\pi} + \ldots, \qquad \ln\frac{1}{x_{max}} = Y(1 - \zeta_{max}) \tag{17}$$

It seems reasonable to start with comparing observed hadron spectra with the limiting distribution. *Infrared stability* of $\overline{D}$, namely the finite limit it gets at $Q_0 = \Lambda$ may serve to justify an attempt to provide the developed cascade not with an expensive increase of total energy E but with decrease of Q_0, thus enhancing the responsibility of PT QCD for jet evolution at recent energies.
Eq. (13) has been derived as asymptotically accurate. However, even for moderate energies it can be believed to give reasonable quantitative predictions because it represents the exact solution of the Evolution Equation (1) which accounts for the main physical ingredients of parton multiplication, namely, colour coherence and energy balance in 2-particle QCD branching.

2.3. Quark Jet versus Gluon Jet

As well known, in the MLLA framework ℓ-distributions of particles from q and g- jets differ only by the normalization factor $C_F/N_c = 4/9$. Account for the Next-to-MLLA effects, namely $\sim O\left(\sqrt{\alpha_s}\right)$ terms in the "preexponential" factor provides a numerically small correction to the multiplicity ratio[6] and leads also to some controllable change in the shape of particle spectra in q and g jets.
Indeed, the residues of the leading "+" trajectory determining the yield of *gluons*-partons in the *quark* and *gluon* jets read:

$$C_q^g = \frac{\Phi_q^g}{\nu_+ - \nu_-} \approx \frac{4C_F/j - 3C_F}{(4N_c/j - a) - (-\frac{2}{3}n_f 2C_F/N_c)} \approx \left[1 + \Delta_q^g \cdot j\right] \frac{C_F}{N_c} \tag{18a}$$

$$C_g^g = \frac{\nu_+ - \Phi_F}{\nu_+ - \nu_-} \approx \frac{4N_c/j - a}{(4N_c/j - a) - (-\frac{2}{3}n_f 2C_F/N_c)} \approx \left[1 + \Delta_g^g \cdot j\right] \tag{18b}$$

with:

$$\Delta_g^g = -\frac{1}{3} n_f \frac{C_F}{N_c^2} \qquad (= -\frac{4}{27} \quad \text{for} \quad n_f = 3)\ , \tag{19a}$$

$$\Delta_q^g = \Delta_0 + \Delta_g^g \qquad (= +\frac{1}{27})\ ; \tag{19b}$$

$$\Delta_0 \equiv \frac{a - 3N_c}{4N_c} \qquad (= +\frac{5}{27})\ . \tag{19c}$$

Additional factor $\Delta \cdot j$ in the integrand of Eq. (10) generates the $\sqrt{\alpha_s}$ correction which can be taken into account by the following trick using $j = \omega + \frac{d}{dY}$ (cf. Eq. (6)):

$$\begin{aligned} \overline{D}_i^g(\ell, Y) &= \int_{\epsilon - i\infty}^{\epsilon + i\infty} \frac{d\omega}{2\pi i} e^{\ell\omega} \left[1 + \Delta_i^g \cdot j\right] \overline{D}^{\lim}(\omega, Y) \\ &= \left[1 + \Delta_i^g \cdot \left(\frac{\partial}{\partial \ell} + \frac{\partial}{\partial Y}\right)\right] \overline{D}^{\lim}(\ell, Y) \approx \overline{D}^{\lim}(\ell + \Delta_i^g, Y + \Delta_i^g)\ . \end{aligned} \tag{20}$$

When calculating the gluon multiplicity, the $\frac{\partial}{\partial}\ell$ term in Eq. (20) vanishes and one is left with,

$$\mathcal{N}_g = \left[1 + \Delta_g^g \frac{\partial}{\partial Y}\right] \mathcal{N}(Y)\ , \tag{21a}$$

$$\mathcal{N}_q = \left[1 + \Delta_q^g \frac{\partial}{\partial Y}\right] \mathcal{N}(Y)\, \frac{C_F}{N_c}\ . \tag{21b}$$

where $\mathcal{N}$ denotes the particle multiplicity corresponding to the limiting spectrum which follows directly from Eq. (13) and can be expressed[3)] in terms of the Modified Bessel function:

$$\mathcal{N}(Y) = \overline{D}^{\lim}(\omega, Y)\Big|_{\omega=0} = \Gamma(B) \left(\frac{4N_c}{b} Y\right)^{-(B-1)/2} I_{B+1}\left(\sqrt{\frac{16N_c}{b} Y}\right)\ . \tag{22}$$

The derivative of $\ln \mathcal{N}(Y)$ is nothing but the value of the anomalous dimension $\gamma_\omega(\alpha_s)$ at $\omega = 0$. Since we are considering now the correction term, γ_0 should be taken in the DLA according to Eq. (9). Thus, the **ratio** of the parton multiplicities in q and g jets takes the form:

$$\frac{N_c}{C_F} \times \frac{\mathcal{N}_q}{\mathcal{N}_g} = \frac{[1 + \Delta_q^g \gamma_0]}{[1 + \Delta_g^g \gamma_0]} \approx [1 + \Delta_0 \gamma_0] = \left[1 + \frac{(a - 3N_c)}{4N_c} \sqrt{4N_c \frac{\alpha_s(Y)}{2\pi}}\right]\ . \tag{23}$$

This next-to-leading correction had been obtained first in Ref. 6).
Notice, that due to the factorization of residues (see Ref. 7) the ratio (23) appears to be the same when one takes quarks instead of gluons as final partons.

Corrections to parton spectra can be described according to Eq. (20) as a **shift** of the limiting distribution to slitely different jet energies. Namely, one gets the spectrum of gluons with fixed energy E_h for a given jet energy E with account of $\sqrt{\alpha_s}$ corrections just by taking the distribution $\overline{D}^{\lim}(\ln(E_{\lim}/E_h), \ln(E_{\lim}/\Lambda))$ given by Eq. (13) at:

$$E_{\lim} = E \exp(\Delta_q) \approx 1.04\, E \quad \text{for } q - jet \tag{24a}$$

and:

$$E_{\lim} = E \exp(\Delta_g) \approx 0.86\, E \quad \text{for } g - jet\ . \tag{24b}$$

The relative "shift" of q and g jets is determined by the same value $\Delta_0 = 5/27$ as the multiplicity ratio. Strictly speaking, only such **relative** features of q and g jets are under the quantitative perturbative control at present. To derive *absolute* predictions with the accuracy including $\sqrt{\alpha_s}$ effects, one needs to go well beyond the MLLA scope to be able to define precisely the energy scale by fixing the normalization scheme.

2.4. Shape of Inclusive Spectrum

One may find the asymptotic shape of limiting distribution not too far from its peak ($\Delta\ell \sim \sigma \sim Y^{3/4}$) by the saddle-point evaluation of Eq. (13):

$$\overline{D}_q(\ell, Y) \approx \mathcal{N}_q(Y) \left(\frac{36N_c}{\pi^2 b Y^3}\right)^{1/4} \exp\left[-\sqrt{\frac{36N_c}{b}}\, \frac{(\ell - \ln 1/x_{\max})^2}{Y^{3/2}}\right] \tag{25}$$

where factor $\mathcal{N}_q$ stands for the averaged multiplicity.

Distribution (25) has a broad Gaussian shape with the peak position given by Eq. (17). In Ref. 8) a transparent expression for $\overline{D}_q(\ell, Y)$ has been obtained in a form of a distorted Gaussian:

$$\overline{D}_q(\ell, Y) \approx \frac{\mathcal{N}_q(Y)}{\sigma 2\pi} \exp\left[\frac{1}{8}k - \frac{1}{2}s\delta - \frac{1}{4}(2+k)\delta^2 + \frac{1}{6}s\delta^3 + \frac{1}{24}k\delta^4\right] \tag{26}$$

with $\delta = (\ell - \bar{\ell})/\sigma$, $\bar{\ell} \equiv \langle \ell \rangle$, $\bar{\ell} = \ln(1/x_{max}) + O(1)$.
Parameters σ, s, k are the width, skewness and kurtosis of the distribution and are predicted to have the following asymptotical values:

$$\sigma = \sqrt{\frac{Y}{3}} d^{1/4} \left(1 - \frac{b}{64} d^{-1/2}\right) + O(Y^{-1/4}) , \tag{27a}$$

$$s = -\frac{a}{16}\sqrt{\frac{3}{Y}} d^{-1/4} + O(Y^{-5/4}) \tag{27b}$$

$$k = -\frac{27}{5Y}\left(d^{1/2} - \frac{b}{24}\right) + O(Y^{-3/2}) \tag{27c}$$

where:

$$d = \frac{bY}{16N_c} = \frac{\pi}{24\alpha_s(Y)} \tag{27d}$$

Whether or not present energies are sufficiently large for parton cascades to determine the shape of hadron spectra is, of course, a question to experiment.

2.5. Hump-backed Plateau at LEP

A strong support to the LPHD ideas came recently from the OPAL data[9] on inclusive momentum distribution of charged particles at the Z^0. To study the expected energy evolution these data were combined with the TASSO results[10].

Figures 1 and 2 show the data on the $\ln(1/x_p)$ distribution ($x_p = 2p/W$) together with the predictions of formulae (13), (25) and (26). Motivating by LPHD, the limiting gluon spectrum (13) was multiplied by an overall fitted normalization factor K. One can easily see from figs. 1 and 2 that the region around the peak is well described. Fit to the OPAL spectrum using Eq. (13) in the interval $2.5 \leq \ell \leq 4.5$ gave the values of the only free parameters: $\Lambda = 0.253$ GeV, $K = 1.28$. For the TASSO data the same Λ value was used; K was determined for each energy separately. Figure 2 shows that the MLLA formulae provide an excellent description of the energy evolution of the shape of the spectrum.

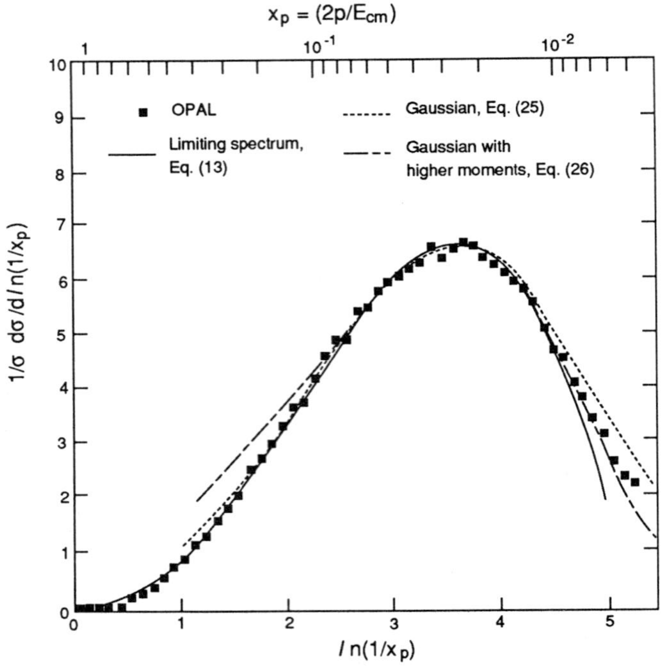

Fig. 1. *($\ln 1/x_p$) distributions at W = 91 GeV compared with analytical formulae.*

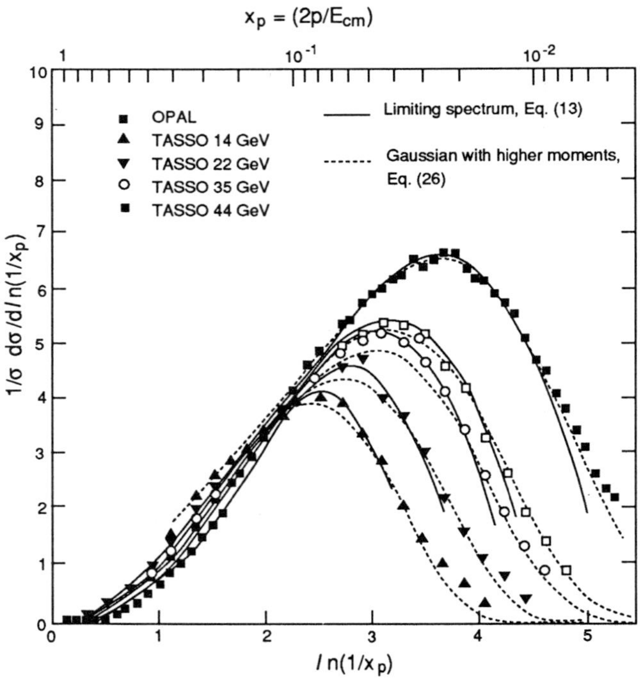

Fig. 2. *($\ln 1/x_p$) distribution at W= 14, 22, 35 44 and 91 GeV versus MLLA limiting spectra for* $\Lambda = 253$ MeV.

A few comments are to be made here. It may look surprising that the limiting spectrum which represents the true solution of the MLLA Evolution Equations (1) agrees with the data outside its formal domain of applicability, namely, at $x_p \sim 1$ where the Gaussian-like approximate formulae break down. This happening can be naturally explained however. Such a coincidence stems from the fact that the approximation (5) for the leading "trajectory" $\nu_+(\omega)$ at $\omega \ll 1$ we made to obtain the limiting spectrum (13) proves to mimic reasonably well behaviour of γ_ω at $\omega \sim 1$ also, which region is responsible for $x \sim 1$ ($\ell \sim 1$). Indeed, at $\omega = 1$ Eq. (5) gives* $\nu(1) = 12/1 - 101/9 = 7/9$ which is not far from the exact value $\nu = 0$ (following from the momentum conservation). At larger ω (5) becomes negative (as the true γ does), thus imitating the scaling violation at $x \to 1$.

Even more intriguing problem comes from the observed[9)] **decrease** by 14%(8%) of $K(Y)$ with the energy increasing from 14(22) GeV to 91. Noticing that the systematic error to the absolute normalization for the OPAL (TASSO) data was ±5%(±7%), one could nevertheless try to take this observation seriously and discuss its consequences.

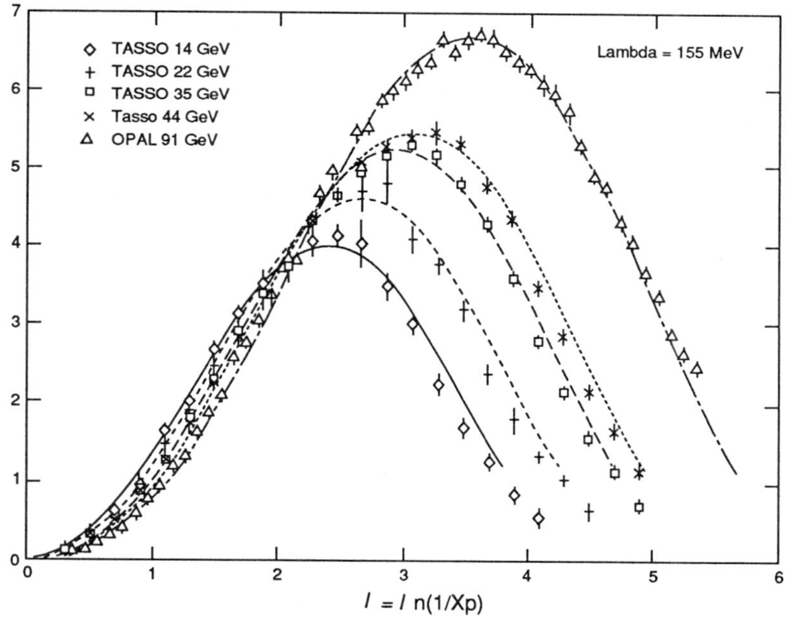

Fig. 3. ($\ln 1/x_p$) *distributions from LEP and TASSO as the sum of the MLLA spectra for* π, K *and* p ($\Lambda = 155$ MeV; $K^\pi = 0.93$; $K^K/K^\pi = 0.22$; $K^p/K^\pi = 0.11$). *For the* $K^\pm$ *and* $p(\bar{p})$ *cases the phase space effects were taken into account.*

* when deriving numerical result we take $n_f = 3$ to account for light quark contributions to parton cascades.

In the LPHD framework parton $\rightarrow$ hadron conversion coefficients must be Y- and x-independent if hadronization occurs *locally* in the phase-space indeed. However, the observed distribution of charged particles is in fact a mixture:

$$\overline{D}^{\text{charged}} = \overline{D}^{\pi^\pm} + \overline{D}^{K^\pm} + \overline{D}^{p\bar{p}} + \dots \tag{28}$$

of different particles, spectra of which prove to be not exactly similar: the position of the hump in $\overline{D}^h$ stiffens with m_h increasing. (The last property can be simulated in perturbative approach by the conjecture[4)] $(Q_0)_{eff} = m_h$.) This can in principal cause some $K^{ch}(Y)$ dependence at fixed values of $K^\pm$, $K^{K^\pm}$, K^p.
A strong influence of massive hadrons contribution is illustrated in fig.3 where experimental data were fitted by (28) with $\overline{D}^{K^\pm}$ and $\overline{D}^{p\bar{p}}$ obtained from the limiting spectrum (28) using definite rescaling procedure [4),11)] accounting for the kinematical effects of finite hadron masses. In this fit Λ_{eff} appeared to be significantly smaller: $\Lambda = 155$ MeV.
Secondly, theoretical explanation of some preasymptotic $K(Y)$ decrease could be connected with the necessity to supply the Eq. (13) with an **extra factor**:

$$\left(1 + \frac{n_f}{3}\sqrt{\frac{4N_c\alpha_s(Y)}{2\pi}} + O\left(\alpha_s(Y)\right)\right) \tag{29}$$

accounting for the yield of the soft sea $q\bar{q}$ pairs which was neglected in the MLLA formula (13) written for the spectrum of *gluons*-partons exclusively.

To clear up this situation as well as for more profound study of LPHD picture conjectures, the measurements of identified hadron spectra are of substantial importance.

3. QCD Bremsstrahlung in Heavy Q Production

Physics of heavy quarks is intensively developing nowadays. Heavy quarks were considered one of the best testing grounds of the strong interaction theory from the very beginning of the QCD epoch.

It is the restriction on the phase space of emitting gluons connected with the kinematics of a heavy quark $Q = c, b, \dots$ which determines the difference of the QCD jet produced by Q from that of the case of an ordinary light (practically massless) quark $q = u, d, s$. Hereafter we shall consider the Q-quark as relativistic ($E_Q \gg M_Q, \Theta_0 \equiv M_Q/E_Q \ll 1$) and ignore small (power) corrections of the order of Θ_0 taking care of logarithmic effects $\sim \alpha_s \ln \Theta_0^{-1} \sim 1$.

3.1. Structure of QCD Cascades in $e^+e^- \rightarrow Q\bar{Q} + \dots$

Primary gluon radiation cross section. Considering the soft gluon emission amplitude one concludes that the large logarithmic contribution to the cross-section:

$$d\sigma_{Q\rightarrow Qg_1} = g_s^2|A^a|^2\frac{d^3k_1}{2(2\pi)^3\omega_1} = \frac{\alpha}{\pi}\,C_F\,\frac{p_\mu d_{\mu\nu}(k_1)p_\nu}{(pk_1)^2 + (M_Q\Gamma_Q/2)^2}\frac{d\Omega_1}{4\pi}\omega_1 d\omega_1\,, \tag{30}$$

where: $d_{\mu\nu}(k_1) = -g_{\mu\nu} + n_{1\mu}\delta_{\nu 0} + n_{1\nu}\delta_{\mu 0}, \quad (n_{1\mu} \equiv k_{1\mu}/\omega_1, \quad \omega_1 \equiv k_1^0)$, comes only from the region of relatively large radiation angles:

$$\Theta_1 \gg \Theta_0 \ . \tag{31}$$

Indeed, $(pk_1) = E\omega_1(n_+n_1) = E\omega_1(1 - v\cos\Theta_1)$, where $n_+ = (1, \vec{v}), \quad v \equiv |\vec{p}|/E \approx 1 - \Theta_0^2/2$ – velocity of Q. Therefore, disregarding Γ_Q for the present,

$$\begin{aligned} d\sigma_{Q\to Qg_1} &= \frac{\alpha_s}{\pi} C_F \left(\frac{2}{\tilde{a}_{+1}} - \frac{n_+^2}{\tilde{a}_{+1}^2} \right) \frac{d\Omega_1}{4\pi} \frac{d\omega_1}{\omega_1} \\ &= \frac{\alpha_s}{\pi} C_F \frac{(2\sin\frac{\Theta_1}{2})^2 \, d(2\sin\frac{\Theta_1}{2})^2}{\left[(2\sin\frac{\Theta_1}{2})^2 + \Theta_0^2\right]^2} \frac{d\omega_1}{\omega_1} \left[1 + O(\Theta_0, \omega_1)\right] \end{aligned} , \tag{32}$$

where we have introduced the notation $\tilde{a}_{+1} \equiv (n_+n_1)$ and made use of the smallness of $\Theta_0 \quad (n_+^2 = 1 - v^2 = \Theta_0^2)$. Eq. (32) describes the angular structure of the primary radiation of the soft gluon g_1 by the massive Q.

"Dead Cone". Universality of large angle emission. For small emission angles $\Theta_1 \ll 1$:

$$d\sigma_{Q\to Qg_1} \sim \frac{\Theta_1^2 \, d\Theta_1^2}{[\Theta_1^2 + \Theta_0^2]^2} \frac{d\omega_1}{\omega_1} , \tag{33}$$

which corresponds to the well famous *double logarithmic* approximation (DLA)[7] :

$$d\sigma_{Q\to Qg_1} \sim \frac{d\Theta_1^2}{\Theta_1^2} \frac{d\omega_1}{\omega_1} = d(\ln\Theta_1^2) \, d(\ln\omega_1) \ . \tag{34}$$

Therefore in the DLA framework structure of the primary g_1 radiation in the region (31) appears to be identical to those for the light q-jet. However, in the angular region:

$$\Theta_1 \leq \Theta_0 \tag{35}$$

angular integration is no longer logarithmic and the yield of final particles originating from the kinematical region (35) proves to give $O(\sqrt{\alpha_s})$ fraction of the total multiplicity.

Since the main logarithmic contribution is absent and the differential cross-section Eq. (32, 33) vanishes in the forward direction, it is natural to call this region the "*Dead Cone*" – relatively depopulated cone around the $\vec{n}_+$ direction with an opening angle $\Theta_1 \sim \Theta_0$.

Notice, that the expression (32) describes adequately the nonlogarithmic region of finite emission angles $\Theta_1 \sim 1$ as well. This part of the radiation phase space contributes as $O(\sqrt{\alpha_s})$ too. Corresponding corrections will be taken into full account later when we shall derive formulae for particle spectra and multiplicities outside the scope of rough DL approximation.

Evolution parameter and running coupling. We draw the conclusion that in the main kinematical region of large gluon emission angles (31) the internal structure of the primary gluon jet proves to be identical to the light q case since an influence of the quark mass on secondary gluon emissions off g_1 is suppressed by the power factor $\sim O(\Theta_0^2/\Theta_{1+}^2)$. This means that the emission angle Θ_{1+} can be taken as the evolution parameter to restrict subsequent cascading of the g_1-jet according to AO. The smaller is radiation angle Θ_{1+}, the less populated with independently emitted partons is the gluon subjet g_1.

The situation changes radically, however, when Θ_{1+} tends to be smaller than Θ_0. Region (35), as we have already mentioned, gives significant (though nonleading) contribution $O(\sqrt{\alpha_s})$ and must be taken into full account. Here the g_1-jet opening angle "friezes" at the value of Θ_0 and no longer decreases with $\Theta_{1+} \to 0$. The reason for this is rather simple. Normally the destructive interference between emission of a soft gluon g_2 by quark and g_1 cancels independent radiation $1 \to 2$ in the angular "disordered" kinematics[12] $\Theta_{21} \geq \Theta_{1+}$. Meantime, in the massive quark case the interference contribution enters the game only when $\Theta_{2+} > \Theta_0$, so that the cancellation of independent $1 \to 2$ term inside the Θ_0-cone does not occur. The opening angle therefore can be chosen as:

$$\tilde{\Theta}_{1+} \equiv \sqrt{\Theta_{1+}^2 + \Theta_0^2} \; . \tag{36}$$

As one can verify by considering maximal invariant mass which decaying partons can get without spoiling logarithmic character of the emission amplitude, similar generalization of the evolution parameter in the region of small gluon angles holds true for the argument of the *running coupling* determining the radiation probability in Eq. (32):

$$\alpha_s \equiv \alpha_s(\tilde{k}_t^{\,2}) \quad , \qquad \tilde{k}_t = \omega_1 \tilde{\Theta}_{1+} \; . \tag{37}$$

To understand better the physical meaning of the introduced new variable $\tilde{k}_t$ let us invoke the space-time picture and discuss how looks the restriction[2) :

$$t_{\text{formation}} < t_{\text{hadronization}} \tag{38}$$

in a case of *heavy* quark radiating a gluon. The last condition is necessary to separate the perturbative stage of emission, and in case of massless partons it gave us the collinear cutoff:

$$t_{\text{form}} \approx \frac{\omega}{k_\perp^{\,2}}, \; t_{\text{hadr}} \approx \omega \mu^{-2} \tag{39a}$$

$$k_\perp^{\,2} > \mu^2 \sim \text{ Few Hundred MeV} \; . \tag{39b}$$

Now the radiating quark Q is massive and one can start from its restframe where the proper time for gluon to be resolved reads: $\tau_{\text{form}} \geq \lambda \sim 1/\omega_0$ with ω_0 the gluon energy. In terms of laboratory frame momenta:

$$2M\omega_0 = 2(Pk) \approx \frac{E_Q}{\omega}\tilde{k_\perp}^2.$$

Therefore, with account of the Lorentz-factor, the formation time in the laboratory frame proves to be:

$$t_{\text{form}} \sim \frac{E_Q}{M} \cdot \tau_{\text{form}} = \frac{\omega}{\tilde{k_\perp}^2}. \tag{40}$$

Thus the restriction (39b) should be substituted by:

$$\tilde{k_\perp}^2 \equiv {k_\perp}^2 + (\omega\Theta_0)^2 > \mu^2 \tag{41}$$

which is in direct accord with the requirement that running α_s must remain small for perturbative considerations to be applicable.

3.2. Multiplicity of Light Hadrons

In this subsection we derive the MLLA expression for the multiplicity of light hadrons accompanying hard production of heavy quark pair. In the LPHD framework it can be obtained via the convolution of the particle multiplicity initiated by *gluon* with the probability of the gluon bremsstrahlung off Q given by Eq. (32) at the hardness parameter $\tilde{k_t}$.

$$N_Q(E) = \int_0^4 \frac{(2\sin\frac{\Theta}{2})^2\, d(2\sin\frac{\Theta}{2})^2}{\left[(2\sin\frac{\Theta}{2})^2 + \Theta_0^2\right]^2} \int_{[k_t>\mu]}^1 dz\, \Phi_F^G(z) \frac{\alpha_s(\tilde{k_t}^2)}{4\pi} N_G(\tilde{k_t}) \tag{42}$$

where:

$$\tilde{k_t}^2 \equiv \omega^2 \left[\left(2\sin\frac{\Theta}{2}\right)^2 + \Theta_0^2\right] \approx \omega^2\, 2\tilde{a}\,. \tag{43}$$

Here we took into account single logarithmic effects connected with "hard" $Q \to Q+g$ splittings at $z \sim 1$ by making use of the exact GLAP kernel instead of its soft limit $4C_F d\omega/\omega$ in Eq. (32):

$$\Phi_F^G(z) = 2C_F \frac{1+(1-z)^2}{z}\,, \qquad z = \omega/E\,. \tag{44}$$

The main contribution comes from the DL phase space region, however expression (42) keeps trace of significant SL effects as well provided the multiplicity factor N_G is calculated with the MLLA accuracy. Introducing convenient variable:

$$\kappa^2 = E^2 \left[\left(2\sin\frac{\Theta}{2}\right)^2 + \Theta_0^2\right], \tag{45}$$

Eq. (42) can be rewritten in a simpler form:

$$N_Q(E) = \int_{M_Q^2}^{W^2} \frac{d\kappa^2}{\kappa^2}\left[1 - \frac{M_Q^2}{\kappa^2}\right]\int_{\mu}^{\kappa} \frac{dk_t}{\kappa}\Phi_F^G(z)\,\frac{\alpha_s(k_t^2)}{4\pi}\,N_G(k_t)\ , \tag{46}$$

where $k_t = z\kappa$ and $W \equiv 2E$ — is the total annihilation energy.
In the massless limit $M_Q \leq \mu$ the contribution of M_Q^2/κ^2 term vanishes and Eq. (46) leads to the known expression for the multiplicity in e^+e^- annihilation into light quarks (22).
Integration can be performed with use of the relation following from the evolution equation for light quark jet multiplicity:

$$N_q' \equiv \frac{\partial}{\partial \ell n Q^2}\,N_q(Q/2)\ = \int_{\mu}^{Q} \frac{dk_t}{Q}\,\Phi_F^G\left(\frac{k_t}{Q}\right)\,\frac{\alpha_s(k_t^2)}{4\pi}\,N_G(k_t)\ . \tag{47}$$

Eq. (46) than takes the form:

$$N_Q(E) = N_q(E) - N_q(\frac{M_Q}{2}) - N_q'(\frac{M_Q}{2}) + O(\alpha_s N_q) + O(\Theta_0^2 N_q')\ . \tag{48}$$

Notice that the factor 2 in the argument of N_q generates $\sqrt{\alpha_s}N_q(M_Q)$ correction and is under the control of present perturbative analysis whereas it could be omitted in the N_q' term as producing $\alpha_s N_q(M_Q)$ terms which we neglected systematically in Eq. (46). Within this accuracy the term $N' \sim \sqrt{\alpha_s}N_q(M_Q)$ can be absorbed by shifting the argument:

$$N_q(\frac{M_Q}{2}) + N_q'(\frac{M_Q}{2}) \approx N_q(M_Q\frac{\sqrt{e}}{2})\quad ,\qquad e \equiv \exp(1)\ . \tag{49}$$

Finally one arrives at the formula expressing the multiplicity in $e^+e^- \to Q\bar{Q}$ in terms of that of $e^+e^- \to q\bar{q}$:

$$\begin{aligned} &N(e^+e^- \to Q\bar{Q}; W) \\ &= N(e^+e^- \to q\bar{q}; W) - N(e^+e^- \to q\bar{q}; \sqrt{e}M_Q) + O(\alpha_s(M_Q^2)N(M_Q)) \end{aligned} \tag{50}$$

The main consequence of Eq. (50) is that the *difference* between particle yields from q- and Q- jets at fixed annihilation energy W depends on the heavy quark mass and remains W–independent. The question arises, to what extent this difference can be quantitatively predicted. The problem is due to ignorance of non-leading corrections $\sim O(\alpha_s(W))$ to the main term.
Strictly speaking, Eq. (50) should look like:

$$\begin{aligned} N(e^+e^- \to Q\bar{Q}; W) = &N^{\rm MLLA}(e^+e^- \to q\bar{q}; W)\left[1 + O(\alpha_s(W^2))\right] \\ &- N^{\rm MLLA}(e^+e^- \to q\bar{q}; \sqrt{e}M_Q)\left[1 + O(\alpha_s(M_Q^2))\right] \end{aligned} \tag{51}$$

where the MLLA multiplicities can be given analytically and account only for exponentiating DL and SL effects ($O(\sqrt{\alpha_s}) + O(\alpha_s)$ in the anomalous dimension) and $\sqrt{\alpha_s}$ terms in the normalization factors ($O(1) + O(\sqrt{\alpha_s})$ in the coefficient function). Rapid energy growth of the multiplicity (faster than any power of $\ln W$) makes the neglected $\sqrt{\alpha_s}\, N(W)$ correction dominating over the subtraction term in Eq. (51) thus endangering the very possibility to discriminate Q and q jets within present theoretical accuracy.

However, considering the origin of subleading corrections to Eq. (51) proportional to $N(e^+e^- \to q\bar{q}; W)$ one can see that all of them prove to be independent of the heavy quark mass M_Q being inherent to the light quark jet evolution itself. For example, first corrections of the order of $O(\alpha_s(W^2)N(e^+e^- \to q\bar{q}; W))$ come from either further improvement of the description of anomalous dimension $\Delta\gamma(\alpha_s) \sim {\alpha_s}^2$ determining intrajet cascades, or from $O(\alpha_s(W^2))$ terms in the coefficient function due to:

1. 3-jet configuration *quark + antiquark + hard gluon with large emission angle*,
2. "dipole correction" to the AO scheme *quark + antiquark + two soft gluons at large angles*

which are insensitive to the Θ_0 value up to *power* correction $\sim \Theta_0 \ll 1$.

Therefore, replacing the approximate MLLA multiplicity factors in the Eq. (51) by real multiplicities one arrives at the Eq. (50) which makes it possible to establish phenomenological relation between light and heavy quark jets with reasonable accuracy.

Generally speaking, it is the QCD coherence which enables this relation. Indeed, according to Eq. (33) the gluon bremsstrahlung off *massive* and *massless* quarks differ only at parametrically small angles $\Theta \leq \Theta_0 \equiv M_Q/E$ where, due to the AO, cascading effects are majorated by the $N(M_Q)$ factor.

3.3. Energy Spectra of Light Hadrons

Effect of the gluon bremsstrahlung suppression (*dead cone*) on the energy spectrum of light particles accompanying Q production can be analyzed in a way similar to the case of mean multiplicity considered above. Resulting particle spectrum $\overline{D}_Q^h(x, \ln(E/\Lambda))$ appears to be *softer* than that in the case of the same energy jet generated by a light quark $\overline{D}_q^h(x, \ln(E/\Lambda))$. To see this let us first make a rough estimate of $\overline{D}_Q^h$.

DL estimate of accompanying hadron spectrum. Due to the *dead cone* phenomenon the difference between $\overline{D}_Q^h$ and $\overline{D}_q^h$ is connected with gluon emission off Q at angles $\Theta \leq \Theta_0$. Radiation concentrated in the restricted angular cone (which is absent in the Q case) can be simply related to the jet production in a process with characteristic hardness M_Q^2 (instead of E^2). Indeed, since we treat M_Q as a parametrically large quantity, one can consider the q-jet production with energy $E_q \sim M_Q$ to be a hard process. This process induces certain amount of gluon radiation and cascading which leads to definite final system of light hadrons. Now, *boosting* this jet with γ-factor $\gamma = E/M_Q$ along its momentum direction one would obtain an exact image of the dead cone: an ensemble of energetic particles (with energies up to E) concentrated inside the cone Θ_0. Thus, we can write a simple formula for the Q-jet particle distribution as:

$$\overline{D}_Q^h\left(x, \ln\frac{E}{\Lambda}\right) = \overline{D}_q^h\left(x, \ln\frac{E}{\Lambda}\right) - \overline{D}_q^h\left(x, \ln\frac{M_Q}{\Lambda}\right) . \qquad (52)$$

Here the structure of Q-jet is expressed in terms of MLLA spectra at different energy scales E^2 and M_Q^2.

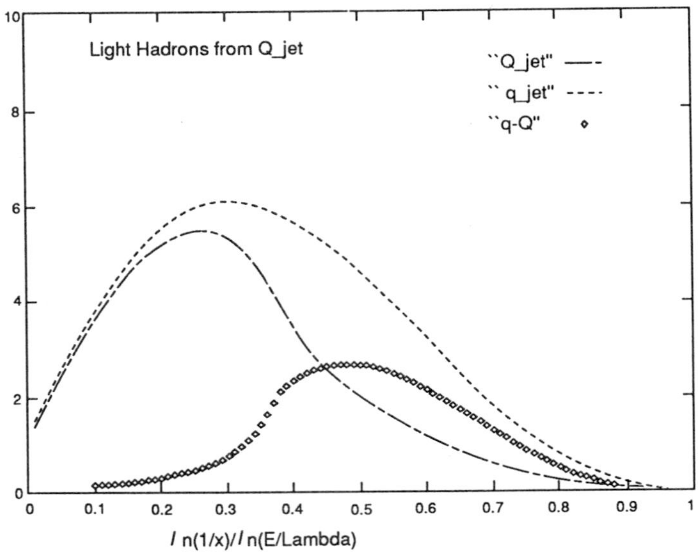

Fig. 4. Light hadrons from light and heavy quark (B) jets according to Eqs.(13), (56). $E_{jet} = 45$ *GeV* , $\Lambda = 150$ *MeV.*

The particle energy fraction x in Eq. (52) is measured with respect to the total Q-jet energy E. But in the subtraction term it can not be smaller than Λ/M_Q which follows from the simple kinematical inequality:

$$\ln\left(\frac{1}{x}\right) < Y \qquad \text{for} \qquad \overline{D}(x, Y) . \qquad (53)$$

Therefore depopulation occures for energetic particles only, whereas in the small energy region $\lambda \leq E_h \leq E\Lambda/M_Q$ spectra should be identical for Q and q jets. This is rather transparent but important consequence of the dead cone physics which is exhibited by Eq. (52). However, if one takes this equation seriously and plots the resulting particle distribution, some unplesant thing happens to happen:

namely, the yield of most energetic particles becomes negative! The reason for this is transparent: according to the *scaling violation* pattern spectrum at large x falls down with increase of the hardness of the process, Q^2, so that the first term in Eq. (52) ($Q^2 \sim E^2$) appears to be smaller for $x \geq 0.2$ than the subtraction term corresponding to $Q^2 \sim M_Q^2$.

MLLA prediction for spectra of light hadrons. This decrease of spectra is due to momentum share which is taken by gluon bremsstrahlung off the leading parton. Thus to recover from this paradox we have to keep trace of the *recoil effects* taking into account the *energy loss* by Q at the first steps of evolution described by the *valence* distribution $\overline{D}_{\text{val}}(x,\xi)$, where $\xi(Q) = \ln(1/\alpha_s(Q^2))$. This can be done with a help of the following representation:

$$\overline{D}_Q^h\left(x, \ln\frac{2E\sin(\Theta/2)}{\Lambda}\right) = \overline{D}_q^h\left(x, \ln\frac{2E\sin(\Theta/2)}{\Lambda}\right) - F_q(x,\Theta_0;E) \qquad (54)$$

with Θ the opening angle of the jet. F_q describes carefully effect of depopulation of the Θ_0-cone and is given by:

$$F_q(x,\Theta_0;E) \;=\; \int_x^1 dy \int^{\tilde{\Theta}_0} \frac{d\tilde{\Theta}}{\tilde{\Theta}} \overline{D}_{\text{val}}(y, \xi - \tilde{\xi}) \int_{x/y}^1 dz \Phi(z) \frac{\alpha_s}{2\pi} \overline{D}_G\left(\frac{x/y}{z}; \ln\frac{zyE\tilde{\Theta}}{\Lambda}\right) \qquad (55a)$$

$$\tilde{\Theta}_0 \;\equiv\; \frac{\Theta_0\sqrt{e}}{\langle y\rangle} \quad , \qquad \tilde{\xi} = \xi(yE\tilde{\Theta}) \qquad (55b)$$

$$\langle y\rangle \;\equiv\; \exp(\langle \ln y\rangle) \quad , \qquad \langle \ln y\rangle = \int_0^1 dy\, \ln y\, \overline{D}_{\text{val}}(y, \Delta\xi). \qquad (55c)$$

Making use of Evolution Equations it is straightforward to convert Eq. (55a) into much more compact form:

$$F_q(x,\Theta_0;E) = \int_x^1 dy \overline{D}_{\text{val}}(y, \xi - \xi_0)\, \overline{D}_q\left(\frac{x}{y}; \ln\frac{yE\tilde{\Theta}_0}{\Lambda}\right) \qquad (56)$$

This equation takes into account the fact that when approaching the evolution stage where characteristic emission angles become as small as Θ_0, radiator has less energy left. Treating $\overline{D}_{\text{val}}$ as $\delta(1-y)$ one would be back to the DLA formula (52). Since characteristic values of y are of the order of unity in Eq. (56) ("leading particle effect"), Eq. (56) can be well approximated for finite x-values ($x > \Lambda/M$) by:

$$\overline{D}_Q^h\left(x, \ln\frac{E}{\Lambda}\right) = \overline{D}_q^h\left(x, \ln\frac{E}{\Lambda}\right) - \overline{D}_q^h\left(\frac{x}{< y >}; \ln\frac{M\sqrt{e}}{\Lambda}\right) \qquad (57)$$

bye where we substituted $\Theta = \pi$ determining the overall jet in Eq. (54) which reminds us subtraction formula Eq. (50) for the multiplicity. In the region of parametrically soft momenta $x \ll \Lambda/M$ the $q-$ and Q-jets should have identical inclusive particle distributions. Figure 4 shows the resulting depopulation of the Q-jet predicted by Eq. (56).

We have presented spectra calculated with use of the MLLA formulae for the valence and $D^{\lim}$ distributions. Meanwhile the content of the heavy quark jet can be predicted by means of known spectra from the "light" jet phenomenologically with use of Eq. (57).

Conclusions

In this lecture we considered two topics of high practical importance: the energy structure of particle spectra from light and heavy quark jets. The study of multiple hadroproduction not only checks the basic structure of parton dynamics at small distances but can be used to gain some actual knowledge about confinement of colour by facing perturbative QCD predictions with properties of observed hadronic systems.

References

1) For reviews see A. Bassetto, M. Ciafaloni, and G. Marchesini, *Phys. Rep.* **C100** (1983) 201; Yu.L. Dokshitzer, V.A. Khoze, A.H. Mueller, and S.I. Troyan, *Rev. Mod. Phys* **60** (1988) 373; V.A. Khoze, Proc. of the Int. Symp. on Lepton and Photon Interactions at High Energies, Stanford, Aug. 1989 (World Scientific) p. 387 ; QCD at LEP, *Proc. of the 1989 Workshop on Z Physics at LEP1*, Z. Kunszt et al., CERN 89-08, **Vol.1**, p. 373.
2) Yu.L. Dokshitzer, Talk given at the International School of Subnuclear Physics, Erice, 1989.
3) Yu.L. Dokshitzer and S.I. Troyan, *Proc. XIX LNPI Winter School*, V.I (1984) 144; preprint LNPI-922 (1984).
4) Ya.I. Azimov et al., *Z. Phys.* **C27** (1985) 65; *Z. Phys.* **C31** (1986) 213.
5) D.Amati and G.Veneziano, *Phys. Lett.* **83B** (1979) 87; G. Marchesini, L. Trentadue, and G. Veneziano, *Nucl. Phys.* **B181** (1981) 335.
6) A.H. Mueller, *Nucl. Phys.* **B213** (1983) 85 and Erratum quoted in *Nucl. Phys.* **B241** (1984) 141.
7) Yu.L. Dokshitzer, D.I. Dyakonov, and S.I. Troyan, *Phys. Rep.* **C58** (1980) 270.
8) C.P. Fong and B.R. Webber, *Phys. Lett.* **B241** (1990) 255.
9) OPAL Collaboration, M.Z. Akrawy et al., CERN EP/90-94.
10) TASSO Collaboration, W. Braunschweig et al., DESY 90-013.
11) Yu.L. Dokshitzer, V.A. Khoze, and S.I. Troyan, in *Perturbative QCD*, ed. A.H. Mueller (World Scientific, Singapore, 1989) p. 241.
12) B.I.Ermolayev and V.S.Fadin, *JETP Lett.* **33** (1981) 285; A.H. Mueller, *Phys. Lett.* **B104** (1981) 161.

Chairman: Y. Dokshitzer

Scientific Secretaries: H.J. Lu, A. Syed, Z. Zhang

Discussion

– *Lu:*

When calculating the cross-section for deep inelastic scattering processes, if one sets the scale of the coupling constant at Q^2, all logarithms will disappear. So, how can one justify the leading logarithm approximation?

– *Dokshitzer:*

Actually that is not an exact statement. Irrespective of the scale that you fix for the running coupling constant, there will still be logarithms. In the diagram:

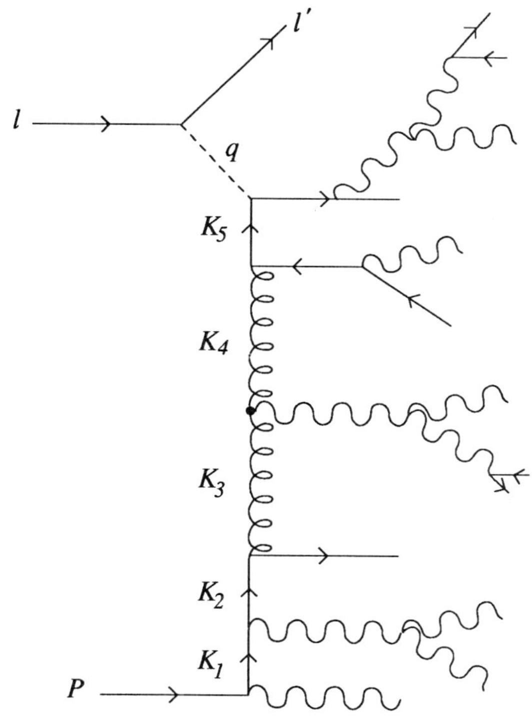

you can fix the renormalization scale for each coupling to be at any point that you like. But due to the fact that the coupling constant is dimensionless and you have logarithms over transverse momenta, you will have a series of these large logarithmic contributions which you will have to consider in the rank of transverse momenta that go from the size of the hadron up to the value of Q^2. So α_s will be in any case multiplied by a large logarithm, irrespective of the scale that you choose for renormalization. If you take the virtuality of the target quark (proton) to be of the order of Q^2, certainly there will be no emissions, but for real proton and large Q^2, there will always be series of big logarithms, you should sum them.

– *Lu:*

But mathematically there must be some μ^2 in the denominator inside the logarithms with $Q^2\ldots$

– *Dokshitzer:*

Sure, Q^2 is divided by the virtuality of the target particle.

– *Lu:*

In my opinion the μ^2 is associated to the scale of the vertex, and if one chooses μ^2 to be equal to Q^2, then all logarithms will diseppear.

– *Dokshitzer:*

Sure, if you take a target particle which is bound to very small distances and that has a virtuality of order of Q^2, there will be no emissions: there will simply be the Born scattering of proton off the quark and no multiple emissions. But in reality this virtuality is fixed because the proton has finite dimension so μ^2 is of order of one fermi^{-2}. This is the reason why these logarithms appear. But mathematically you are right, if you choose a virtuality of the order of Q^2, the problems will disappear.

– *Rosiek:*

There is a 4-gluon vertex in QCD at the tree level, why have you neglected it in your description of the cascade?

– *Dokshitzer:*

I have not neglected anything. In the deep inelastic scattering it is true that there are 4-gluon vertices in addition to the 3-gluon vertices. The 3-gluon contribution is proportional to g^2:

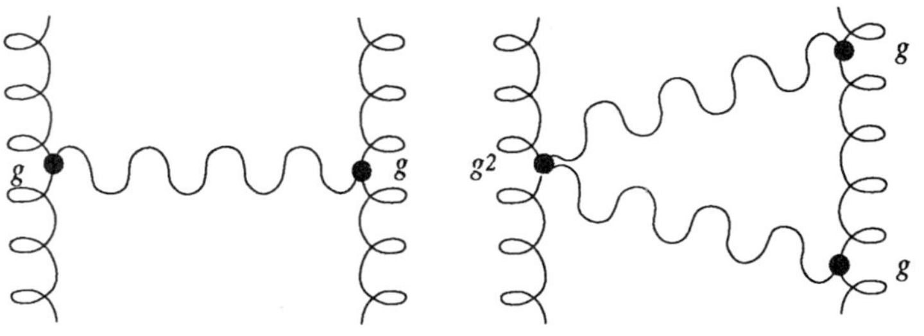

whereas the subdiagram containing 4-gluon coupling is proportional to g^4, which is a higher order correction. If there were an extra logarithm, it will not matter because in the leading log approximation each squared coupling should be compensated by large logarithm. Further, it also depends on the gauge that you are chosing because you are dealing with vector quanta and the amplitudes depend on the gauge. So in principle if you are using Feynman gauge there will be contributions from the 4-gluon vertices, but after you make the summation off all contributing diagrams the 4-gluon vertices will effectively disappear. The 4-gluon vertices are there because of the gauge invariance, just like in QED: when you consider interactions between photons and scalar particles, it is not sufficient to include the processes of the type:

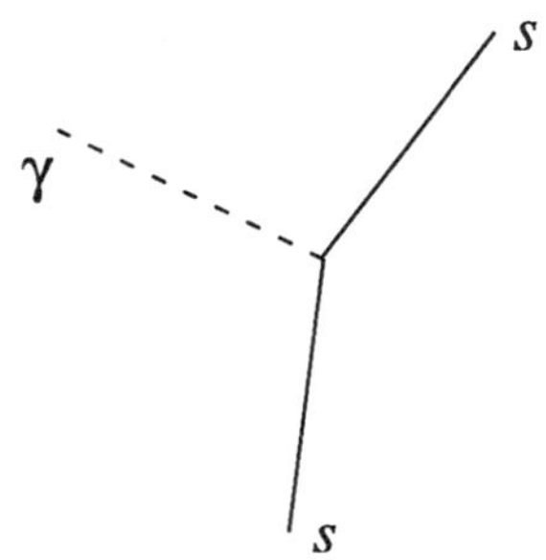

but you are forced to include the point interaction with 2-photons to preserve gauge invariance:

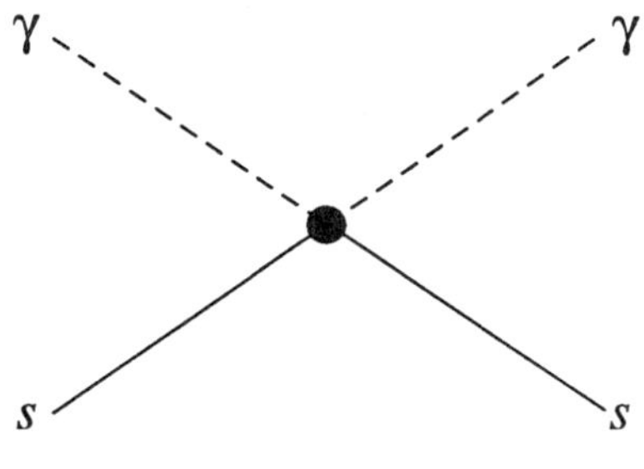

– *Grossmann:*

In your talk you explained the Chudakov effect arguing that there should be interference. What is the reason for these interferences to be destructive and not constructive? Does it depend on the spin of the charged particle?

– *Dokshitzer:*

The reason is that irrespective of the spin of the particle, when you have the emission of a gluon with wavelength larger than the separation between two branches of the fork as below:

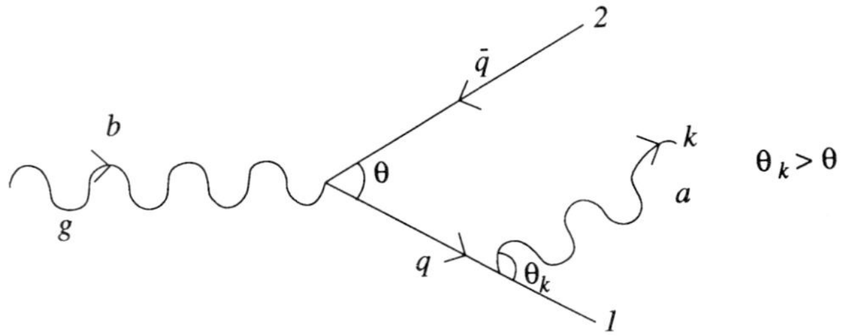

the fork will act as a single emittor of charge equal to the sum of the charges of the particles 1 and 2 and since they have charge sum equal to zero, it is evident that the interference will be destructive.

– *Wexler:*

You said that Gribov and Lipatov showed that all theories with dimensionless coupling constants have Bjorken scaling. What does Bjorken scaling mean in QED?

– *Dokshitzer:*

Let us be a bit more precise, you have approximate Bjorken scaling. So there will be no steep dependence such as powers of Q^2 but there will still be logarithmic dependence of the structure function of Q^2. In QED it is exactly the same story as you are able to study QED structure function of, e.g. a photon. You can organize the same deep inelastic scattering, say a muon scatters off an electron and you produce photons, the photons split into extra pairs which produce, in turn, more photons, etc.

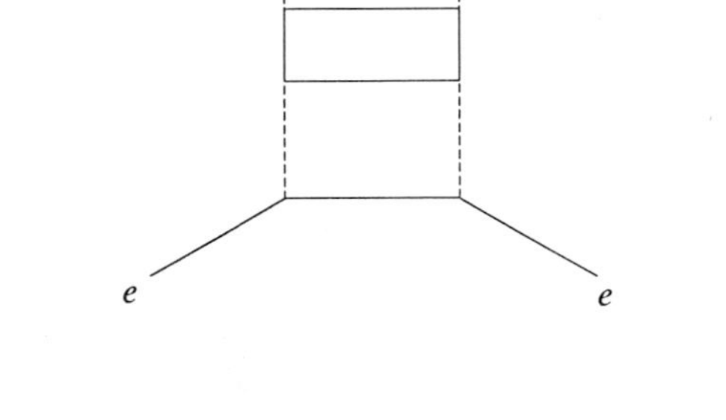

The cross-section of this process can be expressed in terms of structure functions. What you measure is the structure function of an electron. When you are probing an electron at small distances the transverse momenta available are large and you see that inside an electron there are many photons and extra e^+e^- pairs, exactly as in QCD.

– *Hsu:*

You mentioned at the end of your talk that there was no experimental evidence of confinement, could you explain further?

– *Dokshitzer:*

What I meant was that when you are studying the global properties of hadronic jets, that is, multiplicity, energy and angular spectra etc., you find that the experimental characteristics of hadronic final states are proportional to what you calculate from the perturbation theory in terms of partons, completely forgetting about confinement. Confinement is so soft that it does not change these properties, it simply replaces particles which are coloured partons by particles which are "white" hadrons.

– *Ayad:*

You talked about the Marchesini-Webber cascade model which follows the modified leading logarithm approximation evolution. What is the condition when the cascade is stopped?

– *Dokshitzer:*

If you choose the Q_0 value to be large, you will be definitely sure that the coupling constant is numerically small over the region that you are trying to control by the perturbation theory. However if the Q_0 is large you will not have many partons but you will have large relative invariant masses between partons and you will have to invent some hadronization scheme which will work instead of perturbation theory. If you decrease Q_0, which means increasing the domain of responsibility of the perturbation theory, you will find that the invariant masses between partons will be numerically small, of order of 1-2 GeV (the so-called preconfinement). Hadronization will be local, which means that you will be producing a system of partons and after that at the hadronization stage you will convert these partons locally in the phase-space into hadrons. The hadronization scheme used will not affect the result seriously.

Chairman: Y. Dokshitzer

Scientific Secretaries: H.J. Lu, A.Syed, Z. Zhang

Discussion

– *Zichichi:*

The leading effect analysis of the multiplicity distribution induced by hadronic interactions was found to be different in e^+e^- and pp interactions by us , what are your comments on that?

– *Dokshitzer:*

In principle they should be different. If you find the distributions to be similar in e^+e^- and pp, it will be very interesting phenomenologically, but there is no theoretical basis for that. In e^+e^- annihilation we are discussing very hard process of vacuum heating, a quark pair is produced which separate, the phase-space is huge, density of particles is very small and there is no interaction between left and right jet, etc. so it a hard process. What you are discussing in pp is mainly minimum bias events, there is no hard stage in the process; there may be some hard contribution in such processes but the physics is completely different as the protons are hadrons and evolve long before interaction. So the underlying theory in both of these processes is completely different.

– *Brown:*

You say that your analysis depends on only one parameter namely the Λ_{QCD}. Usually this parameter depends on the renormalization scheme used. Which scheme have you used?

– *Dokshitzer:*

When studying soft particle distributions, you have as expansion parameter $\sqrt{\alpha}$ and not α and the anomalous dimension which depends on the running coupling and determines all the story is as follows:

$$\gamma_W^{(\alpha)} = \sqrt{\alpha} + \alpha + \alpha^{3/2} + \dots$$

and $\alpha^{3/2}$ on the third term, it is the first term sensitive to the renormalization because if you multiply Λ_{QCD} two times you get contribution of order of $\alpha^{3/2}\ell n2$. So the double logarithmic approximation and modified leading log approximation are not affected by the normalization scheme used. If we integrate the above expression over virtuality (transverse momentum) you get from the first term $\sqrt{\text{log Energy}}$ the second gives you (log E)$^{\text{some power}}$ and the third one gives you correction of order of $\sqrt{\alpha}$. We would like to include these contributions as well.

– *Lu:*

In e^+e^- annihilation people observe 3-jet events and the usual explanation is that one of the jets is a gluon jet and the others are quark and antiquark jets. It is possible to really tell which jet is coming from which parton?

– *Dokshitzer:*

In principle it is possible to tell the nature of the jets, that is, which jet originates from the quark, the antiquark or the gluon. For example, in the production of $c\overline{c}$ pair you can simply analyse the presence of a fast μ^+ and be sure that it is nothing but the product of c decay. Similarly you can determine the $\overline{c}$ jet and therefore find the origin of all the 3-jets.

This certainly has to be done at LEP now, because the study of these 3-jet events is very important for visualizing QCD coherence phenomena. Also, the ratio between the number of particles in the midway between quark and gluon (N_{qg}) and the number of particles between the quark and antiquark ($N_{q\overline{q}}$) is shown experimentally to be greater than unity:

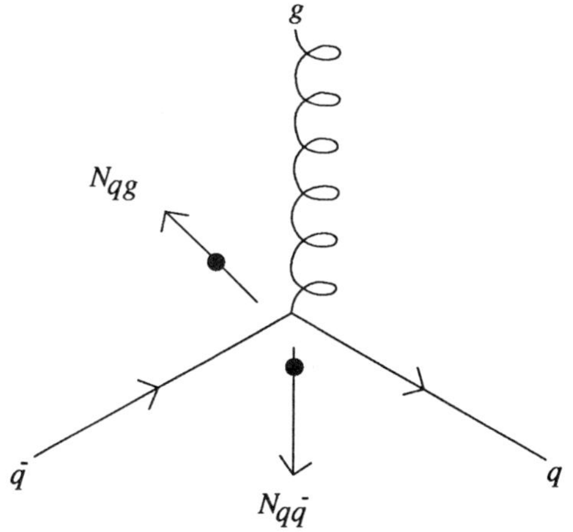

$$\frac{N_{qg}}{N_{q\overline{q}}} \sim 1.2\ldots$$

It is actually a very simple effect of QCD coherence and the QCD prediction for this ratio is

$$\frac{N_{qg}}{N_{q\overline{q}}} = \frac{22}{7} \approx \pi\,!$$

The discrepancy with the experimental value is because of the definition of jets. Usually experimentalists call the softest one the gluon jet, but this has only 60% of chance of being correct. Typically 60% of the gluon jets have the lowest energy, 30% have the intermediate energy and 10% have the largest energy.

What should be done now in LEP is tagging of fast muon in the intermediate jet. If you are sure that this intermediate jet is a quark jet, then you have 90% of confidence that the lowest jet is a gluon jet, because it is very rare that the gluon jet happens to be the fastes one.

– *Ciuchini:*

You have shown an effect on $k's$ that is clear in trend but small in size. Do you think that non-leading correction could be important?

– *Dokshitzer:*

The answer is yes. The reason is that to understand what mean by lambda we certainly need these calculations to be performed. There is an evidence that the picture of strict angular ordering works even better than expected. It takes into account not only leading and next to leading but also next to next to leading correction. There is at least one evidence for that: 5 years ago there was a work by Malaza who showed that two-loop calculation for the ratio of the multiplicity in quark and gluon jets can be obtained by a simple minded cascade with strict angular ordering.

The ratio of quarks multiplicity to gluon multiplicity is $O(\sqrt{\alpha_s})$. This is not a small effect and it represents the ratio of leading contributions.

The function which describes the yield of quarks starts from $\sqrt{\alpha}$ is zero approximation for this function and not a correction.

More studies on what I have spoken about are needed.

Chairman: Y. Dokshitzer

Scientific Secretaries: H.J. Lu, A. Syed, Z. Zhang

Discussion

– *Brown:*

In your question for QED $\Gamma^{TOT} \sim \alpha^2(M^2)$ and $\Gamma_{m\to\gamma\gamma} \sim \alpha^2(0)$, since $\alpha(Q^2)$ increases with Q^2, we have $\Gamma_M^{TOT} > \Gamma_{M\to\gamma\gamma}$ as it should be. In QCD the coupling constant runs the other way, so it would seem that the partial width is greater than the total width. How does one resolve this problem in QCD?

– *Dokshitzer:*

Consider a state x which is built of two heavy quarks $Q\overline{Q}$ and that can decay into two gluons.

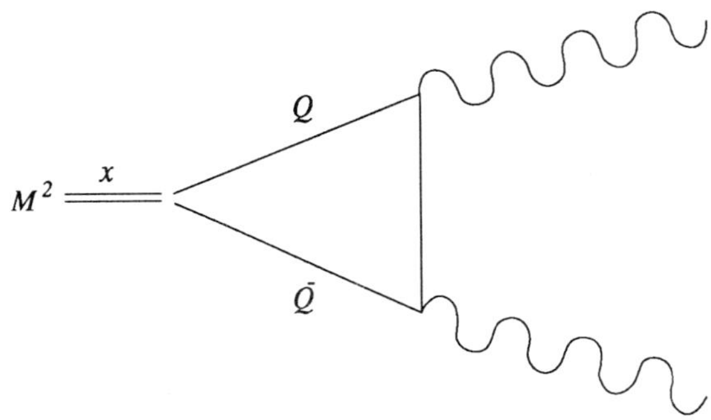

In principle we can do the same analysis and obtain the exclusive gluon decay width and the total hadronic width

$$\Gamma_{\kappa\to 2g} \sim \alpha_s^2(0)$$
$$\Gamma_{\kappa\to anything} \sim \alpha_s^2(M^2)$$

so if the answers were the same as in QED, the total width would appear to be smaller than the partial width.

But in $\Gamma_{\kappa\to 2g}$ there is actually a factor $e^{-\infty}$, because the probability of having two and only two gluons in the final state is exactly zero due to the Sudakov suppression.

Nevertheless, in the Deep Inelastic Scattering experiment we have to put α_s depending on $K_\perp^2$ when studying the multicell structure; $K_\perp^2$ is just the maximum value of mass available in the final state.

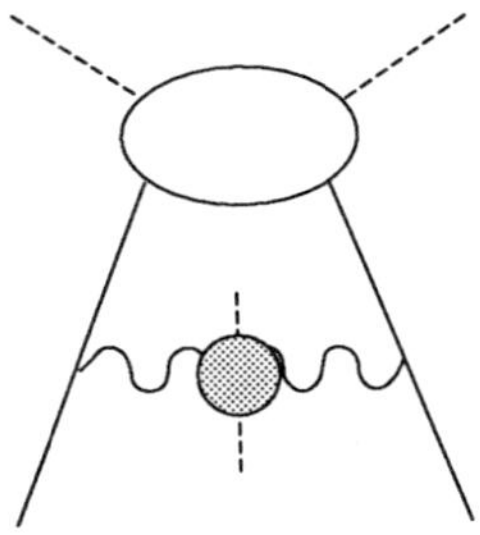

$\alpha_s(K_\perp^2)$

In QED this leads to a total cross-section that is larger than the partial cross-section. (This in fact is the reasoning used by Landau in the early 60's when he made the famous statement that field theory can be forgotten, because a positive photon decay cross-section leads to an increase of the couplings at small distances, so "anti-asymptotically-free" behaviour set-on was thought to be inevitable).

In QCD you have the same contribution due to the quark loop, which is also positive. But there are peculiar QCD contributions that are non-positive, namely, those that look like vertex corrections:

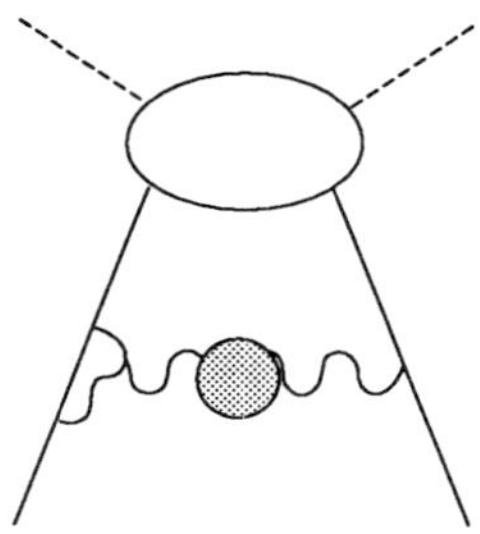

These corrections are negative and the combined result is:

$$B = -\frac{11}{3} N_c + \frac{2}{3} n_f < 0$$

Physically speaking, the gluon propagates not in free space but in the gluon field generated by the quarks. Let us study for example the interaction between two heavy QCD Yang-Mill charges to see that at smaller distances the interaction decreases.

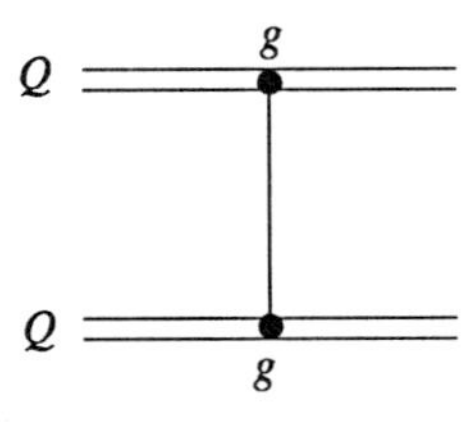

There are two contributions to the renormalization, as it was shown by Gribov:

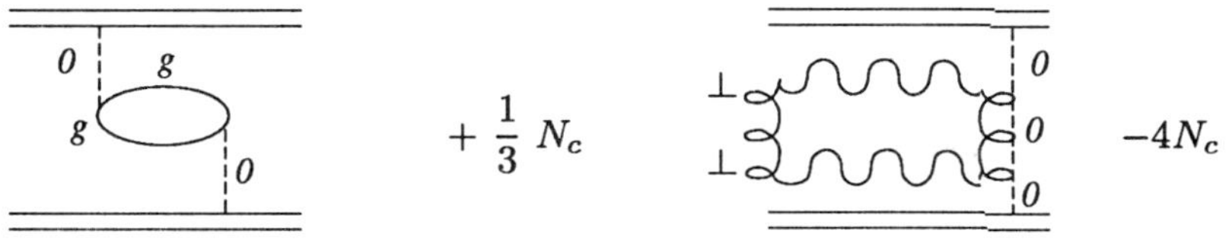

In the first diagram the Coulomb quantum *"0"* decays into physical transversely polarized gluons which then are reabsorbed. It is positive, just like in the QED case. In the second diagram a Coulomb quantum propagates in a space filled with quantum fluctuations of transversely polarized gluons. There is a theorem from quantum mechanics that states that the second order perturbation contribution coming from a small potential leads to a decrease in the bound state energy:

$$\Delta E_o \;=\; -\;|V|^2/E_0$$

This induces large negative contribution to β-function:

$$\overline{b}_{gluon} \;=\Big(-4+\frac{1}{3}\Big)N_c\;=-\frac{11}{3}N_c$$

This is how asymptotic freedom works in QCD.

– *Wexler:*

In your expression for total width, why do you have α^2 and not α^n, which would account for the additional vertices?

– *Dokshitzer:*

When you replace the coupling α by $\alpha(M^2)$

$$\alpha(M^2) \;=\; \frac{\alpha(0)}{1\;-\;\frac{\alpha(0)}{4\pi}\,\frac{4}{3}\,\ell n\Big(\frac{M^2+m_e^2}{m_e^2}\Big)}$$

if you expand it with respect to $\alpha(0)$ then it will generate the extra powers of $\underline{\alpha}$. These powers of $\underline{\alpha}$ will take into account all processes like:

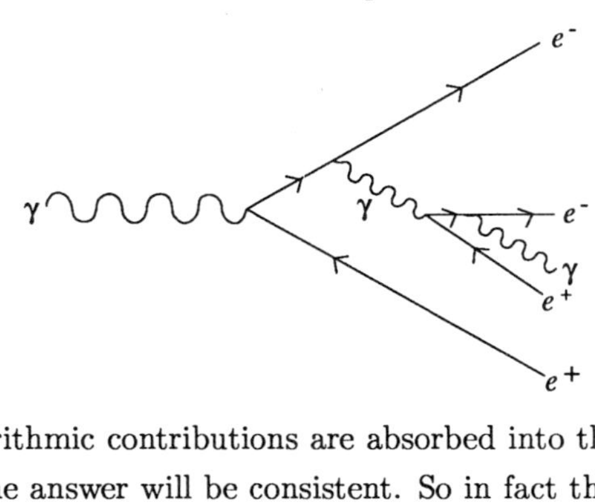

all these large logarithmic contributions are absorbed into the renormalization of the coupling and the answer will be consistent. So in fact this answer is not only a second order result but takes into account all order in the leading logarithmic approximation.

CRUCIAL EXPERIMENTS AT 200 TeV

A. Ringwald*

DESY - Hamburg

1. Introduction

Recently there has been growing interest in non-perturbative effects in the standard electroweak theory. It has been suggested that electroweak interactions become strong at high energies, showing new phenomena like high-multiplicity events involving "weakly" interacting particles and baryon (B) and lepton (L) number violation [1 – 3]. The purpose of this lecture is to present arguments in favour of this possibility and to discuss the implications for future collider experiments.

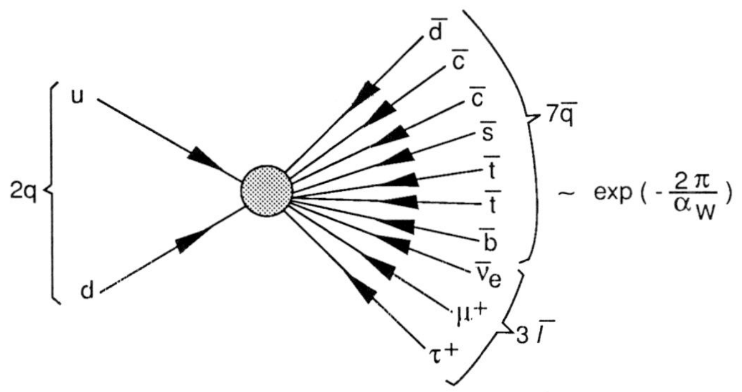

Fig. 1. (B+L)-violating amplitude in the standard model induced by an instanton.

G. 't Hooft was the first to point out B and L are not strictly conserved in the electroweak theory [4]. He showed that topologically non-trivial gauge

* Address after October 1st, 1990: CERN, Geneva.

Physics Up to 200 TeV
Edited by A. Zichichi, Plenum Press, New York, 1991

fields (*instantons*) induce (B+L)-violating vertices (see Fig. 1). These vertices are proportional to $\exp(-S_{\mathrm{inst}}) = \exp(-2\pi/\alpha_w) \sim 10^{-78}$, where S_{inst} is the instanton action, and it was commonly believed, due to the smallness of this number, that the (B+L)-violating processes would be unobservable. However, it has been argued recently that such events might occur frequently in high-energy collisions [5, 1 − 3].

Intuitively one can understand this behaviour through the following reasoning: the factor $\exp(-S_{\mathrm{inst}})$ is associated with a tunneling process [6, 7], which occurs between topologically inequivalent vacua [6] in the electroweak theory, characterized, in the temporal gauge, by the winding numbers n of the vacuum SU(2) gauge fields (see Fig. 2). The transitions between inequivalent vacua have drastic effects due to an anomaly [8] in the B+L current. Whenever the fields make a transition between vacua with different winding numbers, $\Delta n \neq 0$, there occur (B+L)-violating processes according to the selection rule

$$\Delta \mathrm{B} = \Delta \mathrm{L} = -n_g \Delta n, \tag{1.1}$$

where n_g is the number of generations.

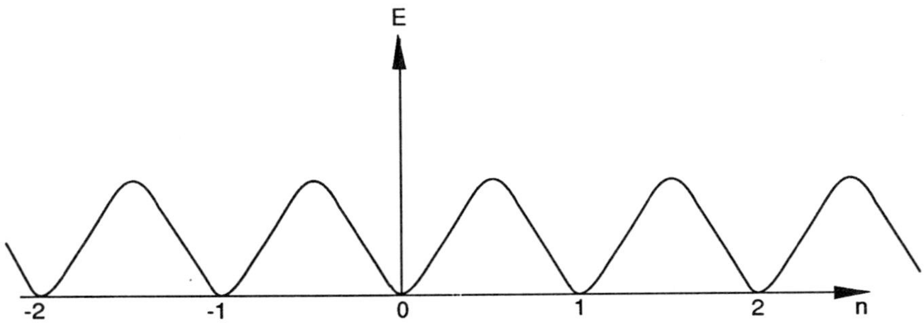

Fig. 2. Schematic plot of the vacuum structure in a SU(2) gauge theory.

However, the transitions with $\Delta n \neq 0$ cost energy: the topologically inequivalent vacua are separated by an energy barrier [6]. In the electroweak theory the minimum barrier height between the different vacua is of order $M_{\mathrm{sp}} \simeq \pi m_w/\alpha_w \simeq$ 10 TeV [9]. The subscript "sp" stands for the *sphaleron*, which is a static saddle-point solution of the bosonic sector of the electroweak theory and corresponds to the configuration which sits at the top of the barrier between topologically inequivalent vacua [9]. Its energy sets the scale above which the nontrivial structure of the vacuum could possibly play a role and where the anomalous (B+L)-violating processes could occur frequently. Below this energy the processes are assumed to be associated with *tunneling* and are therefore exponentially suppressed, $\sim \exp(-2\pi/\alpha_w)$, with the instanton describing the tunneling event [6, 7]. At energies above $\sim$ 10 TeV one may expect from these simple energy considerations that the large suppression disappears because then transitions are possible on the *classical* level. It should be noted that the usual perturbation theory corresponds to the expansion around *one* of the vacua; it ignores $\Delta n \neq 0$ transitions. This is appropriate at low energies, but could be misleading at high energies.

There is by now a consensus that (B+L)-violating effects are important in a hot plasma, in particular in the early universe [10]. Hints of the possible importance of (B+L)-violating processes in high-energy, two-particle collisions have been observed in [1] (and later also in ref. [2]; for an earlier attempt see ref. [5]). It was found that the relevant (B+L)-violating processes are those with the associated production of many, $\mathcal{O}(1/\alpha_w)$, W, Z, and Higgs bosons; this has been conjectured earlier in refs. [9, 11].

The organization of this lecture is as follows. In sect. 2 we review the naive instanton calculation of (B+L)-violating amplitudes in the electroweak theory. It is shown that the amplitudes, in the leading-order semiclassical approximation, are point-like in the sense that there are no form factors suppressing momentum transfers between external legs. Due to this feature the leading-order matrix elements will violate partial-wave unitarity at high energies ($\sim M_{\rm sp}$ for associated production of $n_w \sim \pi/\alpha_w$ W bosons). We discuss the higher-order corrections to the naive instanton result in sect. 3. With the presently-available techniques we can only investigate the total cross-section for B+L violation at energies $\ll M_{\rm sp}$, where it is exponentially growing, but still exponentially small. In sect. 4 we represent speculations about the behaviour of the total cross-section near and above the sphaleron scale. If the 't Hooft suppression factor can be completely overcome at the sphaleron scale, one expects weak interactions above this scale to resemble closely strong interactions, with very interesting phenomenology. Section 5 is devoted to the discussion of this phenomenology. In sect. 6 we present our conclusions.

2. The Naive Instanton Calculation

We will be interested in the *total* (B+L)-violating cross-section. To this end we consider first the *exclusive* processes (see Fig. 3)

$$q_1 + q_1 \Longrightarrow \overline{q}_1 + 3\overline{q}_2 + 3\overline{q}_3 + \overline{l}_1 + \overline{l}_2 + \overline{l}_3 + n_w\ W(Z) + n_h\ H, \tag{2.1}$$

with an arbitrary number of W, Z, and H(iggs) bosons. The subscripts label the different generations ($n_g = 3$). According to the anomaly selection rule (1.1) these are, in fermion content, the simplest anomalous processes ($\Delta n = 1$). We are considering processes which involve also a number of bosons because the sphaleron picture suggests that the *relevant* processes are those with the associated production of many, $\mathcal{O}(1/\alpha_w)$, W, Z, and Higgs bosons in addition to the anti-quarks and anti-leptons. The argument goes as follows [9, 11]: in order to pass over the

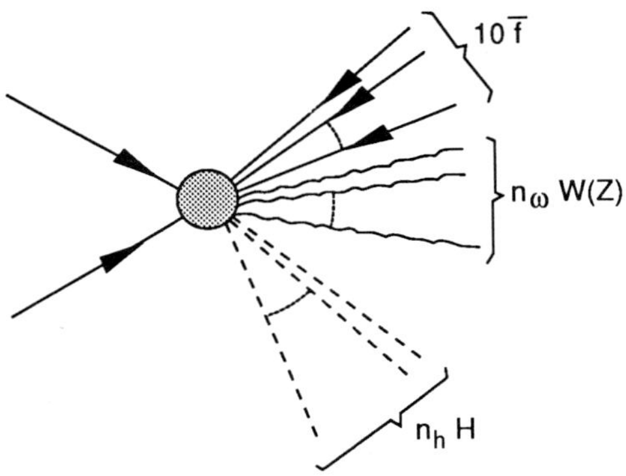

Fig. 3. The fermion-number violating vertex (2.1).

barrier the fields have to configure themselves into "physical" sphaleron-like configurations with masses of order m_w/α_w and radii of order $1/m_w$. The sphaleron configuration then typically decays into $1/\alpha_w$ W(Z) and Higgs bosons, producing the anti-quarks and anti-leptons as a side effect due to the anomaly.

In order to compute the $S-$matrix elements for the exclusive processes (2.1) we use the LSZ approach: calculate the appropriate (connected) Green function, amputate it and obtain the on-shell vertex of the processes. In the following we will neglect fermion masses and set $\sin^2\theta_w = 0$. The Green function associated with the process (2.1) is

$$G^{\text{con}}(\{x_i\},\{y_j\},\{z_k\}) = \langle 0 \mid T\Big[\prod_{i=1}^{12}\overline{\psi}_L^{(i)}(x_i)\prod_{j=1}^{n_w}W_{\mu_j}(y_j)\prod_{k=1}^{n_h}\varphi(z_k)\Big] \mid 0\rangle^{\text{con}}. \tag{2.2}$$

The superscripts on the fermion fields label different doublets. The field $\varphi(x)$ appearing in eq. (2.2) denotes the interpolating field corresponding to the neutral Higgs boson, H.

Since the Green function involves only $\overline{\psi}$'s it vanishes, in every finite order of conventional perturbation theory in α_w (expansion around the topologically trivial vacuum solution). It can be shown [4, 7] that the Euclidean path integral expression corresponding to (2.2) receives contributions only from the sector of gauge fields which have one unit of Euclidean topological (Pontryagin) number

$$q = \frac{g^2}{16\pi^2}\int d^4x \; \text{tr}(F\tilde{F}) = 1, \tag{2.3}$$

where $\tilde{F}$ denotes the dual SU(2) field strength. This is plausible since Euclidean gauge fields with Pontryagin number q correspond to Minkowskian gauge fields which interpolate between topologically inequivalent vacua with $\Delta n = q$ [6, 7]. In the presence of $q = 1$ gauge fields there exists an asymmetry between the number of normalizable right-handed and left-handed zero modes of the Euclidean Dirac operator [12], which is necessary for a non-vanishing fermion-number violating Green function [4, 7].

The path integral over the gauge and Higgs fields can be done in a semiclassical way by expanding around classical gauge and Higgs field configurations which dominate the path integral. It has been shown by 't Hooft [4] and Affleck [13] that the dominant classical gauge fields with $q = 1$ can be represented by a family of configurations, labeled by the position R_μ, a scale size ρ, and a global SU(2)-matrix U, which represents the orientation. The configurations have the following asymptotic form (for $\rho m \ll 1$),

$$W_\mu^{\rm cl}(x-R;\rho,U) = \begin{cases} \frac{2}{g}\frac{\rho^2}{(x-R)^2}\frac{U\overline{\tau}_{\mu\nu}U^\dagger(x-R)_\nu}{(x-R)^2+\rho^2}, & \text{for } \mid x-R \mid < m_w^{-1}; \\ -\frac{4\pi^2\rho^2}{g}U\overline{\tau}_{\mu\nu}U^\dagger\partial_\nu G_{m_w}(x-R), & \text{for } \mid x-R \mid \gg m_w^{-1}, \end{cases} \tag{2.4}$$

$$\Phi^{\rm cl}(x-R;\rho) = \frac{v}{\sqrt{2}}\begin{pmatrix}0\\1\end{pmatrix}\begin{cases}\left(\frac{(x-R)^2}{(x-R)^2+\rho^2}\right)^{1/2}, & \text{for } \mid x-R \mid < m_h^{-1}; \\ \left[1-2\pi^2\rho^2 G_{m_h}(x-R)\right], & \text{for } \mid x-R \mid \gg m_h^{-1}\ . \end{cases} \tag{2.5}$$

Here $\overline{\tau}_{\mu\nu} = \frac{\tau^a}{2}\overline{\eta}_{a\mu\nu}$ (the $\overline{\eta}$'s are 't Hooft's symbols [4]), $v = 246$ GeV is the Higgs field vacuum expectation value, and $G_m(x) = 1/(4\pi^2)(m/r)K_1(mr)$. This configuration has been called a *constrained instanton* [13], since it was obtained by introducing a scale constraint into the path integral. Near the center of the configuration the classical gauge field looks like the pure SU(2) instanton [14]. Far away from the center it decays exponentially to a pure gauge.

We obtain, in the leading-order approximation around the constrained instanton, the following form for the Fourier transform (denoted by tildes) of the Euclidean Green function corresponding to (2.2) [1, 2] (see Fig. 4)

$$\tilde{G}_E^{\rm con}(\{p_i\},\{k_j\},\{q_k\}) = (2\pi)^4\delta^{(4)}\Big(\sum_{i=1}^{12}p_i + \sum_{j=1}^{n_w}k_j + \sum_{k=1}^{n_h}q_k\Big) \tag{2.6}$$

$$\times\int dU\int_0^\infty d\rho\, n(\rho)\prod_{i=1}^{12}\tilde{\chi}_{\rm cl}^{0\dagger}(p_i;\rho,U)\prod_{j=1}^{n_w}\tilde{W}^{\rm cl}(k_j;\rho,U)\prod_{k=1}^{n_h}\tilde{\varphi}^{\rm cl}(q_k;\rho).$$

Here the "instanton density" $n(\rho)$ is proportional to (in lowest order in ρv)

$$n(\rho) \propto \exp(-S_E[W^{\rm cl},\Phi^{\rm cl}]) \sim \exp(-\frac{2\pi}{\alpha_w} - \pi^2 v^2\rho^2), \tag{2.7}$$

times determinants in the background of the constrained instanton. The latter lead essentially to the replacement of α_w by the running coupling constant (in one-loop) $\alpha_w(\rho v)$ [4]. The $\chi_{\rm cl}^0$ are the (right-handed) zero modes in the background of the instanton, $\not{D}(W^{\rm cl})\chi_{\rm cl}^0 = 0$ [4].

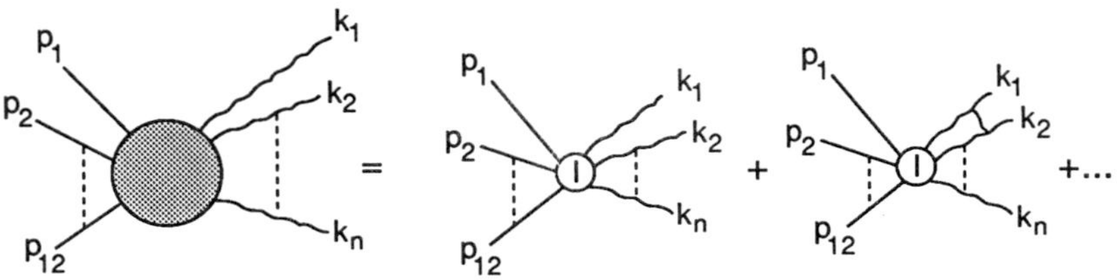

Fig. 4. The semiclassical expansion of the (B+L)-violating Green function in the single-instanton sector. The first term on the right hand side is the leading-order term (2.6). The second term is an $\mathcal{O}(\alpha_w)$ correction: two external lines are connected by a propagator in the instanton background.

It is easily seen that eq. (2.6) has poles only near the mass shells of the individual particles which, in turn, depend solely on the large distance behaviour of the classical fields (see also ref. [15]): the *infrared* behavior of the classical fields determines the behaviour of the leading-order *on-shell* scattering amplitudes. Near the mass-shell the Fourier transforms of the classical fields look like free propagators with non-trivial residua.

Observe that in the leading-order approximation there is no correlation (besides the collective coordinate integrals) between different external legs in the Green function (2.6). The Green function has *form factors* in the *individual virtualities* p_i^2, k_j^2, and q_k^2, but not in the *pair invariants* like $q_i \cdot q_j$. An immediate consequence of this feature is that *off-shell*, deeply in the Euclidean region ($p_i^2, k_j^2, q_k^2 \to \infty$), the Green function (2.6) scales, due to the finite size of the constrained instanton, with the naive scaling dimension. This behaviour is consistent with renormalizability and asymptotic freedom. *On-shell*, however, there are no form factors remaining, even if one does not neglect the finite size of the instanton as 't Hooft did [4].

After amputation of the Green function (2.6), i.e. after multiplication by the inverse free propagators for the external particles, we obtain the *on-shell* vertex function [1, 2]

$$\tilde{\Gamma}_{|\text{on-shell}} = (2\pi)^4 \delta^{(4)}\Big(\sum_{i=1}^{12} p_i + \sum_{j=1}^{n_w} k_j + \sum_{k=1}^{n_h} q_k\Big) \int dU \prod_{i=1}^{12} \Big(U^\dagger \xi\Big)^{a_i}_{\alpha_i} \prod_{j=1}^{n_w} \Big(U \bar{\tau} U^\dagger\Big)^{b_j}_{\mu_j\,\nu} k_j^\nu$$

$$\times \int_0^\infty d\rho\; n(\rho)(2\pi\rho)^{12} \left(i\frac{4\pi^2\rho^2}{g}\right)^{n_w} \left(-2v\pi^2\rho^2\right)^{n_h}. \tag{2.8}$$

Here ξ is a constant matrix acting in spin and weak isospin space. For the momenta k^ν the analytical continuation $k^0 = i\sqrt{\vec{k}^2 + m_w^2} = i\omega_{\vec{k}}$ is implied. The ρ–integral leads to a *factorial growth*, $\sim (n_w + n_h)!$, of the amplitudes. It should be noted however that this is strictly true only as long as long as the dominating ρ in (2.8) is smaller than m^{-1}. Otherwise the large distance normalizations of the classical fields (2.4) and (2.5) as well as the classical action in (2.7) will change. Since the dominating ρ according to (2.8) is given by $\rho_0 \simeq \sqrt{n}/(\pi v)$, our results are only valid for $n_w \ll \pi/\alpha_w$.

The amplitude (2.8) is very similar to a *point-like* interaction: only the gauge fields lead to some momentum dependence. *On-shell* there is no form factor suppressing momentum transfers. In the context of the semiclassical expansion around non-trivial classical background fields there is no way around this paradox. Due to the point-like nature the amplitudes will violate partial-wave unitarity at high energies. The produced particles are concentrated in the lowest partial waves, i.e. they will come out *quasi-isotropically* from the vertex.

In order to obtain the S-matrix element for the process (2.1) we have to multiply the amplitude (2.8) by the wave functions for the external particles*. Disregarding the averaging over the instanton orientation, which unnecessarily complicates the analysis**, we obtain for the squared amplitudes [1, 2, 18]

$$\mathcal{S} \mid \mathcal{M}(qq \to 7\bar{q}\; 3\bar{l}\; n_w W\; n_h H) \mid^2 \simeq$$

$$\left(\frac{\Lambda_W}{v}\right)^{2b} 2^{2(n_w+n_h)} v^{-2(n_w+n_h)} 3^{n_w} \frac{\Big[(n_w+n_h)!\Big]^2}{n_w! n_h!} \prod_{i=1}^{12} \mid \vec{p}_i \mid \prod_{j=1}^{n_w} \left[1 + \frac{4}{3}\left(\frac{\vec{k}_j^2}{m_w^2}\right)\right]. \tag{2.9}$$

Here Λ_W is the energy scale where the SU(2) gauge coupling would become strong if there were no spontaneous symmetry breaking,

$$\Lambda_W = \text{const} \cdot v \cdot \exp\Big(-\frac{2\pi}{b\alpha_w(v)}\Big) \simeq 10^{-23} \text{ GeV}, \tag{2.10}$$

and b is the coefficient of the β-function, $b = \frac{43}{6} - \frac{4}{3} n_g = \frac{19}{6}$.

---

* In ref. [1] the cross-sections for the production of vector particles were underestimated. This was due to the fact that we were not careful enough with the analytic continuation. Since the wave functions of vector particles carry a space-time index they transform non-trivially under the Wick rotation: the zero component of the wave function acquires the factor i. The Euclidean counterparts of the polarization vectors read therefore [16] $\epsilon_\mu^{1,2}(\vec{k}) = (0, \vec{e}^{\,1,2}(\vec{k}))$, $\epsilon_\mu^3(\vec{k}) = \frac{1}{m_w}\left(i \mid \vec{k} \mid, \omega_{\vec{k}} \frac{\vec{k}}{|\vec{k}|}\right)$. In the squared amplitudes one needs the following quantity [16] $\sum_m \epsilon_\mu^m(\vec{k}) \bar{\eta}^a_{\mu\nu} k_\nu \epsilon_\alpha^{*m}(\vec{k}) \bar{\eta}^b_{\alpha\beta} k^*_\beta = \delta_{ab}(\omega_{\vec{k}}^2 + \vec{k}^2) - 2k_a k_b + 2i\omega_{\vec{k}} \epsilon_{abc} k_c$. This formula differs from the corresponding one in ref. [1], eq. (65), in the three-momentum dependence. The origin lies in the factor of i, as explained above.

** In refs. [16, 17] it was found that averaging over the orientation will not affect the final formula for the inclusive (B+L)-violating cross-section, (2.12)-(2.14).

Using eq. (2.9) we obtain [1 − 3, 18] for the (parton) cross-sections of the exclusive (B+L)-violating processes (2.1) (here we use the formula for relativistic phase space)

$$\hat{\sigma}_{n_w,n_h} \simeq \frac{1}{v^2}\left(\frac{\Lambda_W}{v}\right)^{2b}\frac{1}{n_h!}\frac{1}{(n_w!)^3}\left(\frac{\hat{s}}{16\pi^2 v^2}\right)^{13}\left(\frac{\hat{s}}{4\pi^2 v^2}\right)^{n_h}\left(\frac{3\hat{s}^2}{8\pi^2 g^2 v^4}\right)^{n_w}. \quad (2.11)$$

From (2.11) we find that for (parton) cm energies $\sqrt{\hat{s}}$ larger a few times m_w the total (B+L)-violating cross-section rises exponentially

$$\hat{\sigma}_{\Delta(B+L)} = \sum_{n_w,n_h} \hat{\sigma}_{n_w,n_h} \simeq \frac{1}{v^2}\left(\frac{\hat{s}}{16\pi^2 v^2}\right)^{13} \exp\left[-\frac{4\pi}{\alpha_w}F\left(\frac{\sqrt{\hat{s}}}{M_{\text{sp}}}\right)\right], \quad (2.12)$$

where the function F is given by [16, 18]

$$F(x) = 1 - \left(\frac{9}{16\sqrt{8}}\right)^{2/3} x^{4/3}. \quad (2.13)$$

Here we have defined *

$$M_{\text{sp}} \equiv \pi\frac{m_w}{\alpha_w} = 7.2 \text{ TeV}. \quad (2.14)$$

The first term in eq. (2.13) gives rise to the familiar 't Hooft suppression factor, $\sim \exp(-4\pi/\alpha_w)$ [4]. The second term in eq. (2.13), which leads to an exponential growth of the cross-section, comes from W and Z production: for energies below the sphaleron, (B+L)-violating processes are dominated by the associated production of W's and Z's [16 − 18]. The average multiplicity increases according to a power law,

$$\langle n_w \rangle \simeq \frac{4\pi}{3\alpha_w}\left(\frac{9}{16\sqrt{8}}\right)^{2/3}\left(\frac{\sqrt{\hat{s}}}{M_{\text{sp}}}\right)^{4/3}. \quad (2.15)$$

Using eqs. (2.12)-(2.14) we find that the cross-section blows up exponentially, thereby violating S-wave unitarity, at the scale $\sqrt{\hat{s}} \simeq M_0$,

$$M_0 = 2^{-1/4}\frac{8}{3}M_{\text{sp}} \simeq 16 \text{ TeV}, \quad (2.16)$$

where the mean multiplicity is $\langle n_w \rangle \simeq (4\pi/3)\alpha_w^{-1} \simeq 120$**. *The results seem to fit nicely onto the suggestions from the sphaleron picture and related energy considerations: above a threshold of the order of 10 TeV anomalous (B+L)-violating processes are unsuppressed and associated with the multi-particle production of gauge bosons.* However, the leading-order calculation has a very bad disease:

---

* The mass (2.14) is of the order but not identical to the sphaleron mass [9], which depends slightly on the Higgs mass and varies between 7 and 14 TeV if one varies the Higgs mass between 0 and ∞.

** Using the leading order matrix elements, eq. (2.9), but taking into account exact phase space changes M_0 to 17 TeV and n_w at this energy to 85 [19].

partial-wave unitarity will be violated at high energies. Therefore it is necessary to consider higher-order corrections to the amplitudes.

3. Corrections to the Naive Instanton Approximation

The leading-order calculation of the total cross-section for B+L violation certainly breaks down at the energy scale M_0, eq. (2.16). The key question is whether the breakdown does not occur already much earlier, say at m_w? There are many indications that perturbative corrections in the single-instanton sector become important already at $\sqrt{\hat{s}} \sim m_w$ for the *exclusive* cross-sections (2.11) at large n [20 – 24]. However, there are also arguments that the corrections to the leading-order *total* cross-section exponentiate in such a way that their effects are irrelevant up to energies of order M_{sp}; it was argued [16, 21 – 24] that the total cross-section for the processes (2.1) has the generic form, for $\alpha_w \to 0$, $\sqrt{\hat{s}}/M_{\text{sp}}$ fixed,

$$\hat{\sigma}_{\Delta(\text{B+L})} \sim \exp\left[-\frac{4\pi}{\alpha_w} F\left(\frac{\sqrt{\hat{s}}}{M_{\text{sp}}}\right)\{1 + \mathcal{O}(\alpha_w)\}\right]. \tag{3.1}$$

Only if F has a zero is there a chance that electroweak B+L violation in association with multi-W production becomes relevant at future colliders.

For energies much less than the sphaleron energy the function F may be computed in perturbation theory, which is equivalent to a low-energy expansion in terms of $x \equiv \sqrt{\hat{s}}/M_{\text{sp}} \ll 1$. It is possible to show [16], that at low energies, $x \ll 1$, F depends on energy as in eq. (2.13). Unfortunately, the function F is not known at $\sqrt{\hat{s}} \sim M_{\text{sp}}$. At $\sqrt{\hat{s}} \simeq M_{\text{sp}}$ all tree graphs in the background of the instanton seem to contribute substantially to F [16, 21 – 24], indicating a breakdown of perturbation theory around the instanton at this energy. This may signal that it is inappropriate to expand around the instanton and that a better zeroth-order approximation has to be found (the so-called *distorted* instanton [3, 25]).

The arguments leading to the expectation that in general $\hat{\sigma}_{\Delta(B+L)}$ will behave as in eq. (3.1) were mainly combinatorical [16, 21 – 23]. To calculate explicitly the corrections to the exclusive amplitudes (and afterwards to sum over all channels in order to obtain the correction to F) is extremely difficult, since the propagators in the background of the *constrained* instanton - the building blocks of perturbation theory in the single-instanton sector - are not known*. An alternative and perhaps easier approach to the problem of higher-order corrections to F, eq. (2.13), is to consider dispersion relations [17, 18], i.e. to calculate the total cross-section for B+L violation via the imaginary part of the forward elastic scattering amplitude. The main reason why dispersion relations are particularly convenient is that phase

---

* The propagators in the background of the *pure* SU(2) instanton are known explicitly [26]. However, the gauge propagator, which gives probably the most important correction, is singular and must be modified [27, 23].

space and quantum corrections due to W's and H's are automatically taken into account. Also, unitarity is manifestly built in.

The dispersion relation for the total cross-section for fermion-number violating processes (the sum over all W's and H's in eq. (2.1)) reads

$$\hat{\sigma}_{\Delta(B+L)} = \frac{1}{4\hat{s}} \cdot \mathrm{Im}\ \mathcal{T}(p_1, p_2, p_1, p_2) \tag{3.2}$$

$$= \frac{1}{4\hat{s}} \cdot \mathrm{Im}\Big\{ (\mathrm{LSZ}) \Big(\langle 0 \mid T\big[\psi_1(x_1)\psi_2(x_2)\mathcal{P}\psi_1^\dagger(x_3)\psi_2^\dagger(x_4)\big] \mid 0 \rangle \Big) \Big\}.$$

Here $\mathcal{T}$ denotes the elastic scattering amplitude and "LSZ" stands for the usual LSZ procedure. One has to select only those intermediate states which contain exactly 10 fermions (see eq. (2.1)) [17] by means of a projection operator $\mathcal{P}$.

The Euclidean functional integral for the 4-point Green function entering in eq. (3.2)

$$G_E = \int \mathcal{D}W\mathcal{D}\Phi\mathcal{D}\psi\mathcal{D}\psi^\dagger\ \psi_1\psi_2\mathcal{P}\psi_1^\dagger\psi_2^\dagger \exp(-S[W, \Phi, \psi]) \tag{3.3}$$

can be calculated systematically by semiclassical methods; it receives its dominant contribution [28, 29] from an instanton/anti-instanton *valley** in functional space, labeled by the double set of collective coordinates of an instanton/anti-instanton pair. The dominant configuration in Minkowski space is a *saddle-point* [29] - one point of the valley bottom - with one negative mode, which gives rise to the imaginary part of the forward scattering amplitude (see also ref. [17]) . Unfortunately, the valley in functional space is only known for large instanton/anti-instanton separation [28], which in our case means for energies low compared to the sphaleron energy [17, 29]. One finds [29] that the total cross-section of the processes (2.1) behaves like in eq. (3.1), where

$$F(x) = 1 - \left(\frac{9}{16\sqrt{8}}\right)^{2/3} x^{4/3} + \frac{3}{32}x^2 + \mathcal{O}\Big((\frac{\lambda}{g^2} + 1)x^{8/3} \Big). \tag{3.4}$$

We see that indeed the sphaleron energy (remember, $x \equiv \sqrt{\hat{s}}/M_{\mathrm{sp}}$) sets the scale for the energy dependence of the total cross-section. We have therefore verified the conjecture [16, 21 − 24] that the total cross-section for B+L violation has the form (3.1). Note, since we have a complete control over the order of the neglected contributions, we can state explicitly that (3.4) gives the correct behaviour of the

* An instanton/anti-instanton pair at large separations R is not a new solution of the equations of motion, but represents a valley configuration [28]: the action varies only slowly for this pair. In general, the valley can be defined as the gradient line of steepest descent in the zero topological charge sector of functional space [28].

total fermion-number-violating cross-section in the energy range $m_w \ll \sqrt{\hat{s}} \ll M_{\text{sp}}$. The prefactors before the exponential in (3.1) can also be calculated [29].

If we *naively* omit the unknown corrections in eq. (3.4) we find that the cross-section has a maximum at $x = 8\sqrt{2}/3$, where it is still exponentially small, $\hat{\sigma}_{\Delta(B+L)} \sim \exp(-4\pi/\alpha_w)(1/3)) \sim 10^{-52}$. It becomes much larger than the 't Hooft factor but is still unobservably small*. This maximum is of course outside the region of validity of our formula. However, it demonstrates that there is a possibility that the total cross-section, even in the single-instanton sector, will not violate the unitarity bound. In this case one does not need to invoke multi-instanton induced processes [32] in order to save unitarity.

4. High-Energy Electroweak Interactions

To summarize, we know that the total cross-section for electroweak B+L violation grows exponentially at energies small compared to the sphaleron scale. We know nothing about the behaviour of the cross-section near or above this scale. There are still two possibilities for the function F ($\hat{\sigma}_{\Delta(B+L)} \sim \exp[-(4\pi/\alpha_w)F(x)]$): *at high energies*

i) F has a minimum above zero; in this case B+L violation will always be suppressed;

ii) F has a zero or approaches zero; in this case B+L violation will be relatively strong.

---

* In a different approach [30], based on a Minkowski-space calculation, it was claimed that in theories in which the instanton has a fixed size - unlike the electroweak theory - the cross-section will have the same qualitative behaviour as obtained here: it will first grow exponentially, but will never be unsuppressed because the overlap of the inital states with the sphaleron decreases exponentially with energy. One comment on this approach is in order: The Minkowski-space calculation done in ref. [30] relies on the notion of the "most probable escape path in Minkowski space" which is obtained from the instanton by interpreting its Euclidean time variable as a path parameter; the corresponding Minkowskian classical field interpolates between topologically inequivalent vacua and passes the sphaleron. However, it is not clear that for energies approaching the sphaleron it is appropriate to consider just one particular path which exactly passes the sphaleron - there may be many more paths which are equally probable: we do not have to create exactly the sphaleron. That there is indeed the possibility that large coherent fields can be produced at high energies in the electroweak theory has been discussed in several works [31]. In any case, both approaches, the Euclidean and Minkowskian approach, should ultimately converge to give the same answer.

McLerran, Vainshtein and Voloshin [3] gave arguments in favour of the second possibility. They suggested that above a sharp threshold at $\sqrt{\hat{s}} \sim M_{sp}$ weak interactions would become "strong" in the sense that $n_w \sim \pi/\alpha_w = \mathcal{O}(100)$ weakly-interacting particles (mainly W-bosons) will be produced with a relatively large cross-section, $\mathcal{O}(0.1)$ nb. Close to the threshold these particles come out quasi-isotropically from the vertex. At higher energies, the S-wave amplitude would then be bounded by unitarity, and other partial waves, as well as (B+L)-*conserving* amplitudes would become strong, inducing large forward-peaked cross-sections at asymptotic energies $\sqrt{\hat{s}} \gg M_{sp}$.

Possibly the strong interaction above $M_{sp} \sim m_w/\alpha_w$ is related to the long-expected breakdown of conventional (in the topologically trivial sector) perturbation theory at high orders $N > 1/\alpha_w$ (see e.g. [33] and references therein). Indeed, it was argued in ref. [31] that conventional perturbation theory, applied to (B+L)-conserving processes, breaks down at similar energies and multiplicities, at least in the high-energy, fixed-angle regime.

There may even be deeper analogies between the electroweak interactions and QCD at very high energies, as both are non-Abelian gauge theories with similar infrared divergences [34] and topological features. These analogies refer not only to the scattering of a few particles at large angles, which is well described by (parton model) perturbation theory, but also to the large, logarithmically-rising, total inelastic cross-section, where perturbation theory presumably fails.

Based on above considerations it was suggested [3, 34, 35]:

1) At high energies $\sqrt{\hat{s}}$ the total weak inelastic cross-section becomes approximately energy-independent

$$\hat{\sigma}_w = 4\pi c_w m_w^{-2} \approx 0.1 \text{ nb} \quad - \quad 10 \ \mu\text{b}, \tag{4.1}$$

where c_w is a logarithmically-varying coefficient;

2) the asymptotic behaviour sets in abruptly above a certain threshold energy E_w^{crit} of the order of a few TeV to about 20 TeV;
3) the mean constituent multiplicity is high, typically of order $1/\alpha_w \sim 30$;
4) there is a substantial contribution of (B+L)-violating events to the total weak cross-section;
5) the inclusive differential cross-sections decrease rapidly for transverse momenta exceeding a characteristic value of order m_w.

In a simple picture [3, 34, 35] these processes arise from the scattering of the clouds of weak gauge bosons and scalars carried along by colliding quarks or leptons. Both the almost constant multi-particle cross-sections and the characteristic p_t distribution directly reflect the finite size of these clouds ($\propto m_w^{-1}$) which is, of course, related to the finite range of the underlying interactions. We therefore call processes with these characteristics Geometrical Flavour Interactions (GFI) [35].

In collisions with low $p_t < m_w$ the resolution is "bad", $\Delta x > m_w^{-1}$. Therefore, only the gross *geometrical* features are relevant here, namely, the "weak size" of quarks or leptons corresponding to the size of the W, Z, Higgs clouds, and the "opacity" of the colliding objects, which is measured by the factor c_w in (4.1). Within such a geometrical scattering scenario, the most speculative assumptions concern the high multiplicity and the large value of c_w. (The "opaqueness" of the colliding objects is actually related to the large multiplicity.) Intuitively, one may envisage that a non-perturbative behaviour of the multiparticle cross-sections arises from the strong "flavourelectric" and "flavourmagnetic" fields created by the presence of more than α_w^{-1} weak charges in a volume with a characteristic linear size m_w^{-1} [34, 35].

5. Possible Consequences at Future Colliders

There remain considerable theoretical uncertainties in these non-perturbative estimates, notably whether the rapidly-rising (B+L)-violating cross-section flattens out at some energy below $M_{\rm sp}$, and whether it ever reaches a sizeable fraction of the unitarity limit. Nevertheless, the consequences of large non-perturbative electroweak interactions are so far-reaching and dramatic*, that we believe it is worthwhile to study the phenomenology that would be expected at future colliders. As working hypotheses we will adopt points 1)-5) of sect. 4.

We explore the notion that GFI events set in abruptly once the energy is high enough to produce a critical number $n_w^{\rm crit}$ of weakly-interacting particles. As mentioned above, these phenomena are presumably related to the breakdown of perturbation theory for large particle numbers, namely when $n_w \alpha_w$ becomes large. We therefore assume $n_w^{\rm crit}\alpha_w = \nu_w$ with ν_w some constant of order one. The threshold energy can then be roughly guessed by requiring that $\sqrt{\hat{s}}$ must be large enough to produce $n_w^{\rm crit}$ particles with mass m_w without a strong phase space suppression. We therefore parametrize

$$E_w^{\rm crit} = f_w n_w^{\rm crit} m_w = f_w \nu_w m_w / \alpha_w. \tag{5.1}$$

Both instanton estimates [1 – 3] and a comparison with QCD [34, 35] suggest values of $n_w^{\rm crit}$ and $E_w^{\rm crit}$ in the range $n_w^{\rm crit} \simeq 30 - 100$, $E_w^{\rm crit} \simeq (3.5 - 20)$ TeV. We use a conservative value $n_w = 30$ for the phenomenological discussion: all results can be rescaled easily for the experimentally even more advantageous case of higher multiplicity.

---

* For a detailed discussion of the expected phenomenology see ref. [35]. A short version is also contained in ref. [36].

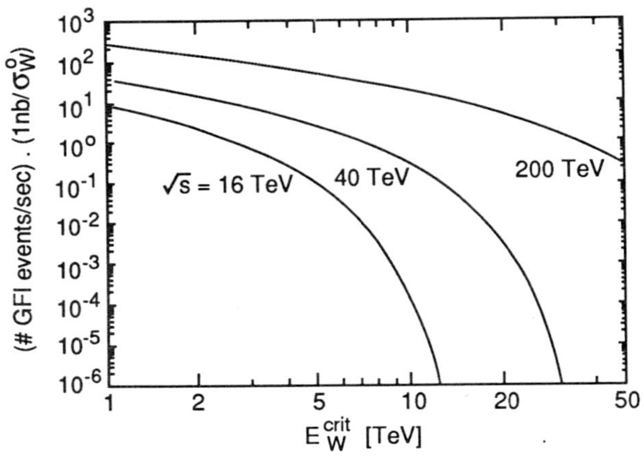

Fig. 5. Number of GFI events per second as a function of the (parton) threshold energy E_w^{crit} (taken from ref. [35]). We assume a constant cross-section $\sigma_w^0 = 1$ nb for weakly-interacting partons with energies above E_w^{crit} and a pp luminosity of $\mathcal{L} = 10^{33}$ cm^{-2}sec^{-1}.

Let us assume an abrupt onset of geometrical QFD cross-sections for $\sqrt{\hat{s}} \geq E_w^{\text{crit}}$, and approximate $\hat{\sigma}_w(\sqrt{\hat{s}}) = \sigma_w^0 \theta(\sqrt{\hat{s}} - E_w^{\text{crit}})$, with $\sigma_w^0 \simeq 0.1$ nb $-$ 10 μb [3, 34, 35]. After folding $\hat{\sigma}_w$ with the parton distributions (no gluons) we obtain the event rate for the production of many weakly-interacting particles in proton-proton collisions, see fig. 5. The limit of detectability should be around 100 events per year. We conclude that GFI events can be seen for $E_w^{\text{crit}} < 11, 28$, and 130 TeV for the LHC, SSC, and ELOISATRON, respectively, if σ_w^0 is around 1 nb. The possibility for detection at the LHC depends crucially on the precise values of E_w^{crit} and σ_w^0. A future collider in the 100 TeV range (ELOISATRON) would be needed for a full exploration of the asymptotic behaviour beyond the threshold energy.

Let us concentrate [35] on the light fermions of mass smaller than the transverse momenta of the produced particles $\sim 30 - 50$ GeV, which may be produced both promptly and as decay products of W, Z bosons, etc. The associated jets and charged leptons are directly accessible experimentally. One finds [35] the following characteristic features of "geometrical flavour interaction" (GFI) events (for $n_w = 30$):

1) For parton kinematics near the threshold ($\sqrt{\hat{s}} \approx E_w^{\text{crit}}$), the events look rather central, i.e. jets and charged leptons are distributed over the whole angular range. Only partons with energies much above E_w^{crit} produce more forward-oriented events. For "threshold" machines with $\sqrt{s} = 16 - 40$ TeV most of the activity will be in the central detector ($10^\circ < \vartheta < 170^\circ$). Even at $\sqrt{s} = 200$

TeV one may expect on the average about one half of the produced particles in this central region.

2) The average transverse momentum per light particle is estimated to be $\overline{p}_t^l \approx 35$ GeV, and the total transverse energy is large: $\langle E_t^{\rm tot} \rangle \approx 1.6 - 2.2$ TeV.
3) We expect on average at least 3.5 "isolated" electrons or muons per event.
4) A similar number of neutrinos is responsible for an average missing transverse momentum $\langle \not{p}_t^{\rm tot} \rangle \approx 100$ GeV.
5) More than 20 jets/event, corresponding to ~ 200 charged hadrons, should be seen in the central detector.
6) In addition, we expect many events with a high charged hadron multiplicity ($n_h > 100$) in the forward and/or backward detectors ($|\eta| > 2.5$).
7) If B+L violation is relatively strong in comparison to the (B+L)-conserving GFI processes, it may be possible to observe the L violation by measuring average lepton charge asymmetries and/or $\langle E_{e^+,\mu^+} \rangle > \langle E_{e^-,\mu^-} \rangle$.

There is essentially no background for these events [35].

Let us enlarge [35] on point 7), since, of course, B+L violation would be the most spectacular consequence of our speculations. If B+L violation becomes strong in the TeV range [1–3], the topology of the (B+L)-violating events will, presumably, resemble roughly the one corresponding to generic GFI events described above. It will be rather central close to the threshold energy and increasingly forward-oriented as the parton energy increases beyond the threshold. A search for B+L violation should therefore directly concentrate on quantum numbers [35] (see also ref. [19]).

A (B+L)-violating interaction

$$q + q \Longrightarrow 7\overline{q} + 3\overline{l} + X \tag{5.2}$$

(see eq. (2.1)) produces on average more positrons and μ^+ than electrons and μ^-. Thus one should measure the average lepton charge asymmetries,

$$\frac{\langle N_{e^+(\mu^+)} - N_{e^-(\mu^-)} \rangle}{\langle N_{e^+(\mu^+)} + N_{e^-(\mu^-)} \rangle}, \tag{5.3}$$

for isolated electrons and muons (with $p_t > 20$ GeV) in the central region ($10^\circ < \vartheta < 170^\circ$). The asymmetry vanishes, in principle, for L-conserving GFI events and gives, therefore, a direct measure of the relative strength of (B+L)-violating interactions.

In addition, one expects the primary e^+'s and μ^+'s from a (B+L)-violating interaction to be more energetic than the decay products of the associated gauge bosons. This leads to an asymmetry in the mean energy of the fastest anti-lepton as compared to that of the fastest lepton,

$$\langle E_{e^+,\mu^+} \rangle > \langle E_{e^-,\mu^-} \rangle. \tag{5.4}$$

As to the usefulness of the average charge asymmetries (5.3) or the energy asymmetries between leptons/anti-leptons (5.4), some comments [35] are in order. i) Since the charges for electrons and muons are almost uncorrelated for a given event, we expect rather large fluctuations of $N_{e^+(\mu^+)} - N_{e^-(\mu^-)}$ on an event by event basis. For (B+L)-conserving interactions the average value of the asymmetry should only vanish for a large number of events, with a statistical error determined by the average size of the fluctuations and the number of events. ii) Lepton number conservation implies $\langle N_{e^+(\mu^+)} - N_{e^-(\mu^-)}\rangle = 0$ only if the experimental cuts (like e.g. angular cuts , isolation cuts or p_t cuts) are completely *symmetric* with respect to the lepton charge distribution. This is far from trivial, since an asymmetry in the kinematic distributions of W^+ and W^-, for example, may reflect itself in an induced effective charge asymmetry of the lepton cuts. To minimize such possible effects it seems advantageous to concentrate on isolated leptons (with $p_t > 30$ GeV) in a central rapidity region. iii) Using special properties of the primordial fermions to filter out the (B+L)-violating events, like the asymmetry (5.4), has to compete always with the "tails" of geometrical flavour production.

Therefore, only L violation that is large compared to the (B+L)-conserving GFI background can be detected by a measurement of (5.3) and/or (5.4).

6. Conclusions

We have discussed in this lecture the possibility that non-perturbative effects in the electroweak theory, like multi-particle production of weakly interacting particles and B+L violation, may become important at energies $\sim M_{\rm sp} \sim \frac{m_w}{\alpha_w} \sim 10$ TeV, which will be reachable at colliders in the near future.

It is by now established that the total cross-section for electroweak B+L violation grows exponentially at energies below the sphaleron scale, $m_w \ll \sqrt{\hat{s}} \ll M_{\rm sp}$, and becomes much larger than the 't Hooft factor, $\exp(-4\pi/\alpha_w)$, would suggest. It is dominated by the associated production of (many) W bosons. Unfortunately, we do not know the behaviour and size of the cross-section near and above the sphaleron scale. The presently-available calculational techniques break down at this scale. Much theoretical work is needed to come to a definitive conclusion whether B+L violation becomes of observable size.

If the rapidly-rising (B+L)-violating cross-sections are qualitatively correct near the sphaleron scale, we expect very interesting phenomenology at future colliders. Multi-particle production of $\mathcal{O}(\alpha^{-1}) \sim 30$ weakly-interacting particles and B+L violation would occur with relatively large cross-sections, $\hat{\sigma}_w \sim 0.1$ nb-10 μb. Pictorially, these processes could be ascribed to the scattering of the clouds of weak gauge bosons and scalars carried along by colliding quarks or leptons.

We have discussed the phenomenology associated with this "geometrical" flavour production in the multi-TeV regime. We found that there is essentially no QCD background for these processes. At least the multi-particle production of weakly-interacting particles would be observed, if the threshold is below 11, 28, or 130 TeV for the LHC, SSC, or ELOISATRON, respectively. The observation of B+L violation would require a lot of statistics and would be a challenge for experimentalists.

References

[1] A. Ringwald, Nucl. Phys. **B330** (1990) 1.

[2] O. Espinosa, Nucl. Phys. **B343** (1990) 310.

[3] L. McLerran, A. Vainshtein and M. Voloshin, Phys. Rev. **D42** (1990) 171.

[4] G. 't Hooft, Phys. Rev. **D14** (1976) 3432.

[5] H. Aoyama and H. Goldberg, Phys. Lett. **188B** (1987) 506.

[6] R. Jackiw and C. Rebbi, Phys. Rev. Lett. **37** (1976) 172; C. Callan, R. Dashen and D. Gross, Phys. Lett. **63B** (1976) 334.

[7] R. Peccei and H. Quinn, Nuovo Cim. **41A** (1977) 309.

[8] S. Adler, Phys. Rev. **177** (1969) 2426; J. Bell and R. Jackiw, Nuovo Cim. **A60** (1969) 47.

[9] F. Klinkhamer and N. Manton, Phys. Rev. **D30** (1984) 2212.

[10] V. Kuzmin, V. Rubakov and M. Shaposhnikov, Phys. Lett. **155B** (1985) 36; P. Arnold and L. McLerran, Phys. Rev. **D36** (1987) 581; A. Ringwald, Phys. Lett. **201B** (1988) 510.

[11] P. Arnold and L. McLerran, Phys. Rev. **D37** (1988) 1020.

[12] M. Atyah and I. Singer, Ann. Math. **87** (1968) 484.

[13] I. Affleck, Nucl. Phys. **B191** (1981) 445.

[14] A. Belavin, A. Polyakov, A. Schwartz and Y. Tyupkin, Phys. Lett. **59B** (1975) 85.

[15] J. Kripfganz and A. Ringwald, Mod. Phys. Lett. **A5** (1990) 675.

[16] S. Khlebnikov, V. Rubakov and P. Tinyakov, *"Instanton induced cross sections below the sphaleron"*, INR preprint (1990).

[17] M. Porrati, Pisa preprint, IFUP-TH **13/90** (1990).

[18] V. I. Zakharov, TPI-MINN-**90/7-T** (1990).

[19] G. Farrar and R. Meng, DESY **90-099** (1990).

[20] L. McLerran, A. Vainshtein and M. Voloshin, Phys. Rev. **D42** (1990) 180.

[21] P. Arnold and M. Mattis, Phys. Rev. **D42** (1990) 1738.

[22] S. Khlebnikov, V. Rubakov and P. Tinyakov, Nucl. Phys. **B347** (1990) 783.

[23] L. Yaffe, *"Scattering Amplitudes in Instanton Backgrounds"*, Proceedings of the Santa Fe Workshop *"Baryon Number Violation at the SSC?"*, eds. M. Mattis and E. Mottola, (World Scientific, Singapore, 1990).

[24] A. Mueller, Orsay preprint, LPTHE-Orsay-**90/19**; Columbia preprint, CU-TP-**481**.

[25] P. Arnold and M. Mattis, Argonne preprint, ANL-HEP-PR-**90-78** (1990).

[26] L. Brown, R. Carlitz, D. Creamer and C. Lee, Phys. Rev. **D17** (1978) 1783.

[27] H. Levine and L. Yaffe, Phys. Rev. **D19** (1979) 1225.

[28] I. Balitsky and A. Yung, Phys. Lett. **168B** (1986) 113; A. Yung, Nucl. Phys. **B297** (1988) 47.

[29] V.V. Khoze and A. Ringwald, CERN-TH-**5912/90**.

[30] T. Banks, G. Farrar, M. Dine, D. Karabali and B. Sakita, Nucl. Phys. **B347** (1990) 581.

[31] J. Cornwall, Phys. Lett. **243B** (1990) 271; H. Goldberg, Phys. Lett. **246B** (1990) 445; J. Cornwall and G. Tiktopoulos, UCLA/90/TEP/**56**; H. Goldberg, Northeastern Univ. preprint, NUB-**3005** (1990).

[32] H. Aoyama and H. Kikuchi, Kyoto preprint KUCP-**23** (1990), unpublished.

[33] J. Zinn-Justin, Phys. Rept. **70** (1981) 109.

[34] A. Ringwald and C. Wetterich, DESY **90-067**.

[35] A. Ringwald, F. Schrempp and C. Wetterich, DESY **90-127**.

[36] J. Ellis, V.A. Khoze, A. Ringwald, F. Schrempp and C. Wetterich, *"Signatures for Geometrical Flavour Interactions and B+L Violation at the LHC"*, CERN Yellow Book on LHC, to appear.

Chairman: A. Ringwald

Scientific Secretary: R.G. Leigh

Discussion

– *Hsu:*

Could you elaborate on the attempts to calculate B and L violating amplitudes in Minkowski space, rather than in Euclidean space? In particular does one see a breakdown in unitarity or perturbation theory or does the amplitude remain small even at large $\sqrt{s}$?

– *Ringwald:*

It is very hard to calculate this in Minkowski space. You have this picture of a potential barrier lying between two inequivalent vacua. But only after you have constructed a Euclidean solution and interpret this as a family of configurations labelled by the Euclidean time do you get an explicit expression for the potential barrier. You don't know through which barrier you go before you make a Euclidean calculation. Approaches at a Minkowski space calculation have been made by Cornwall and by Banks, et al. What was done there was to take the instanton solution to calculate the potential barrier and to consider the Euclidean time as a path parameter. If you look (Figure 1) now at the energy which is swept out by the instanton and vary Euclidean time τ from $-\infty$ to $+\infty$, you can check that this corresponds to a transition from a vacuum with winding number zero to a vacuum with winding number one. Then you have another collective coordinate, ρ, the size of the instanton and we see that the instanton represents a family of configurations which interpolate between vacua. Taking the instanton solution and the barrier that we have to tunnel through and treating the whole as a one dimensional quantum mechanical system, Banks, et al, exactly reproduce my result at low energy. At high energy they claim that one will always have an exponential suppression. They can only show this if the instanton has a fixed size and they explicitly claim that they cannot show that the exponential suppression remains if the instanton has a scale size. The instanton in the electroweak theory has no fixed size; we have to consider all sizes, and one of these has a lowest energy. Really it is not obvious that this calculation describes tunneling. This is just a convenient way to state this: that the amplitude is suppressed by $e^{-S_{cl}}$, but one really doesn't know the potential barrier, you only know it after the calculation. I don't think that these approaches are really fruitful for this problem.

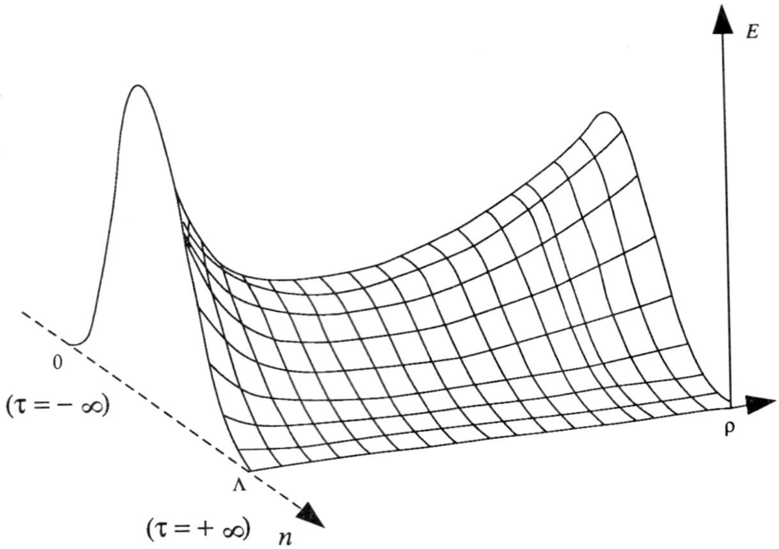

– *Hsu:*

Does the instanton size correspond in any way to the actual size of the coherent fluctuations in Minkowski space?

– *Ringwald:*

In a sense, yes, it could. Here, I have plotted the energy of the instanton (solid line) where I have interpreted Euclidean time as a path parameter (Figure 2): at Euclidean time $\tau = 0$. Then the solid curve has a minimum at ρ, the size of the configuration, of order $1/m_W$. This corresponds to a sphaleron-like configuration. The effective action (dashed line) of the configuration has a minimum at a much smaller value, $1/v$, with v the vacuum expectation value. Really, when we consider a Euclidean calculation, we want to minimize the action. We are integrating over all scale sizes. Small instantons and large instantons are taken into account, and the action does not vary much, as long as the size is less than $1/v$. At this point, the action grows and configurations are suppressed. So actually one was lucky in this calculation that there does not exist instantons of a fixed size in the electroweak theory. It is really very hard to have a physical interpretation of this Euclidean calculation. This was really the first time that S-matrix elements with many external legs had been calculated using instantons.

– *Brahm:*

Why are $7\bar{q}$ and $3\bar{l}$ the expected products of a qq collision?

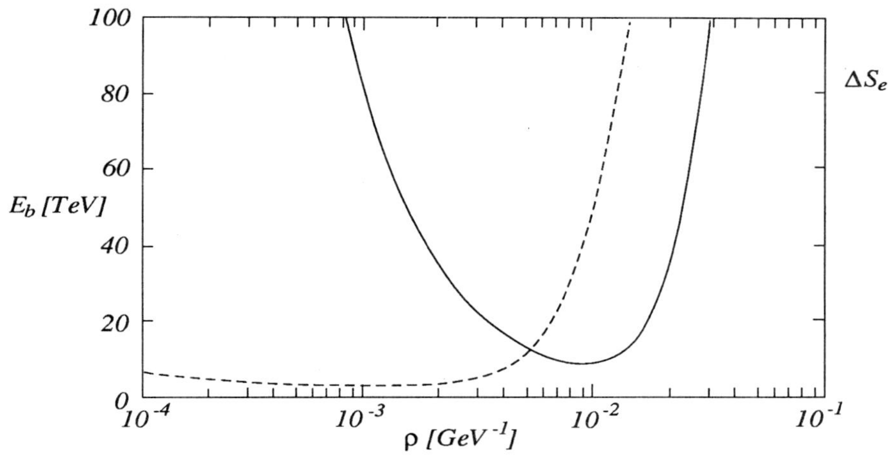

– *Ringwald:*

This comes just from the selection rule. The change in the baryon number, for three generations, is equal to $-3\Delta n$. Each quark has a baryon number 1/3, so this is the simplest process you can have. The quark number has to change by nine units.

– *Zichichi:*

One thing should be pointed out: charge has to be conserved. Seven antiquarks going out and two quarks coming in equals three baryons, which means three leptons must exit.

– *Brahm:*

Are there other possibilities, like having one lepton in the final state along with one antiquark?

– *Ringwald:*

Actually, one must have this violation for each quark flavour. So this is the minimum number. More precisely, for example, I should have written say ud goes to $\bar{d}\bar{c}\bar{c}\bar{s}\bar{t}\bar{t}\bar{b}\bar{\nu}_e\mu^+\tau^+$. If you have two quarks of the same flavour they should have different colours in order that colour is conserved in this process as well.

– *Lu:*

It is very hard for me to understand a Greens function with unbalanced number of fermion fields because it is a Grassman number, and traditionally Grassman numbers are unphysical. Do you have an argument to convince me that fermion number violation calculatons are indeed meaningful?

– *Ringwald:*

The Greens function is not a Grassmann number. If one calculates a path integral with fermion fields, one usually gets the determinant of $\not{D}$, the fermion gauge-kinetic operator. In the presence of instantons however, there are fermionic zero modes; to get a non-zero answer for the path integral we must have insertions of fermion fields. Thus the result of the path integral is the determinant, without zero modes, times the zero mode wavefunctions. These wavefunctions are the coefficients of the (Grassman) zero mode in a mode expansion of the fermion fields that appear as insertions in the path integral.

– *Kastening:*

In calculations of fermion number non-conserving Green functions, the fermion zero modes usually included are associated with massless fermions. Is there a profound reason for considering only massless fermions or is it just that they are easier to deal with?

– *Ringwald:*

Indeed, there are solutions known for non-zero mass. They involve Bessel functions; it's not relevant for this question now, but could be if one needed to consider this problem in more detail.

– *Kastening:*

If, in on-shell calculations, you include the next-to-leading order in the semi-classical approach to the background fields, what effects do the arising form factors have?

– *Ringwald:*

Actually, this is not yet proven. McLerran, Vainshtein, and Voloshin calculated the next-to-leading order corrections which are induced for the scalar field Greens functions. They found no form factor at all. It's still point-like, but it has been pointed out by Arnold and Mattis, Khlebnikov, Rubakov and Tinyakov, and Yaffe, that this is not true generally; for a tree graph there will be corrections, already by exchanging W's, introducing form factor effects. But, what has been argued for by many people, that all these higher order effects become important only at the sphaleron energy and not already at the scale v. This is very important.

– *Kastening:*

There have been criticisms that the dilute instanton gas approximation is not applicable. Does this apply only to the early universe or also to collider experiments?

– *Ringwald:*

I think that this only applies for the early universe calculations. Of course if one gets near the unitarity limit, if the lowest order result in the dilute gas approximation becomes large, then one has to take into account higher order effects. This is very general: if a lowest order amplitude becomes large, then one has to iterate this amplitude in order to save unitarity.

– *Kastening:*

What implications for the early universe would the fact that electroweak interactions become strong, apart from the implications of B and L violation? Is this just a dynamical detail or does one arrive at a very different picture?

– *Ringwald:*

I guess that it is a dynamical detail for the early universe. At very high energies one expects that if the interaction becomes strong, it gets forward peaked and very similar to QCD so there should not be much difference between strong interactions and weak interactions. In the early universe, I think that there is not much difference in the scattering considerations if there is a nearly free weakly interacting gas of particles.

– *Haas:*

You have presented a beautiful piece of work that shows that electroweak theory still holds a lot of mysteries. Unfortunately what you have discussed will not be visible in present day colliders. Could you speculate whether electroweak theory might be surprising us at lower energies?

– *Ringwald:*

At the present machines, I don't think that we will see anything from these processes. There is a very sharp threshold for these processes; they are completely negligible below and could become strong above this threshold. And this threshold at the parton level is 16 TeV according to the naïve instanton analysis, and I guess would generally lie between 10 and 100 TeV.

– *Grossman:*

You have demonstrated your method for calculating the S-matrix for a special reaction. Have you calculated other processes?

– *Ringwald:*

I have never used this for any other calculation. At energies a few TeV below the threshold, the cross section for multi-particle production coming from this instanton configuration are suppressed by 20 orders of magnitude so these are really very tiny effects at low energies. There have been calculations using ordinary perturbation theory to see if there are signs that perturbation theory perhaps breaks down. If the number of external legs becomes of order $1/\alpha_W$, it seems that perturbation theory breaks down in the topologically trivial sector. But there is no other method available for the calculation that I have done. If there was a more reliable method, I would use it instead.

– *Grossman:*

You quoted that $\Delta(B + L) = 0$ would become strong. How was this result obtained?

– *Ringwald:*

This is no result; it is just a speculation. But I have another, very old result, referred to by Prof. Dokshitzer, by Fadin, Kuraev, and Lipatov, who considered the $SU(2)$ gauge Higgs model in the leading log approximation, partially summing perturbation theory. Here the total fermion-fermion scattering cross-section, in the range that the energies are much larger than the W-mass, with $\alpha_W \ell n(s/m_W^2)$ of order one (the point where leading log calculations should be reliable), is nearly constant. It is a slowly rising function that still violates unitarity. So this cannot be the final answer. This is a hint from ordinary perturbation theory that the cross section in the electroweak theory which is supposed to be weak can become strong and indeed it was also found by these people that the reason that this violates unitarity is that one gets very high multiplicities that rise with energy as a small power. I don't know if this is related to the instanton calculation but it has very similar behaviour. The cross-section is of order 1nb and the multiplicities in the leading log calculation are very large. The total cross-section is dominated by processes in the multi-Regge kinematics, that is it describes forward scattering. So this is an indication from partially summing up perturbation theory, that weak interactions could be strong at high energies. Of course this calculation can't really be trusted because its asymptotic behaviour still violates the unitarity bound.

– *Larsen:*

You include non-perturbative contributions in calculating the S-matrix. Shouldn't you also include non-perturbative effects in the coupling constant's evolution? When one gets near the unitarity limit, the coupling constants may not be as large as one thinks. Could you comment?

– *Ringwald:*

The determinant in the background of the instanton leads to an exponent involving the running coupling constant. The result is one-loop renormalization group invariant, so it does not depend so much on the renormalization scale. In any case, you would still get a violation of unitarity since the cross section grows like the exponential of a power of the energy.

– *Rosiek:*

Your lecture was about electroweak instantons and you said that the transitions with the B violation are suppressed by the factor 10^{-78}. But probably it is possible to consider also QCD instantons–the α_s is ten times bigger than α_W so naïvely the suppression factor should be only of order 10^{-7} and phenomena with participation of QCD instantons should be observed at much lower energies. What do you think about this and do you have any ideas for experiments?

– *Ringwald:*

If the instantons in the electroweak theory lead to strong coupling in the sense that cross-sections become asymptotically forward peaked, then the instantons in QCD should play some role, at least in the diffractive region in QCD scattering. That anomalous contributions play a large role in this context has also been pointed out by A. White.

Of course, we also have the problem in QCD that because the coupling is strong, the dilute gas approximation is not valid from the beginning and secondly we are also facing the infrared problems: the integral over all scale sizes diverges because the instanton in QCD is not cutoff at the inverse vacuum expectation value. One should take into account consistently the higher order interactions of instantons and this is a very complicated problem.

– *Bimonte:*

In the context of QCD, one might expect that nucleii might decay by this process. Would you be able to estimate the lifetime? Secondly, what is the dependence of the cross section on the difference of the winding numbers of the vacua?

– *Ringwald:*

First, the QCD case is not completely analogous. There we have a non- chiral vector coupling to the fermions. The QCD instantons induce transitions between different vacua which are chirality violating, but chirality violating processes are always there since there are fermion masses. I would not think that some stable object would decay just because of these processes.

Secondly, for $\Delta n = 2$, the baryon number would change by six units. We know from the beginning that the topological number, q, has to be two and the action grows as the absolute value of q. These processes would be proportional to $e^{-4\pi|q|/\alpha_W}$, so these are more suppressed. However when the dilute gas approximation breaks down for the $q = 1$ calculation, it also breaks down for $q = 2$. Below this scale however only the minimum baryon violating process is important.

RELATIVISTIC ION COLLISIONS AND 200 TeV PHYSICS

T. D. Lee

Department of Physics, Columbia University

1. Symmetry-Breaking Forces

At present, we have QCD for the strong force, the $SU(2) \times U(1)$ standard model for the electroweak force and general relativity for the gravitational force. Together, they seem to give us a satisfactory description of the physical world. However, there remains the puzzle of *missing symmetry*:

These three theories are all based on symmetry. Yet, most of the symmetry quantum numbers are not conserved.

Consequently, there must exist an additional class of forces: the symmetry-breaking force, of which very little is known. In order to describe its low-energy behavior, we need nearly twenty phenomenological parameters. We have no understanding of why these parameters should acquire their observed values. On the other hand, this symmetry-breaking force is known to be closely related to

1. the origin of all masses
2. the structure of the vacuum.

Hence, there is every reason to expect that the underlying principle of this symmetry-breaking force should be as profound as the theoretical foundation of the three known forces.

Let us discuss the symmetry-breaking in the electroweak interaction. In this case, the problem is also tied to

3. the possible existence of H , the Higgs particles.

Suppose the Higgs mass m_H is relatively low, less than or about a few hundred GeV. That means (according to the standard model) the gauge bosons do not interact strongly. The Higgs can be discovered at LHC (8 TeV $\times$ 8 TeV). However, in this case, the determination of m_H only adds one more number to the already very large pool of nearly twenty phenomenological parameters. The mere knowledge of a relatively low Higgs mass does not lead to a deeper understanding of the symmetry-breaking force.

Physics Up to 200 TeV
Edited by A. Zichichi, Plenum Press, New York, 1991

We will still be in exactly the same unsatisfactory situation. In order to achieve a breakthrough from the present bind, we must search for new phenomena at a much higher energy collider, such as the SSC (20 TeV $\times$ 20 TeV) and the Eloisatron (100 TeV $\times$ 100 TeV).

A more exciting scenario is for H to have a much higher mass, near or about 1 TeV. This means the Higgs width would be much too broad for H to appear as a clear resonance. The situation would resemble the σ-model, which is very useful in the description of the low energy hadronic phenomena in the few hundred MeV and the GeV region, but it fails in the prediction of the σ-particle. Even today, while there are some broad spin-0 "resonances" in the 700 MeV-1 GeV range, none can be identified as the σ itself.

A large m_H carries the important implication that there would be new strong interaction between the gauge bosons. To unfold the nature of the symmetry-breaking force, it is imperative to study the gauge boson scatterings, especially for the longitudinally polarized ones. The following table lists the estimate of *signal/noise* ratio made by M. Chanowitz[1]; calculated for a luminosity of $10^{33}/\text{cm}^2/\text{sec}$ and over a time period of 10^7 (one working year):

channel \ beam energy	15 TeV	17 TeV	20 TeV	100 TeV
ZZ	8/8	11/10	17/12	
WZ	5/7	6/8	9/9	>> 1
WW	28/23	37/27	53/34	

Thus, at the SSC (20 TeV $\times$ 20 TeV) energy, we can begin to explore this new class of strong forces; a full investigation of the symmetry- breaking interaction *requires a 100 TeV × 100 TeV collider.*

2. Structure of the Vacuum

The conventional wisdom in dealing with the puzzle of missing symmetry is to put it under the general heading "spontaneous symmetry breaking." In such a scheme, one assumes that there exists some spin-0 field ϕ that can carry the missing quantum number and whose vacuum expectation value is not zero:

$$\phi_{\text{vac}} \equiv < \text{vac} \,|\, \phi \,|\, \text{vac} > \neq 0 \,. \tag{1}$$

Consequently, the observed asymmetry can be attributed entirely to the state vector of our universe, not to the physical law. On the other hand, unless we have other links that connect matter with the vacuum, how can we be sure that this idea is right and not merely a tautology? This then leads to the question[2]: Is the Lorentz-invariant vacuum actually a physical medium? If so, then there must be ways to alter the properties of the vacuum.

Another puzzle is the problem of quark or color confinement, which makes half of the elementary particles, quarks and gluons, non-direct observables. The explanation is again to invoke the vacuum. We assume the QCD vacuum to be a condensate of gluon pairs and quark-antiquark pairs so that it is a perfect color dia-electric[3] (i.e., color dielectric constant $\kappa = 0$). This is in analogy to the description of a superconductor as a condensate of electron pairs in BCS theory, which results in making the superconductor a perfect dia-magnet (with magnetic susceptibility $\mu = 0$). When we switch from QED to QCD we replace the magnetic field $\vec{H}$ by the color electric field $\vec{E}_{\text{color}}$, the superconductor by the QCD vacuum, and the QED vacuum by the interior of the hadron. As shown in Figure 1, the roles of the inside and the outside are interchanged. Just as the magnetic field is expelled outward from the superconductor, the color electric field is pushed into the hadron by the QCD vacuum, and that leads to color confinement, or the formation of bags.[3] This situation is summarized in Table 1.

In the resolution of both puzzles, missing symmetry and quark confinement, the system of elementary particles no longer forms a self-contained unit. The microscopic particle physics depends on the coherent properties of the macroscopic world, represented by the appropriate operator averages in the physical vacuum state.

If we pause and think about it, this represents a rather startling conclusion, contrary to the traditional view of particle physics which holds that the microscopic world can be regarded as an isolated system. To a very good approximation it is separate and uninfluenced by the macroscopic world at large. Now, however, we need these vacuum averages; they are due to some long-range ordering in the state vector. At present our theoretical technique for handling such coherent effects is far from being developed. Each of these vacuum averages appears as an independent parameter, and that accounts for the twenty-some constants in the present theoretical formulation.

Table 1. Analogies between superconductivity and the QCD vacuum

QED superconductivity as a perfect dia-magnet		QCD vacuum as a perfect color dia-electric
$\vec{H}$	$\longleftrightarrow$	$\vec{E}_{\text{color}}$
$\mu_{\text{inside}} = 0$	$\longleftrightarrow$	$\kappa_{\text{vacuum}} = 0$
$\mu_{\text{vacuum}} = 1$	$\longleftrightarrow$	$\kappa_{\text{inside}} = 1$
inside	$\longleftrightarrow$	outside
outside	$\longleftrightarrow$	inside

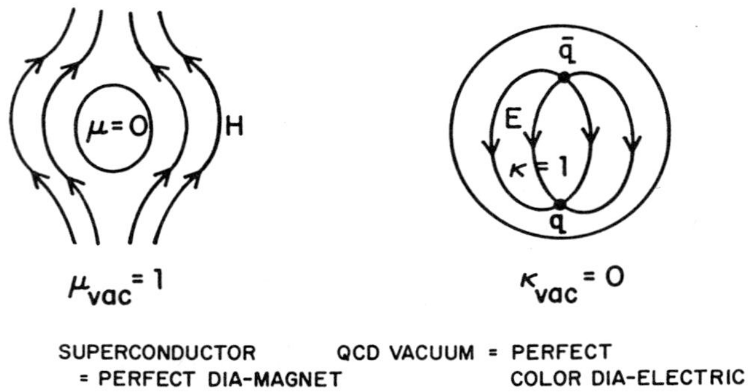

Figure 1. Superconductivity in QED vs. quark confinement in QCD.

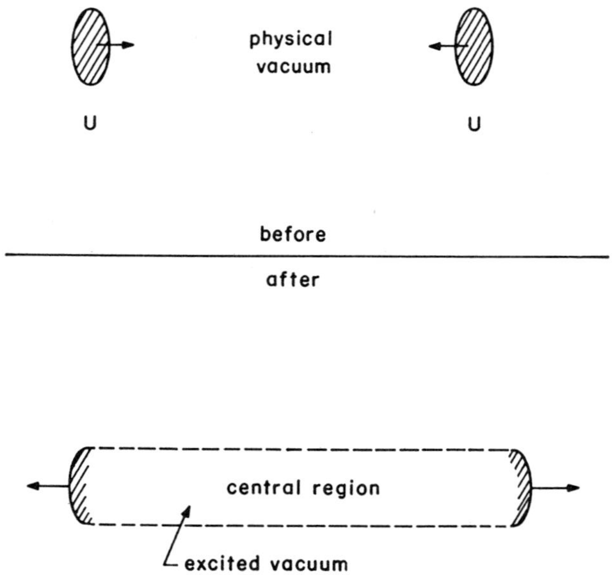

Figure 2

On the experimental side, there has hardly been any direct investigation of these coherent phenomena. This is because hitherto in most high-energy experiments, the higher the energy the smaller has been the spatial region we are able to examine. In order to explore physics in this fundamental area, relativistic heavy ion collisions offer an important new direction.[4] The basic idea is to collide heavy ions, say uranium on uranium, at an ultra-relativistic region. Before the collision, the vacuum between the ions is the usual physical vacuum; at a sufficiently high energy, after the collision almost all of the baryon numbers are in the forward and backward regions (in the center-of-mass system). The central region is essentially free of baryons and, for a short duration, it is of a much higher energy density than the physical vacuum. Therefore, the central region represents the excited vacuum (Figure 2).

As we shall see, we need RHIC (100 GeV $\times$ 100 GeV) to explore the QCD vacuum, but Eloisatron (100 TeV $\times$ 100 TeV) for a full investigation of the electro-weak vacuum.

In this connection, it is important to note the difference between the microscopic excitations of a condensate and its long-range order. For example, the mere knowledge of the He atom and the phonon spectrum does not yield in a simple way the superfluidity theory of liquid He II. Likewise, as mentioned before, the discovery of Higgs (i.e., assuming a relatively low m_H) by itself may not lead to any deeper understanding of the symmetry-breaking forces. Since the symmetry-breaking mechanism is closely tied to the vacuum structure, the study of relativistic heavy ion collisions is crucial to further progress in this field.

3. QCD Vacuum

A normal nucleus of baryon number A has an average radius $r_A \approx 1.2A^{\frac{1}{3}} fm$ and an average energy density

$$\mathcal{E}_A \approx \frac{m_A}{(4\pi/3)r_A^3} \approx 130 MeV/fm^3 . \qquad (2)$$

Each of the A nucleons inside the nucleus can be viewed as a smaller bag which contains three relativistic quarks inside; the nucleon radius is $r_N \approx 0.8 fm$ and its average energy density is

$$\mathcal{E}_N \approx \frac{m_N}{(4\pi/3)r_N^3} \approx 440 MeV/fm^3 . \qquad (3)$$

Consequently, even without any sophisticated theoretical analysis we expect the QCD phase diagram to be of the form given by Figure 3.

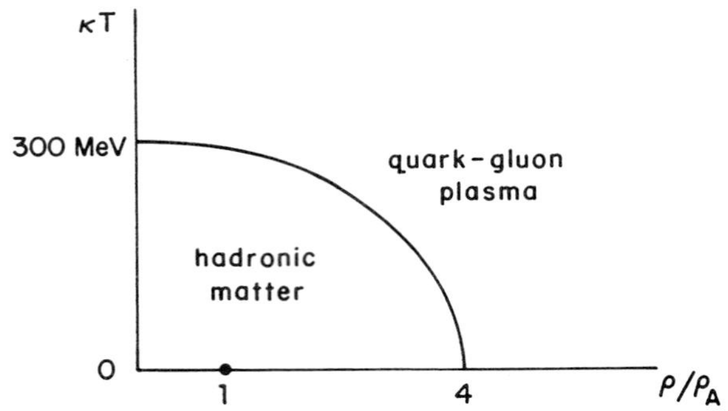

Figure 3. QCD Phase Diagram

In Figure 3, the ordinate is κT ($\kappa =$ Boltzmann constant, $T =$ temperature), the abscissa is ρ/ρ_A ($\rho =$ nucleon density, $\rho_A =$ average nucleon density in a normal nucleus A) and the dot denotes the configuration of a typical nucleus A. The scale can be estimated by noting that the critical $\kappa T \sim 300$ MeV is about the difference of $1 fm^3$ times $\mathcal{E}_N - \mathcal{E}_A$ and the critical $\rho/\rho_A \sim 4$ is just the nearest integer larger than $(1.2/0.8)^3$.

Accurate theoretical calculation exists only for pure lattice QCD (i.e., without dynamical quarks). The result is shown in Figure 4.[5]

If one assumes scaling, then the phase transition in pure QCD (zero baryon number, $\rho = 0$) occurs at $\kappa T \sim 340$ MeV with the energy density of the gluon plasma

$$\mathcal{E}_P \sim 3 GeV/fm^3 . \tag{4}$$

To explore this phase transition in a relativistic heavy ion collision, we must examine the central region. Since only a small fraction of the total energy is retained in the central region, it is necessary to have a beam energy (per nucleon) at least an order of magnitude larger than $\mathcal{E}_P \times (1.2 fm)^3 \sim 5$ GeV; this makes it necessary to have an ion collider of 100 GeV $\times$ 100 GeV (per nucleon) for the study of the QCD vacuum.

The basic energy scale for low energy hadronic physics is the pion mass $m_\pi \sim 140$ MeV, while that for electroweak is the intermediate boson masses: m_W or m_Z. When one scales the RHIC energy (100 GeV $\times$ 100 GeV) by a factor m_W/m_π, one sees that the collider needed for the investigation of the electroweak vacuum is the Eloisatron (100 TeV $\times$ 100 TeV).

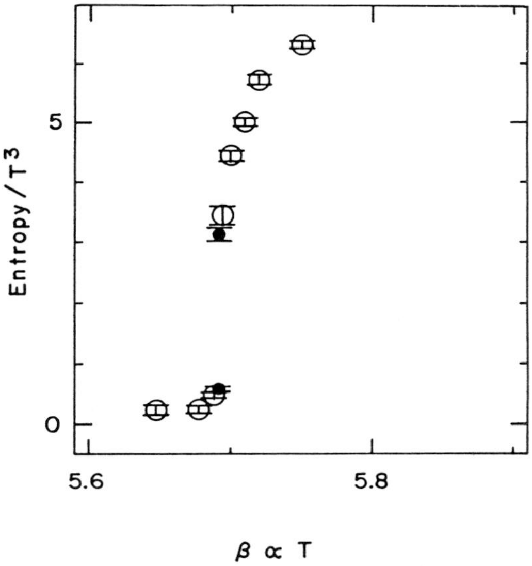

Figure 4

4. Kinematics

(i) *Rapidity*

The rapidity of a particle of energy E and momentum $\vec{p}$ is ($\hbar = c = 1$)

$$y \equiv \frac{1}{2}\, ln\, \frac{E + p_z}{E - p_z} \tag{5}$$

where p_z is the component of $\vec{p}$ along $\hat{z}$, the beam direction. Under a Lorentz transformation along $\hat{z}$, we have $y \rightarrow y+$ constant; hence, it leaves invariant the observed distribution-form in y. For example, in a fixed target experiment, the lab frame and the center-of-mass frame have identical forms in their y-distributions. For a nonrelativistic particle, $y \rightarrow v_z$, the z-component of its velocity. If one wishes, one may regard y as the generalization of v_z for a relativistic particle.

(ii) *Pseudo-rapidity*

For a relativistic particle moving at an angle θ with respect to $\hat{z}$, because $p_z \cong E \cos\theta$, we have $y \cong \eta$ where η is the pseudo-rapidity defined by

$$\eta \equiv -ln \tan\frac{\theta}{2}, \tag{6}$$

and can be readily measured experimentally.

(iii) *Transverse energy vs. y*

Consider the distribution of the total transverse energy

$$E_T \equiv \sum_i E_i \cos\theta_i$$

versus y. An estimate of the energy density $\mathcal{E}$ in the central region can be made (Bjorken):

$$\mathcal{E} \cong \frac{1}{\pi R_{\text{beam}}^2} \frac{dE_T}{\tau dy} \tag{7}$$

where R_{beam} is the radius of the ion and $\tau \sim 1\,\text{fm/velocity of light}$ is the typical time scale for low-energy hadronic physics. Since in the center-of-mass system, $y \sim v_z$ in the central region, $\tau\, dy$ denotes the longitudinal length over which the amount of transverse energy dE_T is deposited. Assuming that the amount of transverse energy deposited indicates the energy density in the central region, we obtain the above estimate.

At present, the CERN SPS experiment gives $dE_T/dy \sim 70$ GeV in the central region; for $R_{\text{beam}} \sim 3$ fm, we estimate $\mathcal{E} \sim 3$ GeV/fm^3, which is marginal for the QCD phase transition.

5. Present Facilities

At present, only *fixed target* experiments have been carried out for relativistic heavy ion collisions:

(i) The AGS at Brookhaven accelerates heavy ions up to $^{28}S_i$ at 14.5 GeV per nucleon. In Figure 5,[6] the differential cross-sections $d\sigma/dE_T$ versus E_T are given[6] for target nuclei A varying from Al, Cu, Ag to Au (Experiment E802).
The distribution for Ag is almost the same as that for Au, even though there is a sizable difference in their radii. Thus, viewed from the projectile, the target nuclei appear black for nuclei heavier than Ag. These AGS experiments can probe physics in the medium energy density and relatively high baryon number density region.

(ii) The SPS at CERN accelerates heavy ions up to ^{32}S at 200 GeV per nucleon. In Figure 6, $d\sigma/dE_T$ versus E_T is given[7] for ^{32}S collisions with Al, Ag, W, Pt, Pb and U.
As can be seen, there is a significant change in the high E_T distributions from Ag to U, showing that at 200 GeV/nucleon the target nucleus still seems transparent, which in turn means a relatively low baryon density ρ in the central region.

(iii) In addition, the Brookhaven RHIC project is under construction; it plans to do heavy ion collider physics at 100 GeV $\times$ 100 GeV (per nucleon) for nuclei up to U on U. This will be the first major thrust to explore the structure of the QCD vacuum.

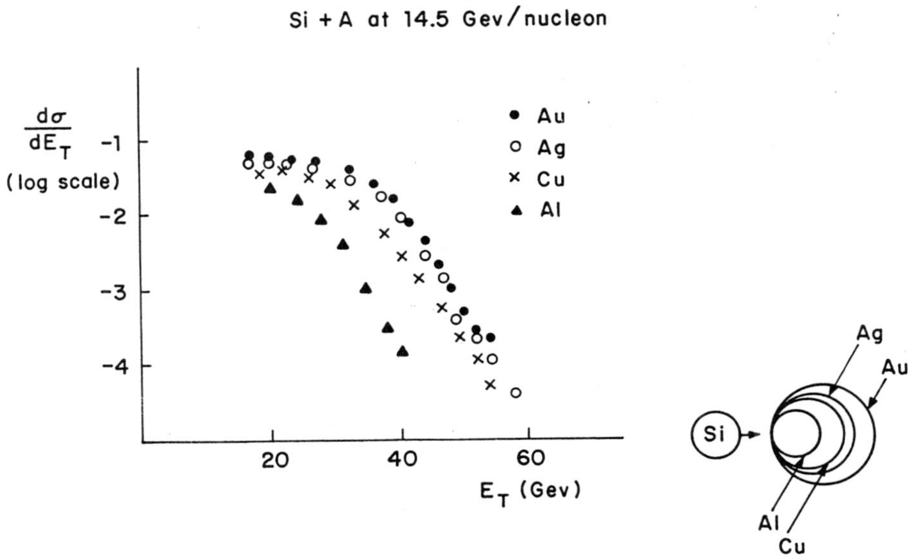

Figure 5

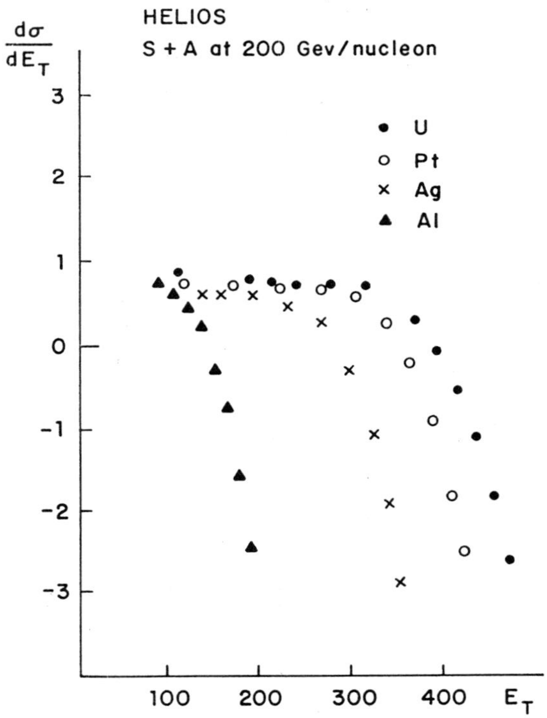

Figure 6

6. High Energy Heavy Ion Experiments

In the following, I shall review three experiments that may serve as precursors (but by no means as evidence) of quark-gluon plasma: The K^+/π^+ enhancement, the J/ψ suppression and the $\pi\pi$ interferometry.

(i) K^+/π^+ *enhancement*

Consider a quark-qluon plasma in a baryon-enriched region (such as the environment that may be approached by the Brookhaven AGS experiments). In addition to the large number of u, d quarks already present, there should be a fair number of $s\overline{s}$ pairs, as well as $u\overline{u}$ and $d\overline{d}$ pairs, produced in the plasma. Both K^+/π^+ and K^-/π^- ratios should increase from those in the pp collision. Since in such a medium it is relatively easy to have $u\overline{s}$, as compared to $\overline{u}s$, we expect the enhancement of the K^+/π^+ ratio to be more dramatic.

The results of Brookhaven AGS experiment E-802 are given in Figure 7.[6]

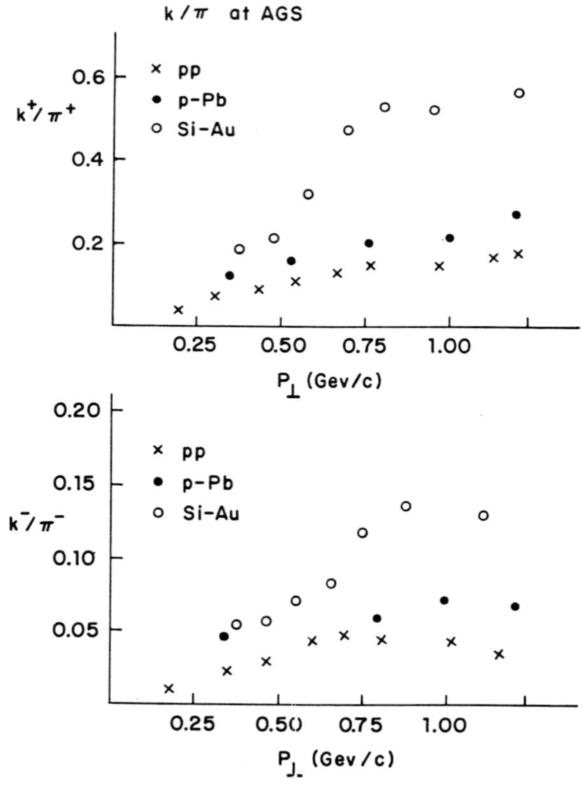

Figure 7

At large transverse momentum, the K^+/π^+ ratio increases $\sim 50\%$ from pp to $p-P_b$, but about a factor of 3 from pp to $Si-Au$. The integrated ratio for K^+/π^+ is ~ 5 to 10% for pp but $\sim$ 20% for $Si-Au$. The corresponding integrated ratio for K^-/π^- is abut 3% for pp and 4% for $Si-Au$.

(ii) *J/ψ suppression*

Consider the production of $c\bar{c}$ in a relativistic heavy ion collision. In the fragmentation region (i.e., forward and backward in the center-of-mass system), because of the quark confinement mechanism it is relatively easy for the $c\bar{c}$ to form a bound J/ψ state. On the other hand, in the central (mid-rapidity) region, if the quark-gluon plasma were formed, then because of de-confinement the $c\bar{c}$ pair tends to drift apart, which results in a J/ψ suppression[8]. The CERN SPC experimental result for

$$S + U \rightarrow \mu^+ + \mu^- + \cdots$$

is given in Figure 8.[9]

One sees that the J/ψ production for $E_T > 125$ GeV (central region) is indeed much suppressed as compared to $E_T < 51$ GeV (fragmentation region).

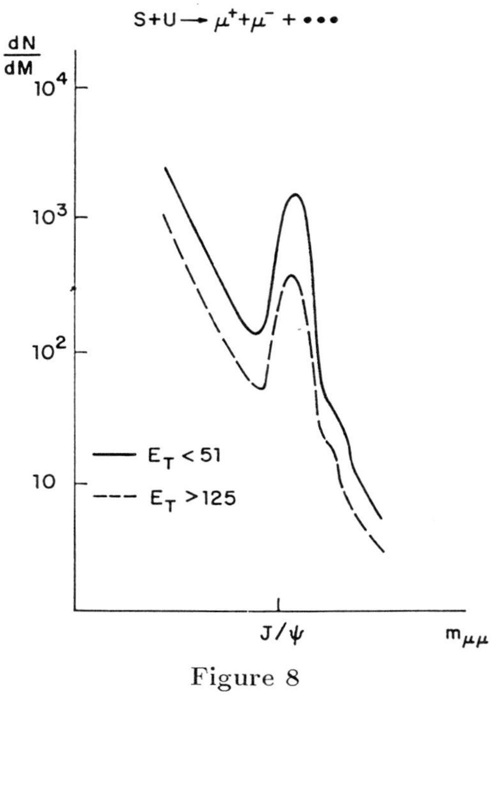

Figure 8

(iii) $\pi\pi$-*interferometry*

The emission amplitude of two pions of the same charge with momenta $\vec{k}_1$ and $\vec{k}_2$ from points $\vec{r}_1$ and $\vec{r}_2$ is proportional to

$$A \equiv e^{i\vec{k}_1\cdot\vec{r}_1+i\vec{k}_2\cdot\vec{r}_2} + e^{i\vec{k}_2\cdot\vec{r}_1+i\vec{k}_1\cdot\vec{r}_2}, \tag{8}$$

because of Bose statistics. Let

$$\vec{q} \equiv \vec{k}_1 - \vec{k}_2 \qquad \text{and} \qquad \vec{r} \equiv \vec{r}_1 - \vec{r}_2 . \tag{9}$$

Since

$$|A|^2 \equiv 1 + \cos\vec{q}\cdot\vec{r} \tag{10}$$

changes from $|A|^2 = 2$ as $\vec{q} \to 0$, to $|A|^2 = 1$ as $\vec{q} = \infty$, a measurement of the $\pi\pi$ correlation gives a determination of the geometrical size R of the region that emits these pions, like the Hanbury-Brown/Twiss determination of the stellar radius. (There is an important difference; here, one measures the amplitude-amplitude correlation, whereas the Hanbury-Brown/Twiss experiment measures the probability-probability correlation; i.e., fourth order in amplitude, not second order. This is because for a classical object the second order correlation vanishes after averaging over the large number of quantum states involved.)

Now, if the central region is a plasma of entropy density S_P occupying a volume V_P, which later hadronizes to ordinary hadronic matter (of entropy density S_H and volume V_H), the total final entropy $S_H V_H$ must be larger than the total initial entropy $S_P V_P$. Since $S_P > S_H$, we have

$$V_H > V_P . \tag{11}$$

The experimental configurations and results are given in Figure 9 and Table 2.[7] One sees that the hadronization radius in the central region is indeed much larger than that in the fragmentation region.

All three examples are, at best, only indicative of the quark-gluon plasma. Much work and higher energy are needed for a more definitive proof. Nevertheless, they do show that relativistic heavy ion is an effective means of exploring the structure of the vacuum.

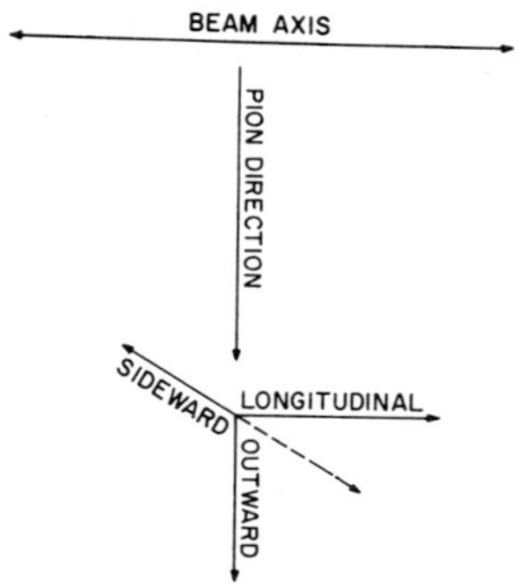

Figure 9

Table 2

200 GeV/nucl $O + Au \rightarrow \pi\pi + \cdots$

Rapidity Interval	Gaussian R_T(fm)	R_L (fm)	Λ
$1 < y < 2$	4.3 ± 0.6	2.6 ± 0.6	$0.34^{+0.09}_{-0.06}$
	$R_T^{side} = 4.0 \pm 1.0$ fm	2.6 ± 0.6	$0.34^{+0.09}_{-0.06}$
	$R_T^{out} = 4.4 \pm 1.0$ fm	2.6 ± 0.6	$0.34^{+0.09}_{-0.06}$
Central region (mid-rapidity) $2 < Y < 3$	8.1 ± 1.6	$5.6^{+1.2}_{-0.8}$	0.77 ± 0.19
	$R_T^{side} = 6.6 \pm 1.8$ fm	$5.6^{+1.2}_{-0.8}$	0.77 ± 0.19
	$R_T^{out} = 11.2 \pm 2.3$ fm	$5.6^{+1.2}_{-0.8}$	0.77 ± 0.19

R (Oxygen) $\cong$ 3 fm

7. Concluding Remarks

To conclude we emphasize, once again, that the most challenging problems in physics are

(1) the symmetry-breaking force, and

(2) the structure of the vacuum.

If the Higgs mass m_H were relatively low (about or less than a few hundred GeV), then the determination of m_H only adds one more member to the very large set of phenomenological parameters. By itself, a low Higgs mass does not lead to a deeper understanding of the problems that face us. Consequently, it is necessary to examine physics at a much higher energy; that means we need SSC (20 TeV $\times$ 20 TeV) and Eloisatron (100 TeV $\times$ 100 TeV). If the Higgs mass is about 1 TeV, then it is necessary to study gauge boson scatterings. At the SSC energy, one can begin to observe these processes, but a full investigation definitely needs 100 TeV $\times$ 100 TeV.

To understand the QCD vacuum, we need RHIC (100 GeV $\times$ 100 GeV). To understand the electro-weak vacuum, Eloisatron is necessary.

References

[1]. M. Chanowitz, private communication.

[2]. T.D. Lee and G.C. Wick, Phys.Rev. D**9**, 2291 (1974).

[3]. R. Friedberg and T.D. Lee, Phys.Rev. D**18**, 2623 (1978).

[4]. T.D. Lee, "Relativistic Heavy Ion Collisions and Future Physics," in *Symmetries in Particle Physics*, ed. I. Bars, A. Chodos and C.H. Tze (New York, Plenum Press, 1984), p. 93.

[5]. F.R. Brown, N.H. Christ, Y.F. Deng, M.S. Gao and T.J. Woch, Phys.Rev.Lett. **61**, 2058 (1988).

[6]. S. Nagamiya, "Nucleus-Nucleus Collisions in the Brookhaven-AGS Energy Domain," in *Relativistic Heavy-Ion Collisions*, ed. R.C. Hwa, C.S. Gao and M.H. Ye (New York, Gordon and Breach Science Publishers, 1990), p. 1.

[7]. W. Willis, "Experimental Studies on States of the Vacuum," in *Relativistic Heavy-Ion Collisions*, ed. R.C. Hwa, C.S. Gao and M.H. Ye (New York, Gordon and Breach Science Publishers, 1990), p. 39.

[8]. T. Matsui and H. Satz, Phys.Lett. B**178**, 416 (1986).

[9]. H. Gutbrod, "Experiments with Highly Relativistic Heavy-Ions," in *Relativistic Heavy-Ion Collisions*, ed. R.C. Hwa, C.S. Gao and M.H. Ye (New York, Gordon and Breach Science Publishers, 1990), p. 105.

Chairman: T.D. Lee

Scientific Secretaries: B. Grossmann, H.R. Hoorani, J. You

Discussion

– *U. Vikas:*

How does the study of WW and ZZ collisions lead to an understanding of Spontaneous Symmetry Breaking?

– *T.D. Lee:*

The study of WW and ZZ is really a way to understand whether electroweak is intrinsically weak or whether electroweak is basically strong. If electroweak is strong then there will be a relatively high probability of the longitudinal modes of gauge bosons scattering on each other. This high possibility in terms of observations is still very difficult as I explained during my lecture.

We are very fortunate that theorists have been rather successful in the past few decades. But in all new discoveries, if you look at all the great accelerators that have been constructed each has done its duty in terms of surprising discoveries. Only in two cases: the discovery of $\bar{p}, \bar{n}$ and the discovery of W and Z, the machine was constructed for that specific purpose, and the discovery was made. But none of the other great discoveries was expected. In other words we are making a move in science it is like playing chess with nature: do not believe that your initial move will make a check mate! Otherwise it would be an overestimation of our abilities.

– *Grossmann:*

You recommended the study of RHIC to investigate the structure of the QCD vacuum. But due to the complicated processes involved in these collisions, it might be rather difficult to decide whether a given signal is due to a new effect, or whether it could also be explained by standard methods (see e.g. Brodsky's explaination for the J/ψ suppression). What is your opinion about this problem?

– *T.D. Lee:*

Life is full of challenges. The question is how smart we are and how we can overcome the difficulties. Take the example of the interferometry method: we can use $\pi\pi, KK$ or lepton lepton. If it is truly a geometrical phenomenon then the geometry (i.e. radius) will be independent of the kind of hadrons that you use. No matter whether you use $\pi\pi$ or KK you get the same radius because it is entirely determined by the size of the object. Any kind of dynamical ideas like ρ decays into $\pi\pi$ or other decays will be different. It will be a long way to sort it out whether it is geometrical or non-geometrical. A good analogy will be with a stellar phenomena: in each of these stellar phenomena we can measure many many photons, but we try to extract out of it the nature of large objects, not particularly the properties of individual photons. I believe we can overcome the challenge.

– *Wexler:*

This is a question about your analogy between the QCD vacuum and the superconductors. By applying a large enough magnetic field, the diamagnetic vacuum of a type II superconductor can be turned into a magnetic flux lattice; by adjusting the temperature, the flux lattice can be melted into a liquid, and maybe even turned into a glass. Do you expect the QCD vacuum to possess an analogous phase structure?

– *T.D. Lee:*

My personal view is that once the vacuum becomes a dynamical variable, then we will discover that since this is a macroscopic object, there is no reason why it should be describable by a scalar field with a quartic renormalizable theory. There is no such restriction. So therefore, once the vacuum becomes a dynamical variable it can be much more complicated. So far, the reason why we need ultra high energies is that we believe that big things are made of small things and small things made of smaller things. By uncertainty principle, the smaller the thing we need the larger the momentum and the larger the energy. But in that approach we concentrate on localization, therefore we miss macroscopic coherence. Investigations of macroscopic coherence compliment the drive to examine small distances. Once the large extension is there available for studies we will discover it will be far more complicated then the current kind of renormalizable scalar theories. I believe that it is important we must pursue this direction.

– *Lu:*

This morning you gave an explanation of the J/ψ suppression based on quark-gluon plasma picture, while Prof. Brodsky gave another scenario based on more conventional QCD. Can you think of any experimental consequences that could allow us to distinguish these two scenarios?

– *T.D. Lee:*

At present I imagine that the J/ψ suppression is a possible indication of the quark gluon plasma. Now if it is truly due to the plasma then it will exist only in heavy ion collisions. If it is due to microscopic physics then it will also be in pp collisions. So by studying the dependence on A, one can probably separate this out.

– *Lu:*

Usually E_T is used to indicate the formation of plasma. But at high E_T the particles generated during the collision practically leave the central region without further interacting with other particles, so how can one justify the formation of the quark gluon plasma? Is there enough time for the plasma to reach equilibrium?

– *T.D. Lee:*

These are the questions we do not know the answers. We need high energy ions about 100 GeV/nucleon. Then the total E_T of interest is way over 1 GeV. Therefore, multiple collisions will happen. The geometrical measurement of volume using $\pi - \pi$ interferometry gives the size of the central region after the hadronization. Because it bases only on the second law of thermodynamics of entropy increase, it should be applicable to non equilibrium phenomena as well.

– *Brahm:*

How would confinement work in the absence of light quarks?

– *T.D. Lee:*

Confinement should work independent of the quark mass. The confinement within lattice gauge which gives the most reliable calculations so far has no light quarks. The dynamical light quarks are really much harder to treat. Therefore, confinement works very well without light quarks. Therefore, confinement works very well without light quarks.

– *Brahm:*

Is there an asymmetry between $< s >$ and $< \bar{s} >$ densities in the quark gluon plasma, analogous to the asymmetry in the proton described by Stan Brodsky?

– *T.D. Lee:*

It depends. Quark Gluon Plasma itself is a general term. The Baryon number or the quark number of that soup can be zero or can be nonzero. In particular for the plasma generated in the heavy ion collisions in RHIC the plasma in the central region should be essentially symmetric. From AGS to SPS the target is more and more transparent, as I described to you that morning. If you extrapolate to the RHIC energies the central region will be essentially a symmetrical one. But the fragmentation region is more Baryon enriched. So by slicing the collision sausage into little pieces you have different varieties.

– *Kastening:*

Assume that QCD is wrong in the sense that there is not the predicted phase transition in heavy ion collisions with the expected properties. Due to the nature of the experiment involving multiparticle initial states, is it still conceivable to extract constructive information out of such an experiment, rather than only being able to falsify the QCD predictions or predictions of other theories?

– *T.D. Lee:*

It is a very big "if" that QCD is wrong. It opens a lot of possibilities. Based on your assumption I would say the sky is the limit. Right now the QCD vacuum should be taken as a first guideline. If you first assume QCD to be right, then you may discover what is wrong with QED. If you make the assumption of it to be wrong then you may never know what you are doing.

– *Pittori:*

Do you think that viewing the vacuum as a physical medium can be related in some way to the view that you explained here in Erice in 1983 that space-time can be seen as dynamical variables in the discrete quantum theory?

– *T.D. Lee:*

That is a different speculation. Since I have not talked about discrete space time during this Erice meeting I will just try to summarize what I have done some ten years ago when I was interested in trying to make quantization of gravity finite. There are two ways to proceed: one way is the Superstring, that is you increase your degrees of freedom. But you assume more symmetry. Then by cancellation of infinities the system becomes finite. The alternative way that I tried to pursue is to decrease the degrees of freedom so that the system becomes a finite one and therefore no divergencies. When you make space time discrete, your fundamental equations will be difference equations. In that case most of the usual difference equations do not admit the continuous group of transformations. The variety that I created about ten years ago is to guarentee that the difference equations written down will have rigorously the full Einstein Poincaré symmetry. Take any action whose density is a function of x(t), in the continuum theory you vary the action with respect to arbitrary x(t) by fixing the initial point 1 and the final point 2:

$$\delta \int_1^2 \mathcal{L}(x(t))dt = 0$$

It is easy to change this into a different postulate: suppose I consider only piecewise continuous curves, that is instead of only fixing the initial and final points 1 and 2 also fix the number of kinks along that path:

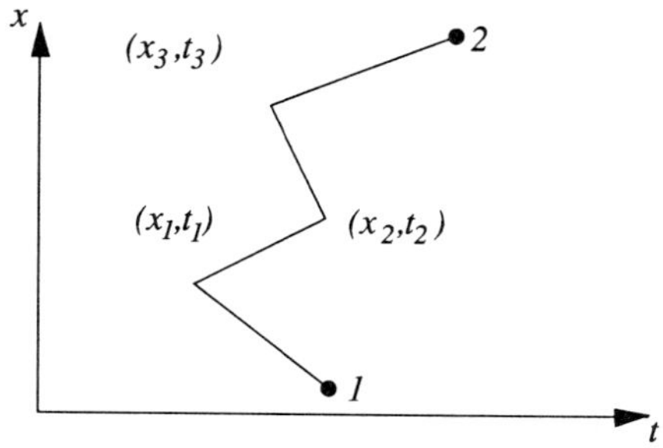

Now it is obvious that if I take this same Lagrangian, apply the Lagrangian to this type of piecewise continuous surface, and vary against its shape, the varying of arbitrary shape is the same as varying the endpoints. And then I will end up with a set of difference equations, and these difference equations by construction have the exact continuum symmetry of the original Lagrangian. In other words the endpoints x_n and t_n will adjust themselves so that from point to point energy and momentum are exactly conserved. This idea can be generalized to the rotation group and to general relativity, and thereby you arrive at a finite theory of discrete space-time points, but with continuous space-time symmetry. The measurement beyond the density of the lattice (in quantum gravity this will be Planck length) has no meaning. So it is a basically different quantum mechanics from what we are used to. Now the vacuum is a long distance phenomenon, so whether our space time at Planck length is discrete or not is of little importance.

Derivation of the Minimal Standard Model Lagrangian

Ronald Kleiss

TH Division
CERN, Genève

Derivation of the model: top-down versus bottom-up approach

The usual derivations of the Minimal Standard Model Lagrangian proceed in a slightly axiomatic way, which can be called a *top-down* approach:

1. Assign the matter fields (leptons and quarks) to particular representations of postulated symmetry groups $SU(2)_L$ and $U(1)_Y$;

2. Add (at least) one scalar field with ϕ^4 self-iteractions and Yukawa coupling to the fermions;

3. Impose local symmetry on the system, which leads to the introduction of gauge fields;

4. Break the symmetry again;

5. Work out the physical fields and the phenomenology of the model;

6. Finally, compare with experiment: the model is seen to work!

In this derivation the actual phenomenology comes in only at the end. I prefer the following, *bottom-up* derivation which remains as close to a constructive phenomenological treatment as possible:

1. Use a number of basic assumptions:

   - QED works extremely well and should be a part of the model;
   - The Fermi model works quite well at low energy, so at low energy our model should reproduce those result;

Physics Up to 200 TeV
Edited by A. Zichichi, Plenum Press, New York, 1991

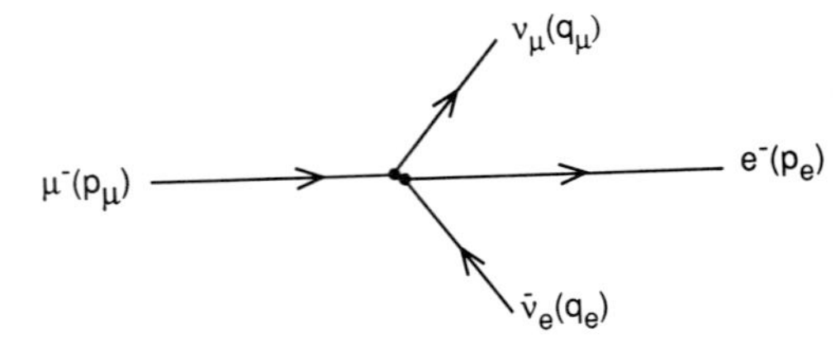

Figure 1. Muon decay in the Fermi model.

- All fields should be really elementary, that is, contain no substructure at any energy scale;

2. Work out the consequences of various Gedanken experiments, at each moment requiring reasonable field-theoretical behaviour of the model;

3. This leads us to introduce several new fields, but we shall only introduce them *where and when they are required*, and not postulate any superfluous fields or interactions.

In this way we can expect to remain closer to physics at each stage of the derivation: in addition a number of useful field-theoretical notions are developed along the way. Obviously a bottom-up derivation cannot supersede a top-down one since there is in principle no limit to the number of possible Gedanken experiments: but for pedagogical purposes it is attractive to show how physical requirements lead to a gauge theory of the Standard Model type almost uniquely.

Muon decay in the Fermi model

Let us start with the primordial weak process, muon decay:

$$\mu^-(p_\mu) \rightarrow \nu_\mu(q_\mu)e^-(p_e)\bar\nu_e(q_e) \ ; \tag{1}$$

in the Fermi model, the amplitude for this process is defined[1] to be

$$\mathcal{M} = -i\frac{G}{\sqrt{2}}\bar u(q_\mu)\left(1+\gamma^5\right)\gamma^\alpha u(p_\mu)\ \bar u(p_e)\left(1+\gamma^5\right)\gamma_\alpha v(q_e) \ . \tag{2}$$

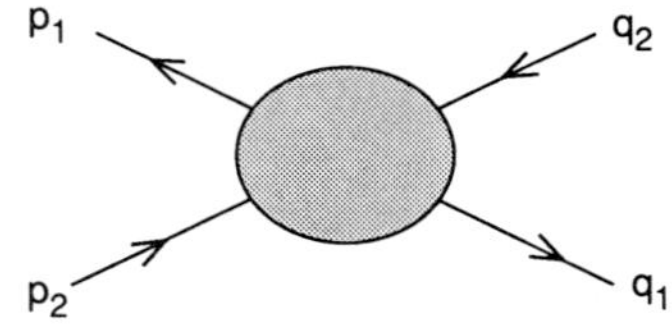

Figure 2. The general four-fermion amplitude

[1]The sign of this amplitude is different from the conventional one: this in order to make the transition to the electroweak model simpler. In general, such an overall sign is irrelevant for the physics content of the model.

The Feynman diagram for this process is depicted in fig.1. Note that both fermion currents have the Lorentz structure $\gamma^\mu + \gamma^5\gamma^\mu = \gamma^\mu - \gamma^\mu\gamma^5$: from the last alternative form, such currents are called V-A currents (V=vector, A=axial vector). The overall constant G is called the Fermi constant. The corresponding matrix element, squared and summed over all spins, is given by

$$\sum_{\text{spins}} |\mathcal{M}|^2 = 128G^2(p_e \cdot q_\mu)(p_\mu \cdot q_e) \quad . \tag{3}$$

Phenomenologically, this Ansatz for the matrix element is in satisfactory agreement with the data on muon decay. If we restrict ourselves to muon decay there would be no reason to doubt the validity of the Fermi model, except for one reason: it is in the Fermi model not possible to calculate higher-order corrections. In the following we shall see how to cure this problem.

The general four-fermion amplitude

Before proceeding let us first present some useful generally valid formulae. In these lectures an important rôle is played by four-fermion interaction processes. The less trivial part of such amplitudes is that which contains the spinors and Dirac matrices for the fermions: it is depicted in fig.2 and its general form can be denoted by

$$A \equiv \bar{u}(p_1)\left(v_p + a_p\gamma^5\right)\gamma^\mu u(p_2)\ \ \bar{u}(q_1)\left(v_q + a_q\gamma^5\right)\gamma_\mu u(q_2) \ ; \tag{4}$$

the masses of the various fermions are given by

$$p_1 \cdot p_1 = m_1^2 \ , \quad p_2 \cdot p_2 = m_2^2 \ , \quad q_1 \cdot q_1 = M_1^2 \ , \quad q_2 \cdot q_2 = M_2^2 \ , \tag{5}$$

and the convention that $m_i < 0$ or $M_i < 0$ denotes an antiparticle. In the Fermi model for muon decay we hence have $p_1 = q_\mu$, $p_2 = p_\mu$, $q_1 = p_e$, $q_2 = q_e$, $m_1 = 0$, $m_2 = m_\mu$, $M_1 = m_e$, $M_2 = 0$, $v_p = a_p = v_q = a_q = (G/\sqrt{2})^{1/2}$. It is straightforward to work out the spin-summed square of this general amplitude:

$$\begin{aligned}\sum_{\text{spins}} |A|^2 = \ & 32\Big\{\left(v_p^2 + a_p^2\right)\left(v_q^2 + a_q^2\right)\left[(p_1 \cdot q_1)(p_2 \cdot q_2) + (p_1 \cdot q_2)(p_2 \cdot q_1)\right] \\ & + 4v_p a_p v_q a_q \left[(p_1 \cdot q_1)(p_2 \cdot q_2) - (p_1 \cdot q_2)(p_2 \cdot q_1)\right] \\ & + (v_p^2 + a_p^2)(v_q^2 - a_q^2) m_1 m_2 (q_1 \cdot q_2) \\ & + (v_p^2 - a_p^2)(v_q^2 + a_q^2) M_1 M_2 (p_1 \cdot p_2) \\ & + 2(v_p^2 - a_p^2)(v_q^2 - a_q^2) m_1 m_2 M_1 M_2\Big\} \ .\end{aligned} \tag{6}$$

To obtain an appropriate matrix element square average, to this we have to add the propagator, colour factors, and a factor for the averaging over incoming spins and colours. The result for muon decay is seen to be a special case of this general formula.

Formulae for cross sections and widths

In addition to the square of the Dirac structure discussed in the previous section it is useful to give here some results pertaining to cross sections and widths. The formula for a $1 \to n$ decay of a massive particle with mass M is given by

$$d\Gamma = \frac{1}{2M}\frac{1}{(2\pi)^{3n-4}} <|\mathcal{M}|^2> d\mathrm{LIPS}_n \quad , \tag{7}$$

whereas for the cross section in a $2 \to n$ scattering process with effectively massless incoming particles we have

$$d\sigma = \frac{1}{2s}\frac{1}{(2\pi)^{3n-4}} <|\mathcal{M}|^2> d\mathrm{LIPS}_n \quad . \tag{8}$$

Here s is the total invariant mass squared. The notation $<|\mathcal{M}|^2>$ denotes the appropriate spin/colour average of the matrix element squared. The Lorentz-invariant phase space integration element $d\mathrm{LIPS}_n$ is defined by

$$\delta^4\left(P - \sum_{i=1}^{n} q_i\right) \prod_{i=1}^{n} d^4 q_i \delta(q_i^2 - m_i^2) \quad ; \tag{9}$$

here, P^μ it the total initial momentum, and q_i^μ are the outgoing momenta with masses m_i. For massless particles the 2- and 3-body phase spaces allow for relatively simple formulations:

$$d\mathrm{LIPS}_2 = \frac{1}{8} d\Omega \quad , \tag{10}$$

where Ω is the $\vec{q}_1 = -\vec{q}_2$ solid angle; and

$$\begin{aligned} d\mathrm{LIPS}_3 &= \frac{1}{8} dq_1^0 dq_2^0 d\Omega_1 d\phi_{12} \quad , \\ \frac{\sqrt{s}}{2} - q_1^0 \leq q_2^0 \leq \frac{\sqrt{s}}{2} \;;, &\quad 0 \leq q_1^0 \leq \frac{\sqrt{s}}{2} \quad . \end{aligned} \tag{11}$$

Here, Ω_1 is the solid angle of $\vec{q}_1$, and ϕ_{12} is the azimuthal angle of $\vec{q}_2$ around $\vec{q}_1$. The last significant point is that for massless particles the total phase space volume is given by

$$\int d\mathrm{LIPS}_n = \left(\frac{\pi}{2}\right)^{n-1} \frac{s^{n-2}}{(n-1)!(n-2)!} \quad . \tag{12}$$

If we define U to be an arbitrary unit of either mass or energy, cross sections have dimension U^{-2}, and widths dimension U. Since the n-body phase space has dimension U^{2n-4} a $2 \to n$ amplitude necessarily has dimension U^{2-n}, and (by crossing) a $1 \to n$ amplitude must have dimension U^{3-n}.

The Fermi constant

We now return to the Fermi model of muon decay. The spin-averaged matrix element squared is

$$<|\mathcal{M}|^2> \;=\; 64G^2(p_e \cdot q_\mu)(p_\mu \cdot q_e) \;=\; 32 m_\mu^2 q_e^0 \left(m_\mu - 2q_e^0\right) \quad , \tag{13}$$

where the dot products have been evaluated in the muon rest frame. The total muon decay width is therefore given by the integral

$$\Gamma_\mu = \frac{m_\mu G^2}{\pi^3} \int_0^{m_\mu/2} dq_e^0 \, (q_e^0)^2 \left(\frac{m_\mu}{2} - q_e^0\right) = \frac{G^2 m_\mu^5}{192\pi^3} \quad . \tag{14}$$

Since the muon mass is $m_\mu = 0.10566$ GeV and its width $\Gamma_\mu = 2.996\ 10^{-19}$ GeV we derive the value of the Fermi coupling to be

$$G = 1.16\ \times\ 10^{-5}\text{GeV}^{-2}\ . \tag{15}$$

The Fermi model at high energies

We now start to examine the cosequences of the Fermi model. If it is to be a good general theory of weak interactions and the muon, electron and neutrino fields are truly elementary, the Fermi model should also provide a description for the crossed process

$$e^-(p_e)\ \bar{\nu}_e(q_e)\ \rightarrow\ \mu(p_\mu)\ \bar{\nu}_\mu(q_\mu) \tag{16}$$

at all energies. The amplitude is of course still given by (a crossed version of) Eq.(2) and its average square by Eq.(6). In the $e^-, \bar{\nu}_e$ centre-of-mass frame, the cross section is therefore

$$d\sigma = \frac{G^2 s}{16\pi^2}(1+\cos\theta)^2 d\Omega\ , \tag{17}$$

where θ is the angle between the incoming e^- and the produced μ^-: the total cross section reads

$$\sigma(e^-\bar{\nu}_e \rightarrow \mu^-\bar{\nu}_\mu) = \frac{G^2 s}{3\pi}\ . \tag{18}$$

As usual for weak interactions, the cross section rises linearly with s: this is also clear from simple dimensional arguments, since the Fermi constant has dimension mass^{-2}.

The optical theorem and diagram cutting

At this point, and several other instances in what follows, it will be very useful to apply notions connected with the *optical theorem*. Consider a quantum world with a number of possible states labelled by indices $i, j, \ldots$ (for simplicity we assume discrete states only). The time evolution in this world can be described by the S (scattering) matrix, which connects initial and final states. The S matrix element between an initial state k and a final state j can be written as

$$S_{jk} = \delta_{jk} + iT_{jk}\ , \tag{19}$$

where the Kronecker delta represents the time evolution without any interaction taking place ('nothing happens'), and the T matrix is the interaction part ('something happens'): the matrix element for the transition $k \rightarrow j$ is given by T_{jk}. Conservation of total probability implies that the S matrix be unitary:

$$\begin{aligned} 1 &= SS^\dagger \\ &= (1+iT)\left(1-iT^\dagger\right) \\ &= 1 + i\left(T - T^\dagger\right) + TT^\dagger\ . \end{aligned} \tag{20}$$

In the special case where the initial and final state are the same, we therefore have

$$2\text{Im}T_{kk} = \left(TT^\dagger\right)_{kk} = \sum_n |T_{kn}|^2\ , \tag{21}$$

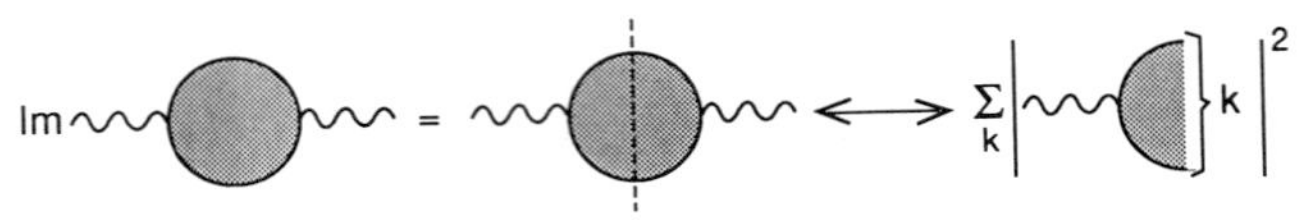

Figure 3. The cutting rule for the photon propgator.

where the sum runs over all intermediate states n. The optical theorem thus relates the imaginary part of the forward amplitude $k \to k$ to the total cross section $k \to$ any n. Note that this relation is *nonlinear*: it connects higher-order imaginary parts to lower-order total cross sections. A tremendously useful diagrammatic application of this is given by the *cutting rules*: consider diagram (a) of fig.3: it depicts the amplitude for a photon to propagate without scattering. Its imaginary part is obtained by cutting through the diagram (in principle this has to be done in all possible ways). The two pieces can then be interpreted as a product of diagrams and conjugate diagrams, forming the total decay cross section for a (off-shell) photon. A specialization of this principle is provided in fig.4 where the contribution of an $f\bar{f}$ intermediate state to the forward amplitude in $e^+e^- \to e^+e^-$ is related to the total cross section for $e^+e^- \to f\bar{f}$:

$$\sigma_{\text{tot}}(e^+e^- \to f\bar{f}) = \frac{1}{s}\text{Im}\mathcal{M}^{(f\bar{f})}_{e^+e^-\to e^+e^-} \quad . \tag{22}$$

Unitarity bounds

In addition to providing a connection between amplitudes, the optical theorem also turns out to put limits on them, owing to its nonlinear character. In the sum over initial states in Eq.(21), the initial state of course appears as a term: so we must have

$$2\text{Im}T_{ii} \geq |T_{ii}|^2 \quad . \tag{23}$$

Writing T_{ii} explicitly as the complex number $\alpha + i\beta$ we therefore have $2\beta \geq \alpha^2 + \beta^2$, which can only be satisfied if

$$\alpha^2 \leq \beta^2 - 2\beta \quad , \quad \beta \leq 2 \ . \tag{24}$$

The allowed region in the Argand plot (the (α, β) plot) is depicted in fig.5. We find that due to the unitarity of the S matrix, total cross sections must be limited, since we have

$$\sum_n |T_{in}|^2 \leq 4 \ . \tag{25}$$

A very explicit example of the above is the simple case of a quantum world with only 2 states. The S matrix is therefore a 2×2 unitary matrix, with the general form

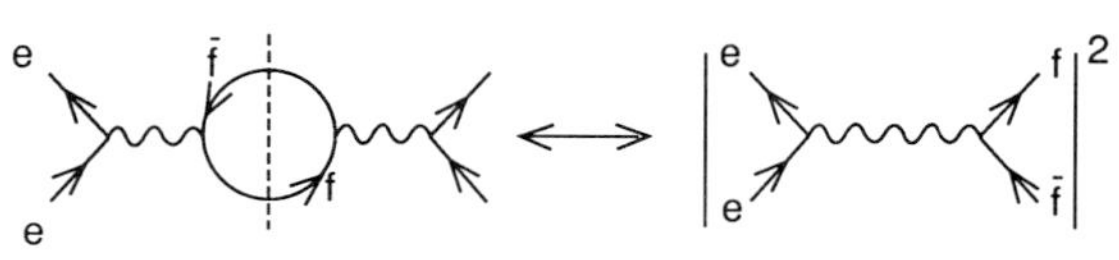

Figure 4. The cutting rule for an $f\bar{f}$ loop diagram.

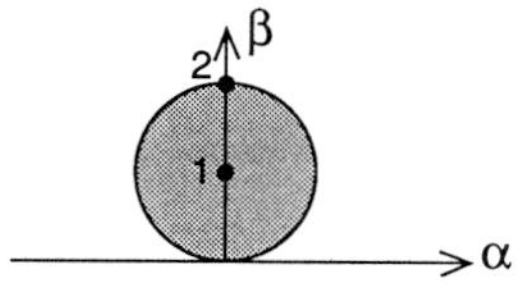

Figure 5. The allowed region in the Argand plot.

$$S = \begin{pmatrix} \cos\theta e^{i\phi_1} & \sin\theta e^{i\phi_2} \\ -\sin\theta e^{i(\phi_1-\phi_2+\phi_3)} & \cos\theta e^{i\phi_3} \end{pmatrix} . \tag{26}$$

Some trivial algebra leads to

$$TT^\dagger = \begin{pmatrix} 2(1-\cos\theta\cos\phi_1) & -\sin\theta e^{i\phi_2}\left[1+e^{-i(\phi_1+\phi_3)}\right] \\ -\sin\theta e^{-i\phi_2}\left[1+e^{i(\phi_1+\phi_3)}\right] & 2(\cos\theta\cos\phi_3) \end{pmatrix} ; \tag{27}$$

so indeed the total $1 \to 1$ probability is

$$\left(TT^\dagger\right) = 2(1-\cos\theta\cos\phi_1) \le 4 . \tag{28}$$

In actuality one has to be a bit more careful in deriving upper bounds on cross sections: for the scattering of two scalar particles the total possible cross section depends on the total angular momentum J as

$$\sigma_J(s) = \frac{16\pi}{s}(2J+1) , \tag{29}$$

while for incoming spinning particles we of course have to add spin-averaging factors: for instance, for unpolarized incoming $f\bar{f}$ we have

$$\sigma_J(s) = \frac{4\pi}{s}(2J+1) . \tag{30}$$

In fact since in a general scattering process arbitrarily high angular momenta can be involved the total cross section could still be infinite. The even more general *Froissart bound* puts a limit on the sum of the $\sigma_j(s)$:

$$\sum_{J=0}^{\infty} \sigma_J(s) \le C\,(\log s)^2 , \tag{31}$$

at least when (as in strong interactions) no massless particles are exchanged.

Unitarity violation in the Fermi model

Let us now return to $e\bar{\nu}_e$ scattering, armed with the unitarity bound. Under the assumption that only left-handed neutrino's ever occur in nature (and therefore the spin average only affects the electron) the unitarity bound implies the following condition:

$$\sigma_{\text{tot}}(e^-\bar{\nu}_e \to \mu^-\bar{\nu}_\mu) \le \frac{24\pi}{s} , \tag{32}$$

where we have used $J = 1$ since the fermion currents are (axial) vector currents. The unitarity condition is seen to be violated at energies

$$\sqrt{s} \geq \left[\frac{6\pi\sqrt{2}}{G}\right]^{1/2} = 1516 \text{ GeV} \ , \tag{33}$$

which implies that the cross section of Eq.(18) can not be correct at such energies[2]. Although experiments involving $e^-\bar{\nu}_e$ scattering at several TeV do not seem likely in the near future, one can imagine such processes taking place in the very early universe. At any rate a model with such a built-in limitation is a poor candidate for a generally valid elementary particle theory. Note that the particular energy at which the unitarity bound is violated is not very important: it is the rising of the cross section with s that leads to trouble.

It should be remarked here that, since the Hamiltonian corresponding to the Fermi interaction is hermitian, the *full* cross section must of course be unitary: what we have established here is only a unitarity violation at the tree level of perturbation theory. However, a theory that fails already at this level is not an attractive one. One way out would be to postulate a Fermi constant that depends on energy so as to satisfy unitarity at all energies. However, such a form-factor-like behaviour indicates that the lepton fields are not really elementary, in contradiction with our assumptions.

Introducing the W

A better improvement on the Fermi model is the assumption of an intermediate vector particle, the W. This means that the point-like four-fermion interaction of fig.1 is split into two fermion-fermion-boson vertices, connected by a propagator as in fig.6. The corresponding amplitude reads

$$\mathcal{M} = i\frac{g_{\rm W}^2}{(p_e+q_e)^2 - m_{\rm W}^2}\bar{u}(q_\mu)\left(1+\gamma^5\right)\gamma^\alpha u(p_\mu)\ \bar{u}(p_e)\left(1+\gamma^5\right)\gamma_\alpha v(q_e) \ , \tag{34}$$

where $m_{\rm W}$ denotes the mass of the W and $g_{\rm W}$ its coupling to the lepton-neutrino currents. Since at low energies the Fermi model is quite succesful we require

$$\frac{g_{\rm W}^2}{m_{\rm W}^2} = \frac{G}{\sqrt{2}} \ . \tag{35}$$

The absence of noticeable form-factor-like effects in muon decay implies $m_{\rm W} \gg m_\mu$, or $g_{\rm W} \geq 10^{-3}$ or so: on the other hand, if we want perturbation theory to be applicable (the very reason why we dropped the Fermi model!) we would rather

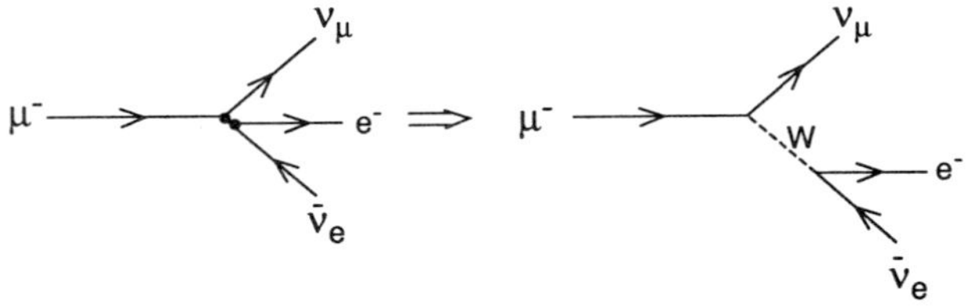

Figure 6. Muon decay in the electroweak model.

[2]An even lower bound of 735 GeV can be obtained by considering $e^-\nu_\mu \to \mu^-\nu_e$, and assuming $J = 0$.

have $g_W \leq 1$ hence $m_W \leq 350$ GeV or at the very least $g_W^2/(4\pi) \leq 1$, implying $m_W \leq 1238$ GeV.

Unitarity restoration and limits on the W mass

As before the total cross section for $e^-\bar{\nu}_e \to \mu^-\bar{\nu}_\mu$ is given by a trivial crossing: in the W model we have instead of Eq.(18):

$$\sigma(e^-\bar{\nu}_e \to \mu^-\bar{\nu}_\mu) = \frac{2g_W^4}{3\pi}\frac{s}{(s-m_W^2)^2} \quad . \tag{36}$$

At low energies, $s \to 0$, this cross section coincides with the Fermi-model result (18), whereas for $s \to \infty$ it has indeed the right $1/s$ behaviour. Moreover, we can derive two interesting statements on the value of m_W:

- unitarity for $s \to \infty$ will be satisfied if

  $$\frac{g_W^2}{6\pi} \leq 1 \quad \Rightarrow \quad m_W \leq 1516 \text{ GeV} \quad , \tag{37}$$

  which is precisely the value of the theory breakdown found before. This is called the *unitarity limit* on the value of m_W.

- On the other hand, if m_W were larger than 1516 GeV, the theory would break down at around those energies, with the possible onset of what would then be called new physics: this is called the *inverse unitarity limit.*

These two limits are not interesting by themselves as m_W is much smaller: but they are instructive because for the Higgs boson similar limits wil arise.

The only question remaining is what happens around the W resonance, since even if $g_W^2/(6\pi) < 1$ the cross section formula will violate unitarity if $|1 - m_W^2/s| \leq g_W^2/(6\pi)$. This problem is solved by considering the quantum mechanics of massive spin-one particles in more detail.

Some field-theoretical properties of spin-1 particles

At the classical level a massive spin-1 field $W^\mu(x)$ is described by the Proca equation:

$$\partial^\alpha\partial_\alpha W^\mu(x) - \partial^\mu\partial_\alpha W^\alpha(x) + m_W^2 W^\mu(x) = 0 \quad . \tag{38}$$

The ***propagator*** of a massive spin-1 field is determined by going to the momentum representation: by definition, the propagator $S_{\alpha\beta}(k)$ satisfies

$$\left[-k^2 g^{\mu\alpha} + k^\mu k^\alpha + m_W^2 g^{\mu\alpha}\right] S_\alpha^{\ \beta}(k) = i g^{\mu\beta} \quad , \tag{39}$$

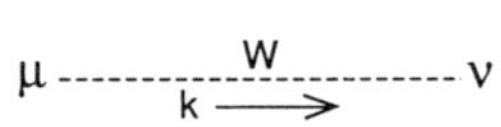

Figure 7. The W propagator.

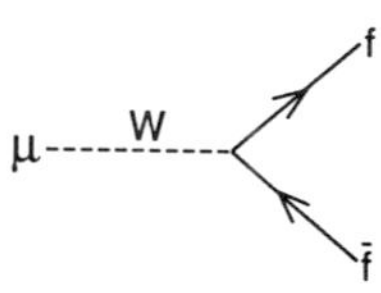

Figure 8. The W-fermion-fermion vertex.

of which the solution[3] is

$$S^{\mu\nu}(k) = \frac{i}{k^2 - m_W^2}\left(-g^{\mu\nu} + \frac{k^\mu k^\nu}{m_W^2}\right) \ . \tag{40}$$

The Feynman diagram corresponding to Eq.(40) is depicted in fig.7. In addition we have already proposed the Wff vertex, depicted in fig.8: it reads

$$ig_W\left(1+\gamma^5\right)\gamma^\mu \ . \tag{41}$$

The *polarization vectors* of the field are defined by the expansion of the field in creation and annihilation operators of plane waves:

$$W^\mu(x) = \int \frac{d^4k}{(2\pi)^4}\delta(k^2 - m_W^2)\theta(k^0) \sum_{\lambda=1,2,3} \left\{a_\lambda(k)\epsilon_\lambda(k)e^{-i(kx)} + a_\lambda^\dagger(k)\epsilon_\lambda^\star(k)e^{i(kx)}\right\} \ ; \tag{42}$$

they are by definition orthonormal, $\epsilon_\lambda\cdot\epsilon_{\lambda'} = -\delta_{\lambda\lambda'}$, and must satisfy the Proca equation:

$$\left[-k^2 g^{\mu\alpha} + k^\mu k^\alpha + m_W^2 g^{\mu\alpha}\right](\epsilon_\lambda)_\alpha(k) = 0 \ . \tag{43}$$

Multiplying this equation with k_μ we immediately get

$$\epsilon_\lambda(k)\cdot k = 0 \quad \text{and} \quad k^2 - m_W^2 = 0 \ . \tag{44}$$

If we take the momentum k^μ to be along the positive z-axis, there are two possible transverse polarizations, as well as one longitudinal one:

$$\begin{aligned} k^\mu &= (k^0, 0, 0, |\vec{k}|) \ , \quad (k^0)^2 = |\vec{k}|^2 + m_W^2 \ , \\ \epsilon_1^\mu(k) &= (0,1,0,0) \ , \quad \epsilon_2^\mu(k) = (0,0,1,0) \ , \\ \epsilon_3^\mu(k) &= \left(\frac{|\vec{k}|}{m_W}, 0, 0, \frac{k^0}{m_W}\right) \ . \end{aligned} \tag{45}$$

If the energy becomes large with respect to the mass, the transverse polarizations ϵ_1 and ϵ_2 remain unchanged, but the longitudinal one, ϵ_3, grows with energy:

$$\epsilon_3^\mu(k) \stackrel{k^0 \gg m_W}{\longrightarrow} \frac{k^\mu}{m_W} + \mathcal{O}\left(\frac{m_W}{k^0}\right) \ . \tag{46}$$

In cross sections the spin sum of ϵ_λ and its conjugate occurs: this can covariantly be written as

$$\sum_{\lambda=1,2,3} \epsilon_\lambda^\mu(\epsilon_\lambda^\nu)^\star = -g^{\mu\nu} + \frac{k^\mu k^\nu}{m_W^2} \ . \tag{47}$$

Note that this corresponds exactly with the denominator of the propagator: this is a general feature since the propagator and the spin sum satisfy the same Proca equation. Alternatively it can be considered as a consequence of the optical theorem, since an external line in a cross section corresponds to an internal line in a higher-

[3]Note that in Eq.(34) we have only used the first term in the propagator, because the second one is suppressed by the electron-to-W mass ratio when contracted with the fermion current: we systematically neglect such terms here.

order diagram[4], in which all physical polarization states must propagate.

Massless spin-1 particles

The result of subsection leads to a complication in the case of massless spin-1 particles (photons) since in that case $\epsilon_3^\mu(k)$, the longitudinal polarization vector, diverges (*cf* Eq.(46)). There is a way out: consider an amplitude $\mathcal{M}$ in which an external photon occurs with polarization vector $\epsilon^\mu(k)$ so that we can write

$$\mathcal{M} = \mathcal{M}_\mu \epsilon^\mu(k) \ . \tag{48}$$

If we now require that all amplitudes in the theory strictly obey

$$\mathcal{M}_\mu k^\mu = 0 \ , \tag{49}$$

then the longitudinal polarization states are effectively decoupled from the physical world, and we can take the limit $m_W \to 0$ without difficulty. The requirement (49) is called *current conservation*: the nomenclature becomes clear if one considers the Fourier transform of Eq.(49) which reads $\partial^\mu \mathcal{M}_\mu = 0$: the $\mathcal{M}_\mu$ is a current which acts as the source of the photon. In what follows we shall require Eq.(49) for all amplitudes with external photons.

Current conservation is related to the Lorentz invariance of the theory as can be seen from the following. Consider a photon with momentum k^μ and polarization ϵ^μ in some Lorentz frame, in which the polarization is purely transverse:

$$k^\mu = (k^0, \vec{k}) \quad , \quad \epsilon^\mu = (0, \vec{\epsilon}_{\rm tr}) \ , \tag{50}$$

with $\vec{\epsilon}_{\rm tr}\cdot\vec{k} = 0$. In another frame we then have

$$k^\mu = (k^{0\prime}, \vec{k}') \quad , \quad \epsilon^\mu = (\eta^0, \vec{\eta}) \ , \tag{51}$$

which generally is no longer purely transverse. Defining the transverse part of $\vec{\eta}$ as

$$\vec{\eta}_{\rm tr} = \vec{\eta} - \frac{\eta^0 \vec{k}'}{k^{0\prime}} \tag{52}$$

the boosted polarization vector can be decomposed as

$$\eta^\mu = \frac{\eta^0}{k^{0\prime}} k^\mu + (0, \vec{\eta}_{\rm tr}) \ . \tag{53}$$

Therefore the requirement that *in any Lorentz frame* it is only the transverse polarizations that interact, immediately leads to Eq.(49).

A last point to be mentioned here is that using Eq.(49) we can also simplify the spin sum of an external photon to

$$\sum_{\rm spins} \epsilon^\mu (\epsilon^\nu)^* = -g^{\mu\nu} \ , \tag{54}$$

since the $k^\mu k^\nu$ term always drops out. In a consistent way we may therefore also use a simpler propagator for internal photons, namely $-ig^{\mu\nu}/k^2$.

---

[4]Strictly speaking, this of course holds only for the imaginary part of the propagator.

The width of the W at tree level

Since the W^- couples to, say, $e^-\bar{\nu}_e$ it can decay into them. The Feynman diagram is given in fig.8 and the amplitude reads

$$\mathcal{M} = i g_{\rm W} \bar{u}(p_e)\left(1+\gamma^5\right)\gamma^\mu v(p_\nu)\epsilon_\mu \quad . \tag{55}$$

In the limit $m_e \ll m_{\rm W}$ the partial width is

$$\Gamma(W^- \to e^-\bar{\nu}_e) = \frac{g_{\rm W}^2 m_{\rm W}}{6\pi} \quad . \tag{56}$$

In N_f of such light families are allowed in W decay (in actuality we have $N_f = 9$, if we count 3 quark colours) we have

$$\Gamma_{\rm W} = \frac{N_f g_{\rm W}^2 m_{\rm W}}{6\pi} = \frac{N_f G m_{\rm W}^3}{6\pi\sqrt{2}} \quad . \tag{57}$$

An interesting consequence is that *in any phenomenologically viable model* (i.e. with the correct prediction for muon decay) the total W width grows as $m_{\rm W}^3$! This leads to yet another mass constraint: an elementary-field interpretation of the W is only really acceptable as long as $\Gamma_{\rm W}$ is much smaller than $m_{\rm W}$. For $N_f = 1$ we find that $\Gamma_{\rm W} = m_{\rm W}$ for $m_{\rm W} = 1516$ GeV (again!), while for $N_f = 9$ the limit becomes $m_{\rm W} = 505$ GeV. As before these limits are not by themselves significant but they are similar to those obtained by the same arguments for the mass of the Higgs boson.

Handwaving quantum mechanics (HQM) of unstable particles

The following argumentation will teach us how the effects of a finite W width can be incorporated in our treatment. Consider a stable particle of mass m at rest. The Schrödinger equation with Hamiltonian $\hat{H}$ is

$$i\frac{\partial}{\partial t}\psi(t) = \hat{H}\psi(t) = m\psi(t) \quad , \tag{58}$$

with the solution

$$\psi(t) = e^{-imt}\psi(0) \quad , \quad |\psi(t)|^2 = |\psi(0)|^2 \quad , \tag{59}$$

so that the probability to observe the particle is time-independent. If the particle is unstable with inverse lifetime (width) Γ we should instead have an exponentially decreasing probability to find the particle at time t:

$$|\psi(t)|^2 = |\psi(0)|^2 \, e^{-\Gamma t} \quad , \tag{60}$$

so that the wave function itself must read

$$\psi(t) = \left[e^{-imt}\psi(0)\right] e^{-\Gamma t/2} = e^{-i(m-\frac{\Gamma}{2})t}\psi(0) \quad , \tag{61}$$

or equivalently obey a Schrödinger equation

$$i\frac{\partial}{\partial t}\psi(t) = \hat{H}\psi(t) = \left(m - i\frac{\Gamma}{2}\right)\psi(t) \quad . \tag{62}$$

Note that this implies a non-hermitian Hamiltonian. Proceeding in this handwaving manner we go from the probability density to find the particle at a given time (assuming it to have been created at $t = 0$) to that for finding it at a given energy E by Fourier transformation:

$$\psi(E) \sim \int_0^\infty dt e^{iEt}\psi(t) = \int_0^\infty dt e^{i(E-m+i\frac{\Gamma}{2})t}\psi(0)$$

$$\propto \frac{1}{E-m+i\frac{\Gamma}{2}} \sim \frac{2m}{E^2-m^2+im\Gamma}\ , \qquad (63)$$

where in the last line we have assumed E to be close to m. One recognizes (as expected) the propagator which now has picked up an additional term $im\Gamma$.

Propagators for particles with finite width, and the $i\varepsilon$ prescription

The last paragraph provides a prescription for handling the W (or any other unstable particle) close to the mass shell. The propagator must be modified as follows:

$$S^{\mu\nu}(k) = \frac{i}{k^2-m_{\rm W}^2}\left(-g^{\mu\nu}+\frac{k^\mu k^\nu}{m_{\rm W}^2}\right) \Rightarrow \frac{i}{k^2-m_{\rm W}^2+im_{\rm W}\Gamma_{\rm W}}\left(-g^{\mu\nu}+\frac{k^\mu k^\nu}{m_{\rm W}^2}\right)\ . \qquad (64)$$

For stable particles we have $im\Gamma \to i\varepsilon \to 0$ so that we recover the famous '$i\varepsilon$ prescription':

$$\frac{1}{k^2-m^2} \Rightarrow \frac{1}{k^2-m^2+i\varepsilon} \qquad (65)$$

for stable internal particle lines in the Feynman diagrams. It is seen that the *sign* of the $i\varepsilon$ is directly connected with the direction of causality: a positive sign indicates that particles tend to *decay* exponentially instead of arising spontaneously with an exponentially increasing density, in accordance with everyday practice.

The W line shape in $e^-\bar{\nu}_e \to \mu^-\bar{\nu}_\mu$

Implementing the above in the result for $e^-\bar{\nu}_e \to \mu^-\bar{\nu}_\mu$ we now find the modified cross section formula

$$\sigma(e^-\bar{\nu}_e \to \mu^-\bar{\nu}_\mu)(s) = \frac{2g_{\rm W}^4}{3\pi}\frac{s}{\left(s-m_{\rm W}^2\right)^2+m_{\rm W}^2\Gamma_{\rm W}^2}\ . \qquad (66)$$

This Breit-Wigner resonance is finite at all energies and reduces to the previous results for $s \to 0$ and $s \to \infty$: around the place of what was in Eq.(36) a singularity, Eq.(66) gives instead

$$\sigma(e^-\bar{\nu}_e \to \mu^-\bar{\nu}_\mu)(m_{\rm W}^2) = \frac{2}{3\pi}g_{\rm W}^4\frac{1}{\Gamma_{\rm W}^2}$$

$$= \frac{24\pi}{m_{\rm W}^2}\left(\frac{\Gamma(W\to e^-\bar{\nu}_e)}{\Gamma_{\rm W}}\right)\left(\frac{\Gamma(W\to \mu^-\bar{\nu}_\mu)}{\Gamma_{\rm W}}\right)\ , \qquad (67)$$

where we have used Eq.(56) to relate $g_{\rm W}$ to the partial W widths. The representation (67) is especially attractive: the first factor is precisely the maximum cross section allowed in the $J=1$ channel by unitarity, while the two other factors are branching ratios. The branching ratio for $W \to \mu^-\bar{\nu}_\mu$ must of course be there since a produced W has only so much chance of going into a $\mu\bar{\nu}_\mu$ final state. The other branching ratio is then compulsory from the crossing symmetry argument that $\sigma(e^-\bar{\nu}_e \to \mu^-\bar{\nu}_\mu)$ should be equal to $\sigma(\mu^-\bar{\nu}_\mu \to e^-\bar{\nu}_e)$. It appears that the peak cross section is *the largest possible* under the constraints of unitarity and crossing symmetry. In our discussion of the Z line shape precisely the same will occur.

The $W^+W^-\gamma$ vertex: first try

If the W couples to $\ell\bar{\nu}_\ell$ it must be charged, so it is clear that also the photon couples to a W^+W^- pair. We now try to construct the form of this vertex. To this end it is useful to study the *Gordon decomposition* of currents. Consider a photon coupling to an outgoing $f\bar{f}$ pair with mass m and charge Q_f, as in fig.9. This vertex can be written as

$$\begin{aligned} iQ_f\bar{u}(p_1)\gamma^\mu v(p_2) &= \frac{iQ_f}{2m}\bar{u}(p_1)\left[(\not{p}_1 + m)\gamma^\mu + \gamma^\mu(\not{p}_2 - m)\right]v(p_2) \\ &= \frac{iQ_f}{2m}\bar{u}(p_1)\left[(p_1^\mu - p_2)^\mu - \sigma^{\mu\nu}(p_1+p_2)_\nu\right]v(p_2) \ , \end{aligned} \tag{68}$$

where we have used the Dirac equation in the first line, and the commutation rule $\gamma^\alpha\not{p} = 2p^\alpha - \not{p}\gamma^\alpha$ in the second line: $\sigma^{\mu\nu}$ stands for $\frac{1}{2}(\gamma^\mu\gamma^\nu - \gamma^\nu\gamma^\mu)$. The first term in Eq.(68) is a spin-independent *convection* term, which equals the vertex one would have if the fermions were actually scalars: it only depends on the charge flow along the charged particle line. The second term depends on the spin of the produced particles. In any vertex for charged particles coupling to a photon, at least the first term should be present. The simplest reasonable choice for the vertex in fig.10 is therefore

$$iQ_{\rm W}(p_1 - p_2)^\mu g^{\alpha\beta} \ . \tag{69}$$

Note that we have contracted the Lorentz indices of the W fields with each other since they have to be present in the vertex and this is the simplest possibility. Let us now study some consequences of this Ansatz.

Current conservation in $u\bar{d} \to W^+\gamma$

The W always couples to two fermions that differ by one unit in charge: in the following we shall call these 'up' (u) and 'down' (d) fermions, where the 'up' fermions can be ν_e, ν_μ, ν_τ, u quarks or c quarks: then the 'downs' are e, μ, τ, d quarks and s quarks, respectively. Let us now consider the Gedanken process

$$\bar{d}(p_1)\ u(p_2) \to W^+(q,\epsilon_+)\ \gamma(k,\epsilon) \ , \tag{70}$$

where we have specified all momenta and polarizations. The available Feynman graphs are given in fig.11. Applying the Feynman rules including the current Ansatz for the $W^+W^-\gamma$ vertex we have

$$\begin{aligned} \mathcal{M} &= \mathcal{M}_1 + \mathcal{M}_2 + \mathcal{M}_3 \ , \\ \mathcal{M}_1 &= -iQ_{\rm u}g_{\rm w}\bar{v}(p_1)\left(1+\gamma^5\right)\not{\epsilon}_+\frac{\not{p}_2 - \not{k}}{(p_2-k)^2}\not{\epsilon}u(p_2) \ , \\ \mathcal{M}_2 &= -iQ_{\rm d}g_{\rm w}\bar{v}(p_1)\not{\epsilon}\frac{\not{k} - \not{p}_2}{(k-p_2)^2}\left(1+\gamma^5\right)\not{\epsilon}_+u(p_2) \ , \end{aligned}$$

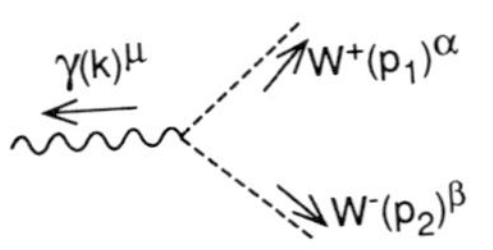

Figure 9. a photon coupling to an outgoing $f\bar{f}$ pair.

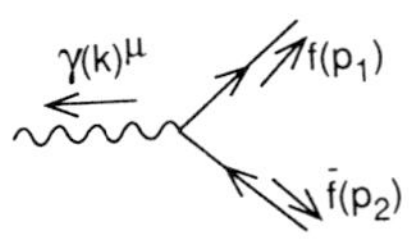

Figure 10. a photon coupling to an outgoing W^+W^- pair.

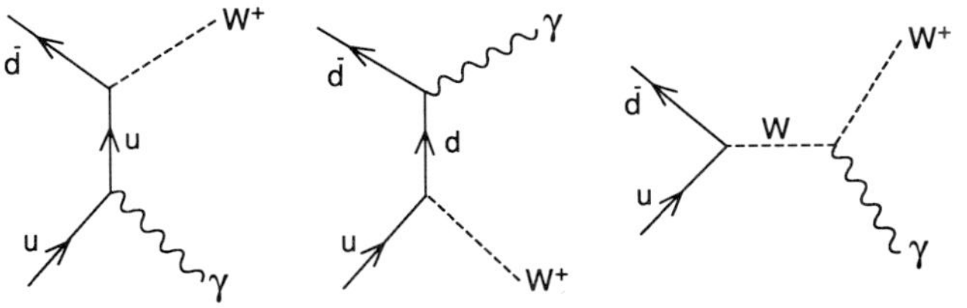

Figure 11. Feynman diagrams for $u\bar{d} \to W^+\gamma$.

$$\mathcal{M}_3 = iQ_{\rm W}g_{\rm W}\bar{v}(p_1)\left(1+\gamma^5\right)\not{q}_+ u(p_2)\frac{(2q+k)^\mu\epsilon_\mu}{(q+k)^2-m_{\rm W}^2} , \tag{71}$$

where we assume that $\Gamma_{\rm W}$ can be neglected if the energy is large enough. We also neglect for the moment the masses $m_{u,d}$.

One of the fundamental requirements for the amplitude (71) is that of electromagentic current conservation as in Eq.(49): we must have

$$\mathcal{M}]_{\epsilon\to k} = 0 . \tag{72}$$

Proceeding very explicitly, we have in the relevant part of $\mathcal{M}_1$:

$$\begin{aligned}\left.\frac{\not{p}_2-\not{k}}{(p_2-k)^2}\not{\epsilon}u(p_2)\right]_{\epsilon\to k} &= \frac{(\not{p}_2-\not{k})\not{k}}{-2k{\cdot}p_2}u(p_2) = \frac{\not{p}_2\not{k}}{-2k{\cdot}p_2}u(p_2)\\ &= \frac{2(k{\cdot}p_2)-\not{k}\not{p}_2}{-2k{\cdot}p_2}u(p2) = -u(p_2) ;\end{aligned} \tag{73}$$

here we have first used that $\not{k}\not{k} = k^\mu k_\mu = 0$, then anticommuted the $\not{p}_2$ and $\not{k}$, and finally the massless Dirac equation $\not{p}_2 u(p_2) = 0$. Similarly we have in $\mathcal{M}_2$:

$$\left.\bar{v}(p_1)\not{\epsilon}\frac{\not{k}-\not{p}_1}{(k-p_1)^2}\right]_{\epsilon\to k} = \bar{v}(p_1) , \tag{74}$$

and in $\mathcal{M}_3$

$$\left.\frac{(2q+k)^\mu\epsilon_\mu}{(q+k)^2-m_{\rm W}^2}\right]_{\epsilon\to k} = 1 . \tag{75}$$

Putting everything together we find

$$\mathcal{M}]_{\epsilon\to k} = i\left(Q_{\rm u}-Q_{\rm d}+Q_{\rm W}\right)\bar{v}(p_1)\left(1+\gamma^5\right)\not{q}_+ u(p_2) , \tag{76}$$

which vanishes provided that

$$Q_{\rm W} = Q_{\rm d} - Q_{\rm u} . \tag{77}$$

This is of course just charge conservation: if we use $u = \nu_e$ with $Q_{\rm u} = 0$ and $d = e^-$ with $Q_{\rm d} = -e$ (where e is the positive unit charge) we have $Q_{\rm W} = -e$. Note that the magnitude of $Q_{\rm W}$ would trivially be expected to be e: but its sign is more difficult to get since it depends on whether one considers W^+ to be the particle and W^- the antiparticle, or vice versa. This ambiguity is resolved by imposing current conservation.

High-energy failure of the first try

In addition to conservation of the electromagnetic current there is another condition to be imposed on the amplitude (71). Since we are dealing with a $2 \to 2$ process, the amplitude must for high s values go to a constant, energy-independent, value (which may be zero) if the cross section is to go as $1/s$. For simplicity we consider here only the limit $s \to \infty$ while we keep all scattering angles fixed. Some simple power counting applies: since

$$\sum_{\text{spins}} u(p)\bar{u}(p) = \not{p} + m \quad , \tag{78}$$

each fermion spinor behaves as $\sqrt{E}$, where E is of order $\sqrt{s}$; fermion propagators behave as $1/E$; the photon polarization is transverse and has therefore components of order $\mathcal{O}(1)$. This means that the energy dependence of, say, $\mathcal{M}_1$ is that of ϵ_+. If the W is transversely polarized, this means that $\mathcal{M}_1$ goes to a constant value as $s \to \infty$, as required; however for longitudinal W's we will get $\mathcal{M}_1 \sim \mathcal{O}(E/m_{\rm W})$ and the cross section will not go as $1/s$. Unitarity will therefore again be violated at high s unless there is some cancellation between $\mathcal{M}_1$, $\mathcal{M}_2$ and $\mathcal{M}_3$ as before. Proceeding as in the e.m. current conservation case, replacing ϵ_+^μ by $q^\mu/m_{\rm W}$ and neglecting $m_{\rm W}/E$ wherever this is safe, we find the analogue of Eq.(76):

$$\mathcal{M}]_{\epsilon_+ \to q/m_{\rm W}} \sim iQ_{\rm W}/m_{\rm W}\bar{v}(p_1)\left(1+\gamma^5\right)\gamma_\mu u(p_2)\left\{\epsilon^\mu - \frac{q\cdot\epsilon}{q\cdot k}k^\mu\right\} \quad . \tag{79}$$

Apparently, no cancellation is generally possible: our Ansatz for the $W^+W^-\gamma$ vertex is not viable.

The $W^+W^-\gamma$ vertex: second try

An alternative choice for the $W^+W^-\gamma$ vertex is easy to construct. The choice (69) appears to work well if we replace the photon polarization ϵ^μ by its momentum k^μ, but not if we replace the W^+ polarization ϵ_+^μ by its momentum q^μ. We may hope to improve on that situation if we make the W's in the vertex behave more 'photon-like'. Therefore we now propose for the vertex of fig.10 a form in which all the field indices and momenta appear on an equal footing, by permuting the vertex form (69):

$$iQ_{\rm W}V(p_1,\alpha,p_2,\beta,k,\mu) \equiv iQ_{\rm W}\left[(p_1-p_2)^\mu g^{\alpha\beta} + (p_2-k)^\alpha g^{\beta\mu} + (k-p_1)^\beta g^{\mu\alpha}\right] \quad , \tag{80}$$

where all momenta are going out of the vertex. We recognize the Yang-Mills vertex of gauge theory.

Of course we have to verify that also this vertex leads to the electromagnetic current conservation of Eq.(72); this is done along the same lines as above, and although the cancellations are now slightly more complicated they still go through if $Q_{\rm W} = Q_{\rm d} - Q_{\rm u}$ as before. The crucial test is to see whether we have improved on the production of longitudinal polarized W bosons. Indeed we find, in the limit $m_{\rm W} \ll E$:

$$\mathcal{M}]_{\epsilon_+ \to q/m_{\rm W}} \sim i\frac{Q_{\rm W} g_{\rm W}}{m_{\rm W}} \bar{v}(p_1)\left(1+\gamma^5\right) \not{q} u(p_2) \left[1 - \frac{q+k)^2}{(q+k)^2 - m_{\rm W}^2}\right] = \mathcal{O}\left(\frac{m_{\rm W}}{E}\right) \ , \tag{81}$$

so that also the longitudinal amplitude is well-behaved (and even vanishing) in the high-energy limit. We therefore adopt this vertex as a part of our theory.

Pair production of W's

Let us now consider another Gedanken process, that of W pair production:

$$\bar{u}(p_1)\, u(p_2) \ \to \ W^+(q_+, \epsilon_+)\ ^-(q_-, \epsilon_-) \ , \tag{82}$$

where again the momenta and polarizations are indicated. The two possible Feynman graphs that we can draw so far are depicted in fig.12, and the amplitude reads

$$\begin{aligned}
\mathcal{M} &= \mathcal{M}_1 + \mathcal{M}_2 \ , \\
\mathcal{M}_1 &= -2ig_{\rm W}^2 \bar{v}(p_1)\left(1+\gamma^5\right) \not{\epsilon}_- \frac{\not{p}_2 - \not{q}_+}{(p_2 - q_+)^2} \not{\epsilon}_+ u(p_2) \ , \\
\mathcal{M}_2 &= iQ_{\rm u} Q_{\rm W} \bar{v}(p_1)\gamma_\mu u(p_2) \frac{1}{s} V(q_+, \epsilon_+, q_-, \epsilon_-, -q_+ - q_-, \mu) \ .
\end{aligned} \tag{83}$$

We now take W^+ to be longitudinally polarized, with ϵ_+^μ replaced by $q_+^\mu/m_{\rm W} + \mathcal{O}\left(m_{\rm W}/q_+^0\right)$, and W^- transversely polarized. The individual diagrams are then again of order $\mathcal{O}\left(\epsilon_+\right) = \mathcal{O}\left(E/m_{\rm W}\right)$: explicitly,

$$\begin{aligned}
\mathcal{M}_1 &\to 2ig_{\rm W}^2 \bar{v}(p_1)\left(1+\gamma^5\right) \not{\epsilon}_- u(p_2) \ , \\
\mathcal{M}_2 &\to iQ_{\rm u} Q_{\rm W} \frac{s - m_{\rm W}^2}{s m_{\rm W}} \bar{v}(p_1) \not{\epsilon}_- u(p_2) \ .
\end{aligned} \tag{84}$$

Again, we see that no cancellation is generally possible, since $\mathcal{M}_1$ contains a pure $V - A$ coupling and $\mathcal{M}_2$ a purely vector one: we have to introduce new fields so that we may hope to obtain a cancellation by adding another Feynman diagram to those of fig.12.

An abortive trial: new fermion fields

A possible way to cure the unitarity problem of the last subsection is to postulate an additional fermion field, U', with electromagnetic charge $Q_{U'} = Q_{\rm u} - Q_{\rm W}$ (that is, $2e$ for $u = \nu$, $5e/3$ for u = up quark, and so on). There would then be a third

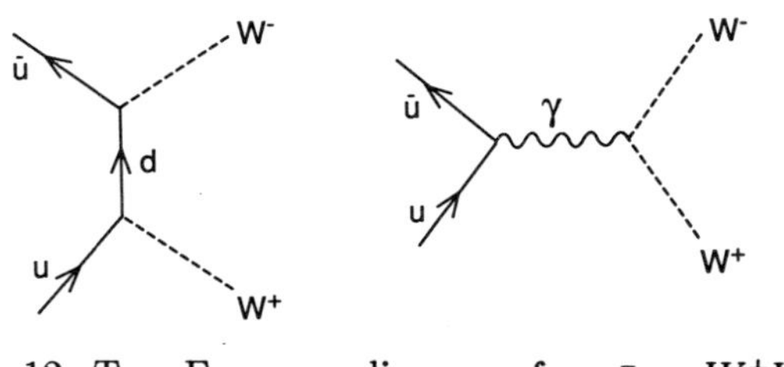

Figure 12. Two Feynman diagrams for $u\bar{u} \to W^+W^-$.

Feynman diagram, that of fig.13, in which not the d fermion is exchanged but the U' fermion:

$$\mathcal{M}_3 = -i\bar{v}(p_1)(V' + A'\gamma^5)\not{\epsilon}_+ \frac{\not{q}_+ - \not{p}_1}{(q_+ - p_1)^2}(V' + A'\gamma^5)\not{\epsilon}_- u(p_2) \ , \tag{85}$$

where V' and A' are the couplings of the W to a $U'\bar{u}$ combination. In the case of longitudinal W^+ and high energy this becomes

$$\mathcal{M}_3 \to -i\bar{v}(p_1)(V' + A'\gamma^5)^2\not{\epsilon}_- u(p_2) \ . \tag{86}$$

Combining this $\mathcal{M}_3$ with the $\mathcal{M}_{1,2}$ of the previous subsection we see that unitarity can in principle be restored provided

$$V'^2 + A'^2 = 2g_{\rm W}^2 + Q_{\rm u}Q_{\rm W} \quad \text{and} \quad 2V'A' = 2g_{\rm W}^2 \ , \tag{87}$$

where the first relation is required by cancellation of the vector part and the second one by cancellation of the axial part. It is seen, however, that this solution for unitarity restoration is unworkable because:

- From Eq.(87) it follows that $(V' - A')^2 = Q_{\rm u}Q_{\rm W}$ which is negative if $Q_{\rm u} > 0$: this implies couplings with imaginary parts, leading to CP violation and related problems.
- Having introduced a U' with charge $Q_{\rm u} - Q_{\rm W}$ we would then have to study also $U'\bar{U}' \to W^+W^-$: this would force us to introduce yet another field, U'', with charge $Q_{U'} - Q_{\rm W} = Q_{\rm u} - 2Q_{\rm W}$, and so on: an infinite sequence of fields with higher and higher charge. Surely the theory would break down at some point.

Introducing the Z^0

If extra fermion fields do not solve the unitarity problem in $u\bar{u} \to W^+W^-$ then we may look for an extra boson field. Let us therefore introduce another, neutral, boson field Z (or Z^0) so that the Feynman graph of fig.14 becomes available. Since the two diagrams of fig.12 have unequal V and A structure, the couplig of the Z to fermions cannot be purely vector, nor purely V-A. We therefore take for this vertex the form

$$i(v_f + a_f\gamma^5)\gamma^\mu \ , \tag{88}$$

with the vector and axial part to be determined. For the ZWW vertex of fig.16 we reasonably take the same form that worked well in the γWW case:

$$ig_{\rm WWZ}V(p_1, \alpha, p_2, \beta, k, \mu) \ , \tag{89}$$

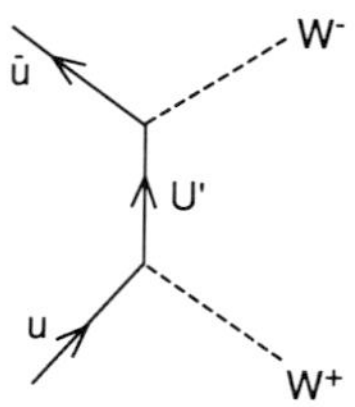

Figure 13. A possible additional diagram for $u\bar{u} \to W^+W^-$.

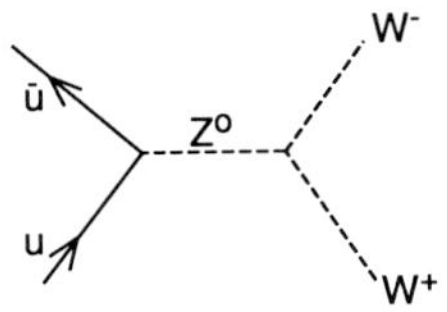

Figure 14. An additional diagram for $u\bar{u} \to W^+W^-$, with exchange of a Z^0.

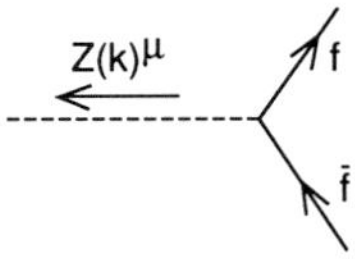

Figure 15. The $Zf\bar{f}$ vertex.

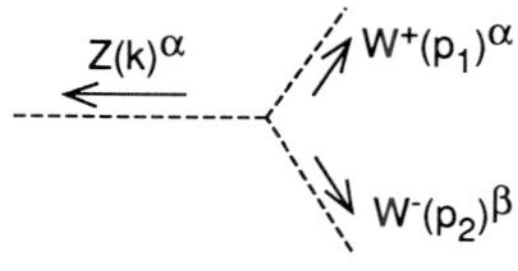

Figure 16. The ZW^+W^- vertex.

where again the unknown coupling constant g_{WWZ} must still be determined. The diagram of fig.14 now reads

$$\mathcal{M}_3 = ig_{\mathrm{WWZ}}\bar{v}(p_1)(v_u + a_u\gamma^5)\gamma_\mu u(p_2)\frac{1}{s - m_{\mathrm{Z}}^2}V(q_+, \epsilon_+, q_-, \epsilon_-, -q_+ - q_-, \mu) \ , \quad (90)$$

where m_{Z} is the Z mass. In the case of longitudinal W^+, transverse W^- and high energy we have

$$\mathcal{M}_3 \to ig_{\mathrm{WWZ}}\frac{s - m_{\mathrm{W}}^2}{s - m_{\mathrm{Z}}^2}\bar{v}(p_1)(v_u + a_u\gamma^5)\not{\epsilon}_- u(p_2) \ . \quad (91)$$

Coupling constants in the fermion-vector boson sector

The previous subsection resulted in the following requirement on the Z couplings to WW and ff in order that the process $u\bar{u} \to W^+W^-$ remain unitarity at high energy, even for longitudinal polarization of one of the W's:

$$\begin{aligned} 2g_{\mathrm{W}}^2 + Q_{\mathrm{u}}Q_{\mathrm{W}} + v_u g_{\mathrm{WWZ}} &= 0 \ , \\ 2g_{\mathrm{W}}^2 + \qquad\qquad a_u g_{\mathrm{WWZ}} &= 0 \ . \end{aligned} \quad (92)$$

Considering the process $d\bar{d} \to W^+W^-$ similarly leads to:

$$\begin{aligned} -2g_{\mathrm{W}}^2 + Q_{\mathrm{d}}Q_{\mathrm{W}} + v_d g_{\mathrm{WWZ}} &= 0 \ , \\ -2g_{\mathrm{W}}^2 + \qquad\qquad a_d g_{\mathrm{WWZ}} &= 0 \ . \end{aligned} \quad (93)$$

Moreover, from $u\bar{d} \to W^+\gamma$ we had already obtained

$$Q_{\mathrm{W}} = Q_{\mathrm{d}} - Q_{\mathrm{u}} \ , \quad (94)$$

while from the analogous process $u\bar{d} \to W^+Z^0$ we have

$$g_{\mathrm{WWZ}} = v_d + a_d - v_u - a_u \ . \quad (95)$$

Other conceivable processes like $u\bar{u} \to d\bar{d}$, $u\bar{u} \to Z^0Z^0$ do not lead to similar constraints since their amplitudes are already unitary. We now proceed to solve the system of equations (92) through (95).

First of all we immediately derive that

$$g_{\mathrm{WWZ}}^2 + Q_{\mathrm{W}}^2 = 8g_{\mathrm{W}}^2 \ . \tag{96}$$

We can therefore define an angle θ_w, with $s_w \equiv \sin\theta_w$ and $c_w \equiv \cos\theta_w$ such that

$$Q_{\mathrm{W}}^2 = e^2 = 8g_{\mathrm{W}}^2 s_w^2 \quad \text{and} \quad g_{\mathrm{WWZ}}^2 = 8g_{\mathrm{W}}^2 c_w^2 \ . \tag{97}$$

This also leads to the following constraint:

$$g_{\mathrm{W}}^2 = \frac{e^2}{8s_w^2} \quad \Rightarrow \quad m_{\mathrm{W}} = \frac{1}{s_w}\left[\frac{\pi\alpha}{G\sqrt{2}}\right]^{1/2} \ : \tag{98}$$

we have established a *lower limit* of

$$\left[\frac{\pi\alpha}{G\sqrt{2}}\right]^{1/2} = 37.281 \ \ \mathrm{GeV} \tag{99}$$

on the mass of the W! Furthermore, implementing Eq.(97) into the system (92-95), we express all the couplings uniquely[5]:

$$\begin{aligned}
g_{\mathrm{W}} &= e\frac{1}{s_w\sqrt{8}} \ , \\
Q_{\mathrm{W}} &= -e \ , \\
g_{\mathrm{WWZ}} &= -e\frac{c_w}{s_w} \ , \\
a_u &= -a_d = e\frac{1}{4s_w c_w} \ , \\
v_u &= e\frac{1}{4c_w s_w}\left[1 - 4\frac{Q_{\mathrm{u}}}{e}s_w^2\right] \ , \\
v_d &= e\frac{-1}{4c_w s_w}\left[1 + 4\frac{Q_{\mathrm{d}}}{e}s_w^2\right] \ .
\end{aligned} \tag{100}$$

Again, these are precisely the couplings as predicted by the Minimal Standard Model as a spontaneously broken gauge theory. Note however that in this derivation so far no gauge symmetry, let alone its breakdown, has been introduced. In particular one sees that the precise form of the Z coupling constants must be independent of the symmetry breaking mechanism, provided the low-energy phenomenology of muon decay is correctly predicted.

Finally, the mass m_{Z} is still undetermined: the above holds as long as $s/m_{\mathrm{Z}} \to \infty$ whenever $s/m_{\mathrm{W}} \to \infty$. Therefore the definition of s_w^2 in terms of coupling constants is 'more fundamental' than that in terms of the W/Z mass ratio which we shall encounter later on.

Four-boson vertices

So far we have introduced couplings between a W^+W^- pair and γ or Z. Note that couplings between neutral bosons (vertices ZZZ, $ZZ\gamma$, etcetera) are nowhere

[5] In fact, uniquely up to an overall sign for all couplings with one Z involved, which is of course physically irrelevant.

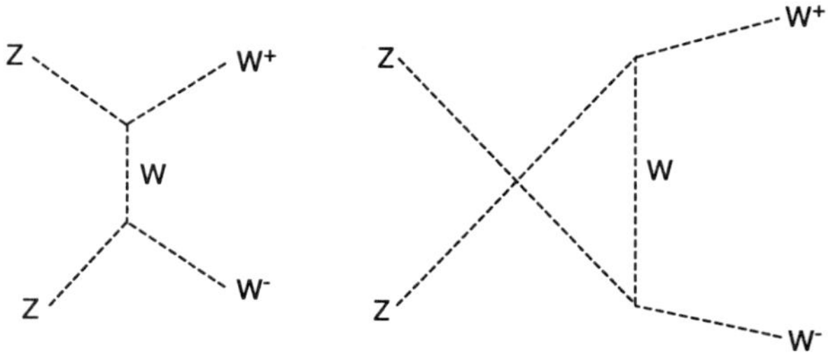

Figure 17. Two diagrams for the process $Z^0Z^0 \to W^+W^-$.

necessary and in fact would spoil the unitarity of processes like $u\bar{u} \to ZZ$. We can now also perform scattering of two bosons into two bosons, as in fig.17 where the two available diagrams for the process

$$Z^0(p_1,\epsilon_1)\,Z^0(p_2,\epsilon_2) \;\to\; W^+(q_+,\epsilon_+)\,W^-(q_-,\epsilon_-) \tag{101}$$

are depicted. The first of these reads

$$\begin{aligned}\mathcal{M}_1 \;=\;& ig^2_{\rm WWZ}V(q_+,\epsilon_+,p_1-q_+,\mu,-p_1,\epsilon_1)V(p_2-q_-,\nu,q_-,\epsilon_-,-p_2,\epsilon_2)\\ &\times\frac{1}{(p_1-q_+)^2-m^2_{\rm W}}\left(g^{\mu\nu}-\frac{1}{m^2_{\rm W}}(p_1-q_+)^\mu(p_1-q_+)^\nu\right)\ ,\end{aligned} \tag{102}$$

while $\mathcal{M}_2$ is obtained from $\mathcal{M}_1$ by interchanging p_1,ϵ_1 and p_2,ϵ_2. As before we take W^+ to be longitudinal, and neglect $m_{\rm W}/E$ wherever possible. We find

$$[\mathcal{M}_1+\mathcal{M}_2]_{\epsilon_+\to q_+/m_{\rm W}} \sim i\frac{g^2_{\rm WWZ}}{m_{\rm W}}\left[2(q_+\cdot\epsilon_-)(\epsilon_1\cdot\epsilon_2)-(q_+\cdot\epsilon_1)(\epsilon_-\cdot\epsilon_2)-(q_+\cdot\epsilon_2)(\epsilon_-\cdot\epsilon_1)\right]\ , \tag{103}$$

which again diverges as $s \to \infty$. Analogous results follow when we take one of the other bosons longitudinal. This unitarity problem can be cured by adding not a new field but a pointlike four-boson vertex, as depicted in fig.18[a]. This $Z^0Z^0W^+W^-$ vertex is defined as

$$-\,ig^2_{\rm WWZ}Y(\alpha,\beta,\mu,\nu)\ , \tag{104}$$

with

$$Y(\alpha,\beta,\mu,\nu) \equiv 2g^{\alpha\beta}g^{\mu\nu}-g^{\alpha\mu}g^{\beta\nu}-g^{\alpha\nu}g^{\beta\mu}\ . \tag{105}$$

Adding this vertex as a third diagram $\mathcal{M}_3$ to $\mathcal{M}_1+\mathcal{M}_2$ trivially restores the correct unitarity behaviour of the amplitude. By similarly considering the processes $Z^0\gamma \to W^+W^-$, $\gamma\gamma \to W^+W^-$, and $W^+W^- \to W^+W^-$, we can also derive the form of the other necessary four-boson vertices of fig.18[b,c,d], which respectively read

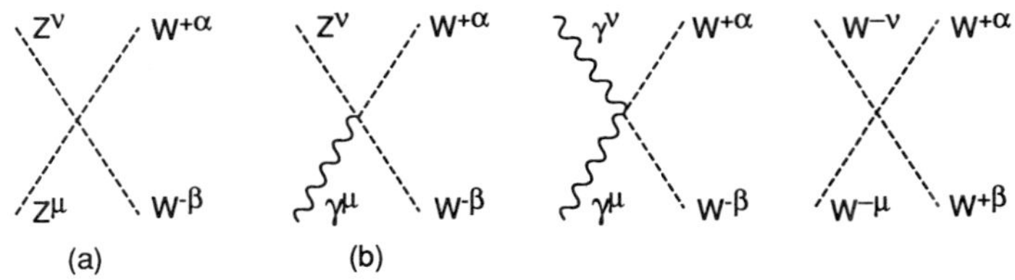

Figure 18. Four-boson vertices.

$$
\begin{aligned}
Z^0\gamma W^+W^- &: \quad -ig_{\rm wwz}Q_{\rm w}Y(\alpha,\beta,\mu,\nu) \ , \\
\gamma\gamma W^+W^- &: \quad -iQ_{\rm w}^2Y(\alpha,\beta,\mu,\nu) \ , \\
W^+W^+W^-W^- &: \quad +i\frac{e^2}{sw^2}Y(\alpha,\beta,\mu,\nu) \ .
\end{aligned} \tag{106}
$$

Note the sign of the last coupling constant: it arises as $(g_{\rm wwz}^2 + Q_{\rm w}^2) = e^2/s_w^2$. Once more, couplings between four neutral bosons are nowhere necessary or even wanted, so we do not introduce any.

Fully longitudinal scattering: introducing the Higgs

Let us now be even more ambitious: consider again the process $ZZ \to WW$ of Eq.(101), and take *all* bosons longitudinal. With only the two diagrams of fig.17 the amplitude would at large energies go as $E^4/m_{\rm w}^2 m_{\rm z}^2$, leading to an s^3 behaviour of the cross section. Including the four-boson coupling of fig.18[a] softens this divergence: we find

$$
\sum_{i=1}^{3}\mathcal{M}_i\Bigg]_{\epsilon_\pm\to q_\pm/m_{\rm W}\ ,\epsilon_{1,2}\to p_{1,2}/m_{\rm Z}} \sim ig_{\rm wwz}^2 E^2\frac{m_{\rm z}^2}{m_{\rm w}^4} + \mathcal{O}(1) \ , \tag{107}
$$

which is less divergent, but of course still not acceptable. Note the occurrence of $m_{\rm w}^4$ in the denominator: two factors of $m_{\rm w}^{-1}$ come from the longitudinal polarization, the other $m_{\rm w}^{-2}$ is due to the $k^\mu k^\nu/m_{\rm w}^2$ term in the internal W propagator, which here appears as a significant part of the amplitude for the first time.

Since there is no room left for introducing another vertex, we have to call in extra fields. There are several possibilities to do so: the simplest one, which corresponds to the Minimal Standard Model, has one additional neutral field, which we call H. We take it to be a scalar field, which will be justified later on. For its couplings to the W and Z we take the simplest possible choice: the vertices depicted in fig.19 are given by

$$
ig_{\rm wwH}g^{\mu\nu} \quad \text{and} \quad ig_{\rm zzH}g^{\mu\nu} \ , \tag{108}
$$

respectively. There is now an additional, fourth diagram available to describe $ZZ \to WW$: it is given in fig.20 and reads

$$
\mathcal{M}_4 = -ig_{\rm wwH}g_{\rm zzH}\frac{1}{s - m_{\rm H}^2}(\epsilon_1\cdot\epsilon_2)(\epsilon_+\cdot\epsilon_-) \ . \tag{109}
$$

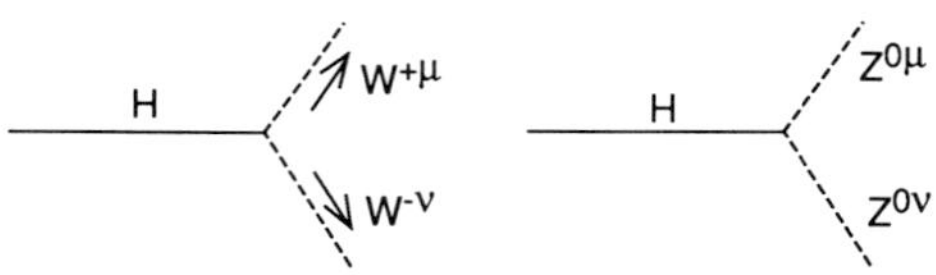

Figure 19. The W^+W^-H and Z^0Z^0H vertices.

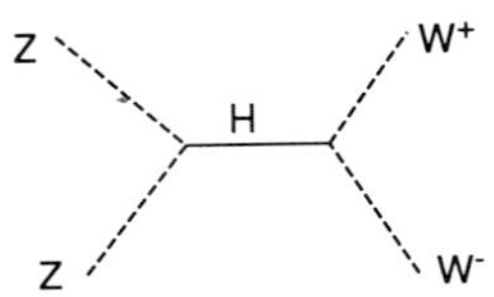

Figure 20. An extra diagram in $Z^0Z^0 \to W^+W^-$, with H exchange.

In the fully longitudinal polarization case this goes to

$$\mathcal{M}_4\Big]_{\epsilon_\pm \to q_\pm/m_W, \epsilon_{1,2}\to p_{1,2}/m_Z} \sim -i g_{WWH} g_{ZZH} \frac{s^2}{4m_W^2 m_Z^2 (s - m_H^2)} \quad . \tag{110}$$

Higgs-vector boson couplings from unitarity

In the last subsection we have introduced the Higgs field H; unitarity now requires that its couplings to W and Z satisfy

$$g_{WWH} g_{ZZH} = g_{WWZ}^2 \frac{m_Z^4}{m_W^2} \quad . \tag{111}$$

Taking the full longitudinal case in $W^+W^- \to W^+W^-$ leads to a second constraint:

$$g_{WWH}^2 = g_{WWZ}^2 \left(4m_W^2 - 3m_Z^2\right) + 4Q_W^2 m_W^2 \quad , \tag{112}$$

while from considering $Z^0 H \to W^+W^-$ we can derive the condition

$$g_{ZZH} = g_{WWH} \frac{m_Z^2}{m_W^2} \quad . \tag{113}$$

Note that four-boson scattering processes involving an external photon are automatically unitarity by e.m. current conservation, so do not lead to an additional constraint. Up to an overall sign in g_{WWH} and g_{ZZH} the equations (111-113) have a unique solution:

$$\begin{aligned} g_{WWH} &= -e\frac{m_W}{s_w} \quad , \\ g_{ZZH} &= -e\frac{m_Z}{c_w s_w} \quad , \end{aligned} \tag{114}$$

and...

$$m_Z = m_Z c_w \quad ! \tag{115}$$

A model with more than one Z boson

There is an interesting observation to be made at this point. When we introduced the Z to cure unitarity in $f\bar{f} \to W^+W^-$ we could of course have used more than just a single neutral field. Suppose we had introduced 2 Z-like bosons, Z_1 and Z_2, that had coupling strengths g_{WW1} and g_{WW2} to the W^+W^-, respectively, and masses m_1 and m_2. We can assume $m_2 > m_1$. We would then have obtained, intead of Eq.(96):

$$g_{WW1}^2 + g_{WW2}^2 + Q_W^2 = 8g_W^2 \quad , \tag{116}$$

so that we could write

$$g_{WW1} = g_{WWZ} \cos\phi \quad \text{and} \quad g_{WW2} = g_{WWZ} \sin\phi \quad , \tag{117}$$

where g_{WWZ} is still the nominal coupling in the 'one-Z' model and ϕ is some undetermined angle. Working through the Gedanken experiments of the last subsection, using processes like $Z_1 Z_1 \to W^+W^-$, $Z_1 Z_2 \to W^+W^-$, etcetera, we would then obtain the analogue of Eq.(115) as

$$m_1^2 \cos^2 \phi + m_2^2 \sin^2 \phi = \frac{m_W^2}{c_w^2} \quad . \tag{118}$$

We immediately find the well-known result that, in a 2-Z model, one of the masses must be smaller, and the other larger, than the nominal value of the Z mass in the one-Z model:

$$m_1 < m_Z < m_2 \quad . \tag{119}$$

Moreover, when m_2 becomes very large, $\sin\phi$ has to go to zero. Note that this does not necessarily imply that m_1 becomes very close to m_Z, though.

A unitarity bound on the Higgs mass

Although we have no relation for m_H in terms of other masses and couplings, we can still obtain an upper bound, by using essentially the same arguments as those for the W mass. In our introduction of the Higgs we saw that the diagram for $Z^0Z^0 \to W^+W^-$ by way of H exchange reads, in the fully longitudinal polarization case:

$$\mathcal{M}_4 = -ig_{WWH}g_{ZZH}(\epsilon_1{\cdot}\epsilon_2)(\epsilon_+{\cdot}\epsilon_-)\frac{1}{s - m_H^2} \to -i\frac{e^2}{4m_W^2 s_w^2}\frac{s^2}{s - m_H^2} \quad . \tag{120}$$

Restricting ourselves for the moment to the divergent (non-unitarity) part of the other three diagrams, we see that this must *necessarily* be such as to cancel Eq.(120) as $s \to \infty$. Therefore, this part of the whole amplitude must read

$$\mathcal{M} \sim i\frac{e^2}{4m_W^2 s_w^2}\left(\frac{s^2}{s - m_H^2} - s\right) = -i\frac{e^2 m_H^2 s}{4m_W^2 s_w^2 (s - m_H^2)} = -i\sqrt{2}Gm_H^2\frac{s}{s - m_H^2} \quad . \tag{121}$$

Equation (121) is quite similar to Eq.(36), and leads of course to a similar *unitarity bound* on m_H, from the unitarity requirement that the amplitude (121) be smaller than 16π as $s \to \infty$[6]:

$$m_H \le \left[\frac{16\pi}{G\sqrt{2}}\right]^{1/2} = 1750 \text{ GeV} \quad . \tag{122}$$

Alternatively, also the *inverse unitarity bound* can be derived, which now states that, in the absence of a Higgs boson, tree-level unitarity will break down around 1.5-2 TeV and one expects either strongly interacting W's and Z's or the onset of other new physics. Note that to obtain this result we did not have to specify exactly the form of the $ZZ \to WW$ amplitude: we have only used the fact that the Higgs exchange diagram must provide a cancellation, and the value of the product $g_{WWH}g_{ZZH} = -4\sqrt{2}Gm_Z^2 m_W^2$. Of course, in Eq.(121) we have neglected that part of the amplitude that is not divergent at high energies, and the exact limit on m_H may be somewhat different. The cancellation behaviour, however, is completely determined.

The Higgs width

Given the WWH and ZZH couplings, also the partial widths $H \to W^+W^-$ and $H \to Z^0Z^0$ are determined: we find

[6]Note that in this case no spin-averaging factor must be included in the unitarity bound, as only one (longitudinal) spin state of each incoming boson is considered.

$$\Gamma(H \to W^+W^-) = \frac{G}{8\pi\sqrt{2}} m_{\rm H}^3 \left(\left(1 - \frac{2m_{\rm W}^2}{m_{\rm H}^2}\right)^2 + \frac{8m_{\rm W}^4}{m_{\rm H}^4} \right) ,$$

$$\Gamma(H \to Z^0Z^0) = \frac{G}{16\pi\sqrt{2}} m_{\rm H}^3 \left(\left(1 - \frac{2m_{\rm Z}^2}{m_{\rm H}^2}\right)^2 + \frac{8m_{\rm Z}^4}{m_{\rm H}^4} \right) . \tag{123}$$

Note that in $\Gamma(H \to ZZ)$ the indistinguishability of the Z's results in an extra factor 1/2. Like the W (Eq.(57)), the total H width goes as $m_{\rm H}^3$: the partial widths for H into fermions are linear in the Higgs mass, and therefore unimportant at high enough mass. For $m_{\rm H} \gg m_{\rm W}, m_{\rm Z}$ we therefore have

$$\Gamma_{\rm H} = \frac{3G}{16\pi\sqrt{2}} m_{\rm H}^3 , \tag{124}$$

which (again, just like in the W case) leads to a breakdown of the elementary-particle interpretation of the Higgs field if $m_{\rm H} \sim \Gamma_{\rm H}$ at $m_{\rm H} \sim 1429$ GeV.

Higgs-fermion couplings

We now reconsider the process $f\bar{f} \to W^+W^-$, but this time we take both W's longitudinal, and do not neglect the fermion mass m_f. The three diagrams of fig.12 and 14 now lead to the following high-energy limit:

$$\mathcal{M}]_{\epsilon\pm \to q\pm/m_{\rm W}} = -2ig_{\rm W}^2 \bar{v}(p_1)u(p_2)\frac{m_f}{m_{\rm W}^2} + \mathcal{O}(1) , \tag{125}$$

where as usual the $\mathcal{O}(1)$ stands for terms that are well-behaved with respect to unitarity. Note that due to the built-in cancellations the amplitude diverges only as $Em_f/m_{\rm W}^2$ and not as $E^2/m_{\rm W}^2$ that naive power counting gives. Still, the behaviour is not good enough: it can be cured by the diagram of fig.21, if the vertex of fig.22 reads

$$ig_{f\bar{f}{\rm H}} \equiv i\frac{e}{2s_w}\frac{m_f}{m_{\rm W}} = \left[G\sqrt{2}\right]^{1/2} . \tag{126}$$

Since, in contrast to the Higgs-vector boson couplings, this is dimensionless, we can derive a meaningful restriction on m_f by requesting that $g_{f\bar{f}{\rm H}} \leq 1$ (or $\sqrt{4\pi}$): this leads to $m_f \sim 247$ GeV (or 875 GeV)as an upper bound, although this has to be borne out by an explicit calculation of some scattering process.

Note that it is the ffH coupling that indicates that the Higgs must be a scalar field: the combination $\bar{v}(p_1)u(p_2)$ of Eq.(125) is a scalar current.

Finally, an even simpler derivation of the coupling is possible, using the process $u\bar{d} \to W^+H$: this gives the uuH and ddH couplings at the same time.

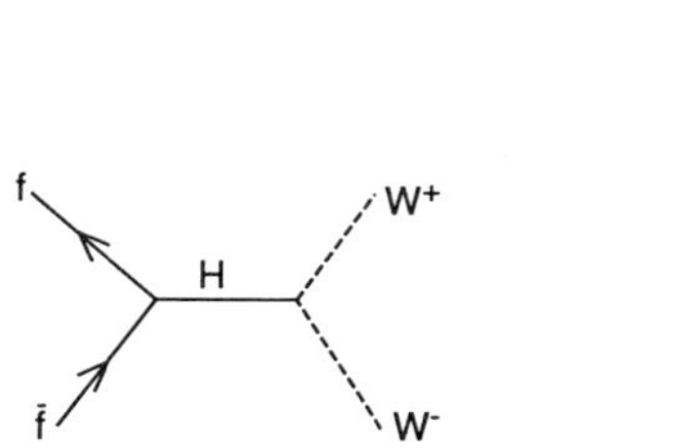

Figure 21. Higgs exchange in to $f\bar{f} \to W^+W^-$

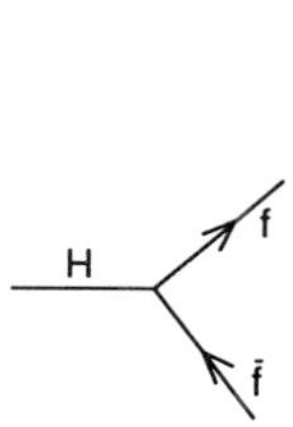

Figure 22. The $Gf\bar{f}$ vertex

Non-minimal Higgs sectors

A very instructive alternative to the process $u\bar{u} \to W^+W^-$ is

$$\bar{d}(p_1)\, u(p_2) \to W^+(q_+, \epsilon_+)\; Z^0(q, \epsilon) \;, \tag{127}$$

as depicted in fig.23. The three diagrams are

$$\begin{aligned}
\mathcal{M}_1 &= -ig_{\mathrm{W}}\bar{v}(p_1)\left(1+\gamma^5\right) \not{\epsilon}_+ \frac{\not{p}_2 - \not{q} + m_{\mathrm{u}}}{(p_2-q)^2 - m_{\mathrm{u}}^2} \left(v_u + a_u\gamma^5\right) \not{\epsilon} u(p_2) \;, \\
\mathcal{M}_2 &= -ig_{\mathrm{W}}\bar{v}(p_1)\left(v_d + a_d\gamma^5\right) \not{\epsilon} \frac{\not{q} - \not{p}_1 + m_{\mathrm{d}}}{(p_1-q)^2 - m_{\mathrm{d}}^2} \left(1+\gamma^5\right) \not{\epsilon}_+ u(p_2) \;, \\
\mathcal{M}_3 &= i\frac{g_{\mathrm{W}}g_{\mathrm{WWZ}}}{s - m_{\mathrm{W}}^2} \bar{v}(p_1)\left(1+\gamma^5\right)\gamma_\mu u(p_2) \left(g^{\mu\nu} - \frac{(p_1+p_2)^\mu(p_1+p_2)^\nu}{m_{\mathrm{W}}^2}\right) \\
&\qquad \times V(q_+, \epsilon_+, -p_1-p_2, \nu, q, \epsilon) \;,
\end{aligned} \tag{128}$$

In this process, no H exchange is possible. Consequently, if we take both W^+ and Z^0 longitudinal in this process, the amplitude must be well-behaved by itself. In fact we find, for nonzero m_{u} and m_{d} that the leading $\mathcal{O}(s)$ divergence cancels, while the subleading $\mathcal{O}(E)$ divergence is given by

$$\left.\sum_{i=1}^{3}\mathcal{M}_i\right]_{\epsilon_+\to q_+/m_{\mathrm{W}},\epsilon\to q/m_{\mathrm{Z}}} \to iG\left(1 - \frac{c_w^2 m_{\mathrm{Z}}^2}{m_{\mathrm{W}}^2}\right)\bar{v}(p_1)\left(m_{\mathrm{d}}\left(1-\gamma^5\right) - m_{\mathrm{u}}\left(1+\gamma^5\right)\right)u(p_2) \;. \tag{129}$$

Unitarity appears to be satisfied, *provided that* $m_{\mathrm{W}} = m_{\mathrm{Z}}c_w$. On the basis of this Gedanken process we can conclude that

- *if* the mass relation Eq.(115) is violated at the tree level (for instance because of a more complicated Higgs sector)
- *then* the theory must contain at least:

    1. *two* neutral scalars coupling to W and Z, since the assumption of only one neutral scalar immediately leads to Eq.(115); and
    2. at least *one* charged scalar field, to repair unitarity in the process (127). Moreover we find that coupling of the charged Higgs to a $(u, \bar{d})$ combination must be proportional to $m_{\mathrm{d}}(1-\gamma^5) - m_{\mathrm{u}}(1+\gamma^5)$.

A unitarity bound on fermion masses?

Yet another instructive example is the scattering of heavy fermions. Let B and T

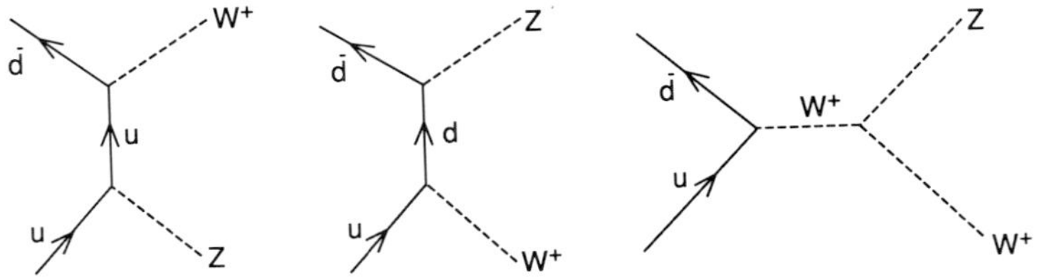

Figure 23. Feynman diagrams for $u\bar{d} \to W^+Z^0$ in the minimal standard model

be 'down' and 'up' fermions, respectively, that have mass $m_{B,T}$ and couplings to the Z $v_{B,T}$ and $a_{B,T}$. Consider the scattering

$$\bar{B}(p_1)\, B(p_2) \to\ Z\ \to T(q_1)\, \bar{T}(q_2)\ , \tag{130}$$

for which the amplitude due to Z^0 and Higgs exchange reads

$$\begin{aligned}
\mathcal{M} &= \frac{i}{s-m_Z^2}\bar{v}(p_1)\left(v_B + a_B\gamma^5\right)\gamma_\mu u(p_2)\left(g^{\mu\nu} - \frac{(p_1+p_2)^\mu(q_1+q_2)^\nu}{m_Z^2}\right) \\
&\times\ \bar{u}(q_1)\left(v_T + a_T\gamma^5\right)\gamma_\nu v(q_2) \\
&+\ \frac{-i}{s-m_H^2} g_{BBH}\, g_{TTH}\bar{v}(p_1)u(p_2)\,\bar{u}(q_1)v(q_2)\ .
\end{aligned} \tag{131}$$

At high enough energies and for large $m_{B,T}$ we can neglect the $g^{\mu\nu}$ part of the Z exchange diagram. Using

$$\bar{v}(p_1)\left(v_1 + a_1\gamma^5\right)\gamma_\mu u(p_2)(p_1+p_2)^\mu = 2a_1 m_1 \bar{v}(p_1)\gamma^5 u(p_2)\ , \tag{132}$$

(and similar for the outgoing $T\bar{T}$), we can write this as

$$\begin{aligned}
\mathcal{M} &\sim \frac{i}{s-m_Z^2}\frac{4a_B a_B m_T m_T}{m_Z^2}\bar{v}(p_1)\gamma^5 u(p_2)\bar{u}(q_1)\gamma^5 v(q_2) \\
&+\ \frac{-i}{s-m_H^2} g_{BBH}\, g_{TTH}\bar{v}(p_1)u(p_2)\,\bar{u}(q_1)v(q_2)\ .
\end{aligned} \tag{133}$$

Note that in general no cancellation between these diagrams is possible: the one is pseudoscalar, and the other one scalar ,which implies that the two contributions do not interfere in the total cross section, which (when we use the expressions (100) and 126 for the couplings) reads

$$\sigma \overset{s\to\infty}{\longrightarrow} \frac{G^2 m_B^2 m_T^2}{\pi s}\ . \tag{134}$$

Counting one $J=1$ and one $J=0$ exchange channel, the unitarity limit is equal to $16\pi/s$ so that we must have

$$m_B m_T\ \le\ \frac{4\pi}{G}\ =\ (1041\ \text{GeV})^2\ . \tag{135}$$

We can tentatively conclude that in the Minimal Standard Model fermion masses should typically be smaller than about 1 TeV.

Two Higgs-two vector boson couplings

Yet another set of vertices can be deduced from looking at the process

$$Z^0(p_1,\epsilon_1)\, Z^0(p_2,\epsilon_2)\ \to\ H(q_1)\, H(q_2)\ . \tag{136}$$

for which the two diagrams available so far are given in fig.24, and read

$$\begin{aligned}
\mathcal{M}_1 &= i g_{ZZH}^2 \frac{1}{(p_1-q_1)^2 - m_Z^2}\left[(\epsilon_1{\cdot}\epsilon_2) - \frac{(p_1-q_1)_\mu \epsilon_1^\mu (p_2-q_2)_\nu \epsilon_2^\nu}{m_Z^2}\right]\ , \\
\mathcal{M}_2 &= \mathcal{M}_1\big]_{q_1\leftrightarrow q_2}\ .
\end{aligned} \tag{137}$$

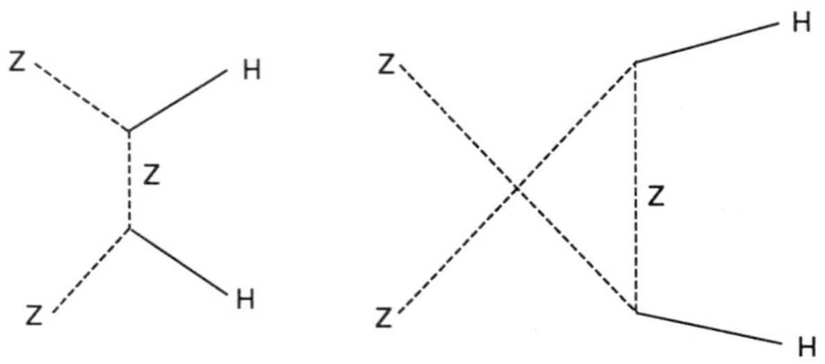

Figure 24. Some diagrams for the process $ZZ \rightarrow HH$

If both Z's become longitudinal, this amplitude goes to

$$\mathcal{M} = \mathcal{M}_1 + \mathcal{M}_2 \rightarrow -ig_{\mathrm{ZZH}}^2 \frac{p_1 \cdot p_2}{m_{\mathrm{Z}}^4} \;, \tag{138}$$

with similar results for the process $W^+W^- \rightarrow HH$. Unitarity can be restored by introducing the four-point vertices depicted in fig.25. The corresponding Feynman rules read

$$\begin{aligned} &ig_{\mathrm{ZZHH}}\,g^{\mu\nu} \quad \text{with} \quad g_{\mathrm{ZZHH}} = \frac{g_{\mathrm{ZZH}}^2}{2m_{\mathrm{Z}}^2} = \frac{e^2}{2s_w^2 c_w^2} \;, \\ &ig_{\mathrm{WWHH}}\,g^{\mu\nu} \quad \text{with} \quad g_{\mathrm{WWHH}} = \frac{g_{\mathrm{WWH}}^2}{2m_{\mathrm{W}}^2} = \frac{e^2}{2s_w^2} \;. \end{aligned} \tag{139}$$

Note that the need for introducing these vertices comes from the $k^\mu k^\nu$ term in the boson propagator: when a Higgs becomes involved, it is usually these 'scalar' parts of the propagators that give rise to such complications[7].

One- and two-loop renormalizability and unitarity

We have now exhausted the $2 \rightarrow 2$ processes of the Minimal Standard Model: other processes like $f\bar{f} \rightarrow HH$, $f\bar{f} \rightarrow H\gamma$, $HH \rightarrow HH$ and so on are unitary by themselves (note that they do not contain external polarization vectors that can become longitudinal and cause trouble) and so do not lead to additional constraints. By the optical theorem, the set of $2 \rightarrow 2$ processes corresponds to the one-loop quantum corrections, as exemplified by fig.26. Therefore, the imaginary parts of these one-loop corrections are well-behaved at all energies: since, as we shall see later, the imaginary parts can be related to the real parts by dispersion integrals, also the real parts will be (sufficicently) well-behaved (although not strictly finite).

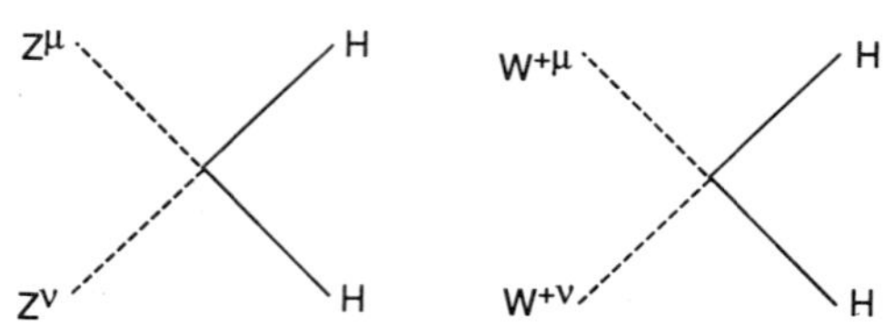

Figure 25. Four-point vertices for $ZZHH$ and W^+W^-HH.

[7]In the gauge-symmetry derivation of the Standard Model this follows from the fact that these propagator parts are in fact remnants of the original Higgs sector, which couple to the Higgs due to its original self-interaction.

This implies that at the one-loop level the theory is *renormalizable*. If we go to the two-loop level, correct energy behaviour of the theory implies that also the

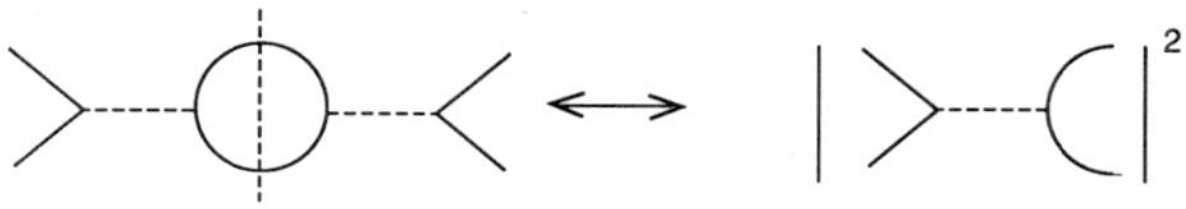

Figure 26. One-loop corrections are related to $2 \to 2$ processes.

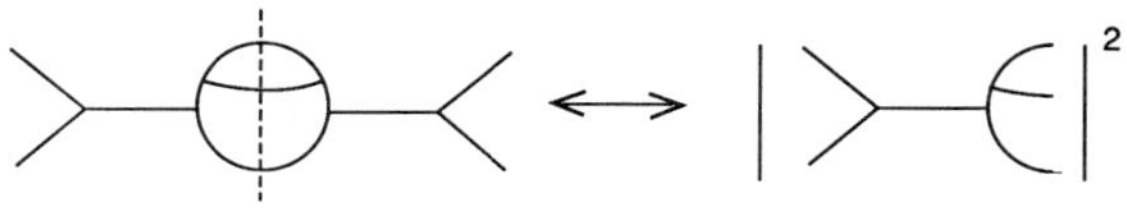

Figure 27. Two-loop corrections are related to $2 \to 3$ processes.

Born level $2 \to 3$ processes must be unitary, as can be seen from fig.27. This implies in principle that we have to examine all possible $2 \to 3$ processes, for which the amplitude must go as $E^{-}1$ for $E \to \infty$.

Higgs self-interactions

Since the $2 \to 3$ processes are more involved than the $2 \to 2$ ones, with both more complicated kinematics and a larger number of Feynman diagrams, we will restrict ourselves to just mentioning the most important result, which is the derivation of Higgs self-interaction vertices with 3 and 4 legs, given in figs. (28) and (29). It should be noted that the best way to obtain these results is numerically, by Monte Carlo generation of random kinematical configurations. The vertices can again be fixed by requiring unitarity: for the HHH vertex we use the process $ZZ \to ZZH$ which has 21 Feynman diagrams depicted in fig.30 The $HHHH$ vertex is subsequently fixed by the process $ZZ \to HHH$ for which we have the 25 diagrams indicated in fig.31. The resulting coupling constants $g_{\rm HHH}$ and $g_{\rm HHHH}$ are

$$g_{\rm HHH} = \frac{3em_{\rm H}^2}{2s_w m_{\rm W}} \quad \text{and} \quad g_{\rm HHHH} = \frac{3e^2 m_{\rm H}^2}{4s_w^2 m_{\rm W}^2} \ . \tag{140}$$

Note that these couplings vanish as $m_{\rm H} \to 0$.

The Lagrangian of the Minimal Standard Model

We now have, as we shall see, collected all the necessary fields and interactions.

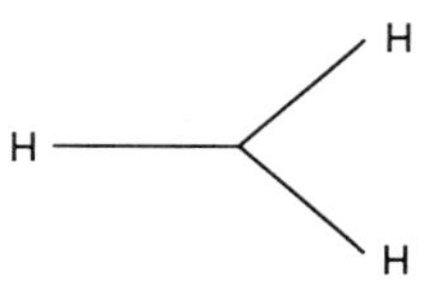

Figure 28. The HHH vertex

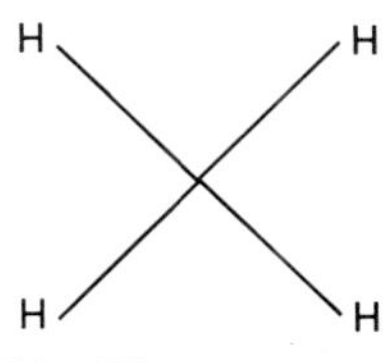

Figure 29. The $HHHH$ vertex

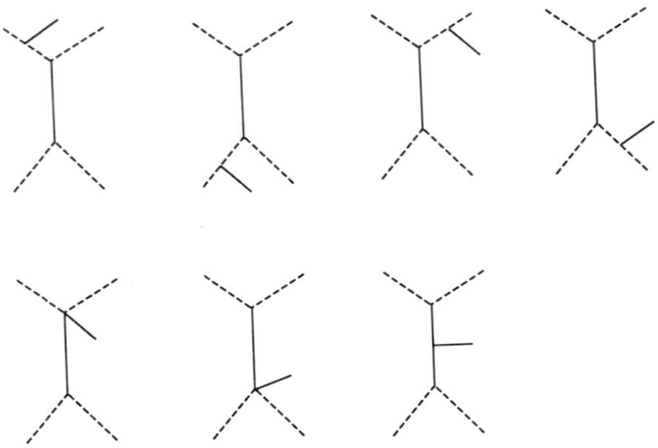

Figure 30. The Feynman diagrams for $ZZ \to ZZH$. Each depicted diagram corresponds to 3 actual ones, related by crossing of the external Z lines.

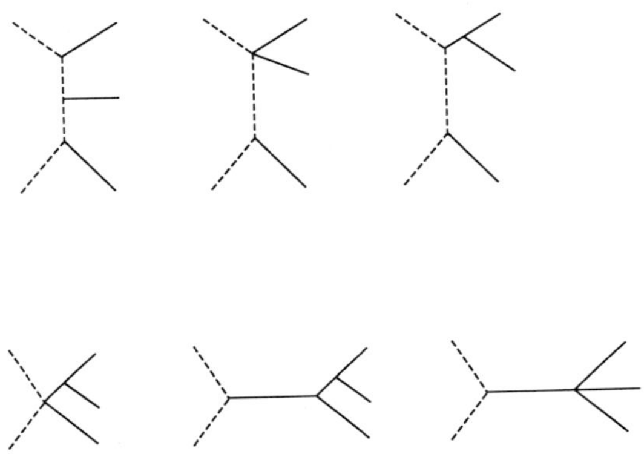

Figure 31. The Feynman diagrams for $ZZ \to HHH$. Again, crossings leading to topologically distinct diagrams are implied.

These results can conveniently be collected into one mathemathical object, the Lagrangian of the Minimal Standard Model in the unitary gauge:

$$\begin{aligned} \mathcal{L}_{\text{MSM}} &= \mathcal{L}_{\text{QED}} + \mathcal{L}_{\text{free V}} + \mathcal{L}_{\text{ffW}} + \mathcal{L}_{\text{ffZ}} + \mathcal{L}_{\text{3V}} + \mathcal{L}_{\text{4V}} \\ &\quad + \mathcal{L}_{\text{free H}} + \mathcal{L}_{\text{ffH}} + \mathcal{L}_{\text{2VH}} + \mathcal{L}_{\text{2V2H}} + \mathcal{L}_{\text{3,4H}} \;, \end{aligned} \tag{141}$$

with the following contributions: the QED Lagrangian

$$\mathcal{L}_{\text{QED}} = -\frac{1}{4}F^{\mu\nu}F_{\mu\nu} + \sum_j \left\{ \bar{f}_j \left(i\partial^\mu \gamma_\mu - m_j \right) f_j - iQ_j \bar{f}_j A^\mu \gamma_\mu f_j \right\} \;; \tag{142}$$

the free vector boson (V=W,Z) Lagrangian:

$$\mathcal{L}_{\text{free V}} = \frac{1}{2}W^*_{\mu\nu}W^{\mu\nu} - m_{\text{w}}^2 W^*_\mu W^\mu - \frac{1}{4}Z_{\mu\nu}Z^{\mu\nu} - \frac{1}{2}m_{\text{z}}^2 Z^\mu Z_\mu \;; \tag{143}$$

the W-fermion interactions:

$$\mathcal{L}_{\text{ffW}} = \frac{ie}{s_w\sqrt{8}} \sum_{j=u,k=d} \left\{ \bar{f}_j \left(1+\gamma^5\right) W^*_\mu \gamma^\mu f_k K_{jk} + \text{h.c.} \right\} \;; \tag{144}$$

the Z-fermion interactions:

$$\mathcal{L}_{\mathrm{ffZ}} = i \sum_j \bar{f}_j \left(v_j + a_j \gamma^5 \right) \gamma_\mu Z^\mu f_j \ ; \tag{145}$$

the three-boson interactions:

$$\mathcal{L}_{3\mathrm{V}} = ie \left(\frac{c_w}{s_w} Z_\nu - A_\nu \right) \left[W_\mu W^*_{\mu\nu} - W^*_\mu W_{\mu\nu} + \partial_\mu \left(W_\mu W^*_\nu - W^*_\mu W_\nu \right) \right] \ ; \tag{146}$$

the four-boson interactions:

$$\begin{aligned} \mathcal{L}_{4\mathrm{V}} \ = \ & e^2 \left(g^{\mu\rho} g^{\nu\sigma} - g^{\mu\nu} g^{\rho\sigma} \right) \left[\frac{1}{2 s_w^2} W^*_\mu W_\nu W^*_\rho W_\sigma \right. \\ & \left. + \left(\frac{c_w}{s_w} Z_\mu - A_\mu \right) \left(\frac{c_w}{s_w} Z_\nu - A_\nu \right) W^*_\rho W_\sigma \right] \ ; \end{aligned} \tag{147}$$

the free Higgs Lagrangian:

$$\mathcal{L}_{\text{free H}} = -\frac{1}{2} \left(\partial_\mu H \right) \left(\partial^\mu H \right) - \frac{1}{2} m_{\mathrm{H}}^2 H^2 \ ; \tag{148}$$

the Higgs-fermion interactions:

$$\mathcal{L}_{\mathrm{ffH}} = -\frac{e}{2 s_w m_{\mathrm{W}}} \sum_j m_j \bar{f}_j H f_j \ ; \tag{149}$$

the interactions between one Higgs and vector bosons:

$$\mathcal{L}_{2\mathrm{VH}} = -\frac{e m_{\mathrm{W}}}{s_w} W^\mu W^*_\mu H - \frac{e m_{\mathrm{Z}}}{s_w c_w} Z^\mu Z_\mu H \ ; \tag{150}$$

the interactions between two Higgses and two vector bosons:

$$\mathcal{L}_{2\mathrm{V}2\mathrm{H}} = -\frac{e^2}{4 s_w^2} W^\mu W^*_\mu H^2 - \frac{e^2}{8 s_w^2 c_w^2} Z^\mu Z_\mu H^2 \ ; \tag{151}$$

and finally the Higgs self-interactions:

$$\mathcal{L}_{3,4\mathrm{H}} = -\frac{e m_{\mathrm{H}}^2}{4 s_w m_{\mathrm{W}}} H^3 - \frac{e^2 m_{\mathrm{H}}^2}{32 s_w^2 m_{\mathrm{W}}^2} H^4 \ . \tag{152}$$

Here we have used the notation

$$\begin{aligned} F^{\mu\nu} &= \partial^\mu A^\nu - \partial^\nu A^\mu \ , \\ W^{\mu\nu} &= \partial^\mu W^\nu - \partial^\nu W^\mu \ , \\ Z^{\mu\nu} &= \partial^\mu Z^\nu - \partial^\nu Z^\mu \ , \end{aligned} \tag{153}$$

while the K_{jk} stands for the (in principle arbitrary) elements of the Cabibbo-Kobayashi-Maskawa matrix. Note that the coefficients in front of the various terms in $\mathcal{L}_{\mathrm{MSM}}$ sometimes differ from the coupling constants deduced in the previous sections by combinatorial factors.

An endless series of Gedanken experiments?

Having concluded that renormalizability of the theory at the n-loop level requires at least unitarity of all $2 \to (n+1)$ processes at tree level, one might wonder whether there is actually a limit to the number of Gedanken experiments that would be necessary to establish renormalizability. In fact, however, we are finished since:

- We have established all fields and vertices of the Minimal Standard Model Lagrangian $\mathcal{L}_{\mathrm{MSM}}$;

- This Lagrangian is the unitary-gauge phenotype of the gauge-symmetric genotype, after symmetry breaking with the minimal possible Higgs sector;
- Such spontaneously broken gauge theories have been proven to be renormalizable[8];
- Moreover such theories lead to tree-level unitarity for all processes.

This admittedly indirect reasoning implies that we do not need to consider additional Gedanken processes: they will all be unitary at tree level, thanks to the gauge symmetry.

From the point of view of our derivation this argumentation is somewhat unsatisfactory since it relies on gauge symmetries instead of direct physical arguments. There *ought* to exist a theorem running in the opposite way, as follows:

- *If* a sufficient number of Gedanken processes satisfy tree-level unitarity; and
- *If* such-and-such additional conditions hold;
- *Then*
    - *All* processes satisy tree-level unitarity;
    - *And* the theory is renormalizable.

Unfortunately I do not know of such a theorem.

[8]This holds provided there are no anomalies. In the Minimal Standard Model this is guaranteed by the fact that the charges in each matter generation sum to zero.

Chairman: R. Kleiss

Scientific Secretaries: K. Adel, N. Brown

Discussion

– *Wexler:*

Is your improved Born approximation consistently accurate to one loop, or to more?

– *Kleiss:*

Partly it is accurate to infinite order, for example in the introduction of the QED running coupling constant. But consider the corrections to the ρ parameter. The one loop corrections go like m_t^2. If this top mass is large then two loop effects will be important, and maybe even higher orders are important. At the moment we have no recipe for summing terms going like m_t^2, we only know how to handle things to all orders with logarithms, and these are taken into account in the Born approximation. For the ρ parameter, however, things are harder.

– *Wexler:*

So consistently it is to one loop?

– *Kleiss:*

Definitely consistent to one loop, and to two loops also, and I think that if someone gets the answer to all orders, it will be numerically indistinguishable from what I wrote down. You should also realize that there are an enormous number of contributions that I did not write, which only give a small contribution in the end. I think at the moment we claim to understand the Z^0 line shape to better than 2 parts per thousand. The accuracy is such that some people are even worrying about the effects of a finite muon mass. I think that everything is pretty much under control.

– *Burke:*

It is known what the effect of a second Higgs doublet would be on the radiative corrections?

– *Kleiss:*

There is obviously the possibility of actually finding the Higgs, but I won't consider that. There is a formula

$$\rho = \frac{V_d^2 + V_t^2}{V_d^2 + 2V_t^2}$$

Here V_d^2 is the sum of the squares of the vacuum expectation values of all the Higgs doublets, and V_t^2 is the corresponding quantity for the Higgs triplets. If you have only doublets and no triplets then the ρ parameter is always going to be 1. If there are triplets then things are different. If there are lots of triplets the ρ parameter will go to 1/2.

– *Zichichi:*

What if there are Higgs singlets?

– *Kleiss:*

The W bosons get their mass from the Higgs. We have a term in the Lagrangian which does this.

$$|D_\mu \phi|^2 = (\partial_\mu \phi)^2 + 2W_\mu \phi \partial_\mu \phi + W_\mu W^\mu \phi^2$$

If the Higgs is an SU(2) singlet then it has no transformation properties under weak SU(2) and so it would not couple to the W or Z: to a large extent it would "disappear" in $\Delta\rho$.

– *Higgs:*

If its symmetry properties are trivial then it is not going to do anything in the way of spontaneous symmetry breaking.

– *Burke:*

What about radiative corrections, are they small for Higgs particles anyway, or would there be any difference for a second Higgs doublet?

– *Kleiss:*

The radiative corrections to Δr with only Higgs doublets are quite small. Even if you have a more complicated Higgs structure the effects are still generally quite small. At the two loop level one can generate terms proportional to m_H^2, but there are suppressed because they are at two loops. Also there are good reasons to expect $m_H < 1\ TeV$, a safe limit. In a more complicated scenario the masses will also not be more than about 1 TeV.

– *Brown:*

You mentioned that it is not known how to resum the corrections to the ρ parameter which go as a power series in m_t^2. An upper limit on m_t is usually provided by measurement of the ρ parameter. It is possible that the top quark mass is large, giving large corrections, but it is just that we do not know how to resum these corrections, in which case we would not be able to put a limit on m_t?

– *Kleiss:*

Yes and no. $\Delta\rho \sim m_t^2$ so if m_t is large then $\Delta\rho$ gets large and you should see the effect. People have worried that in higher order this will be even worse. There were some speculations that because of the resummation I showed earlier, the effect of a really heavy fermion could be suppressed a little bit at higher orders. I do not think it would be good enough to save the ρ parameter. I should say we can allow almost any top mass, by just putting in extra Higgs triplets. If you have a Higgs triplet then the ρ parameter becomes smaller and to get it close to 1 again you need a heavy top quark as I discussed in the lecture, or some supersymmetric doublet with a large mass splitting or so. You can still afford not to believe the limits on the top quark. That will change if we look carefully at the decay $Z \to b\bar{b}$. There we have independent information on m_t and so we may get an upper limit on m_t from LEP, before we find it at the TEVATRON.

– *Hsu:*

As previously stated the corrections to the ρ parameter go like m_t^2. This is hard for me to understand as I naïvely except scalar loops to be quadratically divergent, whilst fermion loops give only logarithmic divergences.

– *Kleiss:*

It is hard to understand intuitively. Yes, you would expect scalar loops to give quadratic divergences, but there is an accidental SU(2) symmetry that tells you these cancel against each other. What is left for the Higgs boson turns out to be a logarithm.

– *Hsu:*

That would be custodial SU(2)?

– *Kleiss:*

Yes.

– *Lu:*

I have a remark about wave function renormalization. You definitely need this when you are calculating cross-sections. What is your opinion about that?

– *Kleiss:*

I do not agree with you. If you look at the Z^0 propagator by itself, then after mass renormalization there are still some divergences left. So, if you want to talk just about the Z^0 propagator it is necessary to perform wave function renormalization to get rid of these divergences. If you are talking about physics, as in some scattering experiment, then the Z^0 propagator is coupled to fermion-Z

vertices. These vertices also have divergences, and because the standard model is a good theory they cancel with the infinities from the Z^0 propagator. Historically people absorb the infinities from diagrams like:

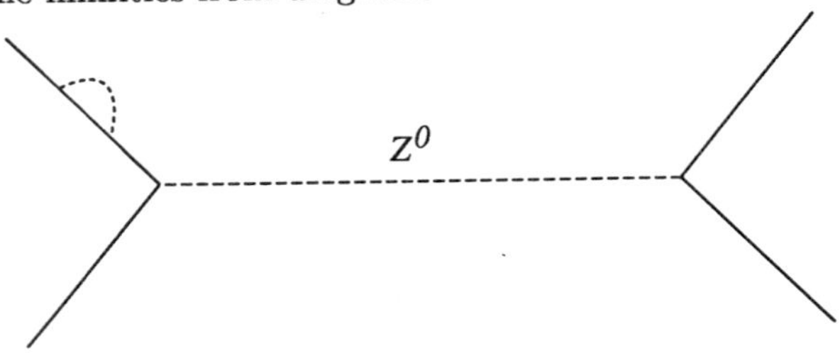

into wave function renormalization of the external legs. If you do that you need to do wave function renormalization on the other parts as well. In order to keep things simple, I only considered the diagrams corresponding to the physics process, and then I maintain that you do not need to worry about wave function renormalization.

– *Lu:*

The diagram you drew has to be multiplied by a factor of 1/2, and also receives contributions at higher orders.

– *Kleiss:*

I completely agree, wave function renormalization is very handy, especially at higher orders, where you have to worry about double counting. This is technical and I did not consider it necessary to go through the whole machinery.

– *Lu:*

I have another question. There are several ways to define the effective $sin^2\theta_W$. What is your opinion on this huge diversity? Is there a definition that is better than the others?

– *Kleiss:*

I was going to comment on that tomorrow. Let me amuse you. We have one s_W^2

$$s_W^2 = 1 - \frac{M_W^2}{M_Z^2}$$

This is the mass ratio. Then there are some others:

$$\bar{s}_W^2$$
$$s_W^{*2}$$
$$s_W^{**2}$$
$$\hat{s}_W^2$$
$$\hat{s}_W^2|_{M_W^2}$$

The first one is the effective $sin^2\theta_W$ I showed you this morning, where you absorb some of the renormalizations. Some people do it differently and get s_W^{*2}. Some arguments arose as to whether this was gauge invariant, and so it was modified to s_W^{**2}. Then some people did MS renormalization to get $\hat{s}_W^2$, where you evaluate this at the scale M_Z, but you may want to consider it at the scale M_W, so we get $s_W^2|_{M_W^2}$. These last five quantities are absolutely identical as far as experiment is concerned, so unless you go to really pathological situations they are all the same at LEP. But people do get angry about which one should be used. All the differences, however, are smaller than α/π. If you worry about that then you must also wory about the box diagrams, and so on... and in the end the results are identical.

I keep on adding to this list. In the end I want to have twelve different definitions, and get the six or seven guys who worry about it together so they can fight it out.

– *Brahm:*

I have a question about the renormalization of the electric charge due to photon-Z mixing, is the coefficient c_W/s_W or s_W/c_W?

– *Kleiss:*

It should be s_W/c_W.

– *Brahm:*

The self energy Σ itself, does that also vanish when $\theta_W \to 0$? What happens if there is no mixing?

– *Kleiss:*

It ought to vanish, because $\Sigma_{\gamma Z}$ only has fermion and boson propagators and they should decouple. If $\theta_W \to 0$ then the U(1) in SU(2) $\times$ U(1) is the U(1) you end up with, and it should completely decouple.

– *Brahm:*

It is also true that this term vanishes in the limit that SU(2) $\times$ U(1) is unbroken?

– *Kleiss:*

I would expect so, yes.

– *Kastening:*

It seems that if you work hard enough you can calculate everything, but if you have hadrons in the final state you will have problems with strong interactions, so how do we deal with that problem?

– *Kleiss:*

You mean the QCD corrections?

– *Kastening:*

Yes.

– *Kleiss:*

Well, we just write them down. People have been working very hard, with not too much work we get

$$1 + \frac{\alpha_s}{\pi} + () \left(\frac{\alpha_s}{\pi}\right)^2$$

where the coefficient at $O(\alpha_s^2)$ is about 1.

– *Kastening:*

So you do not have to introduce some parameterization due to problems in calculation?

– *Kleiss:*

Well, this is perturbative QCD: if you integrate over the whole phase space then you can argue that the problems of hadronization are bound to disappear, because whatever the quarks do after they have been produced you are going to count all those events anyway. In fact this is a theorem. I was going to mention that the corrections above are the only type that are allowed. You cannot have anything like a logarithm of m_Z^2/m_{hadron}^2. So if you really put in a lot of work, you can in fact calculate the $0(\alpha_s^3)$ term, and this was calculated to be about 70, and people got really excited about that. But now it has been discovered that there was a mistake. It is not clear what the number will be, but most people believe it will be a lot smaller. So you do not have to worry about the hadronic R ratio at LEP. If you are asking questions about exclusive jets and things like that, then the situation is completely different, but that will be treated in Prof. Dokshitzer's lecture.

– *Mandelbaum:*

I have two questions. First, you said that the contribution to the ρ parameter could be compensated by a heavy top quark. You mentioned a limit of about 1000 GeV. My question is: how could you calculate anything if the top quark is heavy because the Yukawa coupling becomes large, and how can you speak about the standard model in such a non-perturbative situation?

– *Kleiss:*

I agree, that is unspeakable, what I did was just for fun to show that you can get compensating effects. I am not saying I would believe a top mass of 1000 GeV.

– *Mandelbaum:*

But even that statement does not make sense because it is non-perturbative.

– *Kleiss:*

What are you saying to me is that if I propose a model with one Higgs doublet and one Higgs triplet and $m_t = 1000$ GeV it is going to be wrong. I was not proposing that model, but I could propose a model with 20 Higgs doublets and one triplet, and then I do not need a heavy top quark. But I do agree that if the top quark gets too heavy, then the coupling between the top quark and the Higgs field, which goes as the top mass, gets large, and then what happens to perturbation theory? You are in trouble. Basically no one believes a top quark of much more than 200 GeV in mass. So it is left as an exercise for the reader to work out how many Higgs doublets you need to compensate for one Higgs triplet in order that you do not find an effect with a 300 GeV top quark mass. I would guess about 10 or so.

– *Mandelbaum:*

My second question is related to the top quark being so heavy that in real $t\bar{t}$ production above threshold, you will have a final state interaction mediated by the higgs particle which couples very strongly. If you then have a strong force between the t and the $\bar{t}$, would you expect a "pre-confinement" because at this scale the Higgs coupling may be as strong as α_s?

– *Kleiss:*

Yes, that could indicate "pre-confinement"; you might get funny bound states, not from gluon exchange but from Higgs exchange. So you get toponium back, and that would be nice. If it really is so strong that you get really narrow resonances, people would love it. I cannot definitely say yes to "pre-confinement", but I think that you are right: the coupling becomes very big so that the $t\bar{t}$ will be bound close together.

– *Ferrara:*

I have a remark. Coming back to the question of Higgs singlets, I do not think these are uninteresting because Higgs singlets can have renormalizable couplings with Higgs doublets, and they can communicate with ordinary matter at one loop. These will modify the standard model; they do not have direct tree level couplings, but they couple to the Higgs doublets, and to ordinary quarks and leptons through radiative corrections.

– *Kleiss:*

Yes, I agree, that is a two loop effect, so you can probably think of models where you do have that effect.

– *Ferrara:*

Indeed, in the simple supersymmetric extension of the standard model where you want to get supersymmetry breaking you naturally have these Higgs singlets.

– *Kleiss:*

I agree.

Chairman: R. Kleiss

Scientific Secretaries: K. Adel, N. Brown

Discussion

– *Ciuchini:*

We know that contributions of order $O(\alpha^2)$ must be considered in the calculation of the Bhabha cross-section at the precision required by LEP. Do you think that the full calculation to order $O(\alpha^2)$ is useful and worth doing, or are the sub-leading terms of order $O(\alpha^2)$ enough?

– *Kleiss:*

Look at it numerically: count the correction terms of the form $\alpha^n log^m$. We know that $log \sim 20$, and $\alpha/\pi \sim 0.005$, so $\alpha log \sim 10\%$. At the one loop level you definitely need the αlog term, you also need terms $\sim \alpha$. At higher order we get terms like $\alpha^2 log^2, \alpha^3 log^3, \ldots, \alpha^n log^n, \ldots$ and exponentiation tells us how to deal with these

If you are considering small angle Bhabha scattering, which is used to measure the luminosity, then we must remember that the luminosity monitors used in LEP experiments have their own inherent systematic errors. These are about 0.5%. It is not very useful to go all-out to calculate the α^2 corrections because these are of the order $\sim 10^{-4}$. If we knew the $\alpha^2 log$ term we would be quite happy because at this moment this term is the main source of the uncertainty.

– *Ciuchini:*

How much does the limit on m_t from LEP change if a second Higgs doublet is considered?

– *Kleiss:*

If I consider a second Higgs doublet then the limit on m_t is not changed, because the ρ parameter is 1 if we only have doublets, and the corrections to $\Delta\rho$ are still like m_t^2. If I add a Higgs triplet then the limit on m_t completely disappears, for instance, if we have one Higgs doublet and one Higgs triplet with the same vacuum expectation values, then in order to get ρ in the neighbourhood of 1 the top mass would have to be of the order of 1000 GeV. Any additional assumption on the Higgs structure weakens your limit, and allows much higher top masses. But as I already said yesterday: $\Gamma(Z^0 \to b\bar{b})$ will give us independent information on m_t that depends less on the Higgs structure.

– *Ciuchini:*

So it is independent of the second doublet. However, I thought the CDF limit on m_t depended on whether there was a second Higgs doublet.

– *Kleiss:*

As far as I know that limit depends on the assumption that there is no light charged Higgs into which the top quark could decay, because then the signature of the top events are completely different. If the second Higgs doublet had a mass of 500 GeV, for instance, then this decay would not happen. Then the CDF limit is insensitive to the second Higgs doublet.

– *Ciuchini:*

Which of the three possible LEP future developments that you talked about do you prefer?

– *Kleiss:*

If you are asking for my personal opinion, I think that running LEP at high luminosity is boring, because I am not all that interested in B-physics. However, if you are interested in B-physics then you will be very happy with this B-factory. So far we have not found any Higgs particle, nor any indication of supersymmetry, and the limits on the standard model parameters are already quite good. This does not justify the expense of going to higher luminosity, but again that is a matter of taste.

What is bound to happen is LEP 200, which is more or less "fait accompli"; we are going to go to higher energies. In fact I would be happy if we could go, not to 200 GeV, but a little bit higher, say to 240 GeV - 250 GeV, and there are some plans for that. Then you start to test the non-abelian structure of standard model at the Born level instead of through radiative corrections.

Also if LEP can prove that the beams can be polarized, then polarized beams will be used at some stage in the future. To achieve this we have to build spin rotators, for example. This is relatively expensive and difficult but it can be done.

Of course, this all depends on the plans for the LHC, which would use the same tunnel as LEP and the LHC program should start around 1996, when the normal LEP program should be finishing.

– *Grossmann:*

You mentioned some experimental results on the Higgs mass. There also exist some theoretical estimates on its upper bound. What do you think about the reliability of these estimates and how do they compare to the present experimental ones?

– *Kleiss:*

Let me answer the second question first. At this moment there is no relation between the lower limit from LEP of about 40 GeV and the much higher theoretical upper limits.

If the Higgs is very heavy, then its self-coupling λ is very is very strong and can become larger than 1. This puts a limit on $m_H \sim 1$ TeV. This does not mean that it is forbidden for m_H to be larger than 1 TeV. It is just that perturbation theory becomes unreliable as we enter a strongly interacting region.

There are other similar indications we call "triviality": suppose the Higgs has a large mass and strong self-interactions, then your point of view changes from having the Standard Model with a Higgs sector added on just to give masses to the particles; instead we view the Standard Model as a "small correction" to the strongly interacting Higgs sector. Thus we can consider the original ϕ^4 theory where the runnning coupling constant λ can be evolved from the scale m_H^2 to some other scale M^2:

$$\lambda(M^2) \approx \frac{\lambda(m_H^2)}{1 - \frac{3}{2\pi^2}\, \lambda(m_H^2)\, \ell n\left(\frac{M^2}{m_H^2}\right)}$$

This blows up at at a scale "M" where: $1 - \frac{2}{3\pi^2}\, \lambda(m_H^2)\, \ell n\left(\frac{M^2}{m_H^2}\right) \approx \lambda(m_H^2)$ so that the interaction again becomes strong.

This does not mean that the theory breaks down at such a scale. It just means we can no longer do perturbation theory, and presumably we find compositeness or new physics at that scale. What you can do is choose the scale "M" at which $\lambda(M^2)$ blows up, then you can solve for $\lambda(m_H^2), m_H^2$. For example, $M \sim 10^{19}$ GeV $= M_{Planck}$ would give a reasonable Higgs mass of about 200 GeV. In other words if you want the Higgs sector to be well-behaved up to the Planck scale, then the Higgs had better not be heavier than 200 GeV.

You can continue this analysis for lower and lower values of "M", and if you do not expect things to work above 1 TeV, then you are allowed Higgs masses up to 1 TeV. Again if $m_H > 1$ TeV, you do not expect perturbation theory to hold above this scale. This is more or less the upper limit on the Higgs mass.

This gets more complicated if the top quark is also heavy. What we typically see is a plot like:

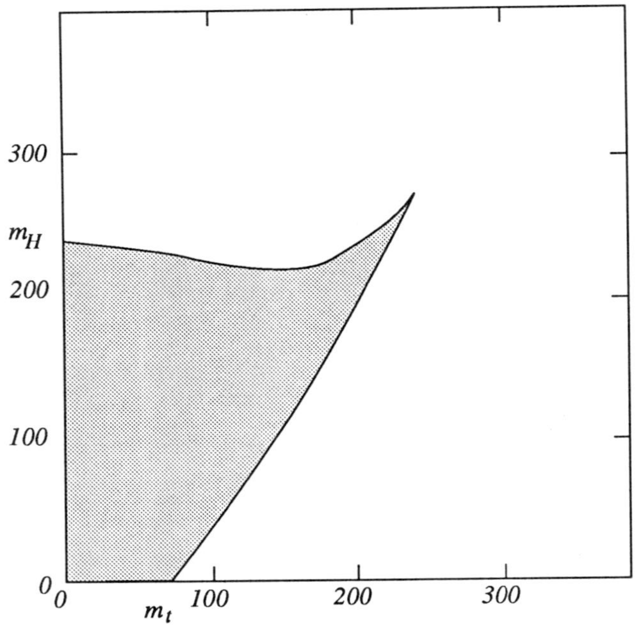

where the shaded region is the allowed region for m_t and m_H. All the values we are looking for at LEP are in the lower left hand corner where the Higgs is relatively light. If the Higgs is heavy, we really need 200 TeV collisions.

– *Hsu:*

You mentioned that LEP II would test, or rather utilize, the trilinear gauge coupling ZW^+W^-. Is there any chance of observing a quartic gauge coupling such as $Z\gamma W^+W^-$, or would this be swamped by the QED background which you discussed?

– *Kleiss:*

The QED background is fairly simple to handle at LEP I because the Z is neutral and the photon does not couple to the Z by itself. If at LEP II you look at the process $e^+e^- \to W^+W^-\gamma$, then indeed we would get very fancy couplings like:

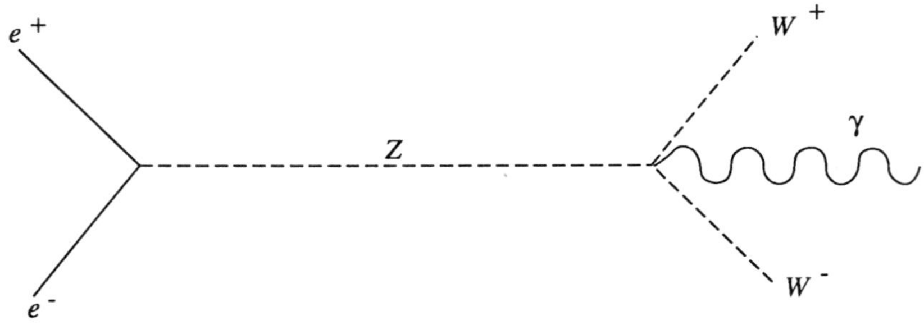

which could be tested. These couplings are *indispensable* if we want to talk about bremsstrahlung, because without them the amplitude will not be gauge invariant.

It is not clear to me that with the limited statistics we will have at LEP 200, we will see many of these events. If these quartic couplings are absent, then we should see crazy effects. So: yes, you would be testing the quartic coupling in some sense.

– *Titard:*

Could you tell us about QCD corrections to the ρ parameter?

– *Kleiss:*

The QCD corrections to the ρ parameter are a two-loop effect from diagrams like:

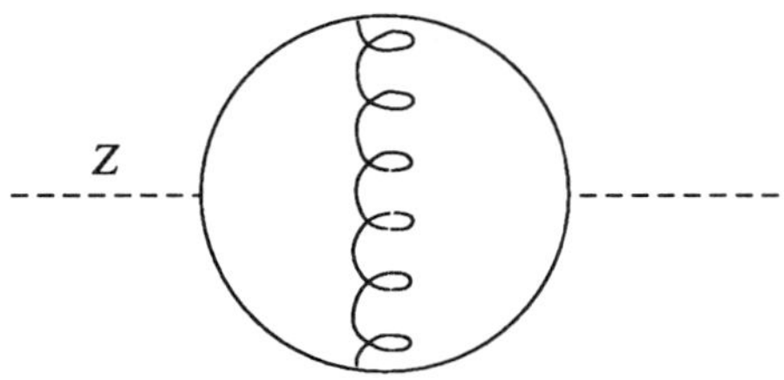

They have been computed sometime ago, in 1983 or so, and to my knowledge, they are generally small. I do not think they are numerically important.

– *Lu:*

In the exponentiation formula, is the second order term exact or is some part of it left out?

– *Kleiss:*

It is not exact. In the formula I wrote down, the dots contain terms which are numerically relevant, but they do not diverge like $\ell n m_\gamma, \ell n E_o$ and so on. What I wrote down is by no means exact; in fact, the whole idea of mimicking the experimental cuts with just some upper limit E_c on the photon energy, is extremely crude. I just did that in order to make clear how the exponentiation works. You should not use the formulae I wrote down, what is hidden in the dots is very important numerically but not for explaining exponentiation.

– *Lu:*

To calculate the final state radiation isn't it simpler to calculate the photon propagator and cut it open, i.e. take the imaginary part of it? This way one does not have to worry about infrared divergences.

– *Kleiss:*

Yes, you can do that. In QCD that is what you typically do, but in QED you do not have to do that. Let me explain further. If you look at final state radiation in QCD, then you have to, for instance, interfere the order $O(\alpha_s)$ diagrams like:

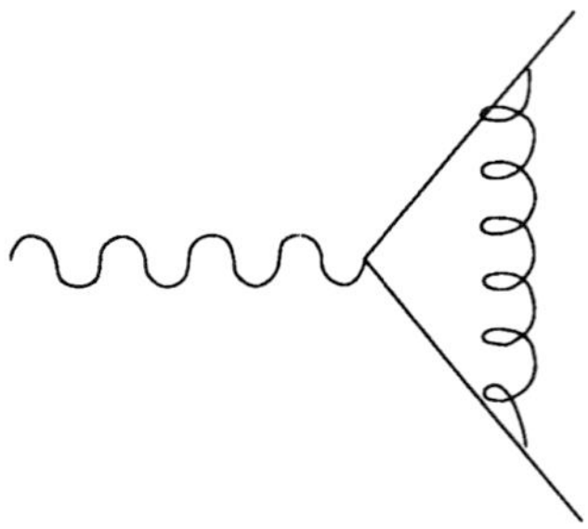

with the lowest order diagram giving us the product

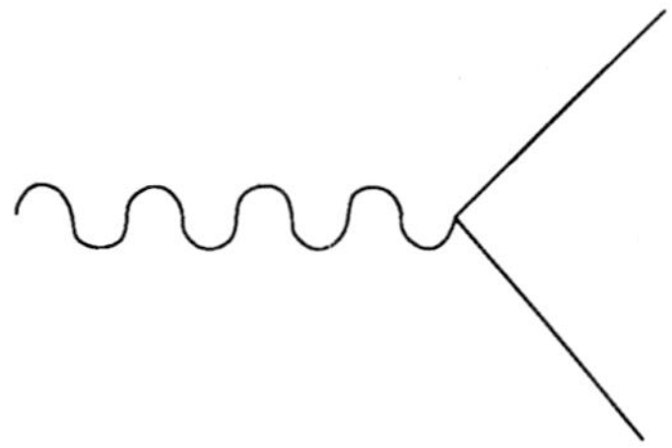

giving us the product:

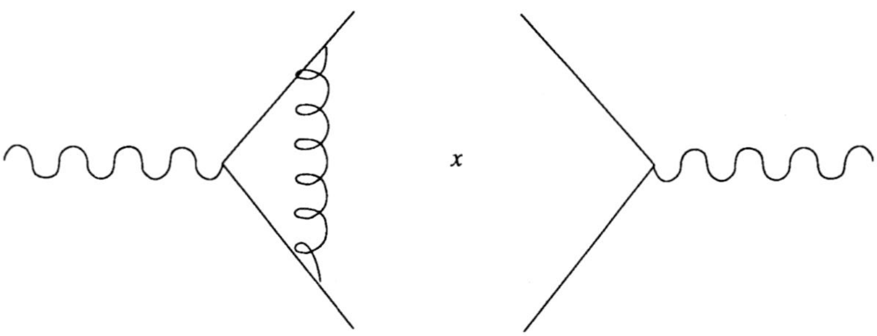

and this is related to the imaginary part of the two-loop correction to the photon propagator:

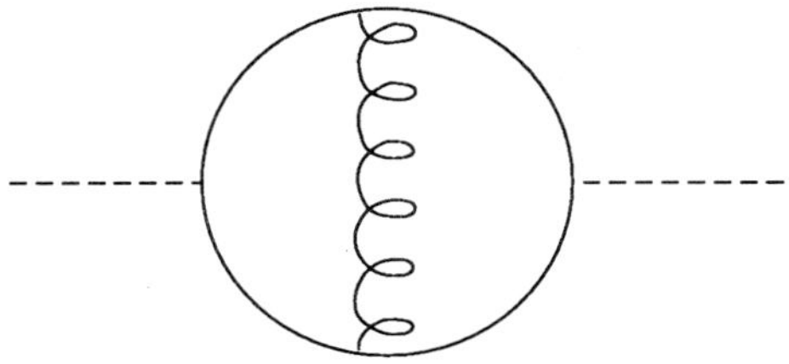

This illustrates the technique which relates the total cross-section to order $O(\alpha_s)$ to the imaginary part of the two-loop photon self-energy.

In practice, however, most QED people calculate the two-loop photon self-energy from the 1-loop final state radiation cross-section rather than the other way around. It is completely equivalent though, you can do it both ways. In QED, with massive charged particles, the infrared problem is somewhat easier to handle than in QCD since the mass acts as a regulator for some of the divergences.

– *Mojzis:*

In measuring the number of families at LEP we measure $\frac{\Gamma_{invisible}}{\Gamma_{neutrino}}$.

As far as I know the quantity is equal to the number of families only if the neutrinos are massless. If they are massive then this is only a lower bound for the number of families. If one wants to have more than a lower bound, one must combine the results from LEP with the results of other experiments on neutrino masses, neutrino oscillations, and so on. My question is: has this been done, and if so, what is the result on the number of families?

– *Kleiss:*

I do not completely agree with you? $\Gamma_{invisible}$ does not give a lower bound on the number of families.

What is true is that in turning $\Gamma_{invisible}$ into a number of neutrino types, you have to assume that the neutrino masses are not very large, although they can be a few GeV.

The point is if a neutrino is very heavy, the width for the Z to decay into that neutrino will be suppressed by phase-space. You cannot rule out a neutrino which has a mass of say 45.5 GeV. You also have to assume the Standard Model couplings of the Z to the neutrinos. Under these assumptions, $\Gamma_{invisible}$ gives you the number of light neutrino types. You cannot rule out a fourth generation neutrino which is very heavy. Thus the statement which has been publicized, that we have proved that there are only three generations, is strictly speaking not true, but there are only three "normal" generations.

– *Zichichi:*

It could be then, that there is a fourth generation of fundamental quarks and leptons, provided that the neutral lepton of the fourth generation is very heavy?

– *Kleiss:*

Yes.

– *Zichichi:*

So it could be that the fourth generation is still there, the quarks are very heavy, obviously, because the top has not been found, and in this case the neutrino must also be very heavy.

– *Kleiss:*

Yes, but I do not consider that a "normal" kind of generation because of the massive neutrino, but it could be there.

– *Brodsky:*

My question is whether or not the radiative corrections from initial state radiation bringing the energy down to the Z pole is a complication for signals at LEP 200. Is it a large cross-section?

– *Kleiss:*

I agree that this signal gives you a sizeable cross-section. The signal for such events would be completely different because hard photon would usually disappear down the beam pipe recoiling against a Z. You would have a tremendous longitudinal momentum imbalance, but at LEP you can measure that. You also know how many events of this type you would expect, so it is pretty much under control.

– *Zichichi:*

I would like to add some information. This morning you quoted the L3 experiment without any τ events. It happens that I am collaborating on L3 and looking at my notes this is the data from L3 which has 2000 τ events.

– *Kleiss:*

I am not saying that L3 does not look at τ's, it is just that in the review by Fernandez at the Neutrino Conference, he did not quote an L3 result.

$\tau\overline{\tau}$	events	2.000
$\mu\overline{\mu}$	events	2.300
$e\overline{e}$	events	2.600

$$N\nu \quad = 3.01 \pm 0.11$$

$$g_v \quad = 0.061^{+0.021}_{-0.016}$$

$$g_A \quad - 0.501 \pm 0.006$$

$$m_{\tilde{\mu}} > 35.7 \quad \text{GeV}$$

$$\Gamma_{b\bar{b}} \quad = \quad 353 \pm 15 \pm 15 \text{ MeV}$$

$$A_{b\bar{b}} \quad \cong \quad - (15.0 \pm 5.0)\%$$

$$\text{mass of} \quad e^*, \ \mu^* > 70 \text{ GeV}$$

$$\text{for} \quad \frac{\lambda}{m_e^*} < 10^{-3}$$

Test of 3 gluon coupling in QCD

$$\alpha_s = 0.115 \pm 0.005 \quad \text{exp}$$

$$\pm 0.012 \quad \text{theory}$$

<u>L3 DATA</u>, July 7, 1990

WEAK SCALE SUPERSYMMETRY

Lawrence J. Hall

Theoretical Physics Group
Lawrence Berkeley Laboratory
Berkeley, California 94720

An introduction to the ideas and current state of weak scale supersymmetry is given. It is shown that LEP data on Z decays has already excluded two of the most elegant models of weak scale supersymmetry.

The LEP data on Z decays and the Fermilab Tevatron data have begun an era of strigent experimental scrutiny of the idea of supersymmetry at the weak scale. In this lecture I discuss the minimal low energy supersymmetric model (MLES) and show why it is not to be particular preferred over several other models having very different experimental signatures. I illustrate the power of the Z data by considering two models which are now excluded in their simplest forms. In one R parity is spontaneously broken by a sneutrino vacuum expectation value,[1] and in the other R parity is promoted to a continuous global $U(1)_R$ symmetry.[2]

One of the goals of the multi-TeV physics of the 1990s is to elucidate the nature of electroweak symmetry breaking and to understand the origin of the weak scale. I will follow the predominant viewpoint that the Planck scale sets the fundamental scale of mass and that all other scales are somehow derivative. This viewpoint is not obviously correct, especially in the case of Brans-Dicke gravity theories, but provides a convenient framework. The QCD scale Λ is of order $10^{-20}M_p$. Such a hierarchy of scales is understood by the gradual logarithmic renormalization of the QCD coupling. If QCD is defined at the Planck scale by the dimensionless parameter α_s, radiative corrections produce the phenomenon of dimensional transmutation: $\Lambda \cong M_p \exp(-1/\alpha_s)$. This is such a plausible and efficient way of generating scales that the idea of dimensional transmutation underlies all of our ideas for the origin of the weak scale. This is most obvious for the case of technicolor.[3] The technicolor idea is that there is a new strong force which gets strong near a TeV and which has techniquark condensates similar to the quark condensates of QCD which break light quark chiral symmetries. If the

Physics Up to 200 TeV
Edited by A. Zichichi, Plenum Press, New York, 1991

techniquarks are chiral under $SU(2) \times U(1)$ their condensates will trigger electroweak symmetry breaking. The weak scale is then identified with the dimensional transmutation scale Λ_{TC}, at which the technicolor coupling gets strong. It is possible that the Higgs boson is composite and that both the binding forces and the forces which give it a vacuum expectation value are due to some new strong gauge force, ultracolor.[4] In this case the weak scale is identified as the dimensional transmutation scale, Λ_{UC} at which the ultracolor coupling gets strong.

There are two viewpoints as to why physics may become supersymmetric above the weak scale. The more modest reason is that supersymmetry protects the Higgs mass against a quadratic divergence. If the Higgs is elementary there should be a symmetry reason explaining why it is so light. All the other particles of the standard model have symmetries protecting their masses: chiral symmetries for fermions and gauge symmetries for vectors. Supersymmetry allows a symmetry to protect scalar masses also. Supersymmetric theories have a bosonic state degenerate with each fermionic state. From the spectrum of the standard model it is obvious that if supersymmetry is relevant to nature it must be broken at least on scales of order M_W. Furthermore, for the Higgs mass to be protected to the weak scale, the supersymmetry breaking scale M_S cannot be much larger than the weak scale. We conclude that M_S must be identified with the weak scale.

A second, more ambitious, viewpoint is that the scale of supersymmetry breaking, and therefore the weak scale, is dynamically generated via a dimensional transmutation.[5] This could occur if the potential for some scalar field were flat at tree level, but had a logarithmically generated minimum via radiative corrections. Such a possibility needs supersymmetry since otherwise the flatness of the potential would be spoiled by quadratic divergences.

The minimal low energy supersymmetric model (MLES) is minimal in several senses. It involves the fewest chiral superfields necessary for a supersymmetric extension of the standard model $Q = (U, D), U^c, D^c, L = (\nu, E), E^c, H_1$ and H_2. Secondly it possesses the fewest interactions for such a model to be consistent with data. These interactions are the supersymmetric gauge interactions of $SU(3) \times SU(2) \times U(1)$, together with supersymmetric Yukawa interactions

$$f = Q\lambda_D D^c H_1 + Q\lambda_U U^c H_2 + L\lambda_E E^c H_1 + \mu H_1 H_2, \tag{1}$$

and a set of interactions which break supersymmetry softly

$$m_{\tilde{g}}\tilde{g}\tilde{g} + ..., \tag{2a}$$

$$m_{\tilde{q}}^2\tilde{q}^*\tilde{q} + ..., \tag{2b}$$

$$A\tilde{q}\tilde{d}^c h_1 + ..., \tag{2c}$$

$$Bh_1h_2 + h.c., \tag{2d}$$

where lower-case letters are used for component fields, and generation indices are suppressed until needed. Finally, MLES is minimal in the sense of having the fewest vevs consistent with data: $\langle h_1 \rangle = v_1, \langle h_2 \rangle = v_2$ and all others vanishing.

Given the above minimality of MLES, it is not surprising that it should receive the most attention. However, from the theoretical viewpoint it does not have a unique minimality, even when we restrict our attention to models with minimal field content. This is because gauge invariance allows

$$f_{\Delta L} = Q\lambda D^c L + L\lambda' E^c L + \mu' L H_2 \tag{3a}$$

and

$$f_{\Delta B} = \lambda'' U^c D^c D^c, \tag{3b}$$

which, if present, would give the proton a weak decay rate. It is necessary to impose some symmetry to remove the interactions of (3a) or (3b). There are four ways of doing this: (3a) may be removed by imposing L, (3b) by imposing B, and both by imposing a Z_2 symmetry (R_p) or a Z_N symmetry ($N > 2$). R parity, R_p, is a multiplicative parity which is $+1$ on all particles and -1 on superpartners. In the case of the MLES model, this is equivalent to matter parity, which changes the sign of all matter superfields. It was originally imposed, in its present form, as a convenient way of constructing a realistic supersymmetric $SU(5)$ grand unified theory.[6)] However, different unified schemes lead to all of the above four cases, and, hence, from the theoretical viewpoint it is difficult to justify a clear preference of one case over the rest. In addition there is the question of whether a sneutrino vev might occur in the R_p invariant case.

If the alternatives to R_p invariance are simply ignored, a convincing argument gives missing energy as a good signature for supersymmetry at colliders. Proton stability leads to R_p invariance which implys the stability of the lightest superpartner (LSP). Cosmology requires the LSP to be neutral and, since its interactions with matter are therefore weak, superpartner pair production at colliders will lead to missing energy signatures. The problem with this litany is that it ignores the alternatives to R_p invariance. These alternatives typically do not lead to events with large missing energy, but have their own distinct and characteristic signatures.

I have recently reviewed the status of these models,[7)] and last year I lectured here on the exciting collider signatures from the models with B and L violation.[8)] This year I will bring you up to date on some more recent developments which have been triggered by the Z data from LEP.

As a prelude, and in case you think supersymmetric model building is a closed subject, let me enumerate the problems encountered by weak scale supersymmetry.

1. Renormalizable B and L violation. The interactions of equations 3 are a major embarrassment. Supersymmetry appears to lead us backwards. At least it forces us to extra symmetries.

2. There is the problem of predictivity. Supersymmetry is justly accused of fitting all data and wriggly out of all constraints. While supersymmetric superpartner interactions do not introduce additional parameters beyond those of the standard model, many extra parameters appear when supersymmetry breaking is included. In the case of breaking supersymmetry in $N = 1$ supergravity models, and imposing the most stringent constraints on the theory at the Planck scale, six new real parameters appear. Renormalized at the Planck scale these are (see equation 2)

i) a degenerate Majorana mass for all gauginos, $\widetilde{m}$

ii) a degenerate mass for all scalars, m^2

iii) a trilinear scalar interaction between squarks and Higgs $|A|e^{i\phi_A}$

iv) a bilinear scalar-scalar mass $(h_1 h_2)|B|e^{i\phi_B}$

3. The weak scale problem. Lets suppose that supersymmetry somehow dynamically generates $\widetilde{m}$ and m to be of order the weak scale. While it is true that

renormalization of the Higgs mass squared leads to a very elegant understanding of $SU(2) \times U(1)$ breaking, there is the question of the supersymmetric parameter μ of equation 1. In the context of the MLES model it is just put in by hand to be of order the weak scale, and this ruins claims to have understood the weak scale. If μ were of order M_p there would be no light Higgs. If μ were absent the form of the supergravity potential forces B to zero, and this ruins electroweak symmetry breaking. This problem has been solved in several ways by complicating the model. None of the problems I am discussing is a true disaster for weak scale supersymmetry. Nevertheless, together they may indicate that MLES is badly off track, and may need radical modification.

4. Ultimately the question of weak scale supersymmetry will be settled experimentally. Experiments over the next decade will teach us a great deal about physics at the TeV scale. However, from this viewpoint I believe that MLES is very awkwardly formulated. It is formulated in terms of some very technical assumptions about the form of the interactions in the $N = 1$ supergravity model at the Planck scale. The reason for this is that when it was formulated theorists were eager to find an elegant understanding for the origin of supersymmetry breaking. However, this formulation is now problematic: if experiments exclude MLES we will learn that this particular supergravity method for breaking supersymmetry is not correct. We will not settle the issue of weak scale supersymmetry. Recently, Lisa Randall and I have formulated weak scale supersymmetry purely in terms of assumptions about physics at the weak scale.[9] We can and should test physics at the weak scale. I will give an example of such a formulation later with the $U(1)_R$ invariant model.

5. In the MLES model a one loop diagram involving either ϕ_A or ϕ_B leads to an electric dipole moment of the quark.[10] In order not to give too large a neutron electric dipole moment these phases are then constrained to be small: $\phi_{A,B} \lesssim 10^{-2} \left(\frac{m_{susy}}{300GeV}\right)^2$. It is not easy to come up with reasons for A and B to have small phases. Of course, numerically the problem is not nearly as bad as the strong CP problem.[11] There are two loop contributions to the CP violating operator $G^2\widetilde{G}$[12] (G is the QCD gluon field strength tensor) which earlier this year looked like constraining ϕ_A to be less than about 10^{-6}. However, an anomalous dimension enhancement sign error had overestimated the effect by 5.10^4. The resulting effect of the operator $G^2\widetilde{G}$ in supersymmetry is less constraining than the electric (and color electric) dipole moment quark operators by about a factor of 10.[13]

One frustration of low energy supersymmetry is that there is any need for a Higgs multiplet at all. The sneutrino has exactly the quantum numbers of the Higgs! However, if the sneutrino is the sole origin of electroweak breaking the corresponding neutrino and charged lepton get large Dirac masses from $\langle\widetilde{\nu}^*\rangle\widetilde{Z}\nu$ and $\langle\widetilde{\nu}\rangle^*\widetilde{w}\ell$. However, it has been pointed out that in the context of the MLES model the tau sneutrino may acquire a vev as well as the two Higgs doublets, provided one studies the $\mu \rightarrow 0$ limit.[1] In this case the physical light tau doublet can be identified as $(\cos\theta\ \nu_\tau + \sin\theta\widetilde{H}, \cos\theta\ \tau + \sin\theta\widetilde{H}^-)$ where θ is a mixing angle which depends on the vevs. This model is interesting for several reasons.

1) The model shows that R parity may be broken even in MLES. This breaking leads to crucial changes in experimental signatures and cosmology.

2) The tau sneutrino vev can easily be triggered by the vevs v_1 and v_2 because the scalar potential contains the D terms: $g^2(h_1^*h_1 - h_2^*h_2 + \nu^*\nu)^2$ and a heavy top quark implies $v_2 > v_1$.

3) Since $\mu = 0$ the model does not require an understanding of why this parameter is of order M_W.

4) Since $\mu = 0$ the model has a higher degree of predictivity that the MLES model. The chargino mass matrix is particularly constrained. Also there is a very light neutralino state $\nu_\perp$, in addition to the three neutrinos.

In fact Z decay data can now exclude this model.[14] The neutralino, chargino, gluino masses are determined in terms of three parameters: $\widetilde{m}$, $\sqrt{v_1^2 + v_\tau^2}$ and v_2. Furthermore $\sqrt{v_1^2 + v_2^2 + v_\tau^2} = 250$ GeV so that there are two free parameters remaining, which I take as $\widetilde{m}$ and $\tan\beta = v_2/\sqrt{v_1^2 + v_\tau^2}$. Figure 1 illustrates the processes which rule out this model in the various regions of parameter space. In region i) there is a chargino light enough to be pair produced in Z decay. Although $\widetilde{m}$ is large, the chargino is light because of a see-saw mechanism. In region ii) the rate for $Z \to v_\perp v_\perp$ is sufficient to give too large a contribution to the invisible Z width. This decay rate depends on tanβ because $v_\perp$ is a superposition of $T_3 = +1/2$ and $-1/2$ components. For tan $\beta = 1$, the $Zv_\perp v_\perp$ coupling vanishes. In region iii) there is an off diagonal coupling $Zv_\perp\chi$ where χ is another neutralino lighter than the Z. It's decay leads to $\bar{b}b+$ missing signatures. These events are similar to those excluded by the Higgs search.

As a second illustration I consider a model proposed recently by Lisa Randall and myself.[2] The idea is to produce as predictive a model of supersymmetry as possible by promoting R parity to a continuous $U(1)_R$ symmetry. In many ways the model is the most elegant model of supersymmetry that I have seen, although I am biased of course. In particular the five problems of the MLES model discussed above are all solved. The $U(1)_R$ symmetry is defined as follows:

All standard model particles:	charge 0
Left handed Higgsinos:	charge -1
Left handed gauginos :	charge +1
Squarks and sleptons (defined as partners of left handed quarks, antiquarks, leptons and anti-leptons):	charge +1

The theory at the weak scale is the most general model with broken supersymmetry which has

A) no quadratic divergences

B) all flavor violation in the usual three Yukawa matrices

C) gauge symmetry $SU(3) \times SU(2) \times U(1)$ and global symmetry $U(1)_R$.

D) minimal fields consistent with A) —C). Since $U(1)_R$ is never broken superpartners must be pair produced, and as for MLES, the lightest is stable.

This model solves the five problems of MLES as follows

1. All the B and L violating interactions of equation 3 are forbidden by $U(1)_R$.

2. The soft parameters $\widetilde{m}, |A|, \phi_A$ are set to zero by $U(1)_R$. Also $\mu = 0$ and ϕ_B is non physical and can be rotated away by a $U(1)_{PQ}$ rotation. The removal of these five parameters leads to a wide set of predictions. In particular it is obvious from the R charges that fermionic superpartners can never acquire Majorana masses. The only allowed masses are Dirac couplings of gauginos to Higgsinos.

3. Since $U(1)_R$ forces μ to vanish, there is no problem in understanding why μ should be of order M_W. Note that it is because (A) allows the most general set of soft operators, not necessarily just those which come from a minimal $N = 1$ supergravity scenario, that B can be non-zero even when $\mu = 0$.

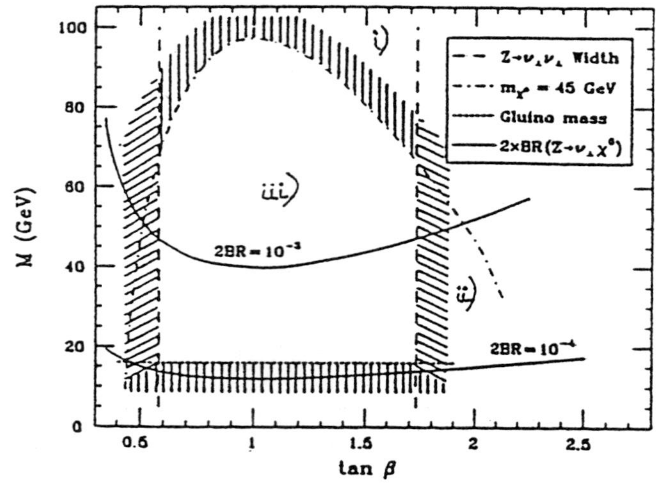

Figure 1. Excluded Regions and BR Contours in the (tan β-*M*) Plane.

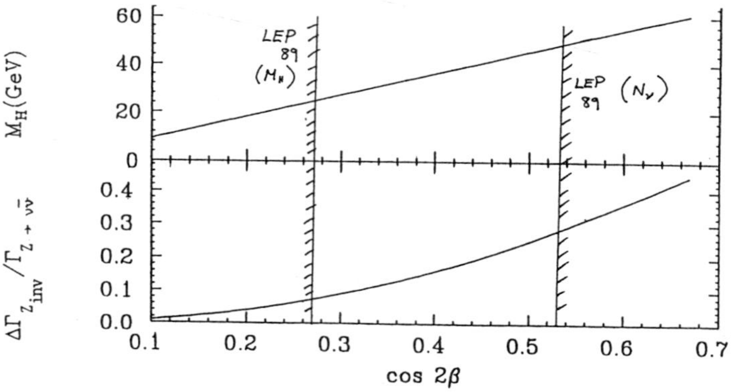

Figure 2. Constraints on cos 2β from Higgs mass and width of *Z* to invisibles.

4. From the nature of the assumptions A) —D) it is clear that this model is formulated at the weak scale. The technical assumptions about $N = 1$ supergravity have been replaced by the mild constraints A) and B). If A) is false there is little motivation for weak scale supersymmetry, while if B is false GIM violation is likely to be problematic

If the model is proved false we learn something about weak scale physics. Some of A) — D) must be wrong. The theory may or may not have a supergravity origin.

5. Since $\phi_A = \phi_B = 0$ there are no one loop diagrams for the neutron electric dipole moment.

Honesty compels me to mention that these problems have been solved at the expense of one new problem: the gluino is massless. This necessitates introducing an extra chiral superfield which is a color octet $(0, \widetilde{0})$. The most general soft operators then include the mass term $\widetilde{g}\widetilde{0}$.[2] Although unpleasant, the positive features of R invariant supersymmetry easily justify its exploration.

The two most exciting predictions are for the electroweak gaugino masses and for the lightest Higgs mass. Since $\mu = \widetilde{m} = 0$ the electroweak gauginos only acquire mass from electroweak symmetry break. The $\widetilde{Z}$ marries $\widetilde{H}_Z = \cos\beta \widetilde{H}_1^0 - \sin\beta \widetilde{H}_2^0$ to become a neutral Dirac fermion of mass M_Z. The photino and $\widetilde{H}_\gamma = \cos\beta \widetilde{H}_2^0 + \sin\beta \widetilde{H}_1^0$ are massless at tree level, and acquire a Dirac coupling of about 1 GeV at one loop. There are two Dirac charginos one heavier and one lighter than the W: $\widetilde{w}^- \widetilde{H}^+$ has a mass $\sqrt{2}\sin\beta M_W$ and $\widetilde{w}^+ \widetilde{H}^-$ a mass $\sqrt{2}\cos\beta M_W$. We follow the usual convention that $\tan\beta = v_2/v_1$. The light Dirac $\widetilde{\gamma}\widetilde{H}_\gamma$ can be produced in Z decay with a width $|\cos 2\beta|^2$ compared to that of a neutrino. Hence the model increases the number of neutrino species measured in Z decay:

$$\Delta N_v = |\cos 2\beta|^2. \tag{4}$$

The potential for the two Higgs doublets h_1 and h_2 is the same as for the MLES model. Hence the Higgs mass relations of that model apply also to the present model. In particular one Higgs is lighter than the Z:

$$m_h < M_Z |\cos 2\beta|. \tag{5}$$

Furthermore we know from (4) that $|\cos 2\beta|$ is quite small, and in this limit h has couplings to the fermions which are the same as the Higgs of the standard model.

In figure 2 I illustrate how the Z decay data on ΔN_v and m_h has rapidly constrained this $U(1)_R$ invariant model. Before LEP it was essentially unconstrained. The 1989 LEP data led to a considerable constraint, but left a sizable allowed region. The data announced in June at the $\nu - 90$ conference has now excluded the model at the 2σ level. This is a powerful illustration of how LEP is bringing us to a new era of understanding of physics at the weak scale.

I conclude with three points:

1) It seems to me unreasonable that the standard model with a single Higgs boson is the whole story for weak scale physics. The weak scale, like all other masses in physics, needs a protection mechanism.

2) Ideas for the protection mechanism, technicolor and supersymmetry for example, are being increasingly constrained by collider data.

3) Could there be an alternative to dimensional transmutations?

1. J. Ellis, G. Gelmini, C. Jarlskog, G. Ross and J. Valle, Phys. Lett. 150B 142 (1985).

2. L. Hall and L. Randall, $U(1)_R$ Symmetric Supersymmetry, LBL preprint LBL-29561 (1990). To be published in *Nucl. Phys. B.*

3. E. Farhi and L. Susskind, Phys. Reports 74 (1981) 277.

4. H. Georgi, Les Houches Summer School Lectures, Ed. P. Ramond and R. Stora, North Holland, 1985.

5. A. Lahanas and D. Nanopoulos, Phys. Rep. **145** (1987) 1.

6. S. Dimopoulos and H. Georgi, Nucl. Phys. **B193** (1981) 150.

7. L.J. Hall, Modern Physics Letters A., 467 (1990).

8. L.J. Hall, Proceedings of the XXIII International School of Subnuclear Physics, Erice, 1989. Ed. A. Zichichi.

9. L.J. Hall and L. Randall, Weak scale effective supersymmetry, LBL preprint 28879 (1990). To be published in PRL.

10. J. Ellis, S. Ferrara and D.V. Nanopoulos *Phys. Lett.* **B114** (1982) 231; J. Polchinski and M. Wise, *Phys. Lett.* **B125** (1983) 393.

11. R. Peccei, CP Violation, C. Jarlskog, World Scientific, 1990.

12. S. Weinberg, *Phys. Rev. lett.* **63** (1983) 2333.

13. R. Arnowitt, J. Lopez and D. Nanopoulos, CTP-TAMU 23/90 (1990).

14. D. Brahm, L.J. Hall and S. Hsu, Phys. Rev. D to be published.

Chairman: L. Hall

Scientific Secretaries: R. Ayad, D. Brahm, M. Ciuchini

Discussion

– *Zichichi:*

Why is the third generation of sneutrino most likely to acquire a vacuum expectation value?

– *Hall:*

Renormalization effects also contribute to the sneutrino mass, and the large mass of the tau tends to lower the tau sneutrino mass the most.

– *Glashow:*

Why can't a fourth generation of sneutrino take the place of the Higgs?

– *Hall:*

There is little distinction in supersymmetry between the Higgs and lepton supermultiplets, so to some degree it is a matter of naming convention. However, one lepton supermultiplet cannot completely replace the Higgses, since only the down-like masses could be generated this way.

– *Wexler:*

How does supersymmetry naturally incorporate gravity?

– *Hall:*

Supersymmetry as a global symmetry at the weak scale need not be connected to gravity. Gauging the supersymmetry, however, and including a graviton supermultiplet lead immediately to the theory of supergravity. In supergravity theories, the scalar potential and the supersymmetry-breaking terms take a very special form. Supergravity is also necessary if supersymmetry breaking is to occur without generating a cosmological constant.

– *Wexler:*

Can you gauge the $U(1)_R$?

– *Hall:*

From the phenomenological viewpoint gauging R is probably OK; after all none of the known particles couple to R. From the theoretical side I see two problems. The model as it stands would have gauge anomalies YR^2, for example from a bino loop. Even if this were fixed up I don't see how to build a globally supersymmetric model with gauged R.

– *Hsu:*

One of the few generic predictions in supersymmetric models with only Higgs doublets is that one Higgs is lighter than M_Z. What implications would the non-observation of such a light Higgs have for supersymmetry? Are there any other testable model-independent predictions?

– *Hall:*

Supersymmetry is very slippery! Even the light Higgs prediction disappears with the introduction of a singlet Higgs field S. Similarly the $U(1)_R$ model can avoid the Higgs mass limit by adding one extra field. It is probably impossible to exclude supersymmetry from any model-independent testable prediction, but at some point the theory becomes so fine-tuned it is no longer of interest. Barbieri and collaborators parameterized the degree of fine-tuning needed to push up the superpartner masses while keeping the weak boson masses at 100 GeV, and it is quite dramatic. An order of magnitude fine-tuning is needed to push all the superpartner masses above 300 GeV, and the degree of fine-tuning increases quadratically with the superpartner masses.

– *Brahm:*

The $U(1)_R$ model uses a supersymmetry-breaking scalar Higgs-Higgs coupling which is not supergravity-inspired. What other supersymmetry-breaking terms are allowed if the constraints from supergravity are lifted?

– *Hall:*

Remember we have imposed another constraint, namely the absence of quadratic divergences. Girardello and Grisaru wrote a paper listing the possible "soft" operators in supersymmetry. I believe they missed one, the dimension-3 coupling of a gaugino to a chiral fermion ($\lambda\psi$), which I used to give the gluino a Dirac mass. They probably left it out because it leads to quadratic divergences in a theory with a gauge singlet. Since the $U(1)_R$ model has no singlets I was able to make use of it.

– *Lu:*

What are the particle charges under R-parity and under $U(1)_R$?

– *Hall:*

Under R-parity, all known particles (q, l, h, $W^{\pm}$, Z^0, etc.) are even, and all superpartners ($\tilde{q}$, etc.) are odd. Under $U(1)_R$, all known particles have charge

0 while the superpartners have charge ±1 : $\tilde{q}$, $\tilde{l}$, and $\tilde{g}$ have charge +1, and the higgsinos $\tilde{h}$ have charge -1. $U(1)_R$ is a rather natural symmetry from the superspace viewpoint, since it corresponds to a phase rotation of the Grassman coordinate θ.

– *Haas:*

Glashow said supersymmetry was dead.

– *Hall:*

No, he did not! He said superstring theory will not connect to weak-scale physics in the near future. But supersymmetry can stand alone as a weak-scale theory, regardless of the status of superstrings. The issue of weak scale supersymmetry will be decided by experimentalists, not by theorists.

– *Kastening:*

If you don't introduce R-parity or $U(1)_R$ to eliminate baryon- and lepton-number violation, could there be sufficient baryon-number violation to explain the cosmological baryon asymmetry, without contradicting experiment?

– *Hall:*

Not if both baryon and lepton number are violated. In this case one expects the proton lifetime to be of order 10^{-12} sec for baryon and lepton number violating couplings λ of order 1.

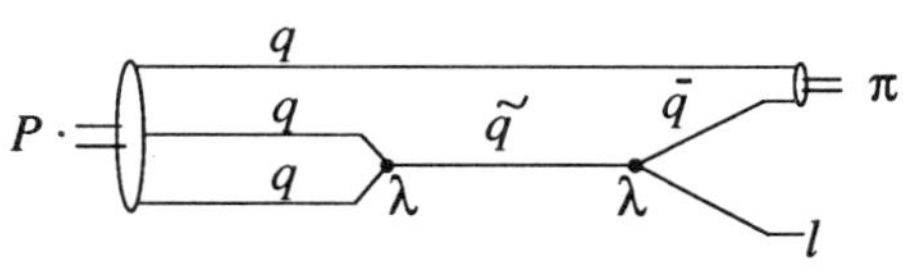

Since we need a suppression of 10^{50}, the couplings need to satisfy $\lambda < 10^{-13}$. Besides requiring incredible fine-tuning, such a coupling would not suffice to give the cosmological baryon excess, since $(n_B - n_{\bar{B}})/n_\gamma \sim \lambda^2$ needs to be about 10^{-10}. In fact, such a coupling would tend to wash out any baryon excess that was produced.

Remember, though, that either baryon or lepton conservation is sufficient to make the proton stable, since the proton must decay to an odd number of leptons. If L is conserved then it is possible to use the baryon violating coupling to generate the cosmological baryon asymmetry, as I showed a couple of years ago with Dimopoulos.

HETEROTIC AND TYPE II SUPERSTRINGS COMPACTIFIED ON CALABI-YAU MANIFOLDS

S. FERRARA

CERN, Geneva
Switzerland

Abstract

Some general properties of superstrings compactified on Calabi-Yau manifolds are reported.

I. Introduction

In this lecture[1)] I will briefly review some general aspects of superstring compactifications on Calabi-Yau manifolds, in particular I will emphasize the point that all low energy interactions of massless fields are "topological" in the sense that all the basic couplings, normalization of kinetic terms or vector couplings are only specified by topological informations and do not depend specifically on the particular Ricci-Flat metric of the "Calabi-Yau" internal space[2)−6)]. This fact is crucial in order to extract model independent properties[7)−22)] of effective supergravity Lagrangians and to study problems like gauge symmetry or supersymmetry breaking in a model independent way.

As is widely known[1)] superstring theories compactified on Calabi-Yau manifolds exhibit N=1 or N=2 space-time supersymmetry depending whether one is considering heterotic[23)−25)] or type II superstrings[26)]. In the world-sheet language these two classes of theories correspond to conformal systems having respectively (1,0) or (1,1) local world-sheet supersymmetry and an "internal" (2,2) superconformal field theory. The existence of space-time supersymmetry is related to the U(1) charge of the internal N=2 superconformal algebra obeying a quantization condition. The basic fact that (2,2) c=9 superconformal systems can be used as "vacua" of heterotic and type II theories corresponds, in the target-space language, to the fact[7)] that a three-dimensional complex manifold with vanishing first Chern class satisfies the gravitational equations of motions where of ten-dimensional N=1 and N=2 supergravity when string corrections are included. In the sequal I will mainly use a target-space language but a corresponding world-sheet language can equally well be used.

An important object of the underlying (2,2) superconformal field theory is the so-called Zamolodchikov metric[6,7] of the moduli fields

$$< V_{\phi_A}(1) V_{\phi_B}(0) > = G_{AB}(\phi) \qquad I-1$$

which is just the (two-point) correlation for truely marginal deformations (moduli) of the superconformal field theory. The couplings ϕ_A play the role, in target space, of background scalar fields and the metric $G_{AB}(\phi)$ is just the normalization of the scalar kinetic term

$$-G_{AB}(\phi)\partial_\mu \phi^A \partial_\mu \phi^B \qquad \mu = 1 \ldots D \qquad I-2$$

describing the low-derivative interactions of these massless excitations in space-time.

From a space-time point of view the metric G_{AB} describes a non-linear σ-model and it defines what is called the geometry of the moduli space, since motion in the coupling constant space "ϕ_A", connecting different superconformal field theories, is specified by the properties of G_{AB}.

In Calabi-Yau compactifications[3),7)] there are two types of moduli, corresponding to the two independent Hodge numbers of a given Calabi-Yau space, $h_{(1,1)}$ and $h_{(2,1)}$. These numbers are just the dimension of the vector spaces $H^{(1,1)}$ and $H^{(2,1)}$ corresponding to the cohomology classes of (1,1) and (2,1) closed forms. Therefore they count the independent harmonic forms existing on the manifold. In the Kaluza-Klein compactifications the existence of these harmonic forms correspond to the existence of massless fields. In particular scalar fields associated to these massless excitations have undetermined vacuum expectation values, since the latter play the role of "deformations" of the Kahler class ((1,1) forms) or of the complex structure ((2,1) forms) and are therefore parameters which are not fixed by the "vacuum condition", i.e. by the 10–dimensional gravitational equations of motion.

In four dimensional language the scalar fields which correspond to "moduli", i.e. to deformations, are said to correspond to "flat" directions of the scalar potential.

This means that

$$\frac{\partial V}{\partial \phi} = 0 \qquad I-3$$

has a solution for any V.E.V. of the moduli fields. Eq. I-3 is a simple consequence of the meaning of moduli as truely marginal deformations of the superconformal field theory since eq. I-3 is nothing but the vanishing of the β function (conformal-anomaly) for the coupling ϕ.

II. Compactification of 10 dimensional N=1 Chiral supergravity on Calabi-Yau manifolds

The massless excitations of critical (10-dimensional) superstrings can be described by 10-dimensional supergravity multiplets. To describe these theories and their compactification it is sufficient to consider bosonic degrees of freedom since the fermionic ones and their couplings are related by space-time supersymmetry.

Let us consider first N=1 D=10 chiral supergravity with Yang-Mills gauge invariance SO(32) or $E_8 \times E_8$. This is the point field limit of 10D heterotic superstrings.

The basic 10-dimensional fields are:

$$\begin{array}{ll} g_{\hat\mu\hat\nu} \text{ (metric)} & \\ B_{\hat\mu\hat\nu} \text{ (antisymmetric tensor)} & A^A_{\hat\mu} \text{ (Yang-Mills potential)} \quad \hat\mu(1\ldots 10) \\ \phi \text{ (dilaton)} & \end{array} \qquad II-1$$

Let us split $\hat\mu = \mu, i, \bar\imath$ with $(\mu = 1, \ldots 4 \; i, \bar\imath = 1, \ldots 3)$

The components of the bosonic fields which correspond to (scalar) massless modes are:

$$g_{i\bar\jmath}, g_{ij}, B_{i\bar\jmath}, \phi, A^I_{ai}, A^I_{a\bar\imath}, b_{\mu\nu} \qquad II-2$$

where a $\in$ SU(3) $\subset$ G and I $\in$ G′ $\subset$ G with G′ commuting with SU(3). For $E_8 \times E_8$ $G' = E_6 \times E_8$.

Tensors with two mixed indices are associated to (1,1) forms since:

$$T_{i\bar\jmath} \;\rightarrow T_{i\bar\jmath} dz^i \wedge d\bar z^{\bar\jmath} \qquad II-3$$

while tensors with two holomorphic indices are associated to (2,1) forms since:

$$\Omega^{\bar\jmath}_{\ell m} \; T_{\bar\imath\bar\jmath} dz^\ell \wedge dz^m \wedge d\bar z^{\bar\imath} \qquad II-4$$

($\Omega^{\bar\jmath}_{\ell m} = \Omega_{\ell m j} g^{j\bar\jmath}$ and $\Omega_{\ell m j}$ is the covariantly constant holomorphic (3,0) form).

Thanks to the identification of the SU(3) holonomy with SU(3) gauge group the A potential which is a one form can be associated to a (1,1) and (2,1) form by contraction with the internal vierbein e^a_i.

$$\begin{array}{ll} A^I_{ai} e^a_j & (2,1) \text{ forms} \\ A^I_{ai} e^a_{\bar\jmath} & (1,1) \text{ forms} \end{array} \qquad II-5$$

This identification also implies:

$$e^a \;=\; \sum_{\alpha=1}^{h_{(1,1)}} C_\alpha A^{\alpha a} \;+\; \sum_{\alpha=1}^{h_{(2,1)}} C'_\alpha \widetilde{A}^{\alpha a} \qquad II-6$$

The physical scalar fields $C^I_\alpha(\alpha = 1, \dots h_{1,1}, I \in 27$ of $E_6)$ $B^I_\alpha(\alpha = 1, \dots h_{(2,1)}, I \in \overline{27}$ of $E_6)$ come from the expansion of the one-form Yang-Mills gauge potential

$$A^I_a = C^I_\alpha(x) A^\alpha_{a\bar{i}}(z,\bar{z}) d\bar{z}^{\bar{i}} + (B_{I\alpha}(x))^* \widetilde{A}^\alpha_{ai}(z,\bar{z}) dz^i \qquad II-7$$

The (1,1) and (2,1) x-dependent moduli correspond to the fluctuation of $\delta g_{i\bar{j}}, \delta B_{i\bar{j}}$ and δg_{ij}. They are defined as follows:[4)]

$$i(\delta g_{i\bar{j}} + i\delta B_{i\bar{j}}) dz^i d\bar{z}^{\bar{j}} = \sum_{\alpha=1}^{h_{(1,1)}} t^\alpha(x) V_\alpha \qquad II-8$$

where V_α is a basis in $H^{(1,1)}$.

$$\varphi_\alpha = \varphi_{\alpha ij\bar{\ell}} dz^i \wedge dz^j \wedge d\bar{z}^{\bar{i}} \qquad \alpha = 1 \dots h_{(2,1)} \qquad II-9$$

with

$$\varphi_{\alpha ij\bar{\ell}} = \frac{\delta g_{\bar{\ell}\bar{p}}}{\delta\psi^\alpha} \Omega^{\bar{p}}{}_{ij} \qquad II-10$$

as a consequence of eq. II-4 with $\delta g_{\bar{i}\bar{j}} \to T_{\bar{i}\bar{j}}$. Here ψ^α are local coordinates on the moduli space of deformations of comples structure.

The structure of the moduli space is related to the homology of the manifold. The complex structure of the manifold can be specified as giving the periods[4),5)] of the holomorphic three form Ω over a canonical homology basis, a^A, b_A (three cycles) $A = 1, \dots b_3/2 = 0, 1, \dots, h_{(2,1)} (b_3 = 2(h_{(2,1)} + 1))$, such that $a^A \cap b_B = \delta^A_B$, $a^A \cap a^B = b_A \cap b_B = 0$. If we call α_A, β^A the dual cohomology basis in $H^3(Z)$ (defined up to a symplectic transformation $Sp(b_3; Z)$:

$$\int_{a^A} \alpha_B = \delta^A_B \quad , \quad \int_{b_B} \beta^A = -\delta^A_B \qquad II-11$$

it follows that:

$$\int_{C_3} \alpha_A \wedge \beta^B = -\int_{C_3} \beta^B \wedge \alpha_A = \delta^B_A, \int_{C_3} \alpha_A \wedge \alpha_B = \int_{C_3} \beta^A \wedge \beta^B = 0 \qquad II-12$$

The holomorphic three-form Ω can be expanded in the (moduli independent) α, β basis:[4),5)]

$$\Omega = Z^A \alpha_A + i F_A \beta^A \qquad II-13$$

The coefficients of Ω are called the periods of Ω and are given, in virtue of eq. II-12 by:

$$\begin{aligned} Z^A &= \int_{a^A} \Omega, = \int_{C_3} \Omega \wedge \beta^A \\ -iF_A &= \int_{b_A} \Omega = \int_{C_3} \Omega \wedge \alpha_A \end{aligned} \qquad II-14$$

In virtue of the fact that Ω is defined up to a rescaling and that $F_A = F_A(Z)$ it follows that Z^A may be regarded as homogeneous coordinates for the moduli space. In a general coordinate basis[13),14)] $\psi, Z^A = Z^A(\psi)$ and $F_A(\psi) = \frac{\partial}{\partial Z^A(\psi)} F(Z(\psi))$ where F is homogeneous of degree 2 in the Z's.

Let us introduce the quantity:

$$\frac{\partial\Omega}{\partial\psi_{\alpha_1}\ldots\partial\psi_{\alpha_n}} = \frac{\partial Z^A}{\partial\psi_{\alpha_1}\ldots\partial\psi_{\alpha_n}}\,\alpha_A + i\,\frac{\partial F_A}{\partial\psi_{\alpha_1}\ldots\partial\psi_{\alpha_n}}\,\beta^A \qquad II-15$$

Since $F_A = \frac{\partial}{\partial Z^A} F$ we have

$$\frac{\partial}{\partial\psi_{\alpha_1}} F_A = \frac{\partial Z^B}{\partial\psi_{\alpha_1}} F_{AB}$$

$$\frac{\partial}{\partial\psi_{\alpha_1}}\frac{\partial}{\partial\psi_{\alpha_2}} F_A = \frac{\partial^2 Z^B}{\partial\psi_{\alpha_1}\partial\psi_{\alpha_2}} F_{AB} + \frac{\partial Z^B}{\partial\psi_{\alpha_1}}\frac{\partial Z^C}{\partial\psi_{\alpha_2}} F_{ABC}$$

$$\begin{aligned}\frac{\partial}{\partial\psi_{\alpha_1}}\frac{\partial}{\partial\psi_{\alpha_2}}\frac{\partial}{\partial\psi_{\alpha_3}} F_A = {} & \frac{\partial^3 Z^B}{\partial\psi_{\alpha_1}\partial\psi_{\alpha_2}\partial\psi_{\alpha_3}} F_{AB} \\ & + 3\,\frac{\partial^2 Z^B}{\partial\psi_{\alpha_1}\partial\psi_{\alpha_2}}\frac{\partial Z^C}{\partial\psi_{\alpha_3}} F_{ABC} \\ & + \frac{\partial Z^B}{\partial\psi_{\alpha_1}}\frac{\partial Z^C}{\partial\psi_{\alpha_2}}\frac{\partial Z^D}{\partial\psi_{\alpha_3}} F_{ABCD}\end{aligned} \qquad II-16$$

it follows that:

$$\int \Omega(\psi)\wedge\Omega(\psi) = \int \Omega(\psi)\wedge\frac{\partial}{\partial\psi_{\alpha_1}}\Omega(\psi) = \int\Omega(\psi)\wedge\frac{\partial^2}{\partial\psi_{\alpha_1}\partial\psi_{\alpha_2}}\Omega(\psi) = 0 \qquad II-17$$

as a consequence of the fact that $\int\Omega\wedge\frac{\partial\Omega}{\partial Z^A} = \int\Omega\wedge\frac{\partial^2\Omega}{\partial Z^A\partial Z^B} = 0$ and that $Z^A F_{ABC} = 0$.

$$\begin{aligned} W_{\alpha_1\alpha_2\alpha_3} &= i\int\Omega\wedge\frac{\partial^3\Omega}{\partial\psi_{\alpha_1}\partial\psi_{\alpha_2}\partial\psi_{\alpha_3}} \\ &= \frac{\partial Z^A}{\partial\psi_{\alpha_1}}\frac{\partial Z^B}{\partial\psi_{\alpha_2}}\frac{\partial Z^C}{\partial\psi_{\alpha_3}} F_{ABC} \\ &= i\,\frac{\partial Z^A}{\partial\psi_{\alpha_1}}\frac{\partial Z^B}{\partial\psi_{\alpha_2}}\frac{\partial Z^C}{\partial\psi_{\alpha_3}}\int\Omega\wedge\frac{\partial^3\Omega}{\partial Z^A\partial Z^B\partial Z^C}\end{aligned} \qquad II-18$$

Since $\frac{\partial\Omega}{\partial\psi_{\alpha_1}} = \frac{\partial Z^A}{\partial\psi_{\alpha_1}}\frac{\partial\Omega}{\partial Z^A}$ it also follows that:

$$\begin{aligned} K_{\psi_{\alpha_1}\overline{\psi}_{\alpha_2}} &= -\frac{\partial}{\partial\psi_{\alpha_1}}\frac{\partial}{\partial\overline{\psi}_{\alpha_2}}\,\ell n(\Omega,\overline{\Omega}) \\ &= -\frac{\partial Z^A}{\partial\psi_{\alpha_1}}\frac{\partial\overline{Z}^{\overline{B}}}{\partial\overline{\psi}_{\alpha_2}}\,\partial_A\partial_{\overline{B}}\,\ell n(\Omega,\overline{\Omega}) = -\frac{1}{(\Omega,\overline{\Omega})}(\varphi_{\alpha_1}\cdot\overline{\varphi}_{\overline{\alpha}_2})\end{aligned} \qquad II-19$$

$$\varphi_\alpha = \frac{\partial}{\partial \psi_\alpha}\,\Omega \;-\; \frac{1}{(\Omega,\overline{\Omega})}\left(\frac{\partial \Omega}{\partial \psi_\alpha}\,,\,\overline{\Omega}\right)\Omega \qquad II-20$$

$$= \nabla_\alpha \Omega \;=\; \frac{\partial Z^A}{\partial \psi_\alpha}\,\varphi_A$$

so that:

$$\int \varphi_\alpha \wedge \overline{\Omega} \;=\; \int \varphi_\alpha \wedge \Omega = 0 \qquad II-21$$

(Since $Z^A \varphi_A = 0$ it also follows that $\varphi_\alpha = -Z^A\, \frac{\partial \varphi_A}{\partial \psi_\alpha}$).

In virtue of eq. II-19 it also follows:

$$\int \frac{\partial \Omega}{\partial \psi_{\alpha_1}} \wedge \overline{\Omega} = \frac{\partial Z^A}{\partial \psi_{\alpha_1}} \int \frac{\partial \Omega}{\partial Z^A} \wedge \overline{\Omega} = \frac{\partial Z^A}{\partial \psi_{\alpha_1}}\,(-4i(N\overline{Z})_A)$$

$$N_{AB} = \frac{1}{4}(F_{AB} + \overline{F}_{AB}) \qquad II-22$$

$$\int \frac{\partial \Omega}{\partial \psi_{\alpha_1}} \wedge \frac{\partial \overline{\Omega}}{\partial \psi_{\alpha_2}} = \frac{\partial Z^A}{\partial \psi_{\alpha_1}}\,\frac{\partial \overline{Z}^B}{\partial \psi_{\alpha_2}}\,(-4iN_{AB})$$

Since $\frac{\partial^2 \Omega}{\partial \psi_{\alpha_1} \partial \psi_{\alpha_2}} \;=\; \frac{\partial^2 Z^A}{\partial \psi_{\alpha_1} \partial \psi_{\alpha_2}}\,\Omega_A \;+\; \frac{\partial Z^A}{\partial \psi_{\alpha_1}}\,\frac{\partial Z^B}{\partial \psi_{\alpha_2}}\,\Omega_{AB}$

it follows that:

$$\left(\frac{\partial \Omega}{\partial \psi_{\alpha_1} \partial \psi_{\alpha_2}}\,,\,\overline{\Omega}\right) \;=\; \frac{\partial Z^A}{\partial \psi_{\alpha_1}}\,\frac{\partial Z^B}{\partial \psi_{\alpha_2}}\,(\Omega_{AB},\,\overline{\Omega}) \;+\; \frac{\partial^2 Z^A}{\partial \psi_{\alpha_1} \partial \psi_{\alpha_2}}\,(\Omega_A,\,\overline{\Omega}) \qquad II-23$$

where $(\Omega_{AB},\,\overline{\Omega}) \;=\; -4iF_{ABC}\overline{Z}^C$.

We note that Ω and φ_α are holomorphic sections of a line bundle[12)] with gauge potential

$$\partial_i K \qquad II-24$$

such that under:

$$K \rightarrow\; K - f - \overline{f}$$

$$\partial_i K \rightarrow\; \partial_i K - \partial_i f \qquad II-25$$

It follows that:

$$\Omega \rightarrow \Omega e^f,\; \varphi_\alpha \rightarrow \varphi_\alpha e^f,\; W_{\alpha_1\alpha_2\alpha_3} \rightarrow W_{\alpha_1\alpha_2\alpha_3} e^{2f} \qquad II-26$$

The latter property being due to the identity:

$$\int \Omega \wedge \frac{\partial^3 \Omega}{\partial \psi_{\alpha_1} \partial \psi_{\alpha_2} \partial \psi_{\alpha_3}} \;=\; \int \Omega \wedge \nabla_{\alpha_1} \nabla_{\alpha_2} \nabla_{\alpha_3} \Omega \qquad II-27$$

On the other hand the Kahler metric is invariant as:

$$K_{\alpha_1\overline{\alpha}_2} = -\frac{1}{(\Omega,\,\overline{\Omega})}\int \nabla_{\alpha_1}\Omega \wedge \nabla_{\overline{\alpha}_2}\overline{\Omega} \qquad II-28$$

It follows from string arguments that the moduli space for complex structure deformations as computed from low-energy Lagrangian considerations actually coincide[21,10] with the "quantum" moduli space of the corresponding abstract conformal field theory as defined by eq. I-1. The reason being that α' corrections correspond in the σ-model language to corrections in α'/T where T is a generic (1,1) modulus. Since the moduli space of (1,1) and (2,1) forms is a product space, as follows from string[10], geometrical[4] and supersymmetry[3] considerations, it follows that the manifold of complex structure deformations is independent of the σ-model coupling constant and therefore is exact both at the perturbative and non-perturbative level. This is not the case for (1,1) forms i.e. for the deformations of the Kahler class. In fact, due to the familiar properties of the antisymmetric tensor $B_{\mu\nu}$ there is in the low-energy effective Lagrangian an unbroken Peccei-Quinn[27),3] symmetry:

$$B \rightarrow B + \Lambda \qquad II-29$$

where Λ is a x independent two form. This amounts to a imaginary constant shift:

$$t^{\alpha}(x) \rightarrow t^{\alpha}(x) + iC^{\alpha} \qquad II-30$$

on the background moduli parameters. This invariance implies that the Kahler potential is a function of $(t+\bar{t})^{\alpha}$ and is not renormalized in σ-model perturbation theory. Actually, due to the fact that the Kahler geometry is "special" (see next section) for Calabi-Yau compactifications this constraint implies[3] that the Kahler potential for (1,1) moduli is of the form:

$$K = -\ell n Y \quad , \quad Y = d_{\alpha\beta\gamma}(t+\bar{t})^{\alpha}(t+\bar{t})^{\beta}(t+\bar{t})^{\gamma} \qquad II-31$$

The $d_{\alpha\beta\gamma}$ coefficients in Calabi-Yau compactifications are given by the low-energy limit (tree level string theory) and coincide with the intersection matrices[2] of (1,1) harmonic forms in $H^{(1,1)}(z)$ (see eq. II-8).

$$d_{\alpha\beta\gamma} = \int V_{\alpha} \wedge V_{\beta} \wedge V_{\gamma}$$

$$Y = \int J \wedge J \wedge J \quad , \quad J = \sum_{\alpha=1}^{h_{(1,1)}} (t+\bar{t})^{\alpha} V_{\alpha} \qquad II-32$$

In string theory non-perturbative effects, i.e. instanton configurations which give a non-trivial map of the world-sheet onto the Calabi-Yau manifolds, modify eq. II-31, since they break the perturbative symmetry given by eq. II-30. Since world-sheet instantons[27),28)] can be generally expressed for large values of the (1,1) moduli (weak coupling regime), therefore eq. II-31 is only an asymptotic formulae for (1,1) moduli.

A common property that the two moduli metrics share and, as we shall see in the next section, is a consequence of (2,2) superconformal invariance or equivalently of the space-time supersymmetry, is that the two metrics:[2),3),4),5)]

$$\frac{1}{V}\int V_\alpha \wedge^* V_\beta \qquad (1,1)\text{ moduli} \qquad II-33$$

$$\frac{1}{\int \Omega\wedge\overline{\Omega}}\int \varphi_\alpha \wedge \overline{\varphi}_\beta \qquad (2,1)\text{ moduli} \qquad II-34$$

are not only Kahler but special Kahler[3),5)]. This means that the Kahler curvature satisfies the additional constraint:[9),10),12),13)]

$$R_{\alpha\overline{\beta}\gamma\overline{\delta}} = g_{\alpha\overline{\beta}}g_{\gamma\overline{\delta}} + g_{\alpha\overline{\delta}}\, g_{\gamma\overline{\beta}} - e^{2K}\, W_{\alpha\gamma\varepsilon}\, g^{\varepsilon\overline{\varepsilon}}\, \overline{W}_{\overline{\beta}\,\overline{\delta}\overline{\varepsilon}} \qquad II-35$$

where $W_{\alpha\beta\gamma}$ is a totally symmetric holomorphic tensor:

$$\partial_{\overline{i}}W_{\alpha\beta\gamma} = 0 \qquad II-36$$

which is actually a section of the line bundle with (1,0) form potential given by eq. II-24. Consequently:

$$W_{\alpha\beta\gamma} \rightarrow e^{2f}W_{\alpha\beta\gamma} \qquad II-37$$

under Kahler transformations.

Given $W_{\alpha\beta\gamma}$ one can construct a covariantly holomorphic tensor:

$$C_{\alpha\beta\gamma} = e^{K}W_{\alpha\beta\gamma} \quad , \quad \overline{\nabla}_i C_{\alpha\beta\gamma} = 0 \qquad II-38$$

which under U(1) gauge transformations transforms as:

$$C_{\alpha\beta\gamma} \rightarrow e^{f-\overline{f}}C_{\alpha\beta\gamma} \qquad II-39$$

The gauge potential

$$A = i(K_i dz^i - K_{\overline{i}} d\overline{z}^{\overline{i}}) \qquad II-40$$

has the property that $dA = J$ where J is the Kahler two form.

For the moduli space of (2,1) forms $W_{\alpha\beta\gamma}$ is defined by eq. II-18 and for the classical moduli space of (1,1) forms $C_{\alpha\beta\gamma}$ is just given by eq. II-32.

It is actually a consequence of eq. II-35 that the "special" Kahler potential can be written as in eq. II-31 with:

$$Y = \overline{Z}^A(\psi)F_A(Z(\psi)) + Z^A(\psi)\,\overline{F}_A(Z(\psi)) \quad , \quad F_A = \frac{\partial}{\partial Z^A}F \qquad II-41$$

Note that under a U(1) transformation:

$$Z^A \rightarrow e^f Z^A \quad , \quad F_A \rightarrow e^f F_A \qquad II-42$$

so that $K \rightarrow K - f - \overline{f}$, as it should be.

In heterotic superstrings to each modulus there is an associated 27 (for (1,1) moduli) and $\overline{27}$ (for (2,1) moduli) representation of E_6 so that $h_{(2,1)} - h_{(1,1)}$ is the net chirality of the manifold. The general form of Yukawa couplings for family and antifamily can be written in a unified way i.e.:

$$W^{1,2}_{\alpha_1\alpha_2\alpha_3} = \frac{\partial Z^A}{\partial\psi^{1,2}_{\alpha_1}} \frac{\partial Z^B}{\partial\psi^{1,2}_{\alpha_2}} \frac{\partial Z^C}{\partial\psi^{1,2}_{\alpha_3}} \partial_A\partial_B\partial_C \, F^{(1,2)}\left(Z(\psi)\right) \qquad II-43$$

where F^1, F^2 are two holomorphic functions which describe the special geometry of (1,1) and (2,1) moduli respectively.

In σ-model perturbation theory, because of eq. II-29, eq. II-31:

$$F^1(Z) = i\, d_{ABC} \, \frac{Z^A Z^B Z^C}{Z^0} \qquad II-44$$

but this formula is modified by non-perturbative (instantons) world-sheet effects.

By the use of eq. II-44, eq. II-43 gives the Yukawa coupling for 27 families in terms of triple intersection of (1,1) forms, as found in ref.2. For $\overline{27}$ families eq. II-43, thank to eq. II-18 reduces to the formula of Candelas et al.[2)].

Let us now consider the trilinear superpotential term, involving 27 scalars (called $C^{\alpha I}$) and $\overline{27}$ scalar (called B^{α}_{I}). The superpotential is:

$$\begin{aligned} W(M,N,C,B) = {} & W^1_{\alpha_1\alpha_2\alpha_3}(M) C^{\alpha_1 I} C^{\alpha_2 J} C^{\alpha_3 K} \, d_{IJK} \\ & + W^2_{\alpha_1\alpha_2\alpha_3}(N) B^{\alpha_1 I} B^{\alpha_2 J} B^{\alpha_3 K} \, d^{IJK} \end{aligned} \qquad II-45$$

Here M, N are the (1,1) and (2,1) moduli respectively (previously called generically ψ_α).

Since under a Kahler transformation the norm:

$$e^G = W W^* e^K \qquad II-46$$

must be invariant it follows that:

$$W \rightarrow W e^{f_1(M)+f_2(N)} \qquad II-47$$

since $K(M,N) = K(M) + K(N)$, which is a consequence of the factorization of the moduli space of (1,1) and (2,1) moduli[3)]. It then follows that:

$$C \rightarrow e^{f_2 - f_1/3} C \quad , \quad B \rightarrow e^{f_1 - f_2/3} B \qquad II-48$$

as a consequence of eq. II-44, II-42 and II-37.

Eq. II-24 gives the relastion between the charges of the U(1) connection of the moduli space and the $U(1)_L$ and $U(1)_R$ charges of the corresponding (2,2) superconformal field theory. Since families and antifamilies correspond to chiral and antichiral primary fields[29] it is obvious that the $U(1)_L$ and $U(1)_R$ charges are proportional to the sum and difference of the U(1) charges defined by the Kahler gauge invariance of the moduli space. The Kahler gauge invariance also fixes the lowest order (quadratic) terms in the family Kahler potential to be:[10]

$$\begin{aligned} K(M,N,C,B) &= e^{K_2-K_1/3}\, G^1_{\alpha_1\overline{\alpha}_2}\,(M)\, C^{\alpha_1 I}\,\overline{C^{\alpha_2 I}} \\ &\quad + e^{K_1-K_2/3}\, G^2_{\alpha_1\overline{\alpha}_2}\,(N)\, B^{\alpha_1}_I\,\overline{B^{\alpha_2}_I} \end{aligned} \qquad II-49$$

Eq. II-49 gives the normalization of the family fields in terms of the moduli metric to be:

$$\frac{\partial K}{\partial C^{\alpha_1 I}\partial\overline{C^{\alpha_2 J}}} = \delta^J_I\, e^{K_2-K_1/3}\, G^1_{\alpha_1\overline{\alpha}_2}\,(M) \qquad II-50$$

$$\frac{\partial K}{\partial B^{\alpha_1}_I\partial\overline{B^{\alpha_2}_J}} = \delta^I_J\, e^{K_1-K_2/3}\, G^2_{\alpha_1\overline{\alpha}_2}\,(N) \qquad II-51$$

This result was obtained using conformal field theory arguments in ref. 10. The G functions defined by eq. II-46 can be exactly computed to all orders in the charged matter field using the 10-D Lagrangian. However one expects superstring corrections[28] for large values of the charged fields. For (2,2) compactifications higher order terms in the charged fields occur in the W superpotential only through non-perturbative effects. Then the trilinear tree-level superpotential given by eq. II-45 produces at least one continuous R-symmetry since under $C \to e^{i\alpha}C, B \to e^{i\alpha}B, W \to e^{3i\alpha}W$ and the G function is invariant.

III. Type II superstrings on Calabi-Yau manifolds

The special geometry of Calabi-Yau moduli space can be explained from the relation between heterotic and type II superstrings[7),3)]. Indeed Calabi-Yau manifolds or equivalently (2,2), c = 9 superconformal field theories can be used as "vacua" of type II theories[2),6)]. The essential point is that in type II theories the number of target space supersymmetries is doubled because of the left-right symmetry of the theory. As a consequence the metric of the moduli in type II theories should be compatible with the stronger requirement of N = 2 space-time supersymmetry. Remarkably the N = 2 assignment of moduli fields led to the discovery of a new class of quaternionic spaces[3,16] called "dual quaternionic" which have the property that the geometry is encoded in a single holomorphic function of the associated special Kahler space.

In type II supergravity, which is the low energy limit of type II superstrings, the moduli fields are associated with the two N = 2 supermultiplets which contain scalar fields, the vector multiplets and the hypermultiplets. The (1,1) moduli are in vector multiplets and the (2,1) moduli are in hypermultiplets in type II A theories. The reverse is true for type II B theories. In addition the quaternionic manifold contains always an additional (quaternionic) coordinate, associated with the dilaton and antisymmetric tensor[7),3)]. Indeed in any type II theory compactified on a (2,2) superconformal field theory, in addition to the $2h_{1,1} + 2h_{2,1} + 2$ degrees of freedom corresponding to the moduli, dilaton and antisymmetric tensor already present in heterotic theories, there are $2h_1 + 2$ additional scalars and $h_2 + 1$ vectors coming from Ramond-Ramond states ($h_1 = h_{1,1}, h_2 = h_{2,1}$ for type II A and $h_1 = h_{2,1}, h_2 = h_{1,1}$ for type II B theory). These additional multiplets enlarge the original scalar manifold[3)]:

$$\frac{SU(1,1)}{U(1)} \times \mathcal{M}(h_{2,1}) \times \mathcal{M}(h_{1,1}) \qquad III-1$$

to the product of a quaternionic manifold and a Kahler manifold:

$$\begin{aligned} &\mathcal{Q}(h_{2,1}+1) \times \mathcal{M}(h_{1,1}) \quad \text{type II A} \\ &\mathcal{Q}(h_{1,1}+1) \times \mathcal{M}(h_{2,1}) \quad \text{type II B} \end{aligned} \qquad III-2$$

where of course $\mathcal{Q}(h+1) \supset \frac{SU(1,1)}{U(1)} \times \mathcal{M}(h)$.

The $\mathcal{Q}$ manifold is the dual quaternionic manifold of the original special Kahler manifold $\mathcal{M}(h)$. Note that because of eq. III-2 the factorization of the Kahlerian moduli space:

$$\mathcal{M}(h_{1,1}) \times \mathcal{M}(h_{2,1})$$

is merely a consequence of the consistency of type II superstrings with (2,2) vacua, giving rise to N = 2 target space supersymmetry.

The map between special Kahler manifolds of complex dimension h and quaternionic manifolds of (quaternionic) dimension h + 1 can be obtaines through three-dimensional duality[3),16)].

The procedure follows the dimensional reduction of D = 4, N = 2 supergravity coupled to h (abelian) vector multiplets to three dimensions. In this way one obtains 3h + 2 scalar degrees of freedom coming from the h vector multiplets, the graviton and the graviphoton. h + 2 additional scalars can be obtained by three dimensional duality from the h + 2 three dimensional vectors coming from h vectors, the graviphoton and the graviton. It is then possible to show that the 3-dimensional Lagrangian for the 4(h + 1) scalar degrees of freedom describes a quaternionic geometry in which the quaternionic metric is entirely determined by the same holomorphic function which determine the special Kahler geometry.

In type II superstring compactifications the 2h coordinates of the original special Kahler geometry correspond to the moduli fields. 2(h + 1) additional scalars come as Ramond-Ramond states and the two extra scalars are the dilaton and antisymmetric tensor (model independent axion).

It is not surprising that fixing the R-R scalars denoted by a complex field $C_i(i = 1, \ldots, h+1)$, the $\mathcal{Q}$ manifold reduces to:

$$\mathcal{Q}(C_i = 0) \rightarrow \frac{SU(1,1)}{U(1)} \times \mathcal{M}(h) \qquad III-3$$

On the other hand if one freezes the moduli fields $Z_i(i = 1, \ldots, h)$ to some value Z_i^P then[16]:

$$\mathcal{Q}(Z_i = Z_i^P) \rightarrow \frac{SU(1, n+2)}{U(1) \times SU(n+2)} \qquad III-4$$

Another property of the $\mathcal{Q}$ manifolds is that they have at least 2h + 4 isometries, which equal the number of the extra coordinates other than the moduli parameters.

For special Kahler manifolds which are symmetric spaces the corresponding quaternionic manifold is also symmetric[18].

In particular there are only two infinite sequences of dual symmetric quaternionic manifolds[14]:

$$\frac{SO(4, n+2)}{SO(4) \times SO(n+2)} \quad , \quad \frac{SU(2, n+1)}{U(1) \times SU(n+1) \times SU(2)} \qquad III-5$$

related to the infinite sequences of special symmetric Kahler spaces[9]:

$$\frac{SU(1,1)}{U(1)} \times \frac{SO(2,n)}{SO(2) \times SO(n)} \quad , \quad \frac{SU(1,n)}{U(1) \times SU(n)} \qquad III-6$$

The exceptional Kahler manifolds:

$$\frac{SU(1,1)}{U(1)} \text{ (with curvature } -\frac{2}{3}\text{), and } \frac{SU(3,3)}{SU(3) \times SU(3) \times U(1)} \qquad III-7$$

correspond to the "exceptional" dual quaternionic spaces:

$$\frac{G_{2(2)}}{SO(4)} \text{ and } \frac{E_{6(2)}}{SU(6) \times SU(2)} \qquad III-8$$

The latter space corresponds to the quaternionic manifold for the untwisted moduli of the Z_3 orbifold ($h_{\mathbf{untwisted}} = 9$). For the Z_3/Z_3 orbifold ($h_{\mathbf{untwisted}} = 3$) the dual quaternionic space is $\frac{SO(4,4)}{SO(4) \times SO(4)}$. In ref. 18 the isometries of $\mathcal{Q}$ induced by additional isometries of $\mathcal{M}(h)$ have been investigated.

IV. Discrete isometries and mirror manifolds

We have seen in the previous sections that the study of the 10-dimensional supergravity theories compactified on Calabi-Yau manifolds define in general a "classical" moduli space which is generally different from the "quantum" moduli space of superstring theories. The difference between the classical and quantum geometry comes from non-trivial topological properties of the world sheet in which superstrings propagate and their map to target space.

On the other hand non-renormalization theorems[27),28)], and more generally consistency with target space supersymmetry allow us to conclude that the classical and quantum geometry of complex structure deformations ((2,1) moduli) coincide while this is not the case, as already discussed, for the geometry of deformations of the Kahler class[21)] ((1,1) moduli).

In the language of conformal field theory quantum Calabi-Yau manifolds correspond to $c = 9$, (2,2) superconformal field theories with a quantization of the U(1) charge. (1,1) and (2,1) moduli are in correspondence with chiral (or antichiral) primary fields. So the only difference between (1,1) and (2,1) moduli is the sign of the right handed U(1) charge[29)].

Recently it has been shown that to a $c = 9$ (2,2) superconformal field theory one can associate an isomorphic conformal field theory in which the role of the chiral and antichiral fields is reversed[29)–31)]. This is not merely a sign convention in the sense that the corresponding theories are physically distinct. This isomorphism led to the introduction of mirror manifolds in which the role of the (1,1) and (2,1) forms is exchanged.

Let us call C a given C-Y threefold and C' its mirror image. Then what is true is that:

$$\mathcal{M}_C(h_{1,1}) = \mathcal{M}_{C'}(h_{2,1}) \qquad IV-I$$

$$\mathcal{M}_C(h_{2,1}) = \mathcal{M}_{C'}(h_{1,1})$$

In heterotic strings because of the properties of special geometry, this implies, for instance, that the Yukawa coupling for families (antifamilies) in C are the same as the Yukawa couplings for antifamilies (families) in C'. In type II superstrings this implies that the scalar manifold $\mathcal{Q} \times \mathcal{M}$ (see eq. III-2) of the type II A (type II B) theory for C coincide with the same manifold of the type II B (type II A) theory in C'. The profound implication of this mapping of C into C' is that the "quantum geometry" of the moduli space for (1,1) moduli in C is completely determined by the classical geometry for (2,1) moduli in C'[30),31)]. Also all the quantum symmetries of string theories, called duality symmetries, are encoded on the action of these symmetries on the (2,1) moduli.

We would like to make some genaral considerations on the action of these symmetries. Since (2,1) moduli in C' correspond to (1,1) moduli in C it is clear that, in some coordinate frame

$$g_{A\overline{B}} \xrightarrow[t^A \text{ Large}]{} -\partial_A \partial_{\overline{B}}\, \ell n\, d_{ABC}\, (t+\overline{t})^A (t+\overline{t})^B (t+\overline{t})^C \qquad IV-2$$

so the asymptotic metric of the moduli space of complex structure deformations has always at least h + 1 isometries:

$$t^A \to t^A + iC^A \quad , \quad t^A \to \lambda t^A \qquad IV-3$$

in the case of a single modulus t eq. IV-3 enlarges to $SL(2,R)$ and $g_{t\bar{t}}$ becomes the metric of the pseudosphere SU(1,1)/U(1) with curvature - 2/3. In some instances (toroidal compactifications) the asymptotic moduli space differ from the exact moduli space only by global properties depending on the fact that some points may be identified by discrete isometries. The identification of these points is made by a discrete group Γ (the duality group) which may have fixed points leading to a singular manifold called an orbifold[23),32),33)]. For instance in the case of a single toroidal modulus $\Gamma = SL(2,Z)$ and the moduli space is the (target space) orbifold $\frac{SU(1,1)}{U(1)}/SL(2,Z)$. This result only applies to moduli which are untwisted i.e. they are Z^h invariant states of the original six dimensional orbifold T_6/Z^h.

Similar arguments apply to the (4,4) moduli space of superconformal field theory giving rise to N = 4 space-time supersymmetry in type II superstrings. The reason for this result can be explained by space-time supersymmetry arguments[7)]. In theories with N = 4 space-time superssyymetry the scalar field kinetic term:

$$G_{AB}(\phi)\ \partial_\mu \phi^A\ \partial_\mu \phi^B \qquad IV-4$$

is constrained to be a non-linear σ-model on the coset space[34)] $\frac{SO(6,n)}{SO(6)\times SO(n)} \times \frac{SU(1,1)}{U(1)}$ where $n+1$ is the number of scalar fields.

This non-linear σ-model is obtained, for example, by extending the moduli space of the K_3 surface (two-dimensional C-Y manifold) and of the T_2 torus[7)]:

$$\frac{SO(4,20)}{SO(4)\times SO(20)} \times \frac{SO(2,2)}{SO(2)\times SO(2)} \qquad IV-5$$

with the scalar massless states coming from the Ramond-Ramond sector.

The moduli space of the c=9, N=4 space-time supersymmetric string is actually[29)]:

$$\frac{SO(4,20)}{SO(4)\times SO(20)}\Big/\ \mathrm{SO(4,20;Z)} \times \frac{SO(2,2)}{SO(2)\times SO(2)}\Big/\ (\mathrm{SL(2,Z)}\times Z_2)^2 \times P_2 \qquad IV-6$$

For Calabi-Yau threefolds let us consider now their moduli spaces for complex structure deformations. If we consider the homology cycles defined through eq. II-14 the action of duality transformations acts as some group Γ embedded in $Sp(b_3, Z)$ acting on the vector given by the periods of three-forms (eq. II-14):

$$\begin{pmatrix} -iF_A \\ Z^A \end{pmatrix} \qquad IV-7$$

with $b_3 = 2(h_{2,1}+1)$ components. The linear action of $b_3 \times b_3$ matrices on the vector defined by eq. IV-7 is induced by some action on the (complex) moduli coordinates $\psi_\alpha(\alpha = 1, \dots h_{2,1})$ of Γ.

For the untwisted (1,1) moduli of orbifold compactifications the moduli space has continuous isometries[18] which are subgroups of $Sp(b_3; \mathcal{R})$. However the twisted state interactions, which are affected by world sheet instantons, break these continuous isometries to some discrete group Γ[32]. In the simplest example of one toroidal modulus, the continuous subgroup of $Sp(4; \mathcal{R})$ is $SL(2, \mathcal{R})$ and its discrete (quantum) isometry is $SL(2, Z) \subset Sp(4; Z)$. Here the number of periods is $2(h_{1,1}+1) = h_{0,0} + h_{3,3} + h_{1,1} + h_{2,2}$. So in mirror manifolds, the equivalent b_3 is $b_0 + b_2 + b_4 + b_6 = 2(1 + b_2)$. For the moduli space of the Z_3/Z_3 orbifold[33] the space is locally $\left(\frac{SU(1,1)}{U(1)}\right)^3$. The duality group Γ of holomorphic transformations is $(SL(2,Z))^3 \times \mathcal{P}_3$ and is generated by four generators which can be given as $Sp(8, Z)$ matrices. There are three additional (non-holomorphic) transformations which give 3 extra Z_2 factors so the entire group Γ is $(SL(2,Z) \times Z_2)^3 \times \mathcal{P}_3$.

In the case of non-trivial C-Y threefolds the continuous isometries only exist for large values of the moduli, where world sheet instanton effects can be neglected (see eq. IV-2). In the case of h moduli corresponding to $H^{(1,1)}$ or $H^{(2,1)}$ let us call G the group of asymptotic continuous isometries, of untwisted moduli, then the "quantum" duality group Γ is a discrete subgroup of G which acts as an element of $Sp(2(h+1), Z)$ on the periods (Z^A, iF_A). For the untwisted moduli of orbifold compactifications the asymptotic continuous isometries are exact (at least locally) for the untwisted moduli space. However this is no longer the case for non-trivial C-Y spaces. It is also not obvious that Γ is related to G.

A recent example has been provided in ref.31 where the exact moduli space for the (2,1) modulus of the mirror manifold of $\mathcal{P}_4(5)$ (or equivalently for the (1,1) modulus of $\mathcal{P}_4(5)$) has been obtained.

Since $h_{(2,1)} = 1$ the asymptotic isometries are uniquely given by $SL(2,R)$ and its asymptotic Kahler potential in some coordinates, is just:

$$K(t) = -3\ell n(t + \bar{t}) \qquad (t \text{ Large}) \qquad IV-8$$

This result is a particular case of eq. IV-2. For small t values instanton effects modify eq. IV-8. In $\mathcal{P}_4(5)$ it would be extremely difficult to compute the small t corrections to the deformations of the Kahler class. However going to the mirror manifold, the very same corrections are given by studying the periods over the homology cycles for the holomorphic three-form $\Omega(\psi)$ under the deformation of the complex structure, related to the ψ parameter.

In Ref. 31 the periods $Z^A(\psi), F_A(\psi)$ have been computed in terms of hypergeometric functions. The Kahler potential is (up to a Kahler transformation) thus obtained as:

$$K(\psi,\overline{\psi}) = -\log i \int \Omega \wedge \overline{\Omega} \qquad IV-9$$

$$= -\log\left(Z^A \overline{F}_A + \overline{Z}^A F_A\right)$$

and the Kahler metric is:

$$\partial_\psi \partial_{\overline{\psi}} K(\psi,\overline{\psi}) \qquad IV-10$$

The "quantum" symmetry of $\mathcal{M}(h=1)$ is a (infinite dimensional) discrete subgroup of the asymptotic isometry $SL(2,R)$ whose generators are two elements of $Sp(4,Z)$:

$$A = \begin{pmatrix} -9 & -3 & 5 & 3 \\ 0 & 1 & 0 & -1 \\ -20 & -5 & 11 & 5 \\ -15 & 5 & 8 & -4 \end{pmatrix} \quad , \quad T = \begin{pmatrix} 1 & 0 & 0 & 0 \\ 0 & 1 & 0 & 1 \\ 0 & 0 & 1 & 0 \\ 0 & 0 & 0 & 1 \end{pmatrix} \qquad IV-11$$

$A^5 = 1$.
These two generators are the analogues of the two elements:

$$S = \begin{pmatrix} 0 & 0 & 2 & 0 \\ 0 & 0 & 0 & -\frac{2}{3} \\ \frac{1}{2} & 0 & 0 & 0 \\ 0 & -\frac{3}{2} & 0 & 0 \end{pmatrix} \quad , \quad P = \begin{pmatrix} 1 & 0 & 0 & 0 \\ 1 & 1 & 0 & 0 \\ \frac{1}{2} & \frac{3}{2} & 1 & -1 \\ -\frac{3}{2} & -3 & 0 & 1 \end{pmatrix} \qquad IV-12$$

$S^2 = 1$.
of $Sp(4,Z)$ which generate the full $SL(2,Z)$ group.

We stress that, although the duality group Γ of the moduli space of $\mathcal{P}_4(5)$ and $SL(2,Z)$ are both subgroups of $SL(2,R)$ they are completely distinct, reflecting the fact that the corresponding C-Y manifolds are distinct.

From the simple example explained in ref. 31 it is remarkable that the periods $Z^A(\psi), F_A(\psi)$ satisfy a linear (fourth order) differential equation and that the related monodromy group is closely related to Γ. If this property extends to more general cases this result gives some insights into how to solve the problem of determining the moduli space and its "quantum" symmetries for general C-Y manifolds solely by using methods of algebraic geometry.

References

1) W. Boucher, D. Friedan, and A. Kent, Phys. Lett. **172B** (1986) 316; A. Sen, Nucl. Phys. **B278** (1986) 289 and **B284** (1987) 423; L. Dixon, D. Friedan, E. Martinec and S.H. Shenker, Nucl. Phys. **B282** (1987) 13; T. Banks and L. Dixon, Nucl. Phys. **B307** (1988) 93; D. Gepner, Nucl. Phys. **B296** (1988) 757, Phys. Lett. **189B** (1987) 380; W. Lerche, D. Lüst and A.N. Schellekens, Nucl. Phys. **B287** (1987) 477.

2) A. Strominger, Phys. Rev. Lett. **55** (1985) 2547, in *"Unified string theory"* eds. M. Green and D. Gross (World Scientific) 1980; A. Strominger and E. Witten, Comm. Math. Phys. **101** (1985) 341; P. Candelas, Nucl. Phys. **B298** (1988) 458; P. Candelas, A.M. Dale, C.A. Lütken and R. Schimmrigk, Nucl. Phys. **B298** (1988) 493.

3) S. Cecotti, S. Ferrara and L. Girardello, Int. J. Mod. Phys. **Vol. 4** (1989) 2475, Phys. Lett. **B213** (1988) 443.

4) P. Candelas, P. Green and J. Hübschz in *"Strings 88"*, eds. J. Gates, C. Preitschopf and W. Siegel (World Scientific) 1989; P. Candelas, T. Hübschz and R. Schimmrigk, Nucl. Phys. **B329** (1990) 582; T. Hübschz, preprint UTT-38-89 (Texas); P. Candelas, X. de Le Ossa, preprint UTT-07-1990 (Texas).

5) S. Ferrara and A. Strominger in *"Strings 89"* eds. Arnowitt, Bryan, Nanopoulos, Pope (World Scientific) 1989 pg. 245.

6) A.B. Zamolodchikov, JETP Lett. **43** (1986) 731.

7) N. Seiberg, Nucl. Phys. **B303** (1988) 206.

8) S. Ferrara and M. Porrati. Phys. Lett. **B216** (1989) 289; M. Cvetic, J. Louis and B. Ovrut, Phys. Lett. **B206** (1988) 239; M. Cvetic, J. Molera and B. Ovrut, Phys. Rev. **D10** (1989) 140.

9) B. de Wit and A. Van Proeyen, Nucl. Phys. **B245** (1984) 89; B. de Wit, P.G. Lauwers and A. Van Proeyen, Nucl. Phys. **B255** (1985) 569; E. Cremmer, C. Kounnas, A. Van Proeyen, J.P. Derendinger, S. Ferrara, B. de Wit and L. Girardello, Nucl. Phys. **B250** (1985) 385; E. Cremmer, A. Van Proeyen, Class and Quan. Gravity **2** (1985) 445.

10) L. Dixon, V. Kaplunovsky and J. Louis, Nucl. Phys. **B239** (1990) 27.

11) S. Ferrara and A. Van Proeyen, Class. and Quan. Gravity **6** (1989) L 243.

12) A. Strominger, Commun. Math. Phys. **133** (1990) 163.

13) L. Castellani, R. D'Auria and S. Ferrara, Phys. Lett. **B241** (1990) 57, Class. Quantum Grav. **7** (1990) 1767.

14) J. Bagger and E. Witten, Nucl. Phys. **B222** (1983) 1.

15) J. Bagger, A. Galpein, E. Ivanhov and V. Ogievetsky, Nucl. Phys. **B303** (1988) 522.

16) S. Ferrara and S. Sabharwal, Nucl. Phys. **B332** (1990) 317.

17) M. Bodner, A. Cadavid, Class. and Quan. Gravity **7** (1990) 829.

18) B. de Wit and A. Van Proeyen, THU-90/18-CERN th-5865/90.

19) P. Candelas, G. Horowitz, A. Strominger and E. Witten, Nucl. Phys. **B258** (1985) 46.

20) E. Calabi in *"Algebraic geometry and topology"*: a Symposium in honour of S. Lefschetz (Princeton University Press) 1957, pg. 38; S.T. Yau, Proc. Natt. Ac. Scie. USA **74** (1977) 5798.

21) M. Dine, P. Huet and N. Seiberg, Nucl. Phys. **B322** 1989) 301.

22) E.S. Fradkin and A.A. Tseytlin, Nucl. Phys. **B261** (1985) 1, Phys. Lett. **158B** (1985) 316, **160B** (1985) 69; C. Callan, D. Friedan, E. Martinec and M. Perry, Nucl. Phys. **B262** (1985) 593.

23) K. Kikkawa and M. Yamasaki, Phys. Lett. **149B** (1984) 357; N. Sakai and I. Senda, Prog. Theor. Phys. **75** (1986) 692; R. Dijkgraaf, E. Verlinde and H. Verlinde, Comm. Math. Phys. **115** (1988) 649.

24) E. Witten, Phys. Lett. **155B** (1985) 151.

25) D. Gross, J. Harvey, E. Martinec and R. Rohm, Phys. Rev. Lett. **54** (1985) 502, Nucl. Phys. **B256** (1985) 253, Nucl. Phys. **B267** (1985) 75.

26) M.B. Green and J.M. Schwarz, Phys. Lett **122B** (1983) 143; J.H. Schwarz, Nucl. Phys. **B222** (1983) 269; J.H. Schwarz and P.C. West, Phys. Lett. **126B** (1983) 301; D. Howe and P.C. West, Nucl. Phys. **B238** (1984) 181; P. Nair, A. Shapere, A. Strominger and F. Wilczek, Nucl. Phys. **B287** (1987) 402; A. Giveon, E. Rabinovici and G. Veneziano, Nucl. Phys. **B322** (1989) 167; A. Shapere and F. Wilczek, Nucl. Phys. **B320** (1989) 669; E.J.. Chun, J. Mas, J. Lauer and H.P. Nilles, Phys. Lett. **B233** (1989) 141; B. Greene, A. Shapere, C. Vafa, S.T. Yau, Nucl. Phys. **B337** (1990) 1; B. Brandenberger and C. Vafa, Nucl. Phys. **B316** (1989) 301; C. Vafa, Mod. Phys. Lett. **A4**

(1989) 1615; S. Ferrara, D. Lüst, A. Shapere and S. Theisen, Phys. Lett. **B225** (1989) 363; S. Cecotti, S. Ferrara and L. Girardello, Nucl. Phys. **B308** (1988) 436, Phys. Rev. **D40** (1989) 1146; H.P. Duff, Phys. Lett. **173B** (1986) 289, Texas preprint TP-Jan. 1990, Nucl. Phys. **B335** (1990) 610; J.H. Schwarz, Caltech preprint CALT-68-1581 (1989); A. Giveon, N. Malkin and E. Rabinovici, Phys. Lett. **238B** (1990) 57; A. Giveon, M. Porrati, Phys. Lett. **246B** (90) 54; Berkeley preprint LBL-29535, UCB-PTH-90/39.

27) M. Dine, N. Seiberg, X.G. Wen and E. Witten, Nucl. Phys. **B278** (1986) 769; Nucl. Phys. **B289** (1987) 319; M. Dine and N. Seiberg, Nucl. Phys. **B301** (1988) 357.

28) M. Cvetic, Phys. Rev. Lett. **59** (1987) 2989; Phys. Rev. **D37** (1988) 2366; M. Dine, C. Lee, Phys. Lett. **B203** (1988) 371.

29) E.J. Martinec, Phys. Lett. **217B** (1989) 431; C. Vafa and N.P. Warner, Phys. Lett. **218B** (1989) 51; W. Lerche, C. Vafa and N.P. Warner, Nucl. Phys. **B321** (1989) 371; K. Intriligator and C. Vafa, Harvard preprint HUTP-89/A054; C. Vafa, Harvard preprint HUTP-89/A057; S. Cecotti, L. Girardello and A. Pasquinucci, Nucl. Phys. **B328** (1989) 701; preprint SISSA 136/89EP (to appear in Journal of Mod. Phys. A).

30) B.R. Greene and R. Plesser, Harvard preprint HUTP-89/B241, HUTP-89/A043; P. Aspinwall, C. Lutken, G. Ross, Oxford preprint OUTP-90-05 P; P. Aspinwall and C. Lutken, Oxford preprint OUTP-90-12 P; M. Cvetic, J. Molera and B. Ovrut, Phys. Lett. **248B** (1990) 83; L. Castellani, M. Monteiro, P. Fré and F. Gliozzi, Torino preprint DFTT 17/90.

31) P. P. Candelas, X. de La Ossa, P. Green and C. Parkes, Texas preprint UTTG-25-1990.

32) S. Ferrara, D. Lüst and S. Theisen, Phys. Lett. **B233** (1989) 147.

33) D. Schevitz, Nucl. Phys. **B338** (90) 283.

34 J.P. Derendinger and S. Ferrara in *"Supergravity 84"* (World Scientific) eds. P. Fayet, B. de Wit and P. Van Nieuwenhuizen.

Chairman: S. Ferrara

Scientific Secretaries: M. Bodner, U. Vikas

Discussion

– *Zichichi:*

What experiment should we perform to discover the superpartners?

– *Ferrara:*

Much of this was covered in the lecture by Larry Hall. Unfortunately, we do not know much about R symmetry. If R symmetry is conserved, probably the best experiments are those in which you produce charged leptons.

– *Zichichi:*

At what mass?

– *Ferrara:*

These should be seen at machines operating at a few TeV. Thus a machine such as the SSC or the Eloisatron should be able to rule out or discover low energy supersymmetry, whereas a machine like LHC is just at the border. Of course you can always think that supersymmetry is above 100 or 200 TeV, but in this case supersymmetry is no longer connected to the origin of the electroweak scale.

– *Bodner:*

It appears that in our compactification scheme, in order to view the theory as a point field theory we must assume that the radius of the internal manifold (althoug small enough to have escaped detection) must be much larger than the string length. Is this true?

– *Ferrara:*

Yes it is.

– *Bodner:*

You mentioned that in the compactified theory our gauge group is $E_6 \times E_8$. Have we broken the original $E_8 \times E_8$ gauge group by embedding the spin connection into one of the $E_8's$? Also you mentioned that Calabi-Yau compactifications preserve $N = 1$ supersymmetry. Are there any other choices for the internal manifold which will give unbroken $N = 1$ supersymmetry in 4-dimensions?

– *Ferrara:*

In the original Kaluza-Klein compactification the gauge group is generated by the isometries of the internal space. However, for the heterotic case the original gauge group is given by $E_8 \times E_8$. The Calaby-Yau six dimensional internal space has no isometries at all. So that compactification destroys part of the gauge group which at the Planck scale in 4-dimensions is in this case $E_6 \times E_8$. The gauge group is not fixed by the isometries of the theory but by consistancy of the theory, for example by solving the equations of motion by embedding the $SU(3)$ holonomy spin connection into the gauge group E_8.

In response to the second question there are a number of different compactification schemes due to the absence of a unique vacuum. The only criteria they must satisfy is anomaly cancelation. However the Calabi-Yau compactifications are more beautiful since the Yukawa couplings are calculable and we have a simple space time interpretation for the vacuum which we do not have in many of the others.

– *U. Vikas:*

Refering to the previous question, you mentioned that we are assuming that the radius of the internal space is much larger than the string length. So there a minimal radius for the compactified manifold at which we can consider the theory as a point field theory.

– *Ferrara:*

In 10-dimensions the volume of the Calabi-Yau manifold must be much larger than the string for it to look like a point particle. We can view our string theory as a point field theory in four dimensions only if α'/R^2 is very small where α'/R^2 is the coupling constant that describes the internal manifold.

This is the same as saying the point field theory in 2-dimensions is weakly coupled, though the non-perturbative aspects of the field theory cannot be obtained from this compactification scheme.

– *Grossmann:*

How do you get a consistent 4-dimensional theory after compactification from 10 dimensions?

– *Ferrara:*

We want to have a Minkowskian metric in 4-dimensions. We also require that $N = 1$ supersymmetry be unbroken in 4-dimensions by the internal manifold.

As I mentioned earlier we work in a regime where α'/R^2 is very small and thus our effective lagrangian corresponds to a point field theory.

These requirements are met by a manifold which is not riemannian but by what is refered to in the literature as Kahler which means its holonomy group is not $SO(n)$ but $U(n)$. These manifolds also have the property of being Ricci flat which means the holonomy of our manifold is $SU(n)$. This is the definition of Calabi-Yau manifolds which are the solutions of the 10 dimensional gravity coupled to Yang-Mills. This was first discovered by Candelas, Horowitz, Strominger and Witten.

– *Leigh:*

How do you calculate the Yukawa couplings in the theory?

– *Ferrara:*

The Yukawa couplings which appear in the effective Lagrangian are at the Planck scale. As I stated the chiral families correspond to the 27 and $\overline{27}$ of E_6 and the Yukawa couplings to $(27)^3$ and $(\overline{27})^3$.

In string theories these fields have flavour indices which correspond to the number of 27, $\overline{27}$ representation in the theory. This index runs from 1 to $h_{1,1}$ for the 27 and 1 to $h_{2,1}$, for the $\overline{27}$ where $h_{1,1}$ and $h_{2,1}$ are topological numbers of the calabi-Yau manifold called the hodge numbers.

– *Zichichi:*

How do you determine the number of families?

– *Ferrara:*

The difference between $h_{1,1}$ and $h_{2,1}$ should be three.

– *Zichichi:*

How do you get three families?

– *Ferrara:*

The point is that there is no basic principle which tells us that there is only 3 families. For example even in the electroweak theory we could have 100 families. Unfortunately in Calabi-Yau manifolds the number of families is less than infinity, but still too large to explain three families. However, the advantage this has over standard field theory is that there is still a finite number of choices.

Continuing with the original question, the coefficients of the 27 and $\overline{27}$ families associated with the Yukawa couplings are topological formulae which are called the triple intersection of forms.In fact all couplings in the Lagrangian can be written in terms of topological properties. For example the kinetic and potential terms can be written in this way.

– *Leigh:*

In principle you must have renormalization problems.

– *Ferrara:*

Of course. To get from the Planck scale to the electroweak scale you have to use the renormalization group to calculate the Yukawa couplings. As in the grand unified scheme the gauge couplings can be calculated from the grand unified to the electroweak scale. Similarly in principle for strings one should be able to calculate not only the gauge couplings but also the Yukawa couplings.

The various masses of particles can also be explained due to this fact in an appropriate model.

– *Ciuchini:*

When you say that $N = 1$ SUSY must be considered to have chiral fermions. Are you assuming that chirality is a fundamental property of fermions up to the Planck scale?

– *Ferrara:*

We have a similar situation in grand unified theory where we assume that chiral symmetry is a low energy phenomenon. Unfortunately no one knows how to formulate a theory which is non-chiral at Planck energies and chiral at low energies.

– *Lu:*

Why do you believe that GUTS alone is not a good theory.

– *Ferrara:*

It is a matter of taste. However, I feel the main problem with GUTS is the fine tuning problem associated with radiative corrections to the Higgs mass.

– *Hsu:*

After compactification, string models typically have one or more massless dilatons which couple to R, the curvature scalar. Unless these dilatons become massive, or the dimensionless coefficient of the dilaton R coupling is extremely small, these models imply long range forces and time variation of G_N which are ruled out by tests of General Relativity. Are there higher order "stringy" corrections which remove the flat direction for the dilaton field?

– *Ferrara:*

The problem of giving a mass to the dilaton is the same problem of breaking supersymmetry. If you could find a viable supersymmetry breaking scenario I think this problem would be solved.

There is no potential for the dilaton in string theory for all orders of perturbation theory. This implies that the dilaton potential goes as $e^{-\phi}$. The dilaton potential can be related to the inverse square of the gauge couplings which would go as e^{-1/g^2} which is a typical instanton effect. Therefore we need an understanding of the non-perturbative aspects of string theory.

We can relate it to the absence of a unique vacuum as I mentioned before. However there have been some recent attempts to understand quantum gravity non-perturbatively in less than 1 dimension which would lead to an understanding of non-perturbative strings.

– *Brahm:*

What are the possible classical trajectories of the string?

– *Ferrara:*

The classical trajectory is the minimal area of the world sheet.

– *Haas:*

Dr. Ferrara, can we attach any physical significance to compactification or do we have to view it solely as a mathematical technicality. Also, what significance do uncompactified superstring theories have?

– *Ferrara:*

If we look at the point field limit of the string theory then compactification has the same space-time interpretation that the Higgs mechanism has in quantum field theory. There a larger symmetry, for example 10 dimensional Lorentz symmetry, breaks to 4-dimensional Lorentz symmetry.

There is another interpretation if we consider for example the bosonic string: in order to maintain unitarity and covariance after quantization we require the dimensionality to be $D = 26$.

– *Bimonte:*

Is there any dynamical explanation of compactification? Also does string theory set any constraint on the space-time metric?

– *Ferrara:*

There is no intrinsic reason coming from string theory as to why we live in 4-dimensional space-time. Though string theory can be compactified to any number of dimensions less than the critical dimensions, but that would have no relation to the world we live in. For example. the concept of chirality would be completely different, or the triangular anomaly would be a penta or hexagonal anomaly depending on the numbers of space dimensions.

The second question can be answered if we take the 2-dimensional analog of the action

$$S = \int d\sigma d\tau \sqrt{-g} g^{ij} \partial_i X^{\mu \partial_j} X^\nu G_{\mu\nu}(X)$$

by replacing $g_{\mu\nu}$ by $\eta_{\mu\nu}$, a 2-D metric. This gives the string moving in a flat space-time, since $\eta_{\mu\nu}$ is the flat Minkowski metric of the ambient space. Replacing $\eta_{\mu\nu}$ with $g_{\mu\nu}$ would give the string in a curved space-time. Now $X(\sigma, \tau)$ is a 2-D field having a non-polynominal interaction with the background metric $g_{\mu\nu}$. This is what corresponds to a non-linear σ model in field theory. The constraint on $g_{\mu\nu}$ comes from the fact that the conformal anomaly should be cancelled at one loop in the 2-D field theory. The Einstein-Hilbert metric $R_{\mu\nu} - \frac{1}{2} g\mu\nu$ is obtained at tree level. But if we take higher order corrections like going to the second loop we get corrections to this metric. This is expected from string theory as it has infinitely many terms which would contribute to the background field. The terms would not only involve the Ricci tensor but higher powers of the Riemannian tensor.

– *Mangano:*

Does the compactificatoion procedure of 6 of the 10 dimension modify the renormalizability properties of the theory?

– *Ferrara:*

The question of renormalizability is not relevant to string theory as in point particle theories. The S matrix enters as a topological expansion over all genus. This is finite and Calabi-Yau manifolds do not effect this property.

SYMMETRIES OF EXTENDED OBJECTS

M. J. Duff†

Center for Theoretical Physics
Physics Department
Texas A&M University
College Station, Texas 77843-4242

ABSTRACT

In analogy with a previous treatment of strings, it is shown that membrane theories exhibit global non-compact symmetries which have their origin in duality tranformations on the three-dimensional worldvolume which rotate field equations into Bianchi identities. However, in contrast to the string, the worldvolume metric also transforms under duality by a conformal factor. In this way the Cremmer-Julia hidden symmetries of supergravity are seen to be a consequence of supermembrane duality. Moreover, the string duality follows from that of the membrane by simultaneous dimensional reduction. Generalization to higher-dimensional objects is straightforward.

1. INTRODUCTION

The purpose of these lectures is to explore the phenomenon of "duality" in string and membrane theories. The word duality has come to acquire many meanings but here a duality transformation will mean a symmetry that rotates field equations into Bianchi identities on the worldvolume of the extended object. In a previous paper [1] devoted to duality rotations in string theory, we saw that in the case of a bosonic string compactified on an n-torus, these continuous transformations were described by an $SO(n,n)$ symmetry of the equations of motion in the presence of the massless background fields. The discrete subgroup, $SO(n,n;Z)$ which survives as a symmetry of the spectrum, contains the typical $R \leftrightarrow \alpha'/R$ transformations whose fixed points correspond to points of enhanced gauge symmetries and which have led to speculations about a minimum length in string theory [2]. As a preliminary to investigating questions of enhanced symmetry or minimum length in membrane theory, therefore, we first wish to discuss the continuous duality transformations.

In fact, our original motivation for studying duality on the worldvolume of extended objects sprang from the old observation that four-dimensional supergravity theories exhibit global non-compact continuous symmetries corresponding to duality symmetries that rotate spacetime field equations into Bianchi identities [3,4].

---

† Work supported in part by NSF grant PHY-9045132.

These were frequently referred to as "hidden" symmetries since their presence is far from obvious by inspection of the four-dimensional Lagrangian. Indeed, they were originally best understood as a consequence of compactifying a higher-dimensional theory [4]. Nowadays, of course, supergravity theories are regarded as being merely the field theory limit of a superstring or supermembrane, and so it is natural to conjecture that these symmetries have their origin on the worldvolume of the appropriate extended object. Some evidence in favour of these ideas may be found in [5,6]. For example, considerations of bosonic strings on group manifolds led us to expect a global $SO(n,n)$ in the lower dimension, where $n = dim\ G$. In the case of the four-dimensional heterotic string this is replaced by global $SO(6,n)$. This latter symmetry is known to be present in four-dimensional $N = 4$ supergravity coupled to a Yang-Mills supermultiplet. The $6n$ scalars in the Yang-Mills sector are described by a non-linear σ-model given by the coset $SO(6,n)/SO(6) \times SO(n)$. [There are also two scalars in the supergravity sector described by $SU(1,1)/U(1)$. The stringy origin of this coset is described in Section 6.] Cosets of this kind were also encounted by Narain [7] in his torus compactification of the heterotic string but where $n = rank\ G$. By group manifold considerations, we were led to the larger symmetry with $n = dim\ G$. It should be emphasized that these larger symmetries with $n = dim\ G$ are broken by gauge interactions but they nevertheless exactly describe the non-linear σ-model of the scalars. Similarly, one might expect that the global $E_{7(+7)}$ Cremmer-Julia [4] symmetry of $N = 8$ supergravity in four-dimensions has its origin on the worldvolume of the eleven-dimensional supermembrane [8] compactified on a seven-torus [4] or a seven-sphere [9]. Once again, in the case of the seven- sphere the $E_{7(+7)}$ will be broken by $SO(8)$ gauge interactions but the $E_{7(+7)}/SU(8)$ coset still describes the non-linear σ-model of the scalars [10].

The proof of these conjectures (in the case of the string) was supplied by Cecotti, Ferrara and Girardello [11] who first pointed out that the two-dimensional worldsheet origin of these symmetries is quite similar to the way they appear in four-dimensional spacetime i.e. through generalized duality transformations of the kind discussed in detail by Gaillard and Zumino [12]. Thus our task in this paper is to generalize the arguments of Cecotti, Ferrara and Girardello to the three-dimensional worldvolume of the membrane. We shall confine our attention in this paper to torus compactification and focus mainly on the bosonic sector. Moreover, we shall follow the approach described in [1] for treating string duality which lends itself to generalization to membranes and other higher-dimensional objects. Let us therefore first recall string duality.

2. REVIEW OF STRING DUALITY

The n-dimensional string is described by a two-dimensional σ-model with worldsheet coordinates $\xi^i = (\tau, \sigma)$, worldvolume metric $\gamma_{ij}(\xi)$ and target space coordinates $x^\mu(\xi)$, with

$$\mathcal{L} = \frac{1}{2}\sqrt{-\gamma}\gamma^{ij}\partial_i x^\mu \partial_j x^\nu g_{\mu\nu} + \frac{1}{2}\varepsilon^{ij}\partial_i x^\mu \partial_j x^\nu b_{\mu\nu} \tag{2.1}$$

We shall consider the case where x^μ correspond to the compactified coordinates with the background fields $g_{\mu\nu}$ and $b_{\mu\nu} (\mu = 1, \ldots, n)$ being x-independent. These backgrounds will admit the interpretation of scalar fields in spacetime and hence will still depend on the spacetime coordinates X^M. Let us define

$$\mathcal{F}^{i\mu} \equiv \sqrt{-\gamma}\gamma^{ij}\partial_j x^\mu \tag{2.2}$$

$$\tilde{\mathcal{F}}^{i\mu} \equiv \varepsilon^{ij}\partial_j x^\mu \tag{2.3}$$

and

$$\tilde{\mathcal{G}}^i{}_\mu \equiv g_{\mu\nu}\mathcal{F}^{i\nu} + b_{\mu\nu}\tilde{\mathcal{F}}^{i\nu} = \frac{\partial \mathcal{L}}{\partial\partial_i x^\mu} \tag{2.4}$$

Then there is a symmetry between the equations of motion

$$\partial_i \tilde{\mathcal{G}}^{i\mu} = 0 \tag{2.5}$$

and the Bianchi identities

$$\partial_i \tilde{\mathcal{F}}^{i\mu} \equiv 0\ . \tag{2.6}$$

The invariance is summarized by the equations

$$\begin{aligned} \delta\tilde{\mathcal{F}}^{i\mu} &= A^\mu{}_\rho \tilde{\mathcal{F}}^{i\rho} + B^{\mu\beta}\tilde{\mathcal{G}}^i{}_\beta \\ \delta\tilde{\mathcal{G}}^i{}_\alpha &= C_{\alpha\rho}\tilde{\mathcal{F}}^{i\rho} + D_\alpha{}^\beta\tilde{\mathcal{G}}^i{}_\beta \end{aligned} \tag{2.7}$$

where A, B, C and D are constant parameters. We must also establish invariance of the equation of motion obtained by varying with respect to γ_{ij}, namely the vanishing of the energy momentum tensor

$$\begin{aligned} \sqrt{-\gamma}\theta_{ij} &= \frac{2\delta S}{\partial\gamma^{ij}} \\ &= \sqrt{-\gamma}\left[\partial_i x^\mu\partial_j x^\nu g_{\mu\nu} - \frac{1}{2}\gamma_{ij}\gamma^{kl}\partial_k x^\mu\partial_l x^\nu g_{\mu\nu}\right] = 0 \end{aligned} \tag{2.8}$$

To fix the group structure more precisely, we introduced in [1] a "dual" σ-model for which the roles of field equations and Bianchi identities are interchanged. One begins by noting that the equations of motion (2.5) may equivalently be derived from a first order Lagrangian with independent variables x^μ and $F_i{}^\mu$,

$$\begin{aligned} \mathcal{L}_x = &-\frac{1}{2}\sqrt{-\gamma}\gamma^{ij}F_i{}^\mu F_j{}^\nu g_{\mu\nu} - \frac{1}{2}\varepsilon^{ij}F_i{}^\mu F_j{}^\nu b_{\mu\nu} \\ &+ \partial_i x^\mu\left(\sqrt{-\gamma}\gamma^{ij}F_j{}^\nu g_{\mu\nu} + \varepsilon^{ij}F_j{}^\nu b_{\mu\nu}\right) \end{aligned} \tag{2.9}$$

Varying with respect to $\partial_i x^\mu$, we have

$$\frac{\partial\mathcal{L}_x}{\partial\partial_i x^\mu} = \sqrt{-\gamma}\gamma^{ij}F_j{}^\nu g_{\mu\nu} + \varepsilon^{ij}F_j{}^\nu b_{\mu\nu} \tag{2.10}$$

and with respect to $F_i{}^\mu$, we have

$$\begin{aligned} \frac{\partial\mathcal{L}_x}{\partial F_i{}^\mu} = &-\sqrt{-\gamma}\gamma^{ij}F_j{}^\nu g_{\mu\nu} - \varepsilon^{ij}F_j{}^\nu b_{\mu\nu} \\ &+ \sqrt{-\gamma}\gamma^{ij}\partial_j x^\nu g_{\mu\nu} + \varepsilon^{ij}\partial_j x^\nu b_{\mu\nu} = 0 \end{aligned} \tag{2.11}$$

This latter equation says

$$F_i{}^\mu = \partial_i x^\mu \tag{2.12}$$

and then (2.10) yields the same equation of motion as in the second order formalism.

Next consider a different first-order Lagrangian with independent variables y_μ and $F_i{}^\mu$,

$$\mathcal{L}_y = \frac{1}{2}\sqrt{-\gamma}\gamma^{ij}F_i{}^\mu F_j{}^\nu g_{\mu\nu} + \frac{1}{2}\varepsilon^{ij}F_i{}^\mu F_j{}^\nu b_{\mu\nu} + \partial_i y_\mu \varepsilon^{ij}F_j{}^\mu \tag{2.13}$$

We have

$$\frac{\partial \mathcal{L}_y}{\partial \partial_i y_\mu} = \varepsilon^{ij}F_j{}^\mu \tag{2.14}$$

and

$$\frac{\partial \mathcal{L}_y}{\partial F_i{}^\mu} = \sqrt{-\gamma}\gamma^{ij}F_j{}^\nu g_{\mu\nu} + \varepsilon^{ij}F_j{}^\nu b_{\mu\nu} - \varepsilon^{ij}\partial_j y_\mu = 0 \tag{2.15}$$

This is an algebraic equation for $F_i{}^\mu$ with the solution

$$F_i{}^\mu = p^{\mu\nu}\frac{1}{\sqrt{-\gamma}}\gamma_{ij}\varepsilon^{jk}\partial_k y_\nu + q^{\mu\nu}\partial_i y_\nu \tag{2.16}$$

where $p^{\mu\nu} = p^{\nu\mu}$ and $q^{\mu\nu} = -q^{\nu\mu}$ are related to $g_{\mu\nu} = g_{\nu\mu}$ and $b_{\mu\nu} = -b_{\nu\mu}$ by

$$\begin{aligned} p_{\mu\nu} &= g_{\mu\nu} + b_{\mu\alpha}g^{\alpha\beta}b_{\nu\beta} \\ p_{\mu\nu}q^{\alpha\nu} &= b_{\mu\beta}g^{\alpha\beta} \end{aligned} \tag{2.17}$$

where $p_{\mu\nu}$ is the inverse of $p^{\mu\nu}$. From (2.14) the equation for y_μ is

$$\partial_i(\varepsilon^{ij}F_j{}^\mu) = 0 \tag{2.18}$$

which implies (2.12), at least locally.

Putting these results together, we find

$$\begin{aligned} \varepsilon^{ij}\partial_j y_\mu &= \frac{\partial \mathcal{L}_x}{\partial \partial_i x^\mu} = g_{\mu\nu}\sqrt{-\gamma}\gamma^{ij}\partial_j x^\nu + b_{\mu\nu}\varepsilon^{ij}\partial_j x^\nu \\ \varepsilon^{ij}\partial_j x^\mu &= \frac{\partial \mathcal{L}_y}{\partial \partial_i y_\mu} = p^{\mu\nu}\sqrt{-\gamma}\gamma^{ij}\partial_j y_\nu + q^{\mu\nu}\varepsilon^{ij}\partial_j y_\nu \end{aligned} \tag{2.19}$$

Thus the field equations of the original Lagrangian $\mathcal{L}_x$ are the Bianchi identities for the "dual" Lagrangian $\mathcal{L}_y$, and vice-versa.

To see the $SO(n,n)$ symmetry explicitly, define

$$\mathcal{G}^i{}_\mu \equiv \sqrt{-\gamma}\gamma^{ij}\partial_j y_\mu \tag{2.20}$$

Then (2.19) may be written

$$\begin{aligned} \tilde{\mathcal{G}}^i{}_\mu &= g_{\mu\nu}\mathcal{F}^{i\nu} + b_{\mu\nu}\tilde{\mathcal{F}}^{i\nu} \\ \tilde{\mathcal{F}}^{i\mu} &= p^{\mu\nu}\mathcal{G}^i{}_\nu + q^{\mu\nu}\tilde{\mathcal{G}}^i{}_\nu \end{aligned} \tag{2.21}$$

or, in compact notation,

$$\Omega_{MN}\tilde{\Phi}^{iN} = G_{MN}\Phi^{iN} \tag{2.22}$$

where $M, N = 1, \ldots, 2n$ and where

$$\tilde{\Phi}^{iN} = \begin{pmatrix} \tilde{\mathcal{F}}^{i\nu} \\ \tilde{\mathcal{G}}^{i}{}_{\beta} \end{pmatrix} \qquad \Phi^{iN} = \begin{pmatrix} \mathcal{F}^{i\nu} \\ \mathcal{G}^{i}{}_{\beta} \end{pmatrix} \tag{2.23}$$

The $2n \times 2n$ matrices Ω and G are given by

$$\Omega_{MN} = \begin{pmatrix} 0 & \delta_{\mu}{}^{\beta} \\ \delta^{\alpha}{}_{\nu} & 0 \end{pmatrix} \tag{2.24}$$

$$G_{MN} = \begin{pmatrix} g_{\mu\nu} + b_{\mu\alpha} g^{\alpha\beta} b_{\nu\beta} & b_{\mu\alpha} g^{\alpha\beta} \\ g^{\alpha\beta} b_{\nu\beta} & g^{\alpha\beta} \end{pmatrix} \tag{2.25}$$

The desired $SO(n,n)$ symmetry is now manifest since the group $SO(n,n)$ may be defined by parameters $A^{M}{}_{N}$ for which

$$\delta\Omega_{MN} = -A^{P}{}_{M}\Omega_{PN} - A^{P}{}_{N}\Omega_{MP} = 0 \tag{2.26}$$

Moreover,

$$G_{MP}\Omega^{PQ}G_{QN} = G_{MN} \tag{2.27}$$

and hence G is an element of $SO(n,n)$. Thus (2.22) is manifestly $SO(n,n)$ invariant with $\tilde{\Phi}^{iN}$ and Φ^{iN} both transforming as the 2n- dimensional vector representation

$$\delta\tilde{\Phi}^{iM} = A^{M}{}_{P}\tilde{\Phi}^{iP}, \quad \delta\Phi^{iM} = A^{M}{}_{P}\Phi^{iP} \tag{2.28}$$

The explicit transformation rules are those of (2.7) with the restrictions

$$\begin{aligned} B^{\mu\beta} &= B^{[\mu\beta]} = A^{\mu\beta} \\ C_{\alpha\nu} &= C_{[\alpha\nu]} = A_{\alpha\nu} \\ D_{\alpha}{}^{\beta} &= A_{\alpha}{}^{\beta} = -A^{\beta}{}_{\alpha} \end{aligned} \tag{2.29}$$

corresponding to the $n(2n-1)$ parameters of $SO(n,n)$. Thus

$$\begin{aligned} \delta\tilde{\mathcal{F}}^{i\mu} &= A^{\mu}{}_{\nu}\tilde{\mathcal{F}}^{i\nu} + B^{\mu\beta}\tilde{\mathcal{G}}^{i}{}_{\beta} \\ \delta\tilde{\mathcal{G}}^{i}{}_{\alpha} &= C_{\alpha\nu}\tilde{\mathcal{F}}^{i\nu} - A^{\beta}{}_{\alpha}\tilde{\mathcal{G}}^{i}{}_{\beta} \end{aligned} \tag{2.30}$$

and similarly for $\mathcal{F}^{i\mu}$ and $\mathcal{G}^{i}{}_{\alpha}$. The transformation rule for G_{MN} is

$$\delta G_{MN} = -A^{P}{}_{M}G_{PN} - A^{P}{}_{N}G_{MP} \tag{2.31}$$

and hence

$$\begin{aligned} \delta g_{\mu\nu} &= -A^{\rho}{}_{\mu}g_{\rho\nu} - A^{\rho}{}_{\nu}g_{\mu\rho} - b_{\mu\alpha}B^{\alpha\beta}g_{\beta\nu} - g_{\mu\alpha}B^{\alpha\beta}b_{\beta\nu} \\ \delta b_{\mu\nu} &= -A^{\rho}{}_{\mu}b_{\rho\nu} - A^{\rho}{}_{\nu}b_{\mu\rho} - b_{\mu\alpha}B^{\alpha\beta}b_{\beta\nu} - g_{\mu\alpha}B^{\alpha\beta}g_{\beta\nu} + C_{\mu\nu} \end{aligned} \tag{2.32}$$

The action of $SO(n,n)$ on the background fields $g_{\mu\nu}$ and $b_{\mu\nu}$ is non-linear. A linear realization may be obtained by rewriting (2.22) as

$$\tilde{\Phi}^{i}{}_{A} = \Phi^{i}{}_{A} \tag{2.33}$$

where

$$\tilde{\Phi}^i{}_A = \Omega_{AB} E_N{}^B \tilde{\Phi}^{iN}$$
$$\Phi^i{}_A = E_{NA} \Phi^{iN} \tag{2.34}$$

and where $E_M{}^A$ is the "vielbein" for which

$$G_{MN} = E_M{}^A E_{NA} \tag{2.35}$$

Multiplying (2.33) by $E_M{}^A$ and using the fact that $E_M{}^A$ is also an element of $SO(n,n)$

$$E_M{}^A \Omega_{AB} E_N{}^B = \Omega_{MN} \tag{2.36}$$

we recover (2.22). As usual, the price to pay for a linear realization of G is a local symmetry H where $H \subset G$. In this case

$$E_M{}^A \to \Lambda^A{}_B E_M{}^B \tag{2.37}$$

where Λ is an element of $SO(n) \times SO(n)$, the maximal compact subgroup of $SO(n,n)$ whose elements commute with Ω. The n^2 physical scalar fields described by G_{MN} parametrize the coset $SO(n,n)/SO(n) \times SO(n)$ and their self-interaction is described by the corresponding non-linear σ-model.

3. MEMBRANE DUALITY

The bosonic sector of the n-dimensional supermembrane is described by a three-dimensional σ-model with worldvolume coordinates $\xi^i = (\tau, \sigma, \rho)$, worldvolume metric $\gamma_{ij}(\xi)$ and target space coordinates $x^\mu(\xi)$, with

$$\mathcal{L} = \frac{1}{2}\sqrt{-\gamma}\gamma^{ij}\partial_i x^\mu \partial_j x^\nu g_{\mu\nu} + \frac{1}{3!}\varepsilon^{ijk}\partial_i x^\mu \partial_j x^\nu \partial_k x^\rho b_{\mu\nu\rho} - \frac{1}{2}\sqrt{-\gamma} \tag{3.1}$$

with background fields $g_{\mu\nu}$ and $b_{\mu\nu\rho} (\mu = 1, \ldots, n)$. Define

$$\mathcal{F}^{i\mu} \equiv \sqrt{-\gamma}\gamma^{ij}\partial_j x^\mu \tag{3.2}$$

$$\tilde{\mathcal{F}}^{i\mu\nu} \equiv \varepsilon^{ijk}\partial_j x^\mu \partial_k x^\nu \tag{3.3}$$

and

$$\tilde{\mathcal{G}}^i{}_\mu = g_{\mu\nu}\mathcal{F}^{i\nu} + \frac{1}{2} b_{\mu\nu\rho}\tilde{\mathcal{F}}^{i\nu\rho} = \frac{\partial \mathcal{L}}{\partial \partial_i x^\mu} \tag{3.4}$$

In the case when the background fields $g_{\mu\nu}$ and $b_{\mu\nu\rho}$ are independent of the coordinate x^μ, its equation of motion is just

$$\partial_i \tilde{\mathcal{G}}^i{}_\mu = 0, \tag{3.5}$$

whereas the Bianchi identity is

$$\partial_i \tilde{\mathcal{F}}^{i\mu\nu} \equiv 0. \tag{3.6}$$

Thus there is a duality symmetry that rotates field equations into Bianchi identities. The invariance is summarized by the equations

$$\begin{aligned}\delta\tilde{\mathcal{F}}^{i\mu\nu} &= \frac{1}{2}A^{\mu\nu}{}_{\rho\sigma}\tilde{\mathcal{F}}^{i\rho\sigma} + B^{\mu\nu\rho}\tilde{\mathcal{G}}^{i}{}_{\rho} \\ \delta\tilde{\mathcal{G}}^{i}{}_{\alpha} &= \frac{1}{2}C_{\alpha\rho\sigma}\tilde{\mathcal{F}}^{i\rho\sigma} + D_{\alpha}{}^{\rho}\tilde{\mathcal{G}}^{i}{}_{\rho}\end{aligned} \tag{3.7}$$

where A, B, C and D are constant parameters. Thus the first major difference from the string is that the duality transformations are non-linear in the "field strength" $\partial_i x^\mu$ by virtue of $\tilde{\mathcal{F}}^{i\mu\nu}$, given in (3.3). We must also establish invariance of the equation of motion obtained by varying with respect to γ_{ij}, namely the vanishing of the energy momentum tensor

$$\begin{aligned}\sqrt{-\gamma}\theta_{ij} &= \frac{2\delta S}{\delta\gamma^{ij}} \\ &= \sqrt{-\gamma}\left[\partial_i\, x^\mu\partial_j x^\nu g_{\mu\nu} - \frac{1}{2}\gamma_{ij}\gamma^{k\ell}\partial_k x^\mu\partial_\ell x^\nu g_{\mu\nu} + \frac{1}{2}\gamma_{ij}\right] = 0\end{aligned} \tag{3.8}$$

This equation is just the statement that γ_{ij} is the induced metric on the worldvolume

$$\gamma_{ij} = \partial_i x^\mu\partial_j x^\nu g_{\mu\nu} \tag{3.9}$$

Here we encounter the second major difference from the string case, where the worldsheet metric is invariant under duality transformations. For the membrane, we must allow the possibility that γ_{ij} also transforms under duality.

The equations of motion (3.4), (3.5) and (3.8) may equivalently be derived from a first-order Lagrangian with independent variables x^μ and $F_i{}^\mu$

$$\begin{aligned}\mathcal{L}_x = &-\frac{1}{2}\sqrt{-\gamma}\gamma^{ij}F_i{}^\mu F_j{}^\nu g_{\mu\nu} - \frac{1}{3}\varepsilon^{ijk}F_i{}^\mu F_j{}^\nu F_k{}^\rho b_{\mu\nu\rho} \\ &+\partial_i x^\mu(\sqrt{-\gamma}\gamma^{ij}F_j{}^\nu g_{\mu\nu} + \frac{1}{2}\varepsilon^{ijk}F_i{}^\nu F_k{}^\rho b_{\mu\nu\rho}) - \frac{1}{2}\sqrt{-\gamma}\end{aligned} \tag{3.10}$$

Varying with respect to $\partial_i x^\mu$, we have

$$\frac{\partial\mathcal{L}_x}{\partial\partial_i x^\mu} = \sqrt{-\gamma}\gamma^{ij}F_j{}^\nu g_{\mu\nu} + \frac{1}{2}\varepsilon^{ijk}F_j{}^\nu F_k{}^\rho b_{\mu\nu\rho} \tag{3.11}$$

and with respect to $F_i{}^\mu$, we have

$$\begin{aligned}\frac{\partial\mathcal{L}_x}{\partial F_i{}^\mu} = &-\sqrt{-\gamma}\gamma^{ij}F_j{}^\nu g_{\mu\nu} - \varepsilon^{ijk}F_j{}^\nu F_k{}^\rho b_{\mu\nu\rho} \\ &+\sqrt{-\gamma}\gamma^{ij}\partial_j x^\nu g_{\mu\nu} + \varepsilon^{ijk}\partial_j x^\nu F_k{}^\rho b_{\mu\nu\rho} = 0\end{aligned} \tag{3.12}$$

This latter equation says

$$F_i{}^\mu = \partial_i x^\mu \tag{3.13}$$

and then (3.11) yields the same equations of motion as in the second order formulation.

Now consider a different first-order Lagrangian with independent variables $y_{\mu\nu}$ and $F_i{}^\mu$

$$\begin{aligned}\mathcal{L}_y = &\frac{1}{2}\sqrt{-\gamma}\gamma^{ij}F_i{}^\mu F_j{}^\nu g_{\mu\nu} + \frac{1}{3!}\varepsilon^{ijk}F_i{}^\mu F_j{}^\nu F_k{}^\rho b_{\mu\nu\rho} \\ &+\varepsilon^{ijk}\partial_i y_{\mu\nu}F_j{}^\mu F_k{}^\nu - \frac{1}{2}\sqrt{-\gamma}\end{aligned} \tag{3.14}$$

We have

$$\frac{\partial\mathcal{L}_y}{\partial\partial_i y_{\mu\nu}} = \varepsilon^{ijk}F_j{}^\mu F_k{}^\nu \tag{3.15}$$

and
$$\frac{\partial \mathcal{L}_y}{\partial F_i{}^\mu} = \sqrt{-\gamma}\gamma^{ij}F_j{}^\nu g_{\mu\nu} + \frac{1}{2}\varepsilon^{ijk}F_j{}^\nu F_k{}^\rho b_{\mu\nu\rho} - 2\varepsilon^{ijk}\partial_j y_{\mu\nu}F_k{}^\nu \tag{3.16}$$

This is an algebraic equation for $F_i{}^\mu$ with the (implicit) solution

$$F_i{}^\mu = 2p^{\mu\nu}\frac{1}{\sqrt{-\gamma}}\gamma_{ij}\varepsilon^{jk\ell}\partial_k y_{\nu\sigma}F_\ell{}^\sigma - 2q^{\rho\sigma\mu}\partial_i y_{\rho\sigma} \tag{3.17}$$

where $p^{\mu\nu} = p^{\nu\mu}$ and $q^{\rho\sigma\mu} = q^{[\rho\sigma\mu]}$ are related to $g_{\mu\nu} = g_{\nu\mu}$ and $b_{\mu\nu\rho} = b_{[\mu\nu\rho]}$ by

$$\begin{aligned} p_{\mu\nu} &= g_{\mu\nu} + \frac{1}{4}b_{\mu\alpha\beta}g^{\alpha\beta\gamma\delta}b_{\nu\gamma\delta} \\ p_{\mu\nu}q^{\alpha\beta\nu} &= \frac{1}{2}b_{\mu\gamma\delta}g^{\alpha\beta\gamma\delta}\ , \end{aligned} \tag{3.18}$$

where $p_{\mu\nu}$ is the inverse of $p^{\mu\nu}$ and where

$$g^{\alpha\beta\gamma\delta} = (g^{\alpha\gamma}g^{\beta\delta} - g^{\alpha\delta}g^{\beta\gamma})\ . \tag{3.19}$$

In proving (3.17), we have made use of the identities

$$\begin{aligned} \varepsilon^{ijk}\varepsilon^{\ell mn} &= \gamma(\gamma^{i\ell}\gamma^{jm}\gamma^{kn} + \gamma^{in}\gamma^{j\ell}\gamma^{km} + \gamma^{im}\gamma^{jn}\gamma^{k\ell} \\ &\quad - \gamma^{i\ell}\gamma^{jn}\gamma^{km} - \gamma^{im}\gamma^{j\ell}\gamma^{kn} - \gamma^{in}\gamma^{jm}\gamma^{k\ell}) \\ \varepsilon^{ijk}\varepsilon^{\ell m}{}_k &= \gamma(\gamma^{i\ell}\gamma^{jm} - \gamma^{im}\gamma^{j\ell}) \\ \varepsilon^{ijk}\varepsilon^{\ell}{}_{jk} &= 2\gamma\ \gamma^{i\ell} \\ \varepsilon^{ijk}\varepsilon_{ijk} &= 6\gamma \end{aligned} \tag{3.20}$$

and have also used the field equation (3.9). From (3.15) the equation for $y_{\mu\nu}$ is

$$\partial_i(\varepsilon^{ijk}F_j{}^\mu F_k{}^\nu) = 0 \tag{3.21}$$

which implies (3.13), at least locally. Putting these results together, we find

$$\begin{aligned} 2\varepsilon^{ijk}\partial_j y_{\mu\nu}\partial_k x^\nu &= \frac{\partial \mathcal{L}_x}{\partial\partial_i x^\mu} = g_{\mu\nu}\sqrt{-\gamma}\gamma^{ij}\partial_j x^\nu + \frac{1}{2}b_{\mu\nu\rho}\varepsilon^{ijk}\partial_j x^\nu\partial_k x^\rho \\ \varepsilon^{ijk}\partial_j x^\mu\partial_k x^\nu &= \frac{\partial \mathcal{L}_y}{\partial\partial_i y_{\mu\nu}} = 2p^{\mu\nu\rho\sigma}\sqrt{-\gamma}\gamma^{ij}\partial_j y_{\rho\sigma} + 2q^{\mu\nu\rho}\varepsilon^{ijk}\partial_j y_{\rho\sigma}\partial_k x^\sigma \end{aligned} \tag{3.22}$$

where $p^{\alpha\beta\gamma\delta}$ is defined by

$$g^{\alpha\beta\gamma\delta} = p^{\alpha\beta\gamma\delta} + q^{\alpha\beta\mu}p_{\mu\nu}q^{\gamma\delta\nu} \tag{3.23}$$

Thus the field equations of the original Lagrangian $\mathcal{L}_x$ are the Bianchi identities for the"dual" Lagrangian $\mathcal{L}_y$, and vice-versa.

4. A SPECIFIC CASE: n=4

In the case of the string, the n $\tilde{\mathcal{G}}^i{}_\mu$ and the n $\tilde{\mathcal{F}}^{i\mu}$ transform as the same 2n-dimensional vector representation of the same orthogonal group, $SO(n,n)$, for all n. In the case of the membrane, however, we shall see that each n tells a different story. The n $\tilde{\mathcal{G}}^i{}_\mu$ and the $n(n-1)/2$ $\tilde{\mathcal{F}}^{i\mu\nu}$ will transform as a $n(n+1)/2$ dimensional representation of some non-compact group. Then the $n(n+1)/2$ $g_{\mu\nu}$ and the $n(n-1)(n-2)/3!$ $b_{\mu\nu\rho}$ will parametrize some $n(n^2+5)/3!$ dimensional coset. (In fact, we shall see in Section 6 that this is valid only for $n \leq 4$ and that the cases $n \geq 5$ require a separate treatment.) The group and the coset will be different for each n.

In this section we shall focus explicitly on $n = 4$. Then the 4 $\tilde{\mathcal{G}}^i{}_\mu$ and 6 $\tilde{\mathcal{F}}^{i\mu\nu}$ will transform as a 10 dimensional representation of the duality group which will turn out to be $SL(5,R)$, and the 10 $g_{\mu\nu}$ and 4 $b_{\mu\nu\rho}$ will parametrize the 14 dimensional coset $SL(5,R)/SO(5)$. To see this explicitly, define

$$\mathcal{G}^i{}_{\mu\nu} \equiv 4\sqrt{-\gamma}\gamma^{ij}\partial_j y_{\mu\nu} \tag{4.1}$$

Then (3.22) may be written

$$\begin{aligned} \tilde{\mathcal{G}}^i{}_\mu &= g_{\mu\nu}\mathcal{F}^{i\nu} + \frac{1}{2}b_{\mu\nu\rho}\tilde{\mathcal{F}}^{i\nu\rho} \\ \tilde{\mathcal{F}}^{i\mu\nu} &= \frac{1}{2}p^{\mu\nu\rho\sigma}\mathcal{G}^i{}_{\rho\sigma} + q^{\mu\nu\rho}\tilde{\mathcal{G}}^i{}_\rho \end{aligned} \tag{4.2}$$

Now define $\tilde{\Phi}^i{}_{MN} = -\tilde{\Phi}^i{}_{NM} (M, N = 1, \ldots, 5)$ via

$$\begin{aligned} \tilde{\Phi}^i{}_{\mu\nu} &= \frac{1}{2g}\varepsilon_{\mu\nu\alpha\beta}\tilde{\mathcal{F}}^{i\alpha\beta} \\ \tilde{\Phi}^i{}_{\mu 5} &= \tilde{\mathcal{G}}^i{}_\mu \end{aligned} \tag{4.3}$$

and $\Phi^{iMN} = -\Phi^{iNM}$ via

$$\begin{aligned} \Phi^{i\mu\nu} &= \frac{1}{2}g^{-1/5}\varepsilon^{\mu\nu\rho\sigma}\mathcal{G}^i{}_{\rho\sigma} \\ \Phi^{i\mu 5} &= g^{-1/5}\mathcal{F}^{i\mu} \end{aligned} \tag{4.4}$$

where g is the determinant of $g_{\mu\nu}$. Then (4.2) may be written

$$\tilde{\Phi}^i{}_{MN} = (G_{MP}G_{NQ} - G_{MQ}G_{NP})\Phi^{iPQ} \tag{4.5}$$

where G_{MN} is the 5×5 symmetric matrix for which

$$\begin{aligned} G_{\mu\nu} &= g^{-2/5}g_{\mu\nu} \\ G_{\mu 5} &= -\frac{1}{3!}g^{-2/5}g_{\mu\alpha}\varepsilon^{\alpha\beta\gamma\delta}b_{\beta\gamma\delta} = G_{5\mu} \\ G_{55} &= g^{3/5}(1 + \frac{1}{3!}b_{\alpha\beta\gamma}g^{\alpha\mu}g^{\beta\nu}g^{\gamma\rho}b_{\mu\nu\rho}) \end{aligned} \tag{4.6}$$

We note that

$$det\ G_{MN} = 1 \tag{4.7}$$

and hence that G_{MN} is an element of $SL(5,R)$. Equation (4.5) is thus manifestly $SL(5,R)$ invariant with $\tilde{\Phi}^{iMN}$ transforming as the 10-dimensional representation

$$\delta\tilde{\Phi}^i{}_{MN} = -A^P{}_M\tilde{\Phi}^i{}_{PN} - A^P{}_N\tilde{\Phi}^i{}_{MP} \tag{4.8}$$

where $A^P{}_M$ is a traceless 5×5 constant matrix. Similarly, we have

$$\delta\Phi^{iMN} = A^M{}_P\Phi^{iPN} + A^N{}_P\Phi^{iMP} \tag{4.9}$$

The explicit transformation rules are those of (3.7) with the restrictions

$$
\begin{aligned}
A^{\mu\nu}{}_{\rho\sigma} =&\delta^\mu{}_\rho A^\nu{}_\sigma - \delta^\nu{}_\rho A^\mu{}_\sigma + \delta^\nu{}_\sigma A^\mu{}_\rho - \delta^\mu{}_\sigma A^\nu{}_\rho \\
&- \delta^\mu{}_\rho \delta^\nu{}_\sigma A^\alpha{}_\alpha + \delta^\mu{}_\sigma \delta^\nu{}_\rho A^\alpha{}_\alpha \\
B^{\mu\nu\rho} &= B^{[\mu\nu\rho]} = -\varepsilon^{\mu\nu\rho\sigma} A^5{}_\sigma \\
C_{\mu\nu\rho} &= C_{[\mu\nu\rho]} = -\frac{1}{g}\varepsilon_{\mu\nu\rho\sigma} A^\sigma{}_5 \\
D_\mu{}^\rho &= -A^\rho{}_\mu + \delta^\rho{}_\mu A^\alpha{}_\alpha
\end{aligned} \tag{4.10}
$$

corresponding to the 24 parameters of $SL(5, R)$. Thus

$$
\begin{aligned}
\delta\tilde{\mathcal{F}}^{i\mu\nu} &= -A^\alpha{}_\alpha \tilde{\mathcal{F}}^{i\mu\nu} + A^\mu{}_\sigma \tilde{\mathcal{F}}^{i\sigma\nu} + A^\nu{}_\sigma \tilde{\mathcal{F}}^{i\mu\sigma} + B^{\mu\nu\lambda}\tilde{\mathcal{G}}^i{}_\lambda \\
\delta\tilde{\mathcal{G}}^i{}_\mu &= A^\alpha{}_\alpha \tilde{\mathcal{G}}^i{}_\mu - A^\sigma{}_\mu \tilde{\mathcal{G}}^i{}_\sigma + \frac{1}{2} C_{\mu\rho\lambda}\tilde{\mathcal{F}}^{i\rho\lambda}
\end{aligned} \tag{4.11}
$$

Whereas,

$$
\begin{aligned}
\delta\mathcal{F}^{i\mu} &= (-\frac{1}{3}A^\alpha{}_\alpha - \frac{2}{3}\frac{1}{3!}B^{\alpha\beta\gamma}b_{\alpha\beta\gamma})\mathcal{F}^{i\mu} + A^\mu{}_\sigma \mathcal{F}^{i\sigma} - \frac{1}{2}B^{\mu\alpha\beta}\mathcal{G}^i{}_{\alpha\beta} \\
\delta\mathcal{G}^i{}_{\mu\nu} &= (\frac{5}{3}A^\alpha{}_\alpha - \frac{2}{3}\frac{1}{3!}B^{\alpha\beta\gamma}b_{\alpha\beta\gamma})\mathcal{G}^i{}_{\mu\nu} - A^\alpha{}_\mu \mathcal{G}^i{}_{\alpha\nu} - A^\alpha{}_\nu \mathcal{G}^i{}_{\mu\alpha} \\
&- C_{\rho\mu\nu}\mathcal{F}^{i\rho}
\end{aligned} \tag{4.12}
$$

The transformation rule for G_{MN} is given by

$$
\delta G_{MN} = -A^P{}_N G_{MP} - A^P{}_M G_{PN} \tag{4.13}
$$

and hence

$$
\begin{aligned}
\delta g_{\mu\nu} =&\frac{4}{3}A^\alpha{}_\alpha g_{\mu\nu} - A^\sigma{}_\mu g_{\sigma\nu} - A^\sigma{}_\nu g_{\mu\sigma} \\
&+ \frac{2}{3}\frac{1}{3!}B^{\alpha\beta\gamma}b_{\alpha\beta\gamma}g_{\mu\nu} - \frac{1}{2}g_{\mu\rho}B^{\rho\alpha\beta}b_{\alpha\beta\nu} - \frac{1}{2}b_{\mu\alpha\beta}B^{\alpha\beta\rho}g_{\rho\nu} \\
\delta b_{\mu\nu\rho} =&2A^\alpha{}_\alpha b_{\mu\nu\rho} - A^\sigma{}_\mu b_{\sigma\nu\rho} - A^\sigma{}_\nu b_{\mu\sigma\rho} - A^\sigma{}_\rho b_{\mu\nu\sigma} \\
&- \frac{1}{3!}B^{\alpha\beta\gamma}b_{\alpha\beta\gamma}b_{\mu\nu\rho} + B^{\alpha\beta\gamma}g_{\mu\alpha}g_{\nu\beta}g_{\rho\gamma} + C_{\mu\nu\rho}
\end{aligned} \tag{4.14}
$$

This action of $SL(5, R)$ on the background fields is nonlinear. A linear realization of $SL(5, R)$, may be obtained by rewriting (4.5) as

$$
\tilde{\Phi}^i{}_{AB} = \Phi^i{}_{AB} \tag{4.15}
$$

where

$$
\begin{aligned}
\tilde{\Phi}^i{}_{AB} &= E^M{}_A E^N{}_B \tilde{\Phi}^i{}_{MN} \\
\Phi^i{}_{AB} &= E_{MA} E_{NB} \Phi^{iMN}
\end{aligned} \tag{4.16}
$$

and where $E_M{}^A$ is the "funfbein" for which

$$
G_{MN} = E_M{}^A E_{NA} \tag{4.17}
$$

and where $E^M{}_A$ is its inverse. Multiplying (4.15) by $E_M{}^A E_N{}^B$ we recover (4.5).

As usual the price to pay for a linear realization of G is a local symmetry H where $H \subset G$. In this case

$$E_M{}^A \to \Lambda^A{}_B E_M{}^B \tag{4.18}$$

where Λ is an element of $SO(5)$, the maximal compact subgroup of $SL(5,R)$. The 14 physical scalar fields described by G_{MN} parametrize the coset $SL(5,R)/SO(5)$ and their self interaction is described by the corresponding non-linear σ-model.

Finally, we should also discuss the transformation properties of the membrane metric. From (3.2) and (3.9) we have

$$-\gamma\gamma^{ij} = g_{\mu\nu}\mathcal{F}^{i\mu}\mathcal{F}^{j\nu} \tag{4.19}$$

Thus from the transformation rules for $g_{\mu\nu}$ of (4.14) and $\mathcal{F}^{i\mu}$ of (4.12) we may deduce

$$\delta\gamma_{ij} = \frac{1}{3}(A^\alpha{}_\alpha - \frac{1}{3!}B^{\alpha\beta\gamma}b_{\alpha\beta\gamma} - \frac{1}{2}\varepsilon)\gamma_{ij} \tag{4.20}$$

where

$$\varepsilon \equiv \frac{1}{\sqrt{-\gamma}}\varepsilon^{lmn}\partial_\ell x^\mu \partial_m x^\nu \partial_n x^\rho g_{\mu\alpha}g_{\nu\beta}g_{\rho\gamma}B^{\alpha\beta\gamma} \tag{4.21}$$

Thus γ_{ij} transforms conformally with a worldvolume coordinate dependent conformal factor. The invariant

$$\tilde{\Phi}^i{}_{IJ}\Phi^{jIJ} = 0 \tag{4.22}$$

as a consequence of (3.9) which just restates, in a manifestly $SL(5,R)$ invariant way, the vanishing of the energy momentum tensor (3.8).

5. COMPARISON WITH d=11 SUPERGRAVITY

By compactifying the membrane on T^4, we discovered in the last section that the duality symmetry is $SL(5,R)$ and that the 14 background fields $g_{\mu\nu}$ and $b_{\mu\nu\rho}$ parametrize the 14-dimensional coset $SL(5,R)/SO(5)$. But this duality symmetry and this coset are precisely those obtained by Cremmer and Julia [4] from compactification of d=11 supergravity on T^4. Once we have accepted that duality symmetries of supergravity can in principle have their origin on the worldvolume of the supermembrane, this correspondence should not be surprising since we know that the κ- symmetry of the d=11 supermembrane [8] forces the background fields of the three-dimensional σ-model to be solutions of the d=11 supergravity equations [13]. With the exception of the graviton, which is a singlet under duality, all the spacetime fields in the supergravity multiplet: gravitinos, vectors, spinors, scalars and antisymmetric tensors will transform under duality. By retaining only the backgrounds $g_{\mu\nu}$ and $b_{\mu\nu\rho}$, we have so far been treating only the spacetime scalars. The scalars alone are nevertheless sufficient to determine the duality symmetries. If we repeat the $n = 4$ analysis of Section 5 for $n < 4$, one finds the duality symmetries listed in Table 1 with the $n(n^2+5)/3!$ $g_{\mu\nu}$ and $b_{\mu\nu\rho}$ parametrizing the corresponding coset. These once again agree with those of Cremmer and Julia. Moreover the $n(n+1)/2$ "field strengths" $\tilde{\mathcal{G}}^i{}_\mu$ and $\tilde{\mathcal{F}}^{i\mu\nu}$ transform as the same representation of G as do the spacetime vector fields.*

For $n \geq 6$, however, there is a mismatch with the number of spacetime scalars as shown in Table 1. The mismatch with the number of spacetime vectors occurs

* We are grateful to Ergin Sezgin for pointing out the importance of this.

Table 1. The hidden global symmetries G and local symmetries H that result from compactifying d=11 supergravity on T^n. For $n \leq 5$, these are compatible with the coset parametrized by the membrane background fields $g_{\mu\nu}$ and $b_{\mu\nu\rho}$ only. Extra spacetime scalars must be included for $n \geq 6$.

n	G	H	$dim\ G/H$	$n(n^2+5)/3!$	
1	R	1	1	1	√
2	$GL(2,R)$	$SO(2)$	3	3	√
3	$SL(3,R) \times SL(2,R)$	$SO(3) \times SO(2)$	7	7	√
4	$SL(5,R)$	$SO(5)$	14	14	√
5	$SO(5,5)$	$SO(5) \times SO(5)$	25	25	√
6	$E_{6(+6)}$	$USp(8)$	42	41	×
7	$E_{7(+7)}$	$SU(8)$	70	63	×
8	$E_{8(+8)}$	$SO(16)$	128	92	×

already for $n \geq 5$ as shown in Table 2. The reason for this is easy to explain and does not present a serious problem. By focussing only on those spacetime scalars arising from $g_{\mu\nu}$ and $b_{\mu\nu\rho}$, we have ignored those arising from other sources [4]. For $n = 6$, we get 1 extra by dualizing the spacetime three-form $b_{MNP}(M, N, P = 1, \ldots, 5)$; for $n = 7$ we get 7 extra by dualizing the spacetime two-form $b_{MN\rho}(M, N = 1, \ldots, 4)$; for $n = 8$ we get 28 extra by dualizing the one-form $b_{M\nu\rho}$ and 8 extra by dualizing $g_{M\nu}$(M = 1,2,3). This correctly accounts for the mismatch in Table 1. Similarly, by focussing on the fields strengths $\tilde{\mathcal{G}}^i{}_\mu$ and $\tilde{\mathcal{F}}^{i\mu\nu}$, we have been ignoring others that couple to spacetime vectors arising from dualization. For $n = 5$, we get 1 extra by dualizing the three-form b_{MNP}, for $n = 6$ we get 6 extra by dualizing the two-form $b_{MN\rho}$. This correctly accounts for the mismatch in Table 2. The cases $n = 7, 8$ are special. The 28 spacetime vectors field strengths combine with 28 dual field strengths to form a 56 of $E_{7(+7)}$ whereas for $n = 8$ all vectors are dual to scalars. Thus we expect that all the hidden symmetries of Cremmer and Julia, including those for $n \geq 5$, will follow from membrane duality provided we start with an enlarged σ-model that includes the couplings to those background fields we could safely ignore for $n \leq 4$. We intend to return to this point elsewhere.

6. STRING DUALITY FROM MEMBRANE DUALITY

The dimension of the extended object (p+1 for a "p-brane") and the dimension of spacetime (D) in which it moves, are severely limited by supersymmetry. Classically, one requires that in a physical gauge there be equal numbers of bosons and fermions on the worldvolume [14]. There are 12 possibilities displayed on the "brane-scan" of Fig. 1. They fall into 4 sequences and the equations of motion for a lower member of the sequence may be obtained from those of a higher member in the same sequence by the process of "simultaneous dimensional reduction"[13]. This is illustrated by the diagonal lines in Fig. 1 which terminate on the strings (p=1) in D=3,4 6 and 10. In particular, the Type IIA superstring in D=10 follows from the supermembrane in D=11. One suspects, therefore, that the string duality

Table 2. Representations of the duality symmetry under which the spin 1 spacetime fields of compactified d=11 supergravity transform. For $n \leq 4$, these are compatible with the representations of the membrane "field strengths" $\tilde{\mathcal{G}}^i{}_\mu$ and $\tilde{\mathcal{F}}^{i\mu\nu}$ only. Extra field strengths must be included for $n \geq 5$.

n	G	spin 1 reps.	$n(n+1)/2$	
1	R	1	1	√
2	$GL(2,R)$	3	3	√
3	$SL(3,R) \times SL(2,R)$	(3,2)	6	√
4	$SL(5,R)$	10	10	√
5	$SO(5,5)$	16	15	×
6	$E_{6(+6)}$	27	21	×
7	$E_{7(+7)}$	56	28	×
8	$E_{8(+8)}$	-	36	×

of Section 2 should follow from the membrane duality of Section 3 by the same simultaneous dimensional reduction. We shall now show that this is indeed the case, using the explicit n=4 example of Section 4.

Le us denote all membrane variables by a carat. Thus the equations of motion (3.5) and Bianchi identities (3.6) now read

$$\partial_{\hat{i}} \hat{\tilde{\mathcal{G}}}^{\hat{i}}{}_{\hat{\mu}} = 0 \tag{6.1}$$

$$\partial_{\hat{i}} \hat{\tilde{\mathcal{F}}}^{\hat{i}\hat{\mu}\hat{\nu}} = 0 \tag{6.2}$$

Similarly, we denote the membrane background fields by $\hat{g}_{\hat{\mu}\hat{\nu}}$ and $\hat{b}_{\hat{\mu}\hat{\nu}\hat{\rho}}$ and the $SL(5,R)$ parameters by $\hat{A}^{\hat{M}}{}_{\hat{N}}$. Here $\hat{i}$ runs over 1 to 3, $\hat{\mu}$ over 1 to 4 and $\hat{M}$ over 1 to 5. In this notation, (3.9) becomes

$$\hat{\gamma}_{\hat{i}\hat{j}} = \partial_{\hat{i}} \hat{x}^{\hat{\mu}} \partial_{\hat{j}} \hat{x}^{\hat{\nu}} \hat{g}_{\hat{\mu}\hat{\nu}} \tag{6.3}$$

Following [13], we now make a two-one split of the worldvolume coordinates

$$\hat{\xi}^{\hat{i}} = (\xi^i, \xi^3), \qquad i = 1,2 \tag{6.4}$$

and a three-one split of the target space coordinates

$$\hat{x}^{\hat{\mu}} = (x^\mu, x^4), \qquad \mu = 1,2,3 \tag{6.5}$$

in order to make the partial gauge choice

$$\xi^3 = x^4 \tag{6.6}$$

which identifies the fourth target space dimension with the third worldvolume dimension. The dimensional reduction is then effected by demanding that

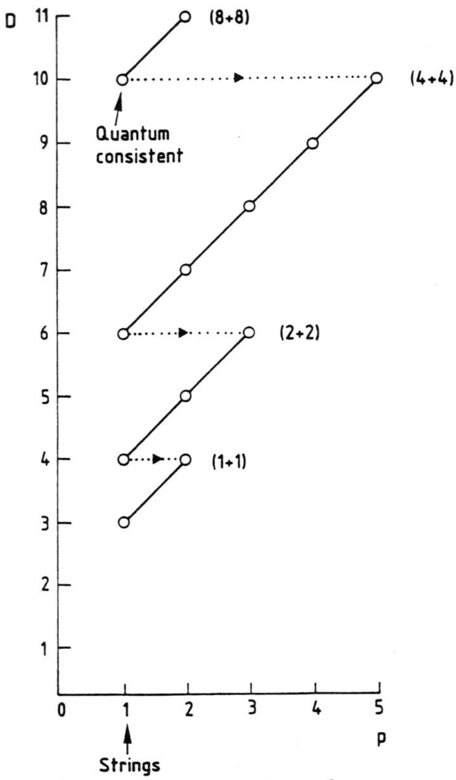

Fig. 1. The "brane-scan".

$$\frac{\partial}{\partial \xi^3} x^\mu = 0 \tag{6.7}$$

(The other requirement of [13], namely

$$\frac{\partial}{\partial x^4} \hat{g}_{\hat{\mu}\hat{\nu}} = 0 = \frac{\partial}{\partial x^4} \hat{b}_{\hat{\mu}\hat{\nu}\hat{\rho}} \ , \tag{6.8}$$

is here superfluous since we are already assuming that the background fields are independent of all $\hat{x}^{\hat{\mu}}$.) A suitable choice of four-dimensional variables is now

$$\hat{g}_{\hat{\mu}\hat{\nu}} = \phi^{-2/3} \begin{pmatrix} g_{\mu\nu} + \phi^2 A_\mu A_\nu & \phi^2 A_\mu \\ \phi^2 A_\nu & \phi^2 \end{pmatrix} \tag{6.9}$$

$$\hat{b}_{\hat{\mu}\hat{\nu}\hat{\rho}} = (\hat{b}_{\mu\nu\rho}, \hat{b}_{\mu\nu 4})$$
$$= (b_{\mu\nu\rho}, b_{\mu\nu}) \tag{6.10}$$

The background fields $g_{\mu\nu}$ and $b_{\mu\nu}$ will be identified with the string backgrounds of Section 2 with $n = 3$, and ϕ will be the string dilaton. The fields A_μ and $b_{\mu\nu\rho}$ are the extra backgrounds that appear in the Type IIA but not the bosonic or heterotic strings. By working in a Green-Schwarz formalism and focussing only on the bosonic sector, we shall see that ϕ, A_μ and $b_{\mu\nu\rho}$ in fact all decouple from the equations of motion. Note, for example, that

$$\widehat{g} = g \tag{6.11}$$

If we now make the following identifications

$$\tilde{\mathcal{F}}^{i\mu} = \widehat{\tilde{\mathcal{F}}}^{i\mu 4} \tag{6.12}$$

$$\tilde{\mathcal{G}}^{i}{}_{\mu} = \widehat{\tilde{\mathcal{G}}}^{i}{}_{\mu} \tag{6.13}$$

and substitute (6.9) and (6.10) into the membrane equations (6.1) and (6.2), we correctly recover the string equations (2.5) and (2.6). The $\widehat{\tilde{\mathcal{F}}}^{\hat{i}\mu\nu}$ and $\widehat{\tilde{\mathcal{G}}}^{\hat{i}}_{4}$ equations are identities, as they must be for consistency. Similarly, if we now make the identifications

$$\begin{aligned} A^{\mu}{}_{\nu} &= \hat{A}^{\mu}{}_{\nu} - \delta^{\mu}{}_{\nu}\hat{A}^{\rho}{}_{\rho} \\ B^{\mu\nu} &= \hat{A}^{5}{}_{\rho}\varepsilon^{\rho\mu\nu} \\ C_{\mu\nu} &= \frac{1}{g}\hat{A}^{\rho}{}_{5}\varepsilon_{\rho\mu\nu}\ , \end{aligned} \tag{6.14}$$

set to zero $\hat{A}^{5}{}_{4}, \hat{A}^{4}{}_{5}, \hat{A}^{4}{}_{4}, \hat{A}^{\mu}{}_{4}$ and $\hat{A}^{4}{}_{\mu}$, and substitute into the membrane transformation rules (4.11), (4.12) and (4.14), we correctly recover the string transformation rules (2.30) and (2.32). Once again, the extra equations not corresponding to string variables are just identities.

Thus in this explicit example, we have seen how the duality symmetry of the string equations of motion for $n = 3$, namely $SO(3,3)$, follows as a consequence of the $SL(5,R)$ duality symmetry of the membrane for $n = 4$. It is not difficult to see, from a group theoretical point of view, how this would work for other values of n. The results are shown in Table 3.

In fact, the string duality symmetries listed in the second column are larger than the $SO(n,n)$ appearing in the third column and discussed in Section 2. The reason is that the $SO(n,n)$ refers to the coset parametrized by $g_{\mu\nu}$ and $b_{\mu\nu}$ only. However, even for the string, spacetime scalars may arise from other sources. First there is the dilaton ϕ which, although decoupling from the bosonic sector, still survives in the fermi-fermi couplings. If we retain the $\hat{A}^{4}{}_{4}$ component of $\hat{A}^{\hat{\mu}}{}_{\hat{\nu}}$, there is an extra $SO(1,1)$ under which $g_{\mu\nu}, b_{\mu\nu}$ and ϕ transform by conformal factors. In four spacetime dimensions we also have the axion b_{MN} (coming from $\hat{b}_{MN11}$ of the D=11 supermembrane) which is dual to a scalar and which, together with ϕ parametrizes the coset $SO(2,1)/U(1)$. In three spacetime dimensions we have 14 more scalars coming from $\hat{g}_{M\hat{\nu}}$ and $\hat{b}_{M\hat{\nu}11}$. These conspire with the dilaton and 49 $g_{\mu\nu}$ and $b_{\mu\nu}$ to parametrize $SO(8,8)/SO(8) \times SO(8)$. For the heterotic string we promote each $SO(n,n)$ to $SO(16+n,n)$ corresponding to the extra 16 left-moving modes. Thus in D=3, we would have $SO(24,8)$.

Of course, we could retain all the spacetime background fields in the dimensional reduction including those that appear only in fermi-fermi couplings and thereby obtain the duality symmetries of the Type IIA superstring. Its duality symmetries would then be given by the first column in Table 3 i.e. the same as those of the d=11 supermembrane.

7. HIGHER EXTENDED OBJECTS

So far we have considered strings (p=1) and membranes (p=2), but similar duality symmetries will be present for other "p-branes" with $p \geq 3$. The Lagrangian takes the form with i = 1,...,(p+1),

$$\mathcal{L} = \frac{1}{2}\sqrt{-\gamma}\gamma^{ij}\partial_i x^\mu \partial_j x^\nu g_{\mu\nu} - \frac{1}{2}(p-1)\sqrt{-\gamma} + \frac{1}{(p+1)!}\,\varepsilon^{ii_1\ldots i_p}\partial_i x^\mu \partial_{i_1}x^{\mu_1}\ldots\partial_{i_p}x^{\mu_p} b_{\mu\mu_1\ldots\mu_p} \tag{7.1}$$

with background fields $g_{\mu\nu}$ and $b_{\mu\mu_1,\ldots,\mu_p} (\mu = 1,\ldots,n)$. Define

$$\mathcal{F}^{i\mu} = \sqrt{-\gamma}\gamma^{ij}\partial_j x^\mu \tag{7.2}$$

$$\tilde{\mathcal{F}}^{i\mu_1\ldots\mu_p} = \varepsilon^{ii_1\ldots i_p}\partial_{i_1}x^{\mu_1}\ldots\partial_{i_p}x^{\mu_p} \tag{7.3}$$

and

$$\tilde{\mathcal{G}}^i{}_\mu = g_{\mu\nu}\mathcal{F}^{i\nu} + \frac{1}{p!}b_{\mu\mu_1\ldots\mu_p}\tilde{\mathcal{F}}^{i\mu_1\ldots\mu_p} \tag{7.4}$$

Then once again there is a symmetry that rotates field equations

$$\partial_i\tilde{\mathcal{G}}^i{}_\mu = 0 \tag{7.5}$$

into Bianchi identities

$$\partial_i\tilde{\mathcal{F}}^{i\mu_1\ldots\mu_p} = 0 \tag{7.6}$$

The dual coordinate now has p indices: $y_{\mu_1\ldots\mu_p}$, and the analogue of (2.19) and (3.22) becomes

$$p!\;\varepsilon^{ii_1\ldots i_p}\partial_{i_1}y_{\mu\mu_2\ldots\mu_p}\partial_{i_2}x^{\mu_2}\ldots\partial_{i_p}x^{\mu_p} = \frac{\partial\mathcal{L}_x}{\partial\partial_i x^\mu}$$

$$= g_{\mu\nu}\sqrt{-\gamma}\gamma^{ij}\partial_j x^\nu + \frac{1}{p!}b_{\mu\mu_1\ldots\mu_p}\varepsilon^{ii_1\ldots i_p}\partial_{i_1}x^{\mu_1}\ldots\partial_{i_p}x^{\mu_p}$$

and

$$\varepsilon^{ii_1\ldots i_p}\partial_{i_1}x^{\mu_1}\ldots\partial_{i_p}x^{\mu_p} = \frac{\partial\mathcal{L}_y}{\partial\partial_i y_{\mu_1\ldots\mu_p}}$$

$$= p!\;p^{\mu_1\ldots\mu_p\;\nu_1\ldots\nu_p}\sqrt{-\gamma}\gamma^{ij}\partial_j y_{\nu_1\ldots\nu_p} + p!\;q^{\mu_1\ldots\mu_p\nu}\varepsilon^{ii_1\ldots i_p}\partial_{i_1}y_{\nu\nu_2\ldots\nu_p}\partial_{i_2}x^{\nu_2}\ldots\partial_{i_p}x^{\nu_p} \tag{7.7}$$

where $p^{\mu\nu} = p^{\nu\mu}$ and $q^{\mu_1\ldots\mu_p} = q^{[\mu_1\ldots\mu_p]}$ are related to $g_{\mu\nu} = g_{\nu\mu}$ and $b_{\mu_1\ldots\mu_p} = b_{[\mu_1\ldots\mu_p]}$ by

$$
\begin{aligned}
p_{\mu\nu} &= g_{\mu\nu} + b_{\mu m} g^{mn} b_{\nu n} \\
p_{\mu\nu} q^{m\nu} &= b_{\mu n} g^{mn}
\end{aligned}
\tag{7.8}
$$

where $p_{\mu\nu}$ is the inverse of $p^{\mu\nu}$. Here we have adopted a condensed notation where the index m means

$$
m \equiv [\mu_1 \ldots \mu_p]
$$

and where a repeated m index means

$$
a_m b^m \equiv \frac{1}{p!} a_{\mu_1 \ldots \mu_p} b^{\mu_1 \ldots \mu_p} \tag{7.9}
$$

The quantity g^{mn} is given by

$$
\begin{aligned}
g^{mn} &\equiv g^{\mu_1 \ldots \mu_p \; \nu_1 \ldots \nu_p} \\
&= \sum_P (-1)^P g^{\mu_1 \nu_1} \ldots g^{\mu_p \nu_p}
\end{aligned}
\tag{7.10}
$$

In this notation, the equations analagous to (2.21) and (4.2) may be written

$$
\begin{aligned}
\tilde{\mathcal{G}}^i{}_\mu &= g_{\mu\nu} \mathcal{F}^{i\nu} + b_{\mu n} \tilde{\mathcal{F}}^{in} \\
\tilde{\mathcal{F}}^{im} &= p^{mn} \mathcal{G}^i{}_n + q^{m\nu} \tilde{\mathcal{G}}^i{}_\nu
\end{aligned}
\tag{7.11}
$$

where

$$
\begin{aligned}
\tilde{\mathcal{G}}^i{}_\mu &= p! \varepsilon^{i i_1 \ldots i_p} \partial_{i_1} y_{\mu \nu_2 \ldots \nu_p} \partial_{i_2} x^{\nu_2} \ldots \partial_{i_p} x^{\nu_p} \\
\mathcal{G}^i{}_n &\equiv \mathcal{G}^i{}_{\nu_1 \nu_2 \ldots \nu_p} = (p!)^2 \sqrt{-\gamma} \gamma^{ij} \partial_j y_{\nu_1 \nu_2 \ldots \nu_p}
\end{aligned}
\tag{7.12}
$$

and where p^{mn} is defined by

$$
g^{mn} = p^{mn} + q^{m\mu} p_{\mu\nu} q^{n\nu} \tag{7.13}
$$

This may be rewritten as

$$
\begin{pmatrix} \tilde{\mathcal{G}}^i{}_\mu \\ \tilde{\mathcal{F}}^{im} \end{pmatrix} = \begin{pmatrix} g_{\mu\nu} + b_{\mu m} g^{mn} b_{\nu n} & b_{\mu m} g^{mn} \\ g^{mn} b_{\nu n} & g^{mn} \end{pmatrix} \begin{pmatrix} \mathcal{F}^{i\nu} \\ \mathcal{G}^i{}_n \end{pmatrix} \tag{7.14}
$$

As remarked in [1] in the context of strings, increasing the dimension of the target space, with coordinates x^μ, to include the extra y_m coordinates is strongly reminiscent of a Kaluza-Klein procedure. This analogy is seen to be even closer when we compare the matrix in (7.14) with the typical Kaluza-Klein decomposition of the metric

$$
G_{MN} = \begin{pmatrix} g_{\mu\nu} + A_\mu{}^m g_{mn} A_\nu{}^n & A_\mu{}^n g_{mn} \\ g_{mn} A_\nu{}^n & g_{mn} \end{pmatrix} \tag{7.15}
$$

Thus the role of the gauge-field $A_\mu{}^m$ is played by the antisymmetric tensor $b_{\mu m}$. Untypical is the fact that the number of "internal" dimensions $\binom{n}{p}$ is determined by the number of "spacetime" dimensions, n, with the curious twist that the role of the "internal" metric g_{mn} is played by g^{mn} which is built out of the inverse of the "spacetime" metric $g^{\mu\nu}$ as in (7.10). Indeed, if we introduce as in [1] a target space with $n + \binom{n}{p}$ dimensions and coordinates

$$Z^M = (x^\mu, y_m) \tag{7.16}$$

the equations of motion and Bianchi identities may be united into a single equation since, after some rearrangement, (7.7) may be written

$$\Omega_{MM_1\ldots M_p} \quad \varepsilon^{ii_1\ldots i_p}\partial_{i_1}Z^{M_1}\ldots\partial i_p Z^{M_p} = G_{MN}\sqrt{-\gamma}\gamma^{ij}\partial_j Z^N \tag{7.17}$$

where G_{MN} is the matrix appearing in (7.14) and $\Omega_{MM_1\ldots M_P} = \Omega_{M[M_1\ldots M_P]}$ is a numerical tensor whose non-vanishing components are given by

$$\begin{aligned}\Omega_\mu{}^{\alpha\beta\ldots\gamma}{}_{\nu\ldots\sigma} &= \frac{1}{p!p}\delta^{\alpha\beta\ldots\gamma}_{\mu\nu\ldots\sigma}\\ \Omega^{\alpha\beta\ldots\gamma}{}_{\mu\nu\ldots\sigma} &= \frac{1}{p!}\delta^{\alpha\beta\ldots\gamma}_{\mu\nu\ldots\sigma}\end{aligned} \tag{7.18}$$

where

$$\delta^{\alpha\beta\ldots\gamma}_{\mu\nu\ldots\sigma} = \sum_P(-1)^P\delta^\alpha{}_\mu\delta^\beta{}_\nu\ldots\delta^\gamma{}_\sigma \tag{7.19}$$

Note that

$$\Omega_{[MM_1\ldots M_p]} = 0 \tag{7.20}$$

and so, multiplying both sides of (7.17) by $\partial_i Z^M$ we learn that

$$\sqrt{-\gamma}\gamma^{ij}\partial_i Z^M G_{MN}\partial_j Z^N = 0 \tag{7.21}$$

as may also be verified explicitly.

Although (7.17) is an elegant way of summarizing the combined field equations and Bianchi identities of an arbitrary p-brane in a target space of arbitrary D, one must not be lulled into thinking that all the hidden symmetries are thus rendered manifest. This is because, with the exception of the string, Ω is not an invariant tensor under the full duality transformations and $\partial_j Z^M$ does not transform as a vector. Rather, the manifest symmetry is only a subgroup of the full duality group, and turns out to be $SL(n,R)\times R^{\binom{n}{p}}$, under which γ_{ij} does not transform. To obtain the complete duality symmetry we must adopt a different route along the lines described in Section 3. What will these p-brane symmetries be?

The first observation to make is that a p-brane couples to a (p+1)-form background $b_{MM_1\ldots M_P}(M=1,\ldots,D)$ whose field strength $F=db$ is a (p+2)-form. But in D spacetime dimensions a (p+2)-form F is dual (in the sense of Poincare duality) to a (D−p−2)-form $\tilde{F}=d\tilde{b}$ where $\tilde{b}$ is (D−p−3) form that couples to (D−p−4)-brane. Hence we expect the duality symmetry of a p-brane in D dimensions compactified on T^n to be the same as that of (D−p−4)-brane in D dimensions compactified on T^n. A good example is provided by the string in D=10 and the 5-brane in D=10. The former couples to the background fields of D=10 supergravity with a 2-form b_{MN}, while the latter couples to the fields of the dual formulation of D=10 supergravity in which the 2-form is replaced by a 6-form $\tilde{b}_{MNPQRS}$. Thus we anticipate that the 5-brane duality symmetries will be exactly the same as those for the string listed in Table 3. Of course, to achieve this it will be necessary, as described in Section 5, to augment the n-dimensional background scalar fields $g_{\mu\nu}$ and

Table 3. String duality from membrane duality via simultaneous dimensional reducton (→). The string symmetries are actually larger than $SO(n,n)$ where n is the number of compactified string dimensions.

n	membrane duality	→	string duality	$SO(n,n)$
1	$GL(2,R)$	→	$SO(1,1)\times SO(1,1)$	$SO(1,1)$
2	$SL(3,R)\times SL(2,R)$	→	$SO(2,2)\times SO(1,1)$	$SO(2,2)$
3	$SL(5,R)$	→	$SO(3,3)\times SO(1,1)$	$SO(3,3)$
4	$SO(5,5)$	→	$SO(4,4)\times SO(1,1)$	$SO(4,4)$
5	$E_{6(+6)}$	→	$SO(5,5)\times SO(1,1)$	$SO(5,5)$
6	$E_{7(+7)}$	→	$SO(6,6)\times SO(2,1)$	$SO(6,6)$
7	$E_{8(+8)}$	→	$SO(8,8)$	$SO(7,7)$

$\tilde{b}_{\mu\nu\rho\lambda\sigma\tau}$ with those arising from other sources and similarly for the field-strengths $\tilde{\mathcal{G}}^i{}_\mu$ and $\tilde{\mathcal{F}}^{i\mu\nu\rho\lambda\sigma}$.

Note, incidentally that whereas in spacetime one must replace the *field-strengths* $F_{MNN_1\ldots N_p}$ of the gauge (p+1)-forms by their duals in order to get equivalent degrees of freedom, for the compactified coordinates (where the (p+1)-form potentials are spacetime scalars) one must replace the *potentials* $b_{\mu\mu_1\ldots\mu_p}$ themselves by their duals. Similar remarks apply when working with the spacetime fields in the light-cone gauge rather than covariantly.

If we consider the full superspace (x^μ,θ^α) couplings and keep all the background fields, then the duality symmetry is preserved by the simultaneous dimensional reduction e.g. as discussed in Section 6, the D=11 membrane duality leads to the D=10 Type IIA superstring duality. Thus there are really just four duality schemes corresponding to the four sequences on the "brane-scan" of Fig. 1. Just as the largest duality group is $E_{8(+8)}$ for the octonionic sequence, a process of counting degrees of freedom and truncating leads to the three groups shown in Table 4 for the quaternionic, complex and real sequences when compactified to D=3 spacetime dimensions. Of course, one might conjecture that these may be enlarged even further to infinite-dimensional symmetries, for example E_9 [4] and E_{10} [5,6,19] in the octonionic case.

8. CONCLUSIONS

We have seen how the Cremmer-Julia hidden symmetries of supergravity have their origin in duality transformations on the three-dimensional worldvolume of the d=11 supermembrane, and how the string duality symmetries follow from those of the membrane by simultaneous dimensional reduction. The duality symmetries for a general p-brane in D spacetime dimensions, will be the same as those of (D−p−4)-brane, to which it is related by Poincare duality, an example being provided by the string and the 5-brane in D=10. Several questions now spring to mind.

First, we have succeeded in writing the equations in a manifestly duality invariant way. Since these symmetries are not symmetries of the p-brane action, however, their presence will never be obvious starting from the σ-model with physical background scalar fields $g_{\mu\nu}$ and $b_{\mu\mu_1\ldots\mu_p}$. However, it would be interesting to

Table 4. Maximal (finite dimensional) duality symmetries for the 4 sequences of extended objects.

Sequence	G	H	$dim\ G/H$
O	$E_{8(+8)}$	$SO(16)$	128
H	$SO(8,8)$	$SO(8) \times SO(8)$	64
C	$SO(4,4)$	$SO(4) \times SO(4)$	16
R	$SO(2,2)$	$SO(2) \times SO(2)$	4

see whether we could write the p-brane action coupled to <u>all</u> the scalars of $E_M{}^A$ which describes both physical and unphysical modes. In this way, at least the local group H, which is the maximal compact subgroup of G, might then be manifest.

Secondly, all the continuous symmetries discussed here follow by demanding that the background fields are independent of the compactified coordinates. This corresponds to a naive dimensional reduction in which it is not even necessary to specify the topology of the extra dimensions. In reality, we must pick a specific topology and geometry e.g. the flat torus T^n, and keep all the Fourier modes. These symmetries will then be broken by the massive states. Morever, there will be quantization conditions imposed by the torus topology. For the string, the τ components of both $\tilde{\mathcal{G}}^i{}_\mu$ and $\tilde{\mathcal{F}}^{i\mu}$ will be quantized, thus

$$\begin{aligned} P_\mu &\equiv \tilde{\mathcal{G}}^\tau{}_\mu = y'_\mu = m_\mu \\ A^\mu &\equiv \tilde{\mathcal{F}}^{\tau\mu} = x'^\mu = n^\mu \end{aligned} \tag{8.1}$$

where m_μ and n^μ are integers. From the point of view of the origial σ-model $\mathcal{L}_x$, m_μ corresponds to the momentum modes and n^μ to the winding modes. Whereas from the point of view of the dual σ-model $\mathcal{L}_y$, the roles are reversed [1]. Thus there is a discrete subgroup of $SO(n,n)$ given by

$$\begin{aligned} \begin{pmatrix} n \\ m \end{pmatrix} &\to S^{-1} \begin{pmatrix} n \\ m \end{pmatrix} \\ G &\to S^T G S \end{aligned} \tag{8.2}$$

where G is given by (2.25). This is the $SO(n,n;Z)$ referred to in the Introduction and which leaves the string spectrum invariant. Thus the question arises whether there are also discrete subgroups of the membrane duality symmetries which leave the membrane spectrum invariant. The equations analogous to (8.1) are

$$\begin{aligned} P_\mu &\equiv \tilde{\mathcal{G}}^\tau{}_\mu = 2\{x^\nu,\ y_{\mu\nu}\} \\ A^{\mu\nu} &\equiv \tilde{\mathcal{F}}^{\tau\mu\nu} = \{x^\mu, x^\nu\} \end{aligned} \tag{8.3}$$

where the Lie bracket is defined by

$$\{X, Y\} = \varepsilon^{ab}\partial_a X \partial_b Y \tag{8.4}$$

Unfortunately, this is a difficult question to answer since the membrane spectra are, as yet, unknown. (This is primarily because the usual light-cone gauge [15] action for a p-brane is highly non-linear except for p=1. An alternatiave gauge which linearizes the equations of motion has been proposed [16], at the expense of introducing a highly non-linear constraint. Incidentally, on the subject of the lightcone gauge, we note that except for p=1, it is inhomogeneous in the components of γ_{ij}, namely

$$\gamma_{ij} = \begin{pmatrix} -h & o \\ o & h_{ab} \end{pmatrix} \tag{8.5}$$

where $h \equiv deth_{ab}$ and $a, b = 1, \ldots, p$. Consequently, the duality symmetry, under which γ_{ij} rescales by a conformal factor (4.20), would be obscured in this gauge. The conformal gauge of [16], which for p=2 looks like

$$\gamma_{ij} = \begin{pmatrix} {A_1}^2 & -A_1A_2 & -A_1A_3 \\ -A_1A_2 & {A_2}^2 & -A_2A_3 \\ -A_1A_3 & -A_2A_3 & {A_3}^2 \end{pmatrix} \tag{8.6}$$

would be much more suitable.) If such discrete subgroups do leave the membrane spectrum invariant, does the $R \to \alpha'/R$ idea continue to apply, and hence does the idea of a "minimum length" also hold for membranes? We would expect so, since as we saw in Section 6, G (string) $\subset G$ (membrane). It would also be interesting to see whether the fixed points correspond to enhanced gauge symmetries.

Finally, is the similarity of the duality symmetry of the string in D=10 and the 5-brane in D=10 indicative of a deeper relationship? The two formulations of D=10 supergravity, one with a 3-form field strength and one with a 7-form field strength, has long been something of an enigma from the point of view of superstrings. As field theories, each seems equally as good. In particular, provided we couple them to $E_8 \times E_8$ or $SO(32)$ Yang-Mills, then both are anomaly-free [17,18]. Since the 3-form version corresponds to the low-energy limit of the heterotic superstring (or Type I superstring) it is naturally to conjecture, as was done some time ago [19], that there exists a "heterotic 5-brane" (or "Type I 5-brane") whose low energy limit is the 7-form version. However, this would require a coupling of the super-5-brane to background Yang-Mills fields which, to date, has not been achieved. It was speculated that such a coupling might exist, making use of the property that a one-time and five-space dimensional worldvolume admits real self-dual three forms (and/or Majorana-Weyl spinors).

This possibility has now become a virtual certainty thanks to the recent remarkable observation by Strominger [20] that the heterotic string admits the heterotic 5-brane as a soliton solution. The soliton interpretation of membranes was the motivation for the original supermembrane paper of Hughes, Liu and Polchinski [21] and has also been pursued by Townsend [22]. Strominger went on to conjecture that the heterotic string and heterotic 5-brane might be "dual"in the sense of Olive and Montonen [23] and that the 5-brane describes the strong coupling limit of the string. Clearly, the many meanings of the word "duality" in theories of extended objects have still not been exhausted.

ACKNOWLEDGMENTS

The work described here was carried out in collaboration with J. X. Lu.

REFERENCES

1. M.J. Duff, Nucl. Phys. B335 (1990) 610.

2. K. Kikkawa and M. Yamasaki, Phys. Lett. 149B (1984) 357. N. Sakai and I. Senda, Prog. Theor. Phys. 75 (1986) 692. V. P. Nair, A. Shapere, A. Strominger and F. Wilczek, Nucl. Phys. B287 (1987) 414. A. Shapere and F. Wilczek, Nucl. Phys. B320 (1989) 669. A. Giveon, E. Rabinovici and G. Veneziano, Nucl. Phys. B322 (1989) 167. M. Dine, P. Huet and N. Seiberg, Nucl. Phys. B322 (1989) 301. P. Ginsparg and C. Vafa, Nucl. Phys. B289 (1987) 414. B. Sathiapaian, Phys. Rev. Lett. 58 (1987) 1597. R. Brandenberger and C. Vafa, Nucl. Phys. B316 (1989) 391, C. Vafa, Mod. Phys. Lett. A4 (1989) 1615. J.M. Molera and B. Ovrut, Phys. Rev. D40 (989) 1146. E. Alvarez and M.A. R. Osorio, Phys. Rev. D40 (1989) 1150. J. Laver, J. Mas and H.P. Nilles, Phys. Rev. Lett. 226B (1989) 251.
3. S. Ferrara, J. Scherk and B. Zumino, Nucl. Phys. B121 (1977) 393. E. Cremmer, J. Scherk and S. Ferrara, Phys. Lett. 68B (1977) 234. E. Cremmer and J. Scherk, Nucl. Phys. B127 (1977) 259. E. Cremmer, J. Scherk and S. Ferrara, Phys. Lett. 74B (1978) 61. N. Marcus and J. H. Schwarz, Nucl. Phys. B228 (1983) 145. J. P. Derendinger and S. Ferrara, in Supersymmetry and Supergravity '84, eds. B. de Wit, P. Fayet and P. van Nieuwenhuizen (World Scientific, Singapore, 1985). M. De Roo, Nucl. Phys. B255 (1985) 515; Phys. Lett. 156B (1985) 331. E. Bergshoeff, I. G. Koh and E. Sezgin, Phys. Lett 155B (1985) 646. L. Castellani, A. Ceresole, S. Ferrara, R. D'Auria, P. Fre and E. Maina, Nucl. Phys. B268 (1986) 317; Phys. Lett. 161B (1985) 91.
4. E. Cremmer and B. Julia, Nucl. Phys. B139 (1979) 141. E. Cremmer, In Supergravity '81, eds. S. Ferrara and J. G. Taylor (C.U.P. 1982). B. Julia, in Superspace and Supergravity, eds. S. W. Hawking and M. Rocek (C.U.P. 1980).
5. M.J. Duff, Phys. Lett. 173B (1986) 289.
6. M. J. Duff in Quantum Field Theory and Quantum Statistics, eds. I.A. Batalin, C.J. Isham and G. Vilkovisky, (Adam Hilger, 1986). M. J. Duff in Physics in Higher Dimensions, eds. T. Piran and S. Weinberg (World Scientific, 1986). M. J. Duff in Architecture of the Fundamental Interactions at Short Distances, eds. P. Ramond and R. Stora (North Holland, 1987).
7. K. S. Narain, Phys. Lett. 169B (1986) 61.
8. E. Bergshoeff, E. Sezgin and P. K. Townsend, Phys. Lett. 209B (1988) 451.
9. M. J. Duff, B.E. W. Nilsson and C.N. Pope, Phys. Rep. 130 (1986) 1.
10. B. de Wit and H. Nicolai, Phys. Lett. 108B (1982) 285.
11. S. Cecotti, S. Ferrara and L. Girardello, Nucl. Phys. B308 (1988) 436.
12. M. K. Gaillard and B. Zumino, Nucl. Phys. B193 (1981) 221.
13. M.J. Duff, P.S. Howe, T. Inami and K. S. Stelle, Phys. Lett. 191B (1987) 70.
14. A. Achucarro, J.M. Evans, P. K. Townsend and D. L. Wiltshire, Phys. Lett. 1988 (1987) 441.
15. J. Hoppe, MIT PhD Thesis, 1982.
16. I. Bars, Membrane Symmetries and anomalies, Preprint USC-89/HEP 14, to appear in Supermembranes and Physics in 2+1 Dimensions, eds. M. J. Duff, C. N. Pope and E. Sezgin (World Scientific, to be published).
17. M. B. Green and J. H. Schwarz, Phys. Lett. 149B (1984) 117.
18. S. J. Gates and H. Nishino, Phys. Lett. 173B (1986) 52.
19. M. J. Duff, Class Quantum Grav. 5 (1988) 189; M. J. Duff in *The Superworld II* (ed. A. Zichichi, Plenum Press, 1990).
20. A. Strominger UCSB-TH-90-16.
21. J. Hughes, J. Liu and J. Polchinski, Phys. Lett. 1808 (1986) 370.
22. P. K. Townsend, Phys. Lett. 202B (1988) 53.
23. C. Montonen and D. Olive, Phys. Lett. 72B (1977) 117.

Chairman: M.L. Duff

Scientific Secretaries: J. Galvez, G. Mangano, M. Wexler

Discussion

– *U. Vikas:*

I don't seem to understand how you intend to get a scenario like duality which is in strings coming from the winding modes in the membranes.

– *Duff:*

Neither do I. In the case of the string, the duality was a swapping of two objects of the same kind. We had these quantities F and G that each had a single index, which when restricted to the torus take on integer values. The duality consisted of an interchange of the two integers. Now in the case of the membrane, we are not rotating a quantity with one index into a quantity with one index, but we are rotating a quantity with one index into a quantity with two indices; and clearly that simple-minded interchange of integers is not going to work in the case of the membrane. That was why I completed my discussion on that issue with a question mark: will there be—as is the case for the string—discrete subgroups of these continuous membrane symmetries, which correspond, in some sense, to the discrete subgroups that we know for the string. And I do not know the answer for that question. All I can say is that since the string symmetries are subgroups of the membrane symmetries, and since we know that there are discrete subgroups of the string symmetries which leave the string spectrum invariant, it is not unreasonable to suppose that a similar phenomenon will happen for membranes. We do not know the answer to the question for the simple reason that we do not know what is the spectrum of the membranes. And the reason we do not know the spectrum is that the equations of motion for the membrane— unlike those for the string— are nonlinear, and we simply do not have any exact solution for those nonlinear equations. In fact, one of the major problems of membrane physics is dealing with this intrinsic nonlinearity that you do not have in the case of the string.

– *U. Vikas:*

I cannot foresee a concept of even a minimum length entering into membranes. I am agreeing that the string case there might be a minimum length, but in the case of membranes can you introduce a concept like a length?

– *Duff:*

That minimum length, that discrete symmetry is a subgroup of SO(1,1) and SO(1,1) is a subgroup of the membrane duality, GL(2,R). So it is not implausible that minimum length would continue to apply to the membrane. But we do not have a formula like

$$M^2 = n^2R^2 + \frac{m^2}{R^2}$$

for the membrane yet because no one knows how to solve the equations.

– *Bodner:*

You mentioned that for $p > 1$ we only have $N = 1$ theories. Can you explain how you obtain this condition?

– *Duff:*

I am not sure that there is a simple hand-waving explanantion; you have to construct the Green-Schwarz superstring action. It is really to do with the requirement that you have equal numbers of bosons and fermions. In my lecture I showed the dimensions of a minimal spinor in the various representations; for example, in ten dimensions the spinor with the fewest number of components is Majorana-Weyl. In the case of eleven dimensions it is Majorana. If you count this number once that corresponds to one supersymmetry; if you wanted two supersymmetries you would have to double the number of fermionic degrees of freedom. It simply turns out that if you look at the cases for $p > 1$, you see that in order to match bosons and fermions, you are only allowed to take the minimal spinor; if you doubled it in order to get two supersymmetries or three, you would find that you would have more fermions than bosons. That logic breaks down in the case of the string, for the simple reason that on the string you have left- and right- movers that can be treated separately, and there you can either have $N = 1$ or $N = 2$, depending on whether you have supersymmetry for left- and right-movers, or just supersymmetry for one of them. It is really a matter of equality of bosons and fermions.

– *Lu:*

Could explain the form of the Lagrangian that you used for the membrane?

– *Duff:*

It consists of three terms: a kinetic term, $\frac{1}{2}\sqrt{-\gamma}\gamma^{ij}\partial_i X^\mu \partial_j X^\nu g_{\mu\nu}$, a so-called world-volume cosmological term, $-\frac{1}{2}\sqrt{-\gamma}$, and a Wess-Zumino term, $\frac{1}{3!}\epsilon^{ijk}\partial_i X^\mu \partial_j X^\nu \partial_k X^\rho b_{\mu\nu\rho}$. It differs from the string into two respects. In the case of the string, we had the Wess-Zumino term, and the usual kinetic term. The kinetic term is so designed that if you vary it with respect to the independent field γ, and then substitute back into the equations of motion, you get the same equations of motion as you do from the Nambu-Goto action,

$$L_{NG} = \sqrt{\det \partial_i X^\mu \partial_j X^\nu g_{\mu\nu}}$$

The Polyakov form of the Lagrangian is where you treat the coordinate X and the metric γ as independent variables. To either of these you can add the Wess-Zumino term, which in the case of the string involves the two-index alternating symbol, and consequently couples to a two-index antisymmetric tensor field, $b_{\mu\nu}$. When we come to the membrane, we have the same kind of kinetic term, but that alone is not equivalent to the Nambu-Goto action in three dimensions because two dimensions is special in this respect, in that if you rescale the metric by some arbitrary function of the coordinates, the Lagrangian remains invariant. That is not true as you go to p-branes for $p > 1$. If you want a Lagrangian that is equivalent to the Nambu-Goto Lagrangian, you have to add the extra world-sheet cosmological term. The net result of that is that if you vary with respect to γ and after a little rearrangement, you learn that its equation of motion is just that γ is the induced metric on the three-dimensional world-volume:

$$\gamma_{ij} = \partial_i X^\mu \partial_j X^\nu g_{\mu\nu}$$

You would not get this equation if you did not have this extra world-volume cosmological term, and in general, when you go to a p-brane, it has some coefficient $p - 1$, so it vanishes only in the case of the string. If you plug this equation back into the Lagrangian, you get the Nambu-Goto Lagrangian back again. As in the case of the string we are free to add a Wess-Zumino term:

$$L_{WZ} = \epsilon^{ijk} \partial_i X^\mu \partial_j X^\nu \partial_k X^\rho b_{\mu\nu\rho}$$

However in the case of the membrane, because it is a three-dimensional surface, the alternating symbol now has three indices, and this is one of the sources of the nonlinearity.

This same procedure works in the bosonic case for any p; you need the cosmological coefficient $p - 1$ and a Wess-Zumino term which has correspondingly more indices as you increase the dimension of the world-volume.

– *Lu:*

Is the Wess-Zumino term the most general term that we can add without spoiling the Nambu-Goto Lagrangian?

– *Duff:*

It is the most general term we can add which preserves the three-dimensional diffeomorphism group without introducing new propagating degrees of freedom. You could always add curvature terms but they would change the number of propagating degrees of freedom.

– *Grossmann:*

By simultaneous dimensional reduction you related theories with different p and D, but you never went to $p = 0$. Would that also be possible?

– *Duff:*

Yes. You can continue down these diagonal lines and obtain superparticles in dimensions 2, 3, 5, and 9. The logic that says you have equal numbers of bosons and fermions no longer applies in one dimension. In fact you can have superparticles in other dimensions as well, which you cannot reach by simultaneous dimensional reduction.

– *Bimonte:*

You said that Strominger was led to the discovery of 5-branes by the consideration of extended soliton solutions of the heterotic string. Would you comment on this?

– *Duff:*

Hughes, Liu, and Polchinski and also Townsend discovered that many of these extended objects—in particular all those in R, C, and H sequences—can be regarded as the solitons of some supersymmetric field theory.

For example consider the quaternion sequence. In four Euclidean dimensions, if you have in a Yang-Mills theory equations

$$\nabla_\mu F^{\mu\nu} = 0$$

we know that there are instanton solutions in which you take self-dual field configurations $F = \pm\tilde{F}$. This is in Euclidean space, where we have four space and no time dimensions. You can do the same thing for super-Yang-Mills. It is known that there are certain moduli associated with this instanton solution—certain zero modes—for example, because of translation invariance, there are four translational zero modes associated with the instanton. In the super-Yang-Mills case there are correspondingly fermionic zero modes as well, coming from the Atiyah-Singer index theorem. Townsend observed that you can do the opposite of this simultaneous dimensional reduction by adding a dimension. Let's begin by adding a time coordinate. From a five-dimensional point of view, this instanton looks like a particle or a monopole solution. If we go up once more, to five space and one time dimensions, it looks like a string. Similarly, 6 space and 1 time looks a membrane; 7 space and 1 time is a 3-brane; 8 space and 1 time is a 4-brane; and finally, 9 space and 1 time is a 5-brane.

– *Bimonte:*

So you get solitons from instantons?

– *Duff:*

That is correct. In the supersymmetric case you get extended objects which have not only bosonic zero modes associated with them, but fermionic ones. So in a supersymmetric theory there is a super-extended object which is the soliton of that theory. All these objects can be explained as the solitons of a supersymmetric Yang-Mills theory. The interesting thing is that this logic also explains why it stops at a 5-brane: because supersymmetric Yang-Mills—as is well known—only exists in spacetime dimensions less than or equal to ten as the result of the supersymmetry. That tells you why you cannot go beyond ten dimensions. You can apply similar arguments to the real and the complex sequences by starting with an Abelian Higgs theory and the nonlinear sigma model, respectively. Curiously, the octonionic sequence does not yet have any soliton interpretation; the only theories of which they could be the soliton are the theories to which they couple, namely ten- and eleven-dimensional supergravity.

The sequence which I described here—the one that Townsend looked at—has in fact 4+4 degrees of freedom. In other words, he looked only at the translation modes associated with the instanton and in the superpartners. What should have been noticed some time ago is that there ought to be a more generalized extended object that takes into account not only the translation modes of the instanton, but also the internal symmetry rotation mode and the dilaton mode. Strominger observed that there ought to be some new 5-brane or a whole sequence of objects which have a different counting of bosons and fermions than the ones that we are used to. He even went one step further, and looked not at pure ten-dimensional Yang-Mills, but ten- dimensional Yang-Mills coupled to ten-dimensional supergravity; in fact he looked at the low energy limit of the ten-dimensional heterotic string. But the trick was very similar: he found an instanton solution in four dimensions, and by the process of reverse double dimensional reduction was able to jazz it up into a 5-brane solution in ten dimensions. But now it is one of the coupled gravity-Yang-Mills system, it is different than the one you get from the pure Yang-Mills system. That is the way Strominger arrived at his heterotic 5-brane as the soliton solution of the low-energy field theory of the heterotic string.

– *Wexler:*

I have been told that for the bosonic string on the circle, duality is broken in a phase transition. Do you think that result might hold also for the superstring, and for higher-dimensional objects?

– *Duff:*

Which phase transition did you have in mind?

– *Wexler:*

There is a Kosterlitz-Thouless phase transition that comes out of matrix models that was shown by Gross and Klebanov to break duality.

– *Duff:*

I know in perturbation theory duality is not broken. Is this a non-perturbative effect?

– *Wexler:*

Yes.

– *Duff:*

I imagine it might very well apply to superstrings.

– *Wexler:*

Do you think it could affect the string-membrane duality as well?

– *Duff:*

I do not know, I suspect that if it is broken for one it may well be broken for the other, but I do not see any evidence that you could spoil the apparent dual relationship that the two theories enjoy. The word "duality" is used in many different contexts which, though sometimes related, should sometimes also be distinguished. There is the individual duality that the string enjoys, like the $R \to 1/R$ symmetry; I showed that the membrane also has similar duality symmetries. These exist whether or not there is a Strominger-type string-membrane duality, because these dualities were concerned with rotating field equantions into Bianchi identities on the world-volume; whereas this duality is more to do with exchanging field equations with Bianchi identities in the ten-dimensional spacetime. So if it were true that there was some non- perturbative breaking of the string duality or even a membrane duality, it does not necessarily say anything about the string-membrane duality.

– *Hsu:*

Can you tell us what happens to the Regge behavior of the string-string scattering in the case of higher p-branes?

– *Duff:*

I do not know the answer to this very interesting question. There is a curious phenomenon: there is the Regge slope between the spin and the square of the mass which involves the slope parameter, α', which has dimensions of length squared. In string language, α' is the inverse of the string tension, and the string tension is the constant that multiplies the action and makes it dimensionless:

$$S = T_1 \int d^2\xi \frac{1}{2}\sqrt{-\gamma}\gamma^{ij}\partial_i X^\mu \partial_j X^\nu g_{\mu\nu}$$

If you go to a p-brane, you have a similar tension that multiplies the action

$$S = T_p \int d^{p+1}\xi \cdots$$

but it has different dimensions because the world-volume of the extended object has a different number of dimensions. So on dimensional grounds—if this were the only dimensionful parameter in the theory—you could never get linear Regge trajectories in the theory. However, if you consider compactifying the p-brane, the very process of compactification introduces a new length scale (for example, if you compactify on the circle, then you inroduce the radius of the circle). Once you have a second dimensional parameter then that argument no longer applies, and in fact, double dimensional reduction consists of identifying the extra dimension in spacetime with the extra dimension on the world-volume, and then compactifying both of them simultaneously and shrinking the radius to zero. In that process you recover the string, hence you recover the string spectrum. So although the membrane spectrum is richer than the string spectrum, in the sense that you have to truncate it to get from one to the other, there is still a string inside that spectrum obeying a Regge-type formula. Whether that means that there is some sort of Veneziano-style scattering amplitude for p-branes, I do not know. It is a very important question.

– *Hsu:*

Minimum distance sounds to me like a discrete spacetime. I recall some work of Klebanov and Susskind on formulating strings on a discrete spacetime. Could you comment?

– *Duff:*

My only comment is that I do not know. Perhaps, since you have raised the subject, I could clarify the question of minimum length. There are really two issues which should be kept separate. One is whether a theory which has a duality symmetry necessarily has a minimum length. The second question is whether string theory has such a symmetry. Now I think it is clear that any theory that has an $R \to \alpha'/R$ symmetry certainly has a minimum length, because symmetry means that the theory cannot distinguish between length and $(\text{length})^{-1}$. If the minimum length is 10^{-33} cm then life at 10^{-34} cm is identical to life at 10^{-32} cm.

– *Wadhwa:*

For $p = 1$, we have one or two strings but for $p > 1$ we have only one string. Could you comment on this?

– *Duff:*

I said that some of the points on the Brane-Scan correspond to more than one theory. These twelve points represent the only values of spacetime dimension and world-volume dimension where you can have a supersymmetric extended object. However, it does not follow that each point corresponds to only one theory, and to illustrate this we only need to look at the ten-dimensional superstring, and we know that there are at least three consistent ten-dimensional superstrings: the type IIA superstring, which has $N = 2$ supersymmetry; the type IIB, which also has $N = 2$; and the heterotic string. So this single point could represent as many as three different string theories. But the case $D = 11$ is unique, there is only one eleven-dimensional supermembrane. What about the other points? Up until recently we believed the ten-dimensional 5-brane is also unique; we have now learned that in fact it is not. In addition to the membrane that we already knew, there is one that carries internal quantum numbers, and in particular there is the heterotic 5-brane which carries the $E_8 \times E_8$ or $SO(32)$ quantum numbers. So there is more than one solution at this point.

– *Wadhwa:*

Can we apply these rules to the physical states of fermions or bosons?

– *Duff:*

The original 5-brane that we knew about had four bosonic and four fermionic degrees of freedom; that came from the translational modes of the four-dimensional instanton. We could already have told two or three years ago that this is not the whole story. We know that the instanton has other zero modes, namely the dilatation and internal symmetry zero modes; the number of these would depend on the gauge group in question. From the Atiyah-Singer index theorem with the corresponding gauge group you would discover many more, depending on the gauge group in question, which would lead to many more fermionic and bosonic degrees of freedom. There are more solutions at this point than we first believed; but they all have equal number of bosons and fermions.

FIRST RESULTS AT THE LEP e^+e^- COLLIDER

J. Steinberger

Scuola Normale Superiore
Pisa, Italy and
CERN, Geneva, Switzerland

ABSTRACT

The first year of LEP running was devoted to the study of Z production and decay on the resonance. Results of the four collaborations, based on $\sim$ 400'000 decays, show that the number of fermion families is three, give a precise measure of the Z mass, furnish tests of the electroweak theory at a new level sensitivity, and constrain the top mass.

LEP

The summer of 1989 saw the first collisions at LEP. Within a few months the four experiments: ALEPH, DELPHI, OPAL and L3 were able to produce results which a) demonstrated that the number of fermion families is three, b) measured the mass of the Z to 1 part in 3000, c) checked the validity of the electroweak theory with a sensitivity of 1%, and d) provided new upper limits on the masses of various postulated new particles, including the Higgs. As of this writing, more than sixty papers have been published, or submitted for publication. In these lectures some of these results are reviewed.

The LEP collider produces collisions between electrons and positrons up to energies of 55 on 55 GeV. This upper energy limit is expected to increase gradually to 90 on 90 GeV in 1994. The 27 km long tunnel is below the plane which separates Geneva, Switzerland from the French Jura, as shown in fig. 1. It is the largest accelerator in the world but not the most energetic, by far. Even its final energy will be ten times less than that of the Tevatron, the present record holder for $p\bar{p}$ colliders, with only one-third the diameter of LEP. This is the price for accelerating electrons and positrons, which, because of their low mass, radiate heavily. The

Physics Up to 200 TeV
Edited by A. Zichichi, Plenum Press, New York, 1991

large radius is a compromise between construction cost and the cost of supplying the R.F. power to replenish the radiated energy. The machine is located in a tunnel 50 to 150 m underground (fig. 2). Four bunches of electrons and positrons produce collisions at four intersection points, 40,000 times per second. The bunch size at collision is $\sim$ 20 mm long, 300 μ transverse horizontally and 10 μ vertically. The luminosity is expected to be limited at $1.6 \times 10^{31} \text{cm}^{-2} \text{sec}^{-1}$ (up to now about one half of this has been reached at 50 GeV).

ALEPH

Figure 3 shows one of the four experimental areas. The experiments are mounted in large underground vaults. On the surface, various buildings house the services, both for the accelerator and the experiment. The vaults measure about 20 m diam. $\times$ 60m in length.

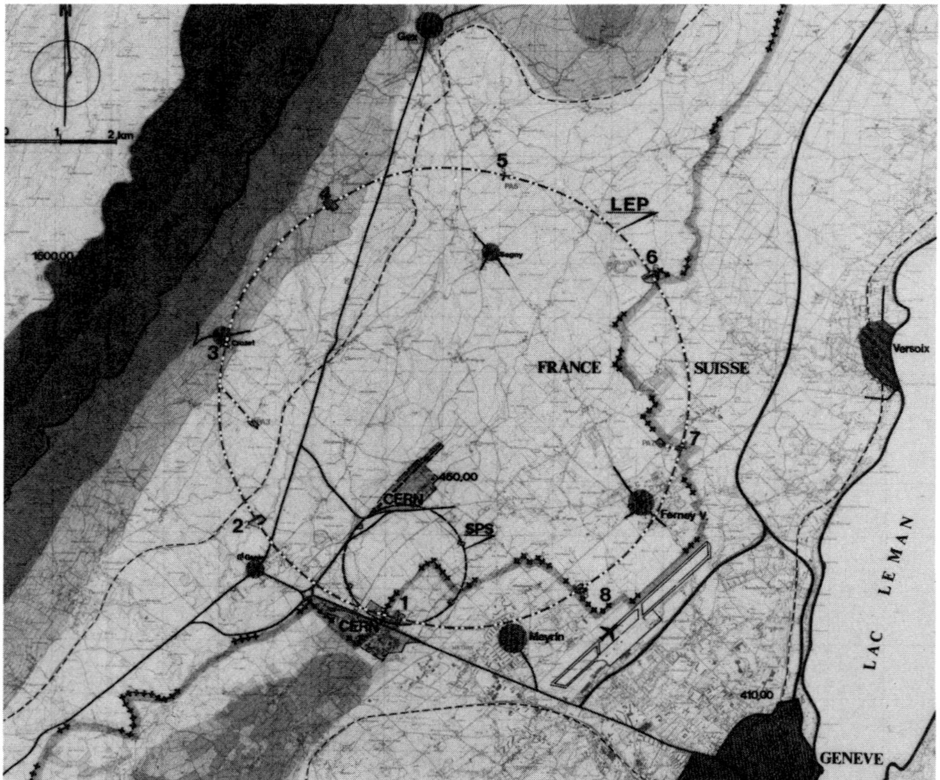

Fig. 1. Siting of the LEP 27 km long tunnel, between the Jura and the Geneva airport.

Fig. 2. View into the LEP tunnel.

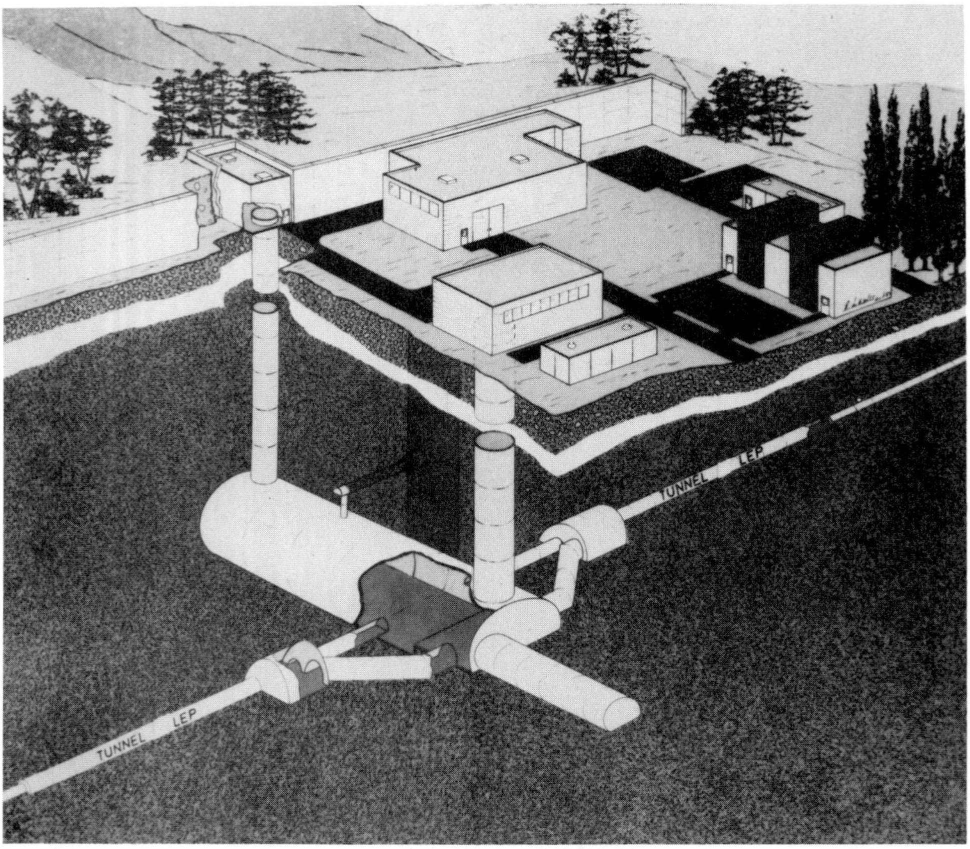

Fig. 3. Cut-away view of one of the four underground experirimental areas.

The four detectors have much in common. They all try to cover as much of the total solid angle as possible, measure the momenta of the charged particles in track detectors, and the energies of all particles, neutral particles as well charged, in calorimeters. The calorimeters are in two layers: electromagnetic and hadronic and serve in a very important way in the identification of the electrons and muons. They are spatially subdivided to achieve as high a "granularity" as possible in order to localize the energy deposit, a feature also essential to the particle identification. The whole is surrounded by 2 or more layers of counters for muon detection and identification. Each experiment groups together upwards of three hundred physicists from thirty or more laboratories. The construction costs have varied between \$50 and 100 M.

In the following I concentrate on the ALEPH detector, and results obtained by the ALEPH collaboration, since I am associated with it; the results of the other collaborations are in general similar. When possible, I will give a combined LEP result. A cutaway view of the ALEPH detector [1] is shown in fig. 4.

Design studies began in 1980; the detector was essentially complete in 1989. Starting inside and moving out, there is the beam pipe, 16 cm in diameter (it will be reduced to 11 cm in 1991) of 0.5 mm Aluminium. This is followed by a 2 layer silicon strip vertex chamber with nominal 10 μm resolution in $r\phi$ and 20 μm in z, covering $\sim$ 70% of the solid angle. This is not yet completely functional. The "Inner Tracking Chamber" (ITC) is a conventional drift chamber with 8 wire layers; it is used also to give tracking information in the first level trigger. The I.D. is 32 cm, the O.D. 52 cm, and the length 2 m. Resolution per wire in $r\phi$ is $\sim$ 100 μm. In z it is $\sim$ 3 cm, on the basis of the time difference of the signals on the two ends of each wire. The ITC is followed by a Time Projection Chamber (TPC) with I.D. 60 cm, O.D. 3.6 m and length 4.4 m. It is a gas filled cylinder. The ionization electrons produced in the traversal of charged particles are drifted to the two ends, in a homogeneous electric field, guided by the strong magnetic field. At the end, avalanches produced in proportional wire chambers are capacitively coupled to circular rows of small pads, in 21 evenly spaced rows, with altogether $\sim$ 50,000 such pads for the two end plates (see fig. 5).

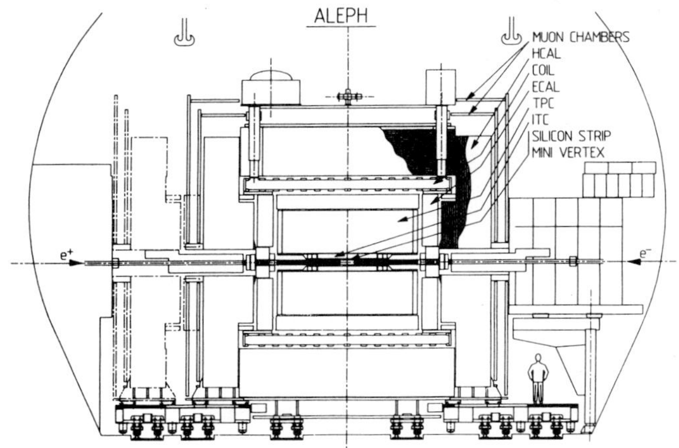

Fig. 4. Cut-away view of the ALEPH detector. From the inside out, there are the silicon surface areas vertex chamber, the inner tracking chamber, the time projection chamber, the electromagnetic calorimeter, the superconducting solenoid, the hadron calorimeter, and the muon chambers.

a

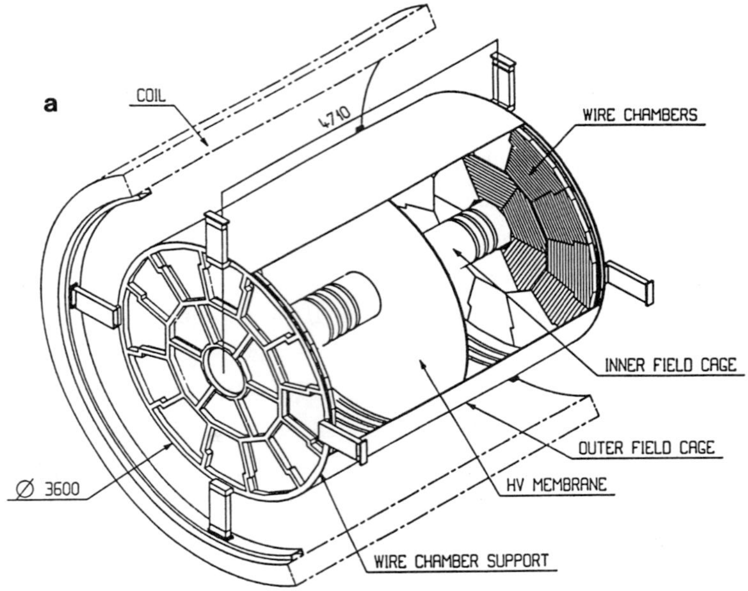

b

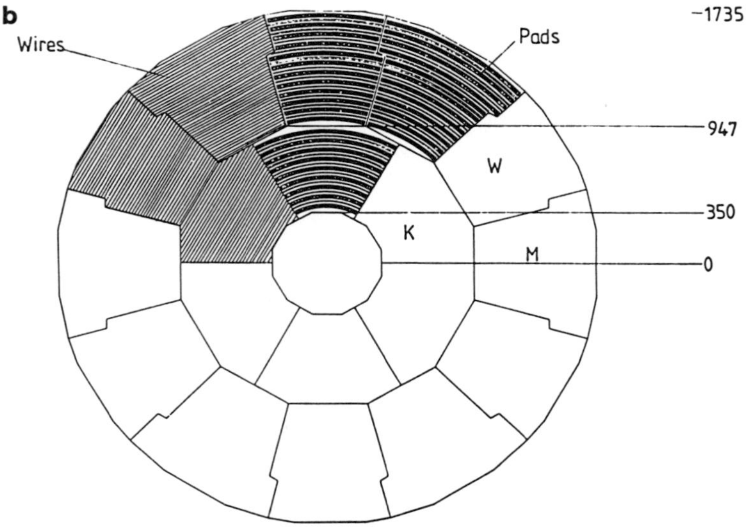

Fig. 5. *The Time Projection Chamber:*

a) Cartoon of overall contsruction.

b) Layout of wire chambers and pads on each end.

The transverse position of the track is measured by comparing the pulse heights of adjacent pads. The $r\phi$ resolution per pad row is $\sim$ 150 μm; and the z position, obtained from the drift time, has a resolution of $\sim$ 1 mm. The momentum resolution which is achieved in the combined tracking in the ITC and TPC is 4% at 45 GeV. For each track, also the pulse heights on the up to 300 traversed wires are recorded and furnish a measure of the specific ionization dE/dx, which is useful in the identification of electrons. The typical resolution in dE/dx is 4.5%. It is illustrated in fig. 6.

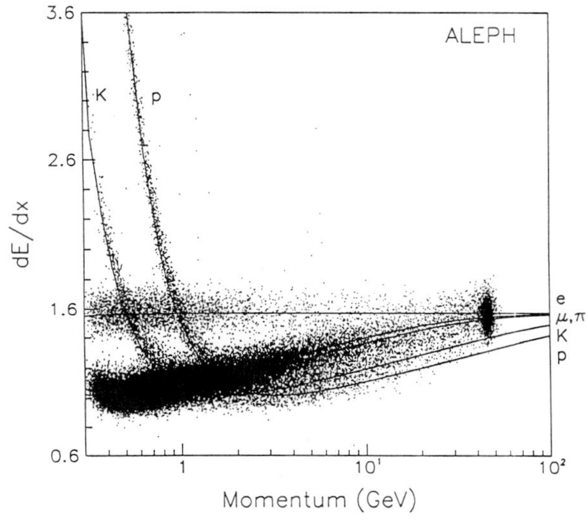

Fig. 6. Ionization density measured in the TPC for a sample of tracks, as function of track momentum.

The barrel of the electromagnetic calorimeter (ECAL) occupies the annular region between the TPC and the magnetization coil. It is a sandwich of lead sheets and multiwire proportional chambers, with 45 such layers, with a total thickness of 23 radiation lengths. ECAL consists of 12 barrel and 2 $\times$ 12 end-cap modules. The signals are read out for each wire plane, but the granularity is obtained from the 3 million capacitively coupled pads. These pads are arranged into 73'000 towers, roughly 3 cm $\times$ 3 cm in cross-section, each tower corresponding to an angular coverage of $0.8^0 \times 0.8^0$. The geometry is a projection to the interaction point; each tower points to the interaction region and is read out in three "stories". The energy resolution for photons and electrons, $\left(1.7 + \frac{19}{\sqrt{E\ (GeV)}}\right)\%$, is not exceptional, but the angular granularity is unusually fine. The high granularity, matched to the size of electromagnetic showers, is useful in the identification of electrons imbedded in hadron jets, The separation achieved on the basis of comparing the energy in the four towers nearest the track with the track momentum, as well as on the longitudinal shower development, is illustrated in fig. 7. The position resolution for an isolated electromagnetic shower is 1.5 mm in each direction.

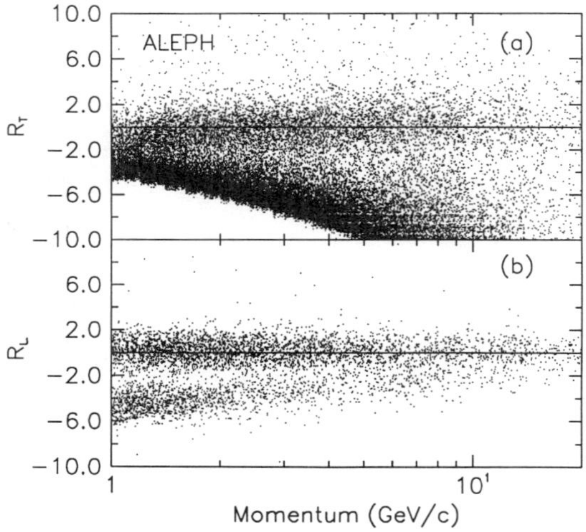

Fig. 7. Separation of electrons and hadrons in ECAL. R_T is a measure based on the ratio of track momentum and energy in the nearest four towers, R_L is a measure based on the longitudinal shower development in the three stories.

The superconducting solenoid is a homogeneous winding of 1712 turns, 6.35 m long, carrying 5000 amperes. On each end there are compensating coils to achieve the field homogeneity imposed by the TPC drift requirements. The field is 1.5 T, the mean diameter 5.3 m, and the weight, including the cryostat is 55 tons. An important feature of the ALEPH design is the location of the ECAL *inside* the coil, so that the electromagnetic showers are not degraded by the material of the coil.

Continuing radially through the detector we find the combined hadron calorimeter (HCAL) and return yoke, with I.D. of $\sim$ 6 m and O.D. $\sim$ 9.4 m. It is a sandwich of iron plates, separated by gaps into which planes of streamer tubes are inserted. There are 22 inner layers of iron, each 5 cm thick, and an outer layer, 10 cm thick. The signals from the tubes are read out in projective towers as in ECAL, but the angular resolution is 4 times courser, so that there are $\sim$ 4500 towers. In addition, the 150,000 streamer tubes are read out digitally (yes or no) at intervals of one cm, so that tracks can be followed through HCAL in one of the two projections. This is useful in the identification of muons. The overall (ECAL + HCAL) energy resolution for hadronic jets is $\sim 60\ \%/\sqrt{E}$ (GEV).

Continuing outwards there are two double layers of streamer tubes, at diameters of $\sim$ 9.5 and $\sim$ 10.6 m for the purpose of identifying muons. These are readout digitally (yes or no) every 5 mm in both projections. Figure 8 shows the assembled barrel, as it was in November of 1988, before the start of the cabling.

Fig. 8. November 1988. The ALEPH barrel is assembled, the cabling of hundreds of thousands of channels is about to begin.

Essential for the determination of absolute cross-sections are the luminosity calorimeters (LCAL). These are located around the beam pipe, at a distance of 2.7 m from the interaction point for the first lead layer, and with outer sensitive diameter $\sim$ 95 cm. They are of the same basic construction as ECAL, with 768 projective towers on each side. An important requirement of the luminosity monitor is the systematic accuracy of the geometrical acceptance. This is accomplished here by means of the projective geometry, the precision of the mechanical construction, and a simple but effective selection which requires that the pulse height of an inner fiducial set of towers is larger than that outside of this boundary. The geometrical precision of the acceptance for Bhabha events is $\sim$ 0.4%.

Physics at LEP

The 100 GeV centre of mass energy available at LEP, eventually 200 GeV in 1994, is clearly much less than the 2000 GeV at the $p\bar{p}$ collider, even after allowance is made for the fact that the quark-antiquark centre of mass energy is only a fraction of the $p\bar{p}$ energy. The strength of the e^+e^- collider is the simplicity of the initial state which is carried over to the final state, so that in general, the rates for different channels can be calculated in the frame of present theories, and can be experimentally sorted out (except that the different quark flavour channels can only rarely be identified).

At collision energies near the Z mass, the cross-section is enhanced through the Breit-Wigner resonance denominator by a factor of the order of $(m_z/\Gamma_z)^2 \sim 10^3$. Since the non-resonant cross-section is very small, of the order of $\frac{4\alpha^2}{s}$, the resonance presents a very important experimental opportunity, by its magnitude alone. All the work at LEP up to now has been on or near the Z peak for this reason. The thrust of the work can be separated into three branches:

a) studies of cross-sections, branching ratios, lineshapes, forward-backward asymmetries, polarization asymmetries. These have given a precise value of m_z, a strong argument that the number of fermion families are three, several checks on the consistency of the electroweak theory much more precise than previously possible and which may still be substantially improved, and some understanding of the top mass through the radiative corrections;

b) searches for new particles, such as Higgs, supersymmetric particles, and others. All these searches have been negative, but they have yielded better limits on the excluded mass domains;

c) work on the properties of heavy flavour quarks, especially b particles. The latter has just begun.

By 1994 it is expected that the LEP energy will exceed the threshold of $\sim$ 160 GeV for W^+W^- pair production. This will open a new domain for the critical experimental check of the electroweak theory, although, because here there is no resonance enhancement, the rates will be painfully low.

The cross-section studies we are concerned with are processes in which, in Born approximation, either a photon or a Z is exchanged, and a fermion pair is emitted:

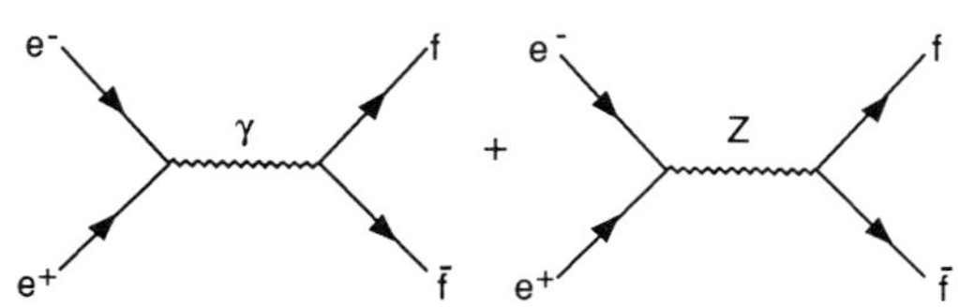

The fermion f may be a charged lepton: e, μ or τ, a neutrino: ν_e, ν_μ, ν_τ, or one of five quark flavours: u, d, s, c or b. No other channels have been observed. For the process $e^+e^- \to f\bar{f}$ the cross-section in lowest order (Born approx.) is:

$$\frac{d\sigma_f}{d\Omega} = \frac{\alpha^2 N_c^f}{4s} \left[F_1(s)\,(1+cos^2\theta) + 2F_2(s)cos\theta\right] \tag{1}$$

where $s = (2\ E_{\rm beam})^2$, N_c^f is the colour factor, 3 for quarks, otherwise 1,

$$F_1(s) = Q_f^2 - 2\ v_e\ v_f\ Q_f\ R_e\ \chi\ +\ (v_e^2 + a_e^2)\ (v_f^2 + a_f^2)\ |\chi|^2$$

$$F_2(s) = -\ 2\ a_e\ a_f\ Q_f\ R_e\ \chi\ +\ 4\ v_e\ a_e\ v_f\ a_f\ |\chi|^2$$

$\chi = \dfrac{s}{s - m_z^2\ +\ i\ m_z\ \Gamma_z}$ is the Breit-Wigner resonance denominator,

$v_f = \dfrac{I_3^f\ -\ 2\ Q_f\ sin^2\theta_w}{2\ sin\theta_w\ cos\theta_w}$ is the weak vector coupling,

$a_f = \dfrac{I_3^f}{2\ sin\theta_w\ cos\theta_w}$ is the weak axial vector coupling,

I_3 is the third component of weak isospin,

and

$\Gamma_z = \sum_{f_i=1}^{n} \Gamma_{f_i}$ is the total width of the Z.

The integrated cross-section at and near the peak is dominated by the Z exchange, $|\chi|^2$ term:

$$\sigma_f = N_f^c\ \frac{4\pi}{3}\ \frac{\alpha^2}{s}\ (v_e^2 + a_e^2)\ (v_f^2 + a_f^2)\ |\chi|^2 \tag{2}$$

The partial widths are:

$$\Gamma_f = \frac{N_f^c}{3}\ \alpha\ m_z\ \cdot\ (v_f^2 + a_f^2) = \frac{N_f^c\ G_F\ m_z^3}{6\ \sqrt{2}\pi}\ (g_{V_f}^2 + g_{A_f}^2)$$

where

$$g_{V_f}^2 = I_3^f - 2\,Q\,sin^2\theta_w \quad \text{and} \quad g_{A_f}^2 = I_3^f \tag{3}$$

Then

$$\sigma_f = \frac{12\,\pi s}{m_z^2} \frac{\Gamma_e \Gamma_f}{|s - m_z^2 + i\,m_z\,\Gamma_z|^2} \tag{4}$$

and at the Z peak, the cross-section for a final fermion pair $f\overline{f}$ is:

$$\sigma_f^{\text{peak}} = \frac{12\,\pi \Gamma_e \Gamma_f}{m_z^2\,\Gamma_z^2} \tag{5}$$

It is useful to note that the peak cross-section are related only to *ratios* of partial to total widths.

The Born approximation is not adequate for the analysis of the LEP experiments; it is essential to include higher order radiative corrections. These include initial state photon radiation, final state photon and gluon radiation, vertex corrections and propagator corrections. They have been the object of considerable study and are summarized and referenced in the publication "Z Physics at LEP" [2]. The initial state radiation can be factored out. The correction is substantial (see fig. 9) but purely electromagnetic, and calculated with adequate precision.

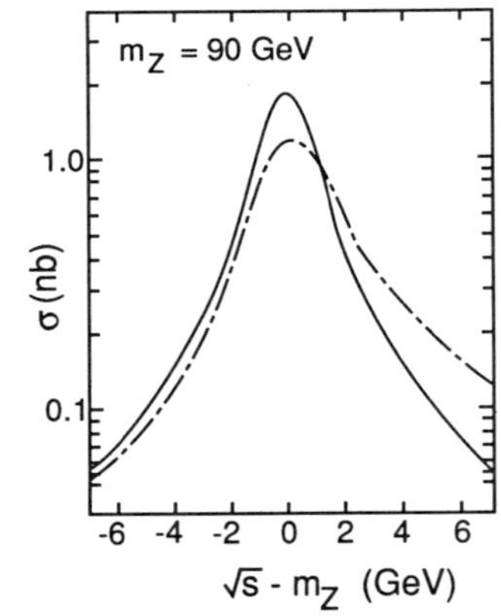

Fig. 9. Effect of initial state radiation on the Z lineshape. Solid line: Born approximation. Dotted line: corrected lineshape.

The remaining corrections are small and calculable in the Standard Model. They depend however on the unknown Higgs and top masses, as well as the not yet very precisely known value of the strong coupling constant α_s. The dominant effect of these corrections can be summarized by replacing $sin^2\theta_w$ by an "effective mixing angle" at the Z mass, $sin^2\theta_w(m_z^2)$, by replacing α by its renormalized value at the Z mass:

$$\alpha(m_z^2) = \frac{\alpha}{1-\Delta\alpha} = 1.064\ \alpha\ ,$$

inserting the factor $\rho = \frac{1}{1-\Delta\rho}$ in the relationships between m_w and m_z and G_F and m_z:

$$m_w^2/m_z^2 = \rho\ cos^2\theta_w(m_z^2)$$

$$\frac{\pi\alpha(m_z^2)}{\sqrt{2}G_F} = \rho m_z^2 sin^2\theta_w(m_z^2)\ cos^2\theta_w(m_z^2)\ ,$$

with

$$\Delta\rho = 3G_F\ m_t^2/8\sqrt{2}\ \pi^2 - 11\ G_F\ m_z^2\ sin^2\theta_w\ \ell n(m_H/m_W)/12\ \pi^2\sqrt{2}$$

$$= 0.0026\ m_t^2/m_z^2 - 0.0015\ \ell n(m_H/m_W)\ ,$$

and for the quark channels, including the second order QCD correction, replacing N_f by:

$$N_c^f(1 + \alpha_s/\pi + 1.4\ \alpha_s^2/\pi^2)\ .$$

Finally, χ in (1) should be replaced by:

$$\chi = \frac{s}{(s-m_z^2) + i\ s\Gamma_z/m_z}\ .$$

These changes give the major radiative corrections for all channels $e^+e^- \to f\overline{f}$, except the $b\overline{b}$ channel, which requires an additional correction which takes into account relatively large t - b amplitudes in the vertex corrections. To give some feeling for the magnitude of the corrections, the top and Higgs mass dependence of Γ_ℓ, Γ_b and Γ_z are shown in fig. 10.

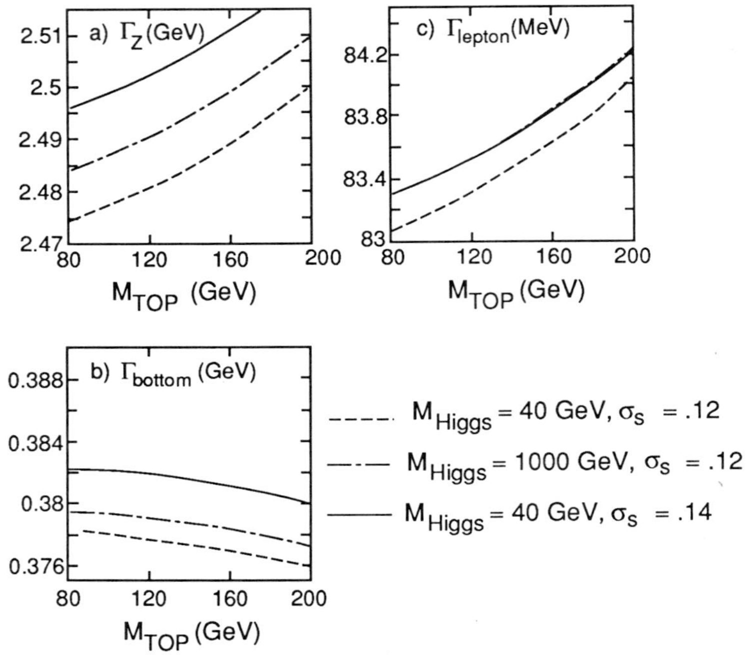

Fig. 10. Dependence on m_z, m_H and α_s of a) Γ_z, b) Γ_b, and c) Γ_ℓ.

The electroweak predictions for the relative rates of the different channels are as follows:

each neutrino	6.7%
each charged lepton	3.35%
up or charm quark	12.0%
down or strange quark	15.4%
bottom quark	15.2%
total, all quarks	70%

The neutrino channel cannot be observed. Of the observed channels the quark (hadronic) channel is by far the most frequent, accounting for 86%; the charged leptons represent only 14%.

LEP Running

The first dozen or two events were observed in each of the four detectors in a five day "pilot run" in August 1989. After one month of improvements, data taking started in earnest in the middle of September. Of course, all four experiments share the same beam energy. A "fill" typically lasts $\sim$ 12 hours before the luminosity has degraded sufficiently to warrant a new fill. For successive fills the energy was rotated in such a way that 1/2 of the data were taken near the top of the Z peak, the other half were more or less equally divided between the points $m_z \pm 1, m_z \pm 2$, and $m_z \pm 3$ GeV. The total number of hadronic events reported by ALEPH in 1989 was 28,000; for all four detectors combined it was 80,000. The luminosity of LEP improved during this period, and ended up with peak luminosities of $\sim 3 \times 10^{30}\text{cm}^{-2}\text{sec}^{-1}$, about one-fifth of design.

After a 3 month shut-down, running resumed in March '90 with peak luminosities of $6 \times 10^{30}\text{cm}^{-2}\text{sec}^{-1}$. At present the luminosity is limited to this value by beam - beam interaction. The integrated luminosity for all running at LEP up to July 1990 was $\sim$ 5 per picobarn. At that time the ALEPH data sample consisted of $\sim$ 100,000 hadronic and leptonic decays.

Lineshape and Number of Families

a) Introduction

The first analysis task is that of classifying the events according to channel. Event classification at LEP is, in general, rather easy (see fig. 11).

The dominant $q\overline{q}$, hadronic events are recognized by the large track multiplicity, the total energy and the wide distribution of calorimetric energy. Electronic and muonic decays both have back to back tracks of opposite electric charge, and momenta equal to one-half the centre of mass energy. The electrons are identified by the energy and shower shape seen in ECAL, the muons by their penetration in HCAL. The τ decays in the beam pipe, so only its decay products are seen. The decays result typically in one or three tracks, not strictly back to back and with missing energy, since in the I decay at least one neutrino is emitted.

The determination of the number of families is based on the total Z width Γ_z. Fermions with masses below $m_z/2$ contribute to the width of Z decay. Their contribution can be calculated in the Standard Model with high precision. The particles which are expected to contribute are those belonging to the three known families, except for the top, since its mass is too high. From these, and other experiments it is known that there are no charged fermions from possible additional families with mass less than $m_z/2$, since these would have been seen. But there may well be quarks and charged leptons with masses greater than $m_z/2$. The neutrinos of these families however, although in the absence of a theory for the masses this cannot be proven, would be expected to have much smaller masses and would then contribute to the Z width:

$$\Gamma_z = \Gamma_h + 3\Gamma_\ell + N_\nu\Gamma_\nu , \tag{6}$$

where Γ_h is the hadronic, $q\overline{q}$ width, Γ_ℓ, the width due to each charged lepton channel, and Γ_ν that due to each neutrino channel and N_ν the number of neutrino families contributing to the "invisible" width, Γ_{inv}. Each additional neutrino channel increases Γ_z by 6.8%. It is therefore interesting to measure Γ_z with much better precision than 6.8%. Γ_z is determined most precisely from a measurement of the cross-section of any of the fermion families at the Z peak, σ_f^{peak}. In practice the total hadronic channel is used since it is the most copious.

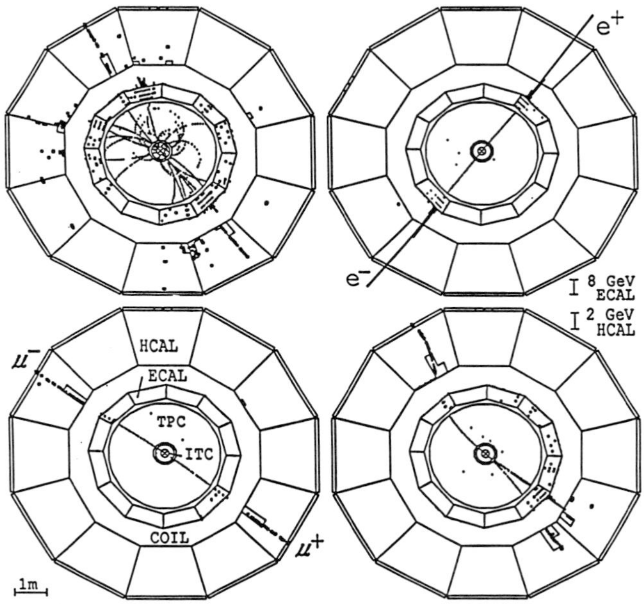

Fig. 11. *Typical events.*

Top left: $e^+e^- \rightarrow q\bar{q}$. *Typically there are two jets of hadrons, each with an average of 10 charged particles. The total observed energy is near to the centre of mass energy.*

Top right: $e^+e^- \rightarrow e^+e^-$. *The electrons are emitted back to back, each with one-half of the centre of mass energy. They are identified by the shape and magnitude of the energy dislribution in ECAL.*

Bottom left: $e^+e^- \rightarrow \mu^+\mu^-$. *Again there are two tracks, each with one-half of the centre of mass energy. They are identified as muons by the small energy deposit in ECAL, and the penetration through HCAL.*

Bottom right: $e^+e^- \rightarrow \tau^+\tau^-$. *The* τ*'s decay within millimetres and are identified on the basis of missing energy and momentum (lost to neutrinos in their decay) and the visible decay products, usually one or three charged tracks in a narrow jet.*

From (5),

$$\Gamma_z = \left[\frac{12\,\pi\,\Gamma_e \Gamma_h}{m_z^2 \cdot \sigma_h^{\text{peak}}} \right]^{1/2} .$$

Here the initial state radiation has been unfolded. α_{peak} refers to the cross-section at the top of the resonance. Then:

$$N_\nu = \frac{\Gamma_\ell}{\Gamma_\nu} \left[\sqrt{\frac{12\pi \cdot \Gamma_h}{m_z^2 \cdot \sigma_h^{\text{peak}} \cdot \Gamma_\ell}} - \frac{\Gamma_h}{\Gamma_\ell} - 3 \right] . \qquad (7)$$

For the evaluation of (7), m_z is measured in the same experiment. Γ_h/Γ_ℓ, is also measured, or it may be taken from the Standard Model. Γ_ℓ/Γ_ν must be taken from the electroweak theory, it cannot be measured. Precise measurement of the absolute cross-section requires precise measurement of the absolute luminosity. This posed an interesting experimental challenge, already in the first weeks of experimentation.

In proceeding now to the experimental data, they are discussed using ALEPH as an illustration [3,4,5], but the results of the three other experiments [6,7,8,9,10,11,12,13] are included in the combined LEP results.

b) Trigger and trigger efficiency

The acquisition of the data must be triggered. The basic problem is the early recognition of e^+e^- annihilation events as distinct from background events, which consist of electrons scattered in the residual gas, cosmic rays, and so called 2 photon processes. The latter are inelastic e^+e^- scatterings with creation of hadrons. It turns out that these backgrounds are sufficiently benign so that triggering at LEP is easy. In order to evaluate the efficiency of the trigger, it is essential to have redundancy in the trigger, that is two or more triggers which are independent and each adequate to trigger on the wanted events. This redundancy is the case in ALEPH for all event types. To trigger on hadronic Z decay, one trigger required only a certain minimum of ECAL energy, another, totally independent, required at least one charged particle with some penetration into HCAL. The trigger efficiency could be shown to be at least 99.9%.

c) Selection of hadronic events

As we have already noted and illustrated in fig. 11, the hadronic and the three charged leptonic channels have very different characteristics in the detector and event selection into the different channels with good efficiency and purity is relatively simple. For the selection of hadronic events there were again two independent methods, one based on the calorimetric properties, requiring no tracks, the other based on having 5 or more tracks, independent of calorimetric energy. The selection efficiencies, respectively 99.4% and 97.5%, are determined on the basis of Monte Carlo simulation of the events and the detector. The near unity of the acceptance ensures also that the uncertainty in the acceptance is small. The detailed agreement of Monte Carlo and data in the distributions of many observables reinforces confidence in the understanding of the events. In fig. 12 the comparison is shown for the distribution in the total observed calorimetric energy. The contamination from background processes and unwanted decay channels is negligible (less than a few tenths of one per cent). Figure 13 shows the distribution in the track energy for track selected events, and fig. 14 the thrust angle distribution, as determined using the charged momenta.

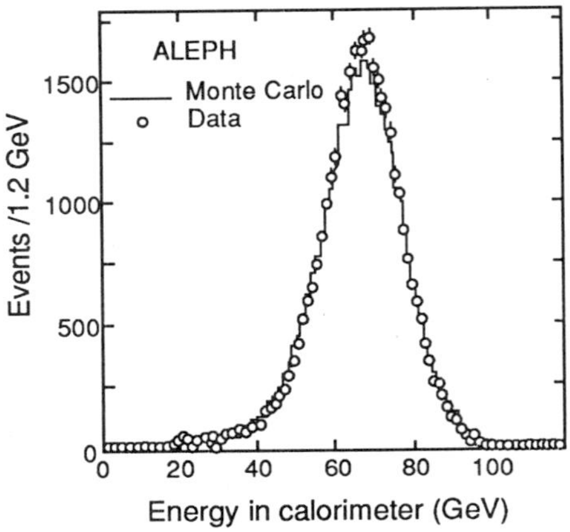

Fig. 12. Comparison of data and MC simulation. Hadronic events selected calorimetrically. Distribution in the ratio of observed energy to beam energy.

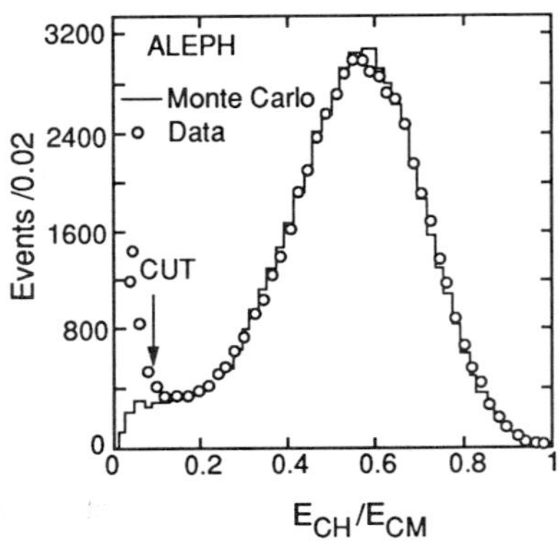

Fig. 13. Distribution in the track energy relative to the centre of mass energy for events with more than five tracks. Acceptance into the hadronic event sample required in addition that this ratio exceeds 0.1. The rise at smaller values is due to e^+e^- inelastic scattering ($\gamma\gamma$ process).

d) Luminosity determination

The luminosity is determined on the basis of the observed small angle e^+e^- (Bhabha) elastic scattering, whose rate can be precisely calculated. The precision of the measurement is limited by the understanding of the acceptance of the luminosity detector. Since the cross section rises steeply with decreasing angle (as θ^{-4}), the precise knowledge of the effective inner cut off is especially critical. The best luminosity measurements reported from earlier, lower energy e^+e^- experiments are at the 3% level. However, the measurement becomes easier with increasing

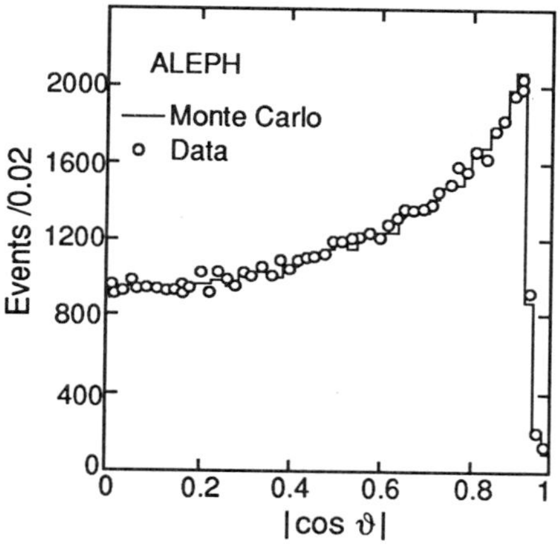

Fig. 14. Hadronic decays selected on the basis of the tracks. Distribution in the angle between beam and thrust axis. Comparison of dala and simulation.

energy and the luminosity detectors of all 4 LEP experiments are substantially better than their predecessors. Most of the LEP luminosity measurements now report errors well below 3%; the systematic experimental luminosity uncertainty in the ALEPH experiments is now 0.4%. This is one of the more beautiful experimental accomplishments of this early work at LEP. Figure 15 shows an ALEPH luminosity event, and fig. 16 compares the observed and simulated polar angle distributions. The agreement is an essential element in the confidence in the precision of the luminosity determination.

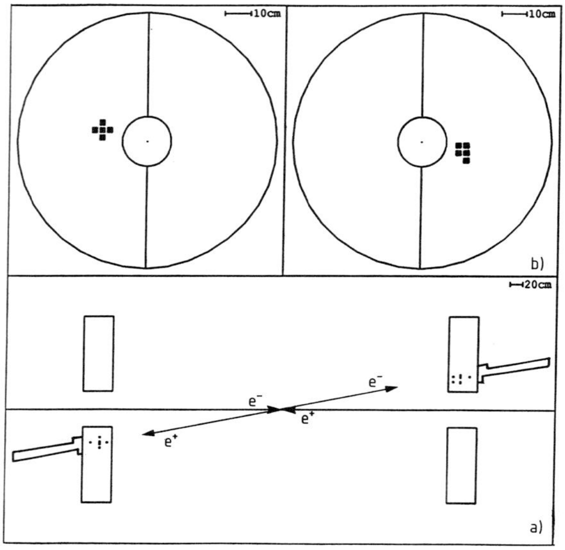

Fig. 15. *Luminosity event in ALEPH.*

a) *Cut including beam direction.*

b) *View along the beam of the two luminosity calorimeters. The two localized energy deposits can be seen to be due to particles 80^0 in azimuth, at small and opposite polar angles, and of comparable energy (since the scattered electron and positron are emitted at opposite angles, each with one half of the total energy.)*

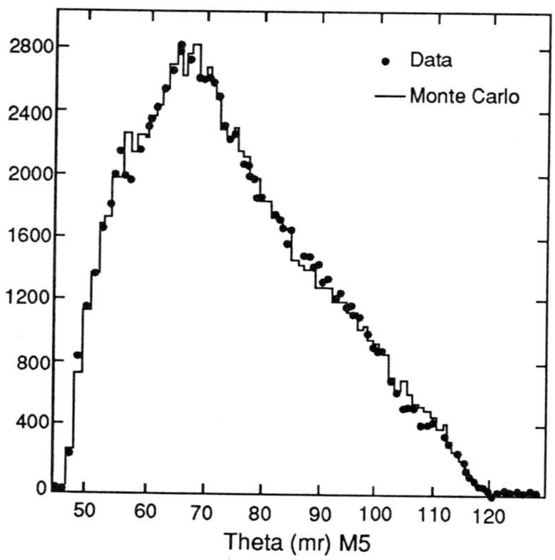

Fig. 16. *Comparison of data and simulation for the polar angle distribution of luminosity events. Agreement is a measure of the systematic understanding of the luminosity.*

e) Results on the Z lineshapes from hadronic decays

A recent ALEPH result on the lineshape of hadronic Z decays is shown in fig. 17. The curves are the electroweak predictions for 2, 3 and 4 neutrino families, with only mz as free parameter. The data can also befitted by a Breit-Wigner distribution of the form:

$$\sigma_h(s) = \frac{\sigma_h^{\text{peak}} \cdot m_z^2 \, \Gamma_z^2}{|(s - m_z^2) + i \frac{s\Gamma_z}{m_z}|^2} \quad ,$$

with three free parameters, m_z, Γ_z and σ_h^{peak}. The most recent results of the four collaborations for these [13] are given in Table 1. They are in very good agreement with each other, as they are with the predictions of the Standard Model, also listed in the table.

The new precise value for the Z mass, $m_z = 91.172 \pm 0.031$ GeV, is now one of the basic parameters of the electroweak theory, together with the fine structure constant a and the Fermi constant, G_F. Its remaining uncertainty is due principally to the 30 MeV uncertainty in the beam energy. Already this level of precision in the beam energy is a great accomplishment. It is based on the comparison of the revolution frequencies of protons and electrons in the same beam orbit. In the future it is hoped to achieve an even smaller error with the use of polarized electrons.

The precise width provides, as we see later, a sensitive (at the moment, the most sensitive) check on the standard model. The peak cross-section is the basis of the argument on the number of neutrino families.

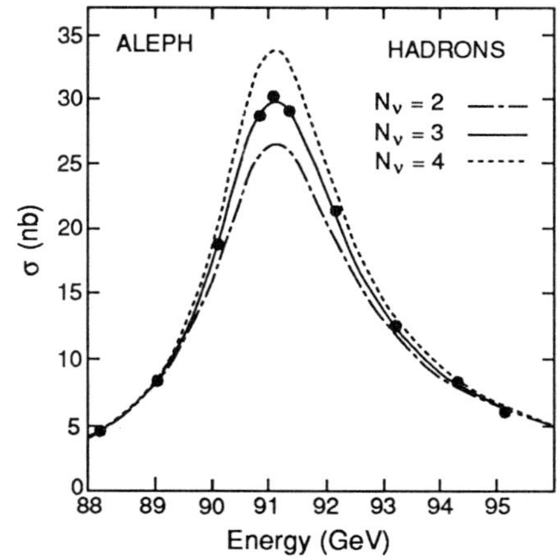

Fig. 17. ALEPH results for the cross section as function of centre of mass energy compared with the expectations of the Standard Model for hadronic events for 2, 3 and 4 neutrino families. The mass of the Z has been left free to achieve the best fit.

Table 1. **Results on the lineshape of the hadronic channel, for each of the four collaborations, and the combined result**

	ALEPH	DELPHI	L3	OPAL	LEP	Standard Model
Number of Hadronic events	84'000	168'000	162'000	112'000	320'000	
m_z, GeV	91'186 ± 0.013 ± 0.030	91'188 ± 0.014 ± 0.030	91'164 ± 0.013 ± 0.030	91'174 ± 0.011 ± 0.030	91'177 ± 0.031	
Γ_z, MeV	2506 ± 26	2476 ± 26	2492 ± 25	2505 ± 20	2498 ± 12.5	2490 ± 15
σ_h^{peak}, nb	41.78 ± 0.63	42.38 ± 1.02	41.38 ± 0.71	41.88 ± 0.74	41.77 ± 0.42	41.42 ± 0.12

f) Results on the Z lineshape from the leptonic channels and partial decay widths of the Z

The selection of the leptonic channels is more complex than that of the hadronic channel. Although, as we saw in fig. 11, they are generally distinct, there are kinematic regions of confusion at the percent level, for instance if both electrons decide to go through cracks in the ECAL, or the muons in HCAL, or τ^+ and τ^- both choose to decay to electron and neutrino or to muon and neutrino. Assignment of these events to the correct channel requires adequate algorithms; checked with the help of simulation. In ALEPH systematic uncertainty in the separation and in the efficiencies is kept below the 1% level. For the e^+e^- channel there is a complication at small angles, the t channel:

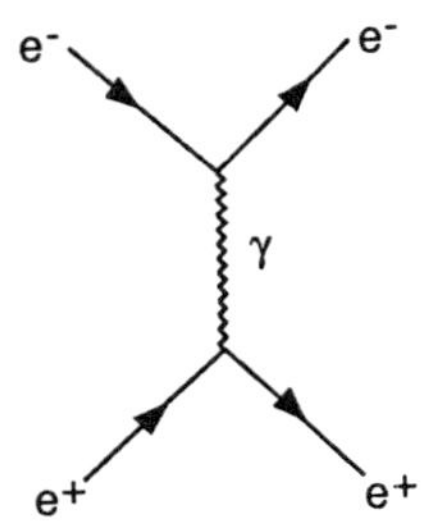

This contributes the large cross-section at small angles used in the luminosity determination. Its interference at larger angles with the s channel is an important and non trivial correction, especially for the asymmetry determination. It is therefore necessary in the case of the e^+e^- channel to exclude the small angles

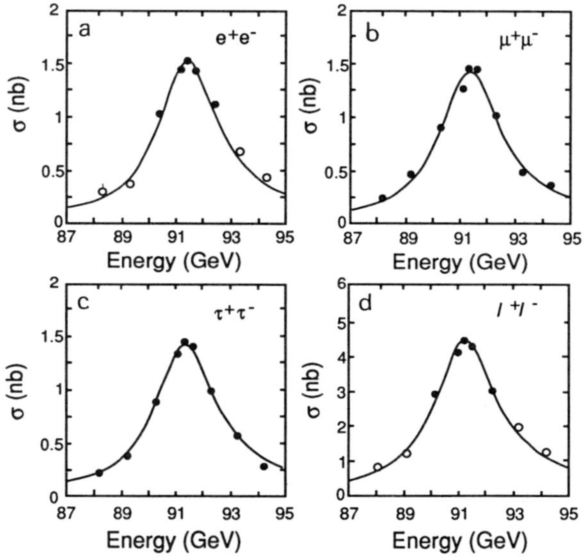

Fig. 18. *ALEPH results for the cross-section:*
a) $e^+e^- \rightarrow e^+e^-$,
b) $e^+e^- \rightarrow \mu^+\mu^-$;
c) $e^+e^- \rightarrow \tau^+\tau^-$ and
d) the total leptonic channel $e^+e^- \rightarrow \ell^+\ell^-$, *as a function of the centre of mass energy. The lines are the electroweak expectations, with* $sin^2\theta_w(m_z^2) = 0.232$.

where the t channel is large, and to subtract it from the remaining angular range on the basis of a QED calculation. In the ALEPH analysis the excluded region is $\cos\theta > 0.7$.

The statistical accuracy of the leptonic channels is limited by the fact that each accounts for only $\sim 5\%$ of the observed events. It is therefore interesting to look also at the combined leptonic channels, assuming universality. This is done in ALEPH using an independent selection procedure which takes advantage of the fact that it is not necessary to distinguish the individual leptonic channels.

Figure 18 shows the ALEPH leptonic cross-sections at the Z resonance for the 3 separate, as well as for the combined leptonic channel [5].

The leptonic channels have been studied extensively at LEP [4,5,10,14,15,16, 17,18,19,20]. The partial widths are obtained from the cross-sections and from Γ_z on the basis of (5):

$$\Gamma_e = \Gamma_z \sqrt{\frac{m_z^2\, \sigma_e^{\text{peak}}}{12 \cdot \pi}}$$

and

$$\Gamma_{\mu(\tau)} = \Gamma_z\, \sigma_{\mu(\tau)}^{\text{peak}} m_z \sqrt{12\, \pi \sigma_e^{\text{peak}}}$$

or for the averaged leptonic channel, assuming universality:

$$\Gamma_\ell = \Gamma Z \sqrt{\frac{m_z^2\, \sigma_\ell^{\text{peak}}}{12\, \pi}} \quad .$$

With:

$$R = \Gamma_h / \Gamma_\ell = \sigma_h^{\text{peak}} / \sigma_\ell \quad ,$$

the partial hadronic width is:

$$\Gamma_h = \Gamma_z \sqrt{\frac{R\, m_z^2\, \sigma_h^{\text{peak}}}{12\, \pi}} \quad .$$

Finally, the invisible (neutrino) width is:

$$\Gamma_{\text{inv}} = \Gamma_z - \Gamma_h - 3\Gamma_\ell \quad .$$

Recent results [13] of the four collaborations, as well as the combined LEP results and the electroweak prediction, are given in Table 2. Again, the experiments are in good agreement with each other and with the Standard Model predictions [2].

Table 2. Results the four collaborations, and the combined result for the partial width R, Γ_h and Γ_{inv}.

	ALEPH	DELPHI	L3	OPAL	LEP	Standard Model
Number of Leptonic events	9'500	4'000	4'000	11'327	29'000	
Γ_e, MeV	84.9 ± 1.1	82.0 ± 1.7	84.3 ± 1.4	82.7 ± 1.3	83.4 ± 0.9	83.6 ± 0.3
Γ_μ, MeV	80.7 ± 2.2	87.2 ± 3.4	82.3 ± 2.9	85.9 ± 2.0	84.1 ± 1.4	83.6 ± 0.3
Γ_τ, MeV	81.8 ± 2.2	86.0 ± 3.9	83.5 ± 3.7	83.9 ± 2.3	83.25 ± 1.5	83.6 ± 0.3
Γ_ℓ, MeV	84.2 ± 1.0	83.7 ± 1.4	83.2 ± 1.4	83.6 ± 0.9	83.7 ± 0.65	83.6 ± 0.3
$R = \Gamma_h/\Gamma_\ell$	20.95 ± 0.31	21.00 ± 0.48	21.02 ± 0.62	21.26 ± 0.32	21.08 ± 0.19	20.80 ± 0.15
Γ_h, MeV	1764 ± 24	1756 ± 32	1748 ± 35	1778 ± 26	1764 ± 16	1739 ± 14
Γ_{inv}, MeV	489 ± 22	469 ± 29	494 ± 32	476 ± 25	482 ± 13	500 ± 2

g) Number of neutrino families

The number of neutrino families N_ν is determined from Γ_{inv} by dividing by the width given in the electroweak theory for any neutrinofamily with mass much smaller than $m_z/2, \Gamma_\mu = 166.7 \pm 0.7$ MeV. N_ν includes, by definition, also other non-detectable decay channels of the Z, such as lowest mass supersymmetric particle, if these exist, are neutral, and their mass is less than $m_z/2$. $\Gamma_{inv} = \Gamma_z - \Gamma_h - 3\Gamma_\ell \cdot \Gamma_z$ is directly measured, but can also be obtained from the peak cross-section

$$\Gamma_z = \sqrt{\frac{12\ \pi \Gamma_\ell \Gamma_h}{m_z^2\ \sigma_h^{\text{peak}}}} \ .$$

The error in N_ν can be can be minimized by using both. For Γ_e and Γ_h either the measured values or the Standard Model values can be used. Given the agreement between these, and the small errors now obtained in the measurements, the results are very similar. The results of the four experiments [Refs. 3 to 13] are given in Table 3.

Table 3. Results the four collaborations for the number of neutrino families with $m_\nu << m_z/2$

	N_ν
ALEPH	2.91 ± 0.12
DELPHI	2.82 ± 0.17
L3	3.01 ± 0.12
OPAL	2.86 ± 0.15

A determination of N_ν, using the results of thc four collaborations for σ_h^{peak} and Γ_z, and the electroweak results for Γ_h and Γ_ℓ, yields:

$$N_\nu = 2.92 \pm 0.08 \quad .$$

The "invisible" contribution to the Z width is in agreement with the Standard Model, with just thrce neutrinos, with an uncertainty of only 0.08 neutrino families. The integer nature of the result is a confirmation of the Standard Model. The fact that it is 3 and not 4 or 5 is a demonstration that, with the extremely unlikely caveat that there may be neutrinos in the frame of the Standard Model with masses greater than 0.45 m_z, the number of families of fermions is 3. Finally, the fact that N_ν is 3 and not 2 shows that the τ ncutrino is distinct from the electron and muon ncutrinos.

Asymmetries

Parity violating asymmetries are among the more sensitive observables for studying the electroweak theory. They can be divided into two classes: forward-backward and polarization. For each of these there are two rather different domains: below and above the Z peak the asymmetry is due to the interference of the electromagnetic vector and the weak axial vector interactions, on the Z peak, and of greater physics interest, it is due to the interference between the weak vector and the weak axial vector interactions.

a) Forward-Backward Asymmetries

This is the asymmetry in the angle between the outgoing fermion and the incident positron. The electroweak prediction is governed by the differential cross-section (1). The asymmetry at the Z peak is proportional to:

$$\frac{(v/a)_e \cdot (v/a)_f}{[1 + (v/a)_{e^2}] \cdot [1 + (v/a)_{f^2}]}$$

Experiments usually report A_{FB}, defined as the difference in the cross-sections integrated over forward and backward hemispheres, divided by their sum:

$$A_{FB} = \frac{\int_0^1 \frac{d\sigma}{dcos\theta}\, dcos\theta - \int_{-1}^0 \frac{d\sigma}{dcos\theta}\, dcos\theta}{\int_0^1 \frac{d\sigma}{dcos\theta}\, dcos\theta + \int_{-1}^0 \frac{d\sigma}{dcos\theta}\, dcos\theta}$$

In terms of F_1 and F_2 of (1), $A_{FB} = \frac{3}{4}\, F_2/F_1 = 3\, v_e a_e v_f a_f/(v_e^2 + a_e^2) \cdot (v_f^2) + a_f^2)$. A_{FB} has been measured at lower energy (see e.g. the review by S.L. Wu [21]), but here, as already mentioned, it is dominated by the interference between photon and Z exchange. At the Z peak the asymmetry is sensitive to the ratio of vector to axial vector couplings v/a. In the electroweak theory, $v/a = 1 - 4|Q|sin^2\theta_w(m_z^2)$. The experimental results are just beginning to emerge, because of the as yet very limited statistics. The clearest channels, experimentally, are the three leptonic channels, however the rates are low. For the e^+e^- channel only, there is the complication of the photon t channel - s channel interference at small angles which would produce huge asymmetries unless corrected. The procedure is to limit the acceptance region in the forward cone (in the case of ALEPH it is limited to $cos\theta < 0.7$) and to subtract the remaining t channel contribution on the basis of the theory. This is shown for the ALEPH data [5] in fig. 19. A_{FB} for leptons is small, very nearly equal to $(v/a)_\ell^2 = (1 - 4|Q|sin^2\theta_w)^2$ which is small because $4\ sin^2\theta_w$ is numerically close to unity. The fact that A_{FB} is quadratic in v/a, as well as small, makes it less sensitive in the measurement of $sin^2\theta_w$.

In fig. 20 the lepton asymmetry observed by ALEPH is shown as function of the centre of mass energy near the Z resonance peak.

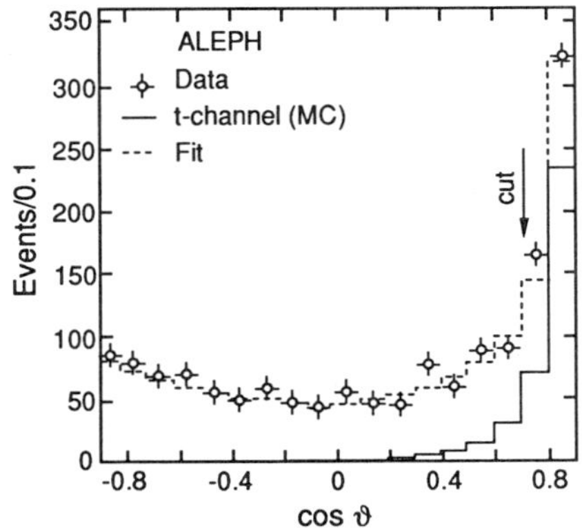

Fig. 19. ALEPH measurement of the angular distribution of the reaction $e^+e^- \rightarrow \ell^+\ell^-$. The solid line is the calculated contribution of the t channel, which is subtracted. Acceptance is limited to $cos\theta < 0.7$.

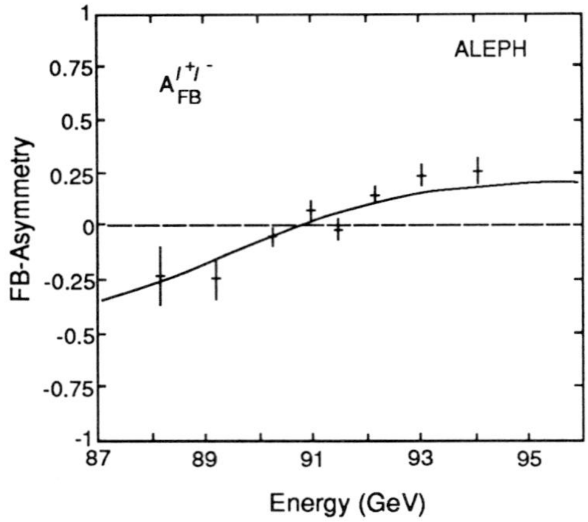

Fig. 20. ALEPH results for lepton forward-backward asymmetry A_{FB} across the Z resonance.

The results for v_ℓ^2/a_ℓ^2 [7,19,20,5,13] are:

	v_ℓ^2/a_ℓ^2
ALEPH	0.0083 ± 0.0034
DELPHI	$0.0l2 \pm 0.009$
L3	0.012 ± 0.007
OPAL	0.038 ± 0.0033
LEP	0.0069 ± 0.0022
E.W.	0.0052 ± 0.0008

The combined LEP results for the forward-backward lepton asymmetry correspond to $sin^2\theta_w(m_z^2) = 0.229 \pm 0.003$.

b) Polarization asymmetries

The asymmetry with respect to longitudinal polarization can, in principle, fumish very precise information on v/a, or equivalently, on $sin^2\theta_w$. The simplest case is the one in which only the polarization of one of the incident particles, or one of the outgoing particles is observed. The basic expressions for these are the same, except for the reversal of the roles of electron and final state fermion. Here they are written for the case of final state polarization. In general, both Z and photon exchange must be considered, as well as their interference. For clarity, only polarization on the Z resonance peak, $s = m_z^2$ is considered, where the photon exchange contribution is negligible.

Then

$$\frac{d\sigma}{dcos\theta}(cos\theta, p) \propto (1 + cos^2\theta)\ F_1 + 2\ cos\theta\ F_2 + p[(1 + cos^2\theta)\ F_3\ +\ 2\ cos\theta\ F_4]$$

where p is the longitudinal polarization, and

$$\begin{aligned} F_1 &= (v_e^2 + a_e^2)\ (v_f^2 + a_f^2) \\ F_2 &= 4v_e a_e v_f a_f \\ F_3 &= 2(v_e^2 + a_e^2)\ v_f a_f \\ F_4 &= 2v_e a_e (v_f^2 + a_f^2)\ . \end{aligned}$$

A_{Pol} and A_{Pol}^{FB} are defined:

$$A_{Pol} \equiv \frac{\int_{-1}^{1} dcos\theta\ \frac{d\sigma(p=1)}{dcos\theta}\ -\ \int_{-1}^{1} dcos\theta\ \frac{d\sigma(p=-1)}{dcos\theta}}{\int_{-1}^{1} dcos\theta\ \frac{d\sigma(p=1)}{dcos\theta}\ +\ \int_{-1}^{1} dcos\theta\ \frac{d\sigma(p=-1)}{dcos\theta}} = < p >$$

Here $p = 1$ and $p = -1$ refer to positive and negative helicities respectively.

$$A_{Pol} = \ -F_3/F_1 = \ -\frac{2\ v_f a_f}{v_f^2 + a_f^2}$$

$$A_{Pol}^{FB} \equiv \frac{\int_{0}^{1} dcos\theta\left[\frac{d\sigma(p=1)}{dcos\theta}\ -\ \frac{d\sigma(p=-1)}{dcos\theta}\right]\ -\ \int_{-1}^{0} dcos\theta\left[\frac{d\sigma(p=1)}{dcos\theta}\ -\ \frac{d\sigma(p=-1)}{dcos\theta}\right]}{\int_{-1}^{1}\left[dcos\theta\ \frac{d\sigma(p=1)}{dcos\theta}\ +\ \frac{d\sigma(p=-1)}{dcos\theta}\right]}$$

$$= \ -\ 3/4\ F_4/F_1 = \ -\ 3/4\ \cdot\ \frac{2\ v_e a_e}{v_e^2 + a_e^2}$$

The polarization asymmetries, being linear rather than quadratic in v/a, provide aventually more sensitive possibilities to measure $sin^2\theta_w$ than A_{FB}.

LEP as yet cannot provide polarized beams. It may in the future be possible to make modifications which will permit longitudinal beam polarization, but this is not yet certain and would be a major effort. At present, at LEP, polarization asymmetries can be studied in the τ channel, where the decays $\tau \rightarrow \nu_\tau + \pi, \tau \rightarrow \nu_\tau + \ell + \nu_\ell$ and $\tau \rightarrow \nu_\tau + \rho$ can serve as analyzers of the polarization. The work is only beginning. The first ALEPH result, presented at the Singapore conference, is [13]: $< p_\tau > = A_{Po\ell,\tau} = -0.151 \pm 0.087$. With $A_{Po\ell,\tau} = \frac{-2v_\tau/a_\tau}{(v_\tau/a_\tau)^2+1} \simeq -2v_\tau/a_{\tau_1} = -2(1-4\ sin^2\theta_w(m_w^2))$ this asymmetry corresponds to $sin^2\theta_w(m_z^2) = 0.231 \pm 0.011$, in agreement with the other measurements, but with substantially larger error.

Comparison with the Standard Model, $\sin^2\theta_w$ Mass of the Top, and α_s.

Table 4 lists the main LEP experimental results relevant to the critical confrontation with the predictions of the Standard Model [2]. For the model predictions, the new measurement of mz, as well as the top mass constraints discussed below, have been used. The agreement is everywhere within the combined uncertainty, some at the 1/2 to 11/2% level of accuracy. Not all of the results are independent. If we consider the measurement of $m_z, \Gamma_z, \sigma_h^{\text{peak}}$ and R as independent, the results for $\Gamma_\ell, \Gamma_h, \Gamma_{\text{inv}}$ and N_ν follow, and so add nothing new. In order to show the precision of the test, the value of $sin^2\theta_w(m_z^2)$ derived from the particular measured quantity and its error are given in the table as well. For comparison, the most precise previous tests of the Standard Model were at best at the 4% level in $sin^2\theta_w$.

The theoretical uncertainty will become more interesting as the experiments get more precise. In $\Gamma_z, \sigma_h^{\text{peak}}$ and R this is dominated by the uncertainty in α_s (assumed here $\pm$ 0.02). The error in Γ_ℓ is dominated by the uncertainty in the top mass. The latter also contributes the bulk of the uncertainty in $sin^2\theta_w(m_z^2)$ as determined from m_z. If $sin^2\theta_w$ were defined directly in terms of m_z, for instance as $sin^2\theta_w cos^2\theta_w = \frac{\pi\alpha}{\sqrt{2}G_F m_z^2}$, instead of the effective $sin^2\theta_w(m_z^2)$ definition used here, the precision in $sin^2\theta_w$ from the m_z measurement would be 0.0002.

The relationship between $sin^2\theta_w(m_z^2)$ and m_z is shown in fig. 21 for the measured values of m_z, Γ_z and Γ_ℓ. Here it is assumed that $m_H = 200$ GeV and $\alpha_s = 0.12$. In the same figure, the information from the measurements of m_w/m_z is also shown. These are: the UA2 measurement of the ratio $m_w = 0.8831 \pm 0.0055$ [22], the CDF measurement $m_w = 79.83 \pm 0.44$ GeV [23] and the result $1 - \frac{m^2 w}{m_z^2} = 0.231 + 0.006$ [24] from the CDHS and CHARM measurements of the neutral to charged current ratio in neutrino deep inelastic scattering. These measurements of fig. 21 taken together furnish the best value of the mass of the as yet undiscovered top, through its effect on the radiative corrections.

$$m_t = 127 \pm 34 \pm 17(m_{\text{Higgs}}) \text{ GeV} \quad .$$

The last error covers an uncertainty in the Higgs mass from 40 to 1000 GeV.

Table 4. **Combined LEP results and Standard Model predictions for $m_z, \sigma_h^{peak}, \Gamma_z, \Gamma_h, \Gamma_\ell, R, N_\nu$ and asymmetries. In the Standard Model predictions it is assumed that $40 < m_H < 1000$ GeV, $\sigma_s = 0.12 \pm 0.02$, and that $90 < m_z < 170$ GeV. The column $\sin^2\theta_w(m_z^2)$ lists the value which follows from the particular measurement**

	Experiment	Standard Model	$sin^2\theta_w(m_z^2)$
m_z, GeV	91.177 ± 0.031		0.2325 ± 0.0010
Γ_z, MeV	2498 ± 12.5	2490 ± 16	0.2313 ± 0.0023
σ_h^{peak}, nb	41.77 ± 0.42	41.42 ± 0.12	0.2350 ± 0.0035
Γ_h, MeV	1764 ± 14	1739 ± 14	0.2290 ± 0.0030
Γ_ℓ, MeV	83.7 ± 0.65	83.6 ± 0.3	0.2318 ± 0.0034
$R \equiv \Gamma_h/\Gamma_\ell$	21.08 ± 0.19	20.8 ± 0.15	0.215 ± 0.015
N_ν	2.92 ± 0.08	Integer	
$(v/a)_\ell^2 = \frac{1}{3}A_{FB,\ell}$	0.0069 ± 0.0022	0.0052 ± 0.0008	0.229 ± 0.003
$(v/a)_\tau = -\frac{1}{2}A_{Po\ell,\tau}$	0.075 ± 0.043	0.072 ± 0.006	0.231 ± 0.011
$A_{FB,b}$	0.172 ± 0.076	0.098 ± 0.008	0.220 ± 0.013

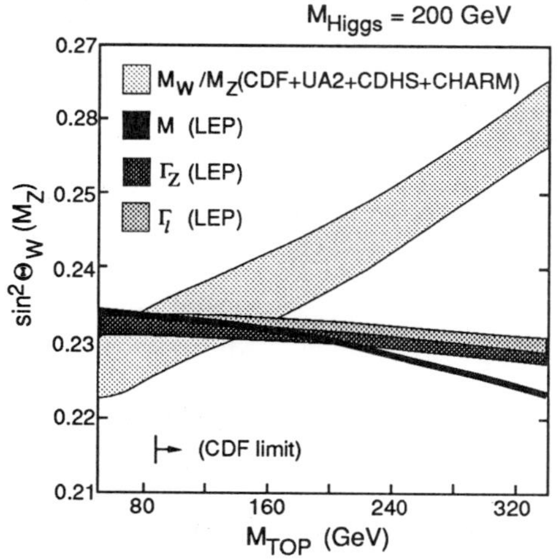

Fig. 21. Implications for $sin^2\theta_w(m_z^2)$ and m_t of the LEP results for m_z, Γ_z and Γ_ℓ, as well as the UA2, CDF, CDHS and CHARM results for m_w/m_z.

The resultant value for $sin^2\theta_w(m_z^2)$ is:

$$sin^2\theta_w(m_z^2) = 0.2325 \pm 0.0003 + 0.0043 \left(1 - \frac{m_t}{127 \text{ GeV}}\right) + 0.0005 \log_{10} \frac{m_H}{200 \text{ GeV}} = 0.2325 \pm 0.0010 \pm 0.0005 \ (m_H) \ .$$

In the determination of m_t and $sin^2\theta_w(m_z^2)$ the m_z and m_w/m_z measurements are the most important ingredients; the Γ_z and Γ_ℓ, measurements contribute less. This can be seen from fig. 21. The agreement in $sin^2\theta_w(m_z^2)$ as determined from Γ_z and Γ_ℓ, as well as from the asymmetries, with the value determined from m_z, illustrate the agreement of all experimental results with thc Standard Model prcdictions.

Finally, the radiativc corrcctions to thc hadronic width [2], $\Gamma_h = \Gamma_{h,ew} \times (l + \alpha_s\pi + 1.3\ \alpha_s^2/\pi^2)$ permit a determination of α_s, already at thc present Ievel of experimental accuracy. Two experimentally independent determinations are possible. From $\Gamma_z = 2498 \pm 12.5$ GeV,

$$\alpha_s(m_z^2) = 0.134 \pm 0.027$$

Here the error includes $\pm$ 0.012 due to the uncertainty in the top mass, taken to be $m_z = 127 \pm 34$ GeV, and ± 0.008 due to that in the Higgs mass, assumed $40 < m_H < 1000$ GeV . From $R = \Gamma_h/\Gamma_\ell = 21.08 \pm 0.19$ GeV,

$$\alpha_s(m_z^2) = 0.162 \pm 0.031 \ ,$$

where the error is dominated by the experimental uncertainty. The combined value is:

$$\alpha_s(m_z^2) = 0.146 \pm 0.020 \ .$$

The unccrtainty is thc experimental one only, thc theoretical uncertainty is unknown to the author.

Summary

Less than a year after the start of LEP operation, of the order of 400,000 Z decays have been recorded. These have already furnished the demonstration that there are just three fermion families, and have provided a number of independent, precise checks of the Standard Model, at the percent level. The effective weak mixing angle has been measured with a new precision, $sin^{\theta}_w(m_z^2) = 0.232 \pm 0.001$.

The mass of the Z has been measured to one part in 3000; together with other LEP results and the results on the W mass, the effects of radiative corrections furnish an indirect, coarse, measure of the top mass: $m_t = 127 \pm 38$ GeV.

Acknowledgements

It is a pleasure to thank all the members of the ALEPH collaboration for the opportunity to participate in this adventure. In particular, I would like to thank E. Fernandez, L. Garrido, J.L. Harton, M. Martinez and A. Roussarie for a great deal of help. Finally many thanks to the ALEPH secretariat staff, A. Mazzari and R. Goidadin for the typing of the manuscript.

References

1. CERN-EP/90-25 ALEPH: *"A Detector for Electron-Positron Annihilation at LEP"*, 22 February 1990; to be published in Nucl. Instr. and Methods.

2. *Physics at LEP*, G. Altarelli, R. Kleiss and C. Verzegnassi editors, CERN 89-08.

3. D. Decamp et al., *"ALEPH Collaboration. Determination of the Number of Light Neutrino Species"*, Phys. Lett. **231B** (1989) 519. 4.

4. D. Decamp et al., *"ALEPH Collaboration. A Precise Determination of the Number of Families with Light Neutrinos and of the Z Boson Partial Widths,.* Phys. Lett. **235B** (1990) 399.

5. S. D. Decamp et al., *"ALEPH Collaboration. Measurement of Electroweak Parameters from Z Decavs into Fermion Pairs.* To be published in Zeit. fur Physik C.

6. B. Adeva et al., L3 Collaboration. *"A Determination of the Properties of the Neutral Intermediate Vector Boson Z*, Phys. Lett. **231B** (1989) 509.

7. M.Z. Akrawy et al., OPAL Collaboration. *"Measurement of the Z^0 Mass and Width with the OPAL Detector at LEP"*, Phys. Lett. **231B** (1989) 530.

8. P. Abreu et al., DELPHI Collaboration. *"Measurement of the Mass Width of the Z^0 from Multihadronic Final States Produced in e^+e^- Annihilations"*, Phys. Lett. **231B** (1989) 539.

9. B. Adeva et al., L3 Collaboration. *"Measurement of Z^0 Decays to Hadrons, and a Precise Determination of the Number of Neutrino Species"*, Phys. Lett. **237B** (1990)136.

10. M.Z. Akrawy et al., OPAL Collaboration. *"A Combined Analy.si.s of the Hadronic and Leptonic Decavs of the Z"*, Phys. Lett. **240B** (1990) 497.

11. P. Abreu et al., DELPHI Collaboration. *A Precise Measurement of the Z Resonance Parameters Through its Hadronic Decays*", Phys. Lett. **241B** (1990) 435.

12. B. Adeva et al., L3 Collaboration. "*A Precision Measurement of the Number of Neutrino Species*", 14 July 1990; to be published in Phys. Lett.B.

13. Reported at the Conference on High Energy Physics, Singapore, August 1990.

14. D. Decamp et al., ALEPH Collaboration. "*Determination of the Leptonic Branching Ratios of the Z*", Phys. Lett. **234B** (1990) 399.

15. P. Abreu et al., DELPHI Collaboration. "*Study of the Leptonic Decavs of the Z^0 Boson*", Phys. Lett. **241B** (1990) 425.

16. M.Z. Akrawy et al., OPAL Collaboration. "*Measurement of the Decay of the Z^0 into Lepton Pairs*", Phys. Lett. **235B**(1990) 379.

17. B. Adeva et al., L3 Collaboration. "*Measurement of g_A and g_V, the Neutral Current Coupling Constants to Leptons*", Phys. Lett. **236B** (1990) 109.

18. B. Adeva et al., L3 Collaboration. "*A Measurement of the Z^0 Leptonic Partial Widths and the Forward-Backward Asvmmetry*", Phys. Lett. **238B** (1990)122.

19. M.Z. Akrawy et al., OPAL Collaboration. "*Analysis of the Z^0 Couplings to Charged Leptons*, 11 June 1990; to be published in Phys. Lett. B.

20. B. Adeva et al., L3 Collaboration. "*A Determination of the Electroweak Parameters from $Z^0 \rightarrow \mu^= \mu^-(\gamma)$*." 21 June 1990; to be published in Phys. Lett. B.

21. S.L. Wu, Phys. Rep. **107** (1984) 60.

22. J. Alitti et al., (UA2 Collaboration), Phys. Lett. **B241** (1990) 150.

23. P. Shalbach et al., (CDF Collaboration), Proceedings of the APS Conference, Washington DC, April 1990.

24. H. Abramowicz et al., (CDHS Collaboration), Phys. Rev. Lett. **57** (1986) 298, and A. Blondel et al., Z. Phys. **C45** (1990) 361. J.V. Allaby et al. (CHARM Collaboration), Phys. Lett. **B177** (1986) 446, and Z. Phys. C36 (1987) 611.

Chairman: J. Steinberger

Scientific Secretaries: P. Maley, M. Wadhwa

Discussion

– *Brahm:*

Does the current data indicate an anomalously high hadronic width for the Z^0 ?

– *Steinberger:*

No.

– *Maley:*

Does the τ polarization affect the measurement of the τ lifetime by changing the shape of the impact parameter distribution?

– *Steinberger:*

The polarization should have some effect, how big it is I don't know but it should be in the Monte-Carlo.

– *Baldini-Celio:*

Once the Top has been discovered (we hope) do you think we shall have enough sensitivity to give results on the Higgs mass ?

– *Steinberger:*

That would be my wildest hope. I believe we will know the mass of the Top eventually. You can see from these results that the mass of the Top should be very close to where the $p\bar{p}$ colliders might be able to find it.

If we assume it's found then I don't know if the small effects which are now predicted from the mass of the Higgs are enough. You still have α_s corrections which you don't know that well but maybe you could imagine measuring also α_s a bit better.

If we could see some effect of the Higgs I think it would be much more important than anything we have done so far but it's very hard. This is my feeling but I've not really got a quantitative answer.

– *Labs:*

What analysis in "B" physics is underway in ALEPH — such as neutral "B" meson mixing ?

– *Steinberger:*

There is a result from ALEPH on mixing which will be given at the Singapore meeting and which is not in disagreement with the previous two results of ARGUS and UA1.

– *Dokshitzer:*

Could you please tell us why ALEPH is not doing "Hadronic Jet Physics ?

– *Steinberger:*

There are a great deal of data on hadronic jets, but it is not obvious how we should analyse them to learn something.

– *Dokshitzer:*

Will it be possible for ALEPH to measure separately yields of πs, κs and ps at momenta around 1 GeV ?

– *Steinberger:*

Yes.

– *Ayad:*

I understand you have used the lepton P_T-cut method to tag the b quark. What is your efficiency ?

– *Steinberger:*

My understanding is that we look at both electrons and muons. Together this would be about 40% of b production because each b can decay, with about 10% probability, into either of them and there are two b quarks in each event so that should give you 30%-40%. It is required that the lepton has a longitudinal momentum of more than 3 GeV and that the P_T is greater than 1 GeV (or 2 GeV in another analysis). In about 25,000 hadronic decays there were about 600 identified Bs, corresponding to an efficiency of $\sim$ 1 %.

– *Hsu:*

Was that a three photon final state you showed us ? It seems to me that you should see only a fraction of a single event with the statistics you have.

– *Steinberger:*

No, I don't think you're right that there is no easy way to get these events. At the level of about 100,000 Z^0 decays there are such events. I would be delighted if it were interesting but I understand it's not.

– *Baba:*

If the mass of the Higgs is close to the Z^0 mass how do you distinguish the Higgs from the Z^0 ?

– *Steinberger:*

It is not obvious. It will be very hard.

A CRUCIAL TEST FOR QCD: THE TIME-LIKE E.M. FORM FACTORS OF THE NEUTRON

Rinaldo Baldini Ferroli Celio

INFN - Laboratori Nazionali di Frascati
P.O. Box 13
I-00044 FRASCATI (Italy)

INTRODUCTION

A very special topic, the neutron e.m. time-like form factors (FF), is considered in the following.

These FF have never been measured. Furthermore the present models of the nucleon, which agree in describing the proton and the neutron space-like FF, are in bad disagreement in providing the cross section for

$$e^+e^- \to n\bar{n} \tag{1}$$

Comparing the ratio $\sigma(e^+e^- \to n\bar{n})/\sigma(e^+e^- \to p\bar{p})$, PQCD foresees a value 0.25, whereas for instance various EVMD predictions range from 1 up to 100.

Actually the nucleon structure still needs to be investigated. This has been proved by different new data, mainly:

- the surprising EMC results[1] about the contributions of the constituents to the spin of the nucleon,

Physics Up to 200 TeV
Edited by A. Zichichi, Plenum Press, New York, 1991

- the new results from the APPLE experiment[2] at LEAR, concerning the proton time-like FF, which are totally unexpected on the basis of the previous data. They show a very steep slope very near threshold, changing suddenly soon after.

The argument of this talk has been already reviewed[3]. In the following an updated, summarized, version is reported and in particular consequences of the new APPLE results are emphasized. Finally a new experiment, FENICE, is collecting data at the renewed storage ring ADONE and preliminary experimental results will be outlined.

DEFINITIONS AND PRESENT EXPERIMENTAL SITUATION OF THE FF

The main properties of the FF are shortly recalled (in the following it is assumed c=h=1 and the exchanged 4-momentum squared Q^2 is defined positive in the time-like region).

The one-photon exchange approximation is a standard assumption in lepton-hadron scattering and lepton-antilepton annihilation into hadron pairs (see Fig.1). Many positive checks have been done in lepton-hadron scattering[4,5]: angular behaviour for a given Q^2, identity among e^- and e^+ scattering, scattering on a polarized target. In e^+e^- annihilation the best check has been the absence of C=+1 final hadronic states.

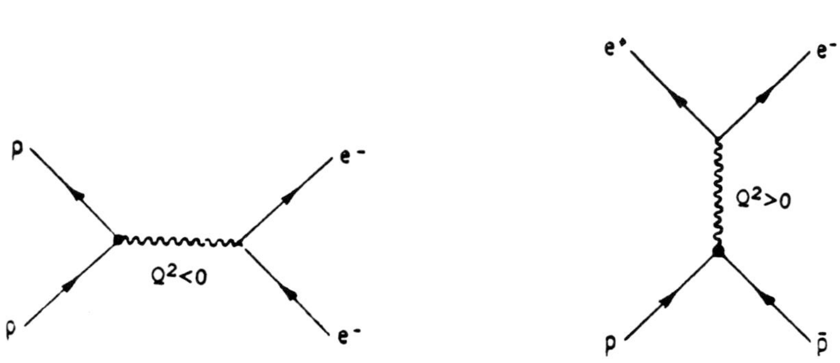

FIG. 1 - One-photon exchange approximation in ep scattering and e^+e^- annihilation into $p\bar{p}$.

In e^+e^- annihilation the virtual exchanged photon is polarized along the beam direction like a real one, if the electron mass is neglected respect to Q. Projecting this virtual photon along the outgoing baryon direction θ, three orthogonal helicity states are available, with different transition amplitudes and different angular behaviours:

$$\mathbf{A} = \mathbf{A}_+ \ (1+\cos\theta)/2 - \mathbf{A}_0 \ /\sqrt{2} \ \sin\theta + \mathbf{A}_- \ (1-\cos\theta)/2 \ .$$

For pointlike fermions with a mass M it is $\mathbf{A}_+:\mathbf{A}_-:\mathbf{A}_0$ = 1:1:2M/Q. P invariance implies $\mathbf{A}_+=\mathbf{A}_-$ for baryons. CPT invariance implies that an outgoing antibaryon with a given helicity corresponds to an ingoing baryon with the same helicity.

More properly $\mathbf{A}_+(Q^2)$ and $\mathbf{A}_0(Q^2)$ are the analytical continuation of the space-like spin-flip (magnetic), $G_M(Q^2)$, and no spin-flip (electric), $(2M/Q)G_E(Q^2)$, form factors. The pion FF behaviour[6] across the space-like and time-like regions brings an empirical evidence for. Unfortunately such a smooth extrapolation is not expected for baryons, due to the presence of vector mesons poles in the unphysical region. In principle this region may be explored by looking at $B\bar{B} \rightarrow \pi^0 e^+e^-$. Unitarity implies that FF are real on the real axis up to the first inelastic threshold in the unphysical region.

The Dirac ff, F_1, and the Pauli ff, F_2, related to the anomalous magnetic moment, are also introduced: $G_E=F_1+Q^2/4M^2F_2$, $G_M=F_1+F_2$.

At last the differential cross section for unpolarized beams for reaction (1) is:

$$d\sigma/d\cos\theta=\pi\alpha^2\beta/2Q^2 \ ((1+\cos^2\theta)\,|G_M(Q^2)|^2+ Q^2/4M^2 \sin^2\theta \,|G_E(Q^2)|^2)$$

S and D waves are allowed. Yet at threshold the S wave is expected to be dominant so that it is expected an isotropic cross section and $G_E(4M^2)=G_M(4M^2)$.

Space-like proton FF measurements, up to $Q^2 \approx 10$ GeV^2, are described by the classical dipole fit[4,5]: $G_M{}^p=\mu_p/(1-Q^2/m_0{}^2)^2$, with m_0=0.84 GeV, and $G_E{}^p=G_M{}^p/\mu_p$.

Time-like data disagree with a straightforward extrapolation of the dipole fit. A very relevant result has been achieved by the APPLE experiment at LEAR[2], reported in Fig.2: new, more accurate measurements have shown a very steep slope very near the threshold, totally unexpected on the basis of the previous data.

The neutron space-like magnetic ff is also well described by the dipole fit. New measurements of the neutron electric ff at low Q^2 are described by the fit[7] $G_E{}^n = a\mu_n\tau/(1+b\tau)$, where $a\approx1.3$, $b\approx7.$ and $\tau=-Q^2/4M^2$.

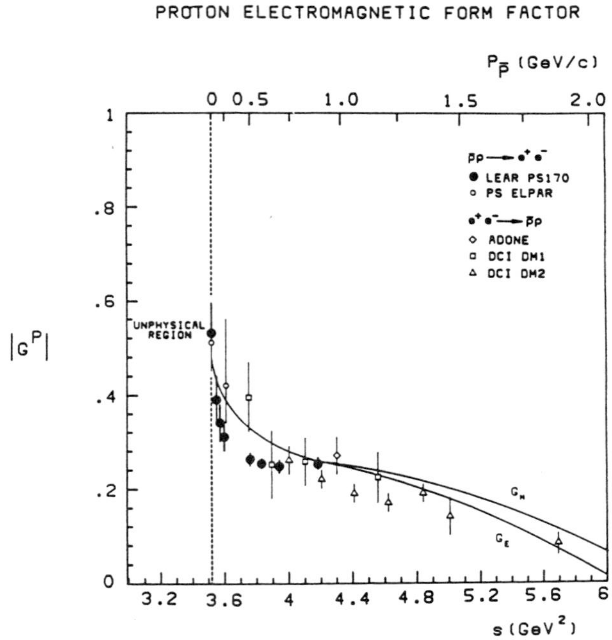

FIG. 2 - Apple data on p FF.

There are no data on the neutron time-like FF.

Finally, only one poor measurement of strange baryon FF exists[8], namely: $\sigma(e^+e^-\rightarrow\Lambda\bar{\Lambda})=110\pm50$ pb at $Q^2=5.76$ GeV^2, which implies $|G_M{}^\Lambda|=0.12\pm0.03$ assuming that $G_E \approx G_M$ is still valid.

MAIN QCD PREDICTIONS ON FORM FACTORS

At present the available QCD predictions on FF mainly concern their asymptotic behaviours and the quark wave functions probed at high Q^2 inside a hadron.

A still open question in QCD is at which Q^2 value an asymptotic prediction is reached. This is the major lack to appreciate Perturbative QCD calculations. Asymptotic expectations, like the Bjorken scaling, are fulfilled already at space-like $Q^2=-4M^2$ in the deep inelastic electron-nucleon scattering[4]. The new APPLE data clearly show that at $Q^2=4M^2$ smooth, asymptotic behaviours are still not reached. The question is: how far is the asymptotic region?

Concerning the asymptotic behaviour of the FF for a hadron made of n pointlike costituents it is foreseen a Q^2 power law[9], modulo terms containing ($\log Q^2$):

$$F \propto (1/Q^2)^{n-1}$$

Actually a power law rests on a more general ground than QCD[9,12]. As a check, results in the time-like region on pion and kaon FF[8] are reported in Fig.3. The overall trend is in good agreement with the expected behaviour (even if not proven) expecially compared to the exponential drop expected in absence of pointlike costituents[10,11].

PQCD predicts for the leading terms in the no helicity flip amplitude T[9,13]:

$$T = \int dy\, dx\, \phi_B^{+}(y,Q)\, T_B(x,y,Q)\, \phi_B(x,Q)$$

where T_B is the scattering or annihilation amplitude between quarks, computed just replacing each baryon with its collinear valence quarks, and ϕ_B is the quark wave function inside the baryon, considering transverse momenta squared up to Q^2.

The power law is recovered taking into account the Q^2 dependence of the various propagators and wave functions according to Fig.4. Additional $\log(Q^2)$ dependence comes from $\sqrt{\alpha_s(Q_i^2)}$ factors at each quark-gluon vertex.

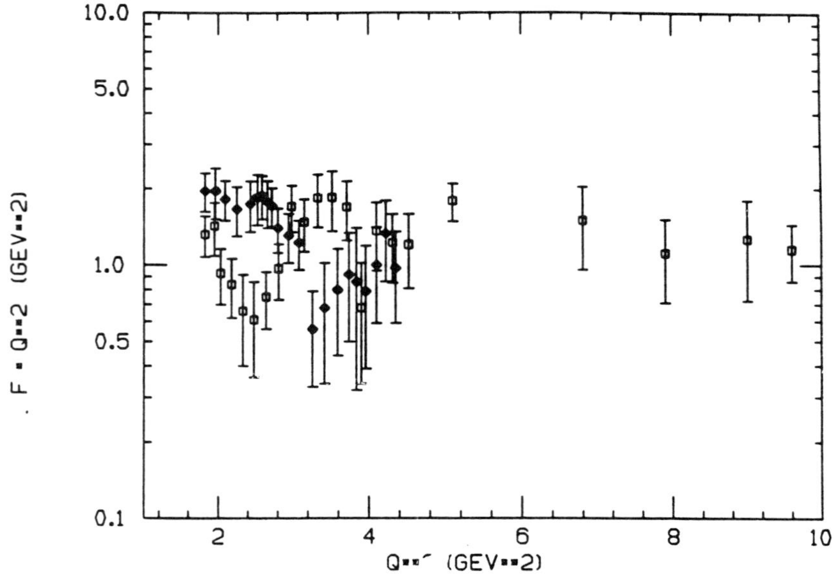

FIG. 3 - Pion (□), charged kaon (◆) time-like FF.

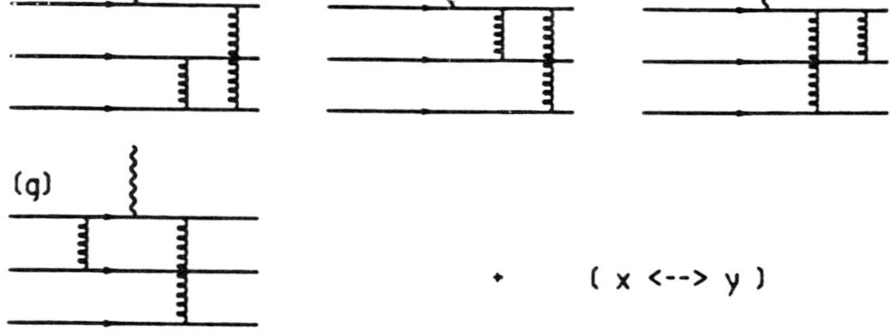

FIG. 4 - Leading contributions to the FF, according to PQCD.

There is no Pauli term in a quark current therefore PQCD naturally predicts the baryon Dirac ff; the Pauli ff scales with an additional $1/Q^2$ factor. Actually, the G_E structure constraints F_1 and F_2 to have different behaviours.

Critical ingredients in the PQCD calculation of the FF are the Q^2 dependence of α_s and of the quark wave function. In the integration on the quark and gluon internal momenta there is a divergence if it is assumed $\alpha_s(Q_i^2) \approx 1/\log(Q_i^2/\Lambda^2)$. A reasonable agreement with the data has been obtained by Ji[14] if a fictitious gluon mass $m_g \approx 0.5$ GeV is introduced, as it is shown in Fig.5.

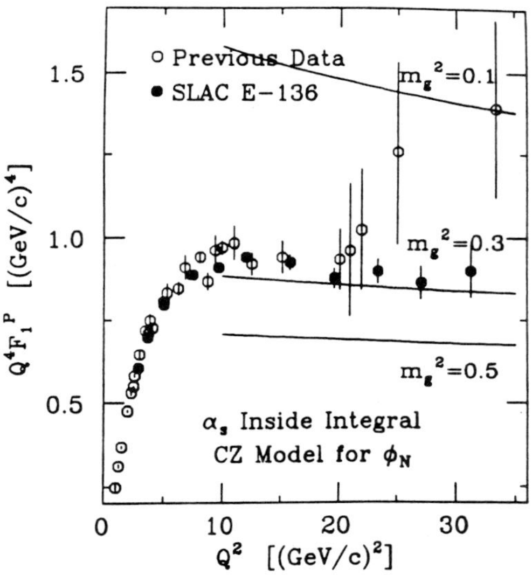

FIG. 5 - Proton Dirac ff, according to PCQD and $m_g \neq 0$.

The quark wave function is also critical because extreme cases, a non-relativistic distribution or an asymptotic one, both give unphysical space-like FF: $G_M^n > 0$, $G_M^p < 0$ or $G_M^p \ll G_M^n$. Chernyak and Zhitnisky[13] have evaluated the wave function according to the S.V.Z. sum rules[15], which allow to know the quark momenta $x_1^{n1}x_2^{n2}x_3^{n3}$ averaged on the nucleon wave function for every n_1 n_2 n_3 values. Very roughly these sum rules connect, according to the uncertainty principle, quantities averaged on the energy to perturbative and short range confining mechanism expectations. Chernyak and Zhitnisky have done an ansatz for the wave function, which agrees with sum rules for $n_1+n_2+n_3 \leq 2$. It is very relevant that in this wave function, see Fig.6, there is a leading u(d) quark in a proton (neutron): $<x_{u(d)}> \approx 2/3$. Such a result is in agreement with a baryon's picture as a diquark-quark bound state. The leading quark should be produced first in e^+e^- annihilation at high Q^2 according to PQCD, so that:

$$\sigma(e^+e^- \rightarrow n\bar{n}) \,/\sigma(e^+e^- \rightarrow p\bar{p}) = (q_u/q_d)^2 = 0.25 \; .$$

A warning about S.V.Z. sum rules applied to heavy $q\bar{q}$ wave functions has been given by Bell and Bertlmann[16]. A $q\bar{q}$ confining potential has been derived, which gives the same quark momenta: $V(r) = -4/3 \; \alpha_S/r + \pi/144 \; <\alpha_S GG> m_q \; r^4$. Yet this potential should be flavour independent and no m_q dependence should be allowed.

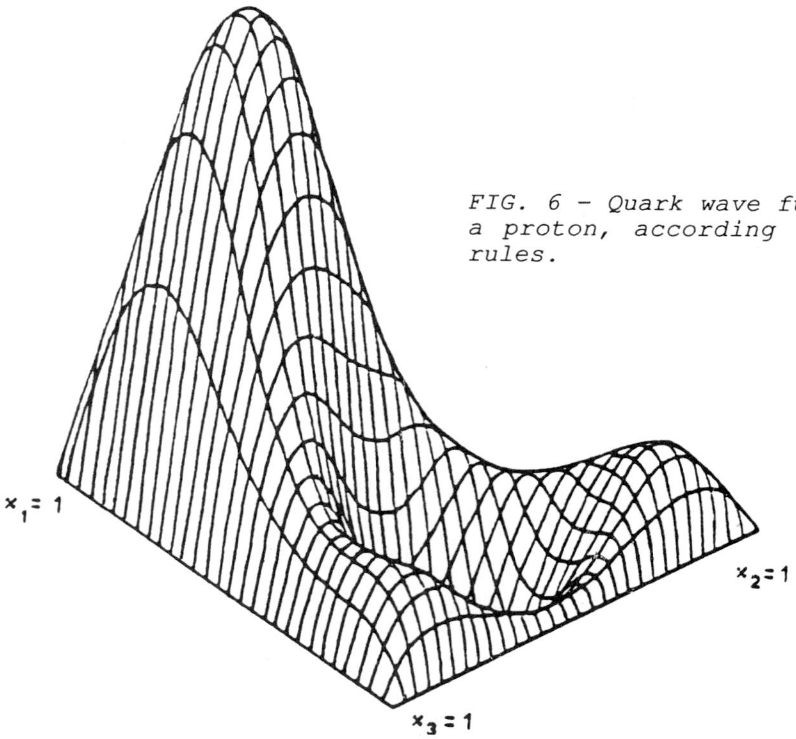

FIG. 6 - Quark wave functions in a proton, according to SVZ sum rules.

High statistics experiments on J/Ψ decays allow to check the theory in the time-like region (even if, actually, at $Q^2 \approx 4M_C^2$). A test of PQCD expectations is done looking at the branching ratios $B(J/\Psi \rightarrow B\bar{B})$. In this case the main decay of J/Ψ in three hard gluons well matches the three quarks of a baryon, as it is shown in Fig.7. This branching ratio is strongly dependent on α_s and ϕ_N [13], namely $B(J/\Psi \rightarrow B\bar{B}) \approx \alpha_s{}^3 \phi_N{}^4$. The experimental value, $B(J/\Psi \rightarrow B\bar{B}) = 0.22 \pm 0.01$ [17], is just recovered if $\alpha_s(M_{J/\Psi}{}^2) \approx 0.2$ and if the wave function is that derived according to sum rules. This expectation decreases by one or two order of magnitude if the asymptotic or non relativistic wave functions are introduced.

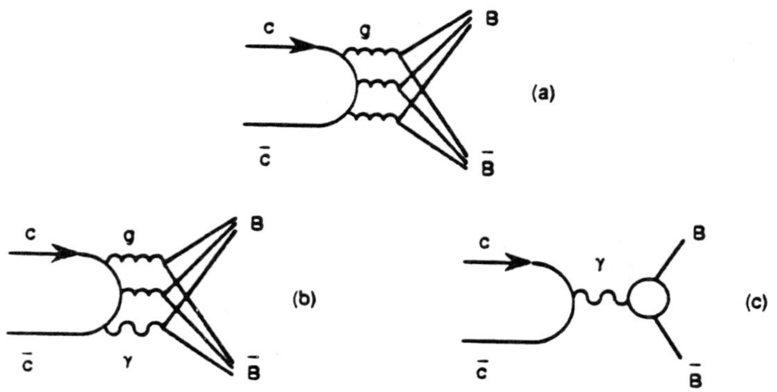

FIG. 7 - a) Leading contribution to $J/\Psi \rightarrow B\bar{B}$, according to PQCD. b,c) e.m. contributions to $J/\Psi \rightarrow B\bar{B}$.

This impressive check conflicts with the helicity non conservation in the sizeable branching ratios $B(J/\Psi \rightarrow \rho\pi + K^*K) = 2.1 \pm 0.2$ % [17]. In fact in the gluon-quark spin matching and helicity conservation only two vector or pseudoscalar mesons should be produced.

The Ψ' agrees with this rule: $B(\Psi' \rightarrow \rho\pi + K^*K) = 0.009 \pm 0.005$%. Indeed the anomaly is the J/Ψ decay. It is expected that J/Ψ and Ψ' branching ratios into the same hadronic channel were in the same ratio as the relative branching ratios into e^+e^- [18]. In Table I many of these ratios are reported, in good agreement with this prediction, only the decays into vector + pseudoscalar meson disagree by a factor as big as $(M_\Psi/M_{\Psi'})^n$, where n=17.

This anomalies have been pointed out many years ago[19]. Recently Brodsky, Lepage and Tuan [18] have proposed a small mixing between the J/Ψ and a "glueball" G nearby, with a difference in mass and width of the order of 100 MeV. In fact G would affect only the J/Ψ and a large effect is expected in channels forbidden to the J/Ψ, if the G width is large. On the other hand the interaction region for G is of the order of $1/m_{u,d,s}$ (to be compared to $1/m_c$ for a $c\bar{c}$ pair). Therefore also many soft gluons contributions are provided for the G decay. Unfortunately there is also a puzzling sizeable branching ratio $B(J/\Psi \rightarrow \omega\pi)$, which would not be expected in this hypothesis.

TABLE I. Ψ/Ψ'decays anomaly.

CHANNEL X	$B(\Psi\rightarrow X)/B(\Psi'\rightarrow X)$
$e^+e^- + \mu^+\mu^-$	7.7
$3(\pi^+\pi^-)\pi^\circ$	8.3
$2(\pi^+\pi^-)\pi^\circ$	11.0
$\pi^+\pi^-\pi^\circ$	167.0
$\pi^+\pi^-K^+K^-$	4.5
$\pi^\circ K^+K^-$	>203.
$\pi^+\pi^- p\bar{p}$	7.5
$\pi^\circ p\bar{p}$	7.9
$\pi^+\pi^-$	1.9
K^+K^-	2.4
$p\bar{p}$	11.6
$3(\pi^+\pi^-)$	26.7
$2(\pi^+\pi^-)$	8.9

Finally it must be noticed that lattice calculations[20], up to now performed in quenched approximation, do not predict a leading quark. However the dipole formula also is not predicted.

VECTOR MESON DOMINANCE PREDICTIONS ON FORM FACTORS

Before QCD, the successful model to interpret the e.m. interactions of hadrons was the Vector Meson Dominance model. According to VMD[21] hadronic and e.m. interactions are mediated by vector mesons as ρ, ω, ϕ and their recurrencies.

It is a curiosity that, just in the early days of the first nucleon FF measurements and of the first vector meson evidences, Cabibbo and Gatto[22] foresaw a very big neutron time-like FF:

$$\sigma(e^+e^-\to n\bar{n}) / \sigma(e^+e^-\to p\bar{p}) \approx 14 \ .$$

At the moment it is assumed that there is a transition from an Extended VMD to a direct photon-quark coupling, passing from low to high Q^2. A duality relationship should exist between these two descriptions[23]. The uncertainty principle should relate cross sections described by PQCD to integrated cross sections described by EVMD. In turn, high Q^2 cross sections are achieved by the convolution of an infinite number of vector mesons.

Actually there is not yet a reliable prediction on the vector meson spectrum and on how to extrapolate their amplitudes out of the resonance. In spite of these uncertainties, most of the EVMD calculations on the time-like neutron FF expect:

$$\sigma(e^+e^- \to n\bar{n}) \gg \sigma(e^+e^- \to p\bar{p})$$

for $Q^2 \leq 10\ \mathrm{GeV}^2$.

The determination of the various vector mesons contributions to the nucleon FF is an important result perse. For instance the size of the Φ contribution is the best measurement of the strange quark content in the nucleon wave function, which is very important[24] to interpret the aforementioned EMC results on the polarized nucleon structure functions.

According to the old Veneziano model[25] the masses of the ρ daugthers are given by $m_n^2 = m_\rho^2 + n/\alpha'$, where $\alpha' \approx 0.9 \div 1.0\ \mathrm{GeV}^2$ is the slope of the ρ trajectory, so that we have $m_{\rho'} \approx 1.25$ GeV, $m_{\rho''} \approx 1.6$ GeV and full degeneracy is expected between ρ and ω recurrences.

Korner and Kuroda [26] have reproduced the baryon space-like FF and predicted the time-like FF (see Fig.8) with a remarkable formula, without free parameters. ρ,ω,ϕ relationships are established according SU_3 flavour symmetry, poles are those provided by the Veneziano model, their number $N_{1,2}$ is fixed by the asymptotic QCD power law requirements ($F_1 \propto 1/Q^4$ and $F_2 \propto 1/Q^6$) and the FF are normalized to $Q^2=0$. Demanding the asymptotic behaviour with a small number of poles should be coherent with the early scaling behaviours in the space-like region. This formula agrees surprisingly well with the bulk of the proton time-like measurements, done many years later, even if do not reproduce the APPLE structure. It is predicted, in the range $3.5 \leq Q^2 \leq 10$ GeV^2 :

$$\sigma(e^+e^- \to n\bar{n}) / \sigma(e^+e^- \to p\bar{p}) \approx 2.$$

Yet a $\rho'(1.25)$ is demanded in this approach, but there is no experimental evidence of this vector meson in $\sigma(e^+e^- \to \text{hadrons})$ and in the pion ff[8].

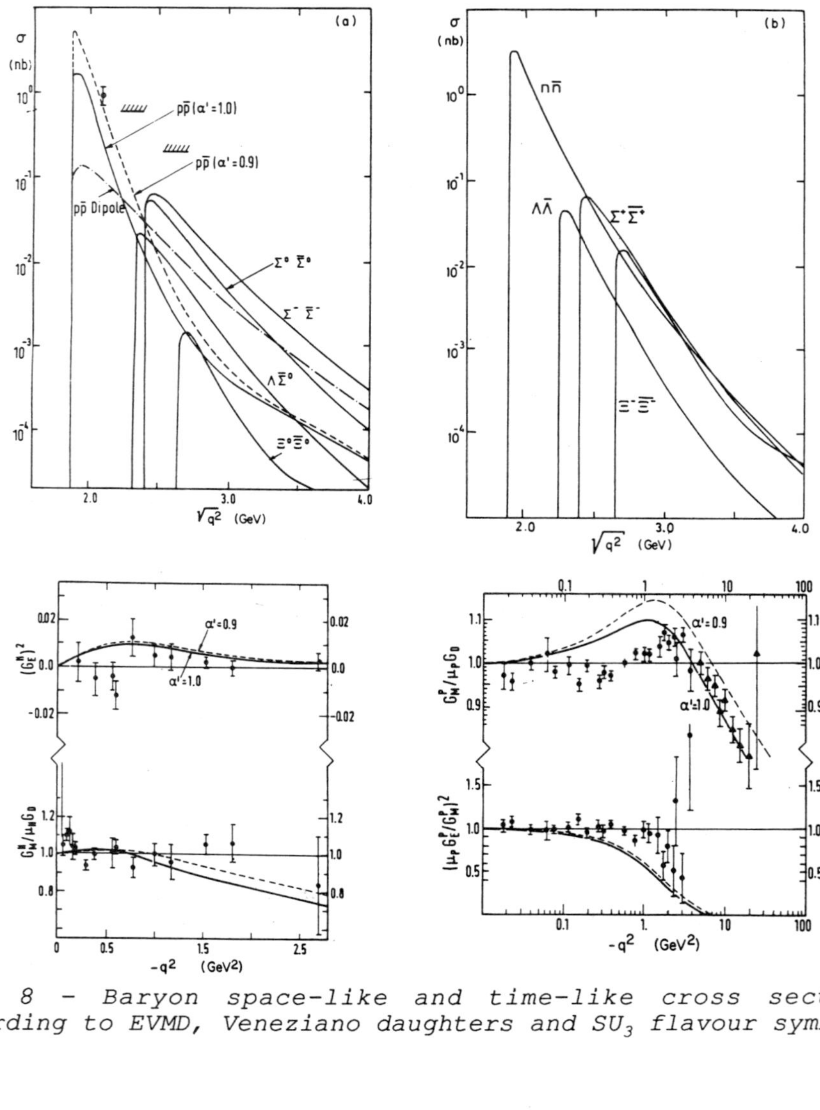

FIG. 8 - Baryon space-like and time-like cross sections, according to EVMD, Veneziano daughters and SU_3 flavour symmetry.

If only even daugthers exist, like the $\rho(1.6)$, good predictions have been given by Etim and Greco[27], assuming simple rules for their widths and coupling constants: for instance $R_\infty = \sigma(e^+e^- \to \text{hadrons}) / \sigma(e^+e^- \to \mu\mu) \approx 2.5$.

According to this, by now standard, scheme, which is in agreement with the bulk of the experimental data in e^+e^- annihilation around the $N\bar{N}$ threshold, it is provided [28] in the range $3.5 \le Q^2 \le 10\ \text{GeV}^2$:

$$\sigma(e^+e^- \to n\bar{n}) / \sigma(e^+e^- \to p\bar{p}) \approx 100$$

if old proton time-like FF data are considered.

Of course the new APPLE results again do not fit in a simple scheme of only large recurrencies. There were already other hints that the spectrum of the ρ recurrencies is much more complicated. The pion ff has a dip at 1.6 GeV and the comparison with the diffractive photoproduction of pion pairs, see Fig.9, definitively shows that photoproduction cannot be related to e^+e^- annihilation according to the simplest version of EVMD [29].

The pion ff has been nicely fitted by Donnachie and his collaborators[29] with two interfering resonances $\rho'(1.4)$ and $\rho'(1.7)$.

The smooth bump around 1.6 of $\sigma(e^+e^- \to \pi^+\pi^-\pi^+\pi^-)$, the so called $\rho'(1.6)$ (see Fig.10a), does not show any interference pattern [30], but many channels may be superimposed in a many bodies process. Indeed in the two body channel $\rho\eta$ there are evidences for these two resonances, but the collected statistics is rather poor [31] (see Fig.10b).

Results from data collected by the DM2 experiment at DCI about $e^+e^- \to \pi^+\pi^-\pi^0\pi^0$ (see fig.11) agree with the two resonances interpretation. In fact $e^+e^- \to \pi^+\pi^-\pi^+\pi^-$ and $e^+e^- \to \pi^+\pi^-\pi^0\pi^0$ are very different, once $e^+e^- \to \omega\pi^0$ (see Fig. 12) has been removed. These two resonances, if confirmed, will have very deep implications on the whole hadron spectroscopy.

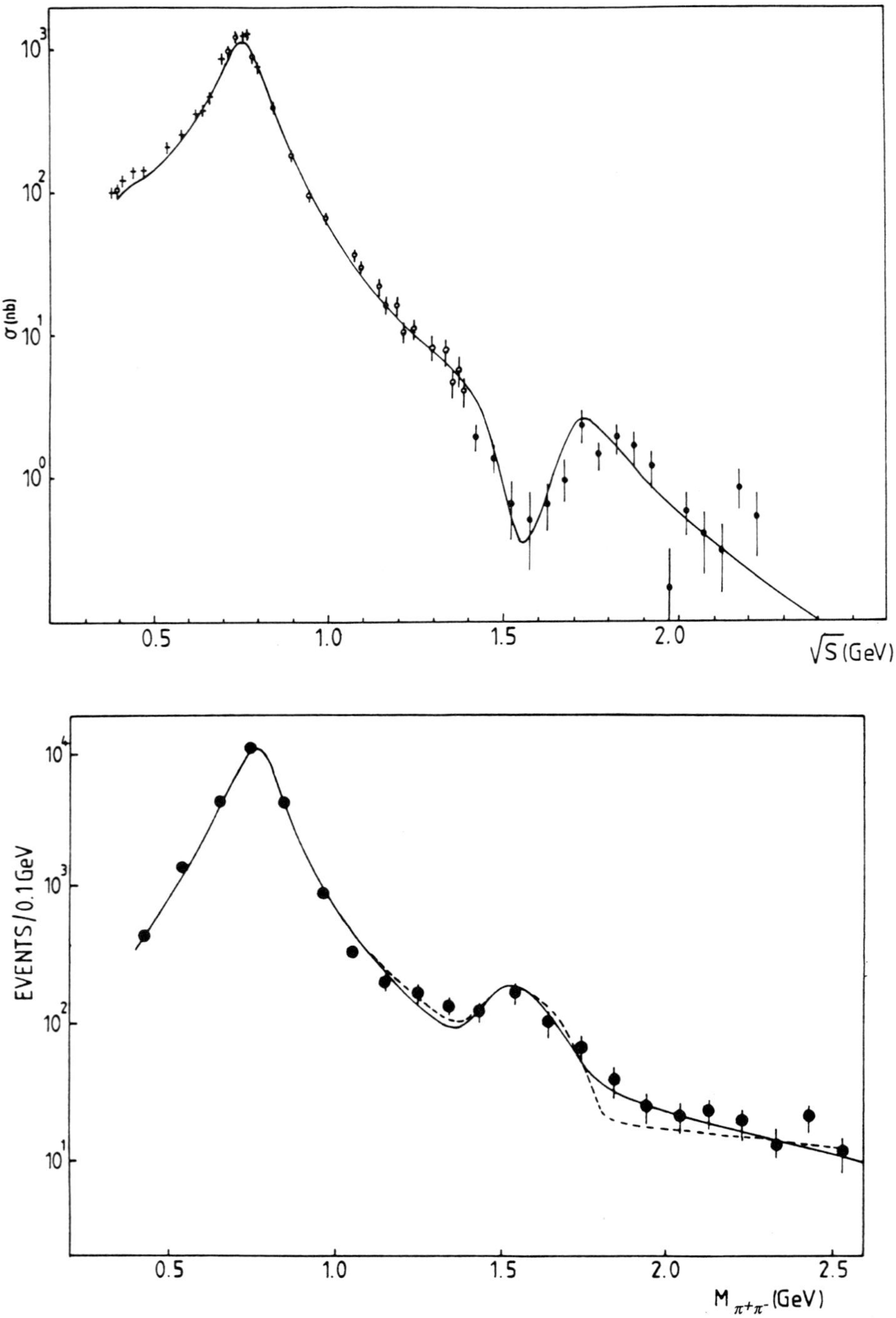

FIG. 9 - Pion ff and diffractive pion pairs photoproduction.

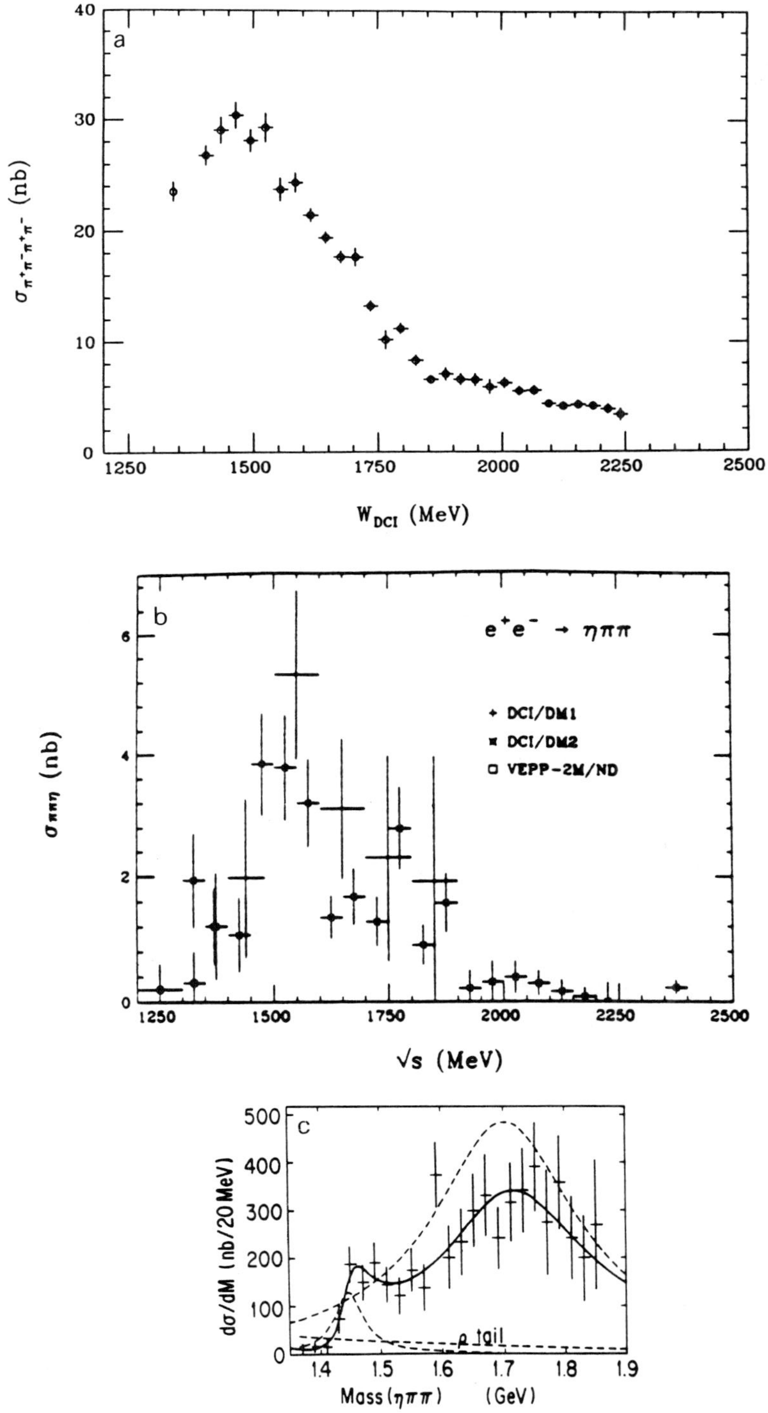

FIG. 10 - a) the $\rho'(1.6)$ bump in $e^+e^- \rightarrow \pi^+\pi^-\pi^+\pi^-$.
b) cross section $e^+e^- \rightarrow \rho'\eta$.
c) cross section as $\pi^-p \rightarrow (\rho\eta)n$, as a function of $M_{\rho\eta}$ for $J^P_{\rho\eta} = 1^-$.

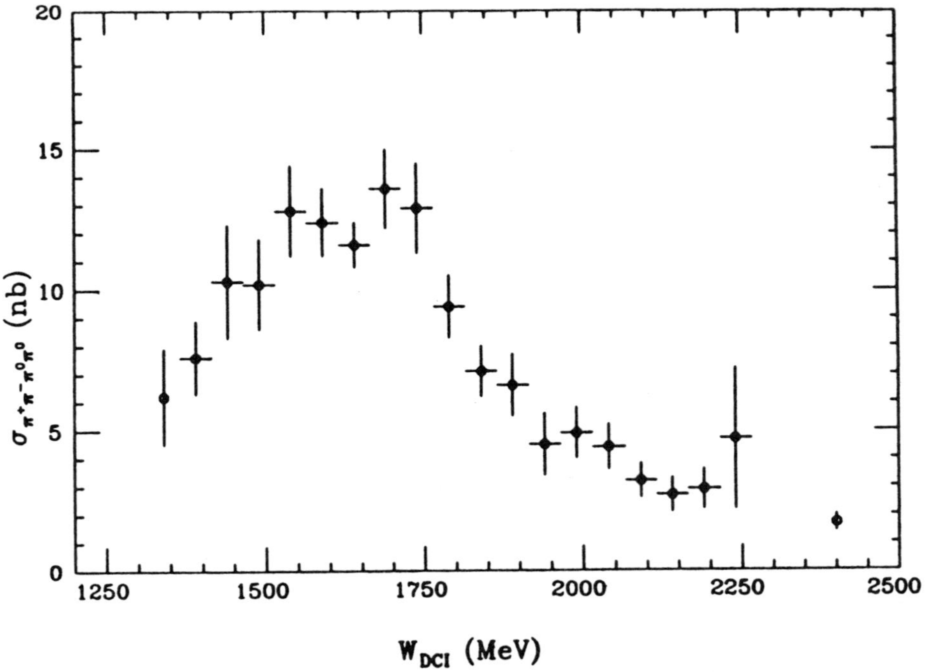

FIG. 11 - $e^+e^- \to \pi^+\pi^- \pi^0\pi^0$ cross section ($\omega\pi^0$ contribution subtracted).

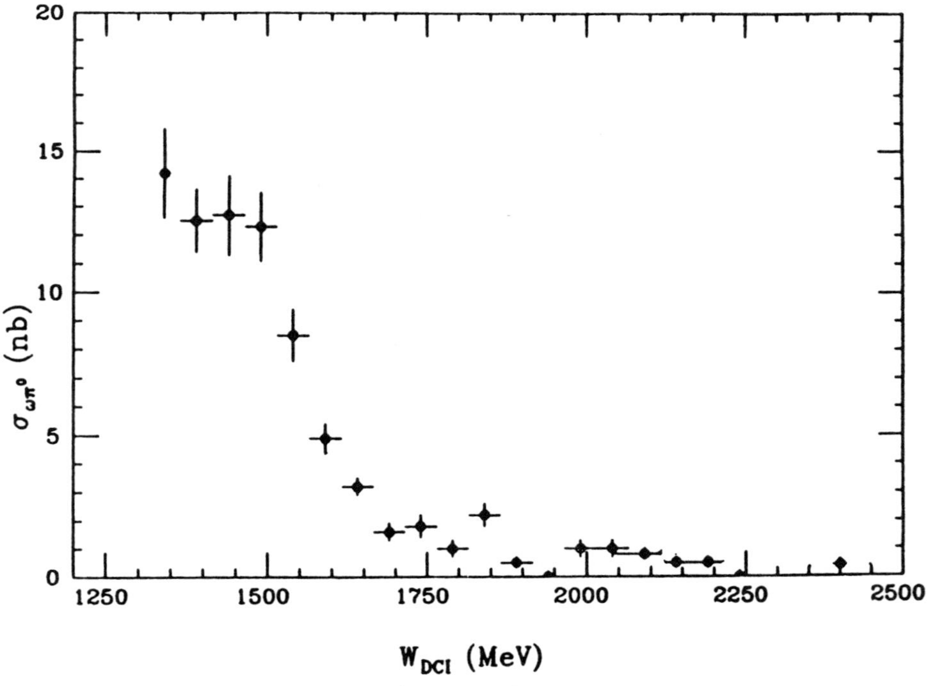

FIG. 12 - $e^+e^- \to \omega\pi^0$ cross section.

IMPROVEMENTS UPON VMD

All the aforementioned EVMD approaches extrapolate resonant amplitudes assuming that the vector meson-nucleon coupling constants remain real and do not vary with Q^2. Many years ago it was shown by Massam and Zichichi [32] that the space-like nucleon dipole fit is recovered by VMD if a further vector meson form factor $F = 1/(1-Q^2/\Lambda^2)$, with $\Lambda \approx 1.$ GeV, is introduced in the vector meson-nucleon coupling constants. Nowadays this procedure is embodied in the Skyrme model and the same cutoff is also used in practice in the One-Boson-Exchange model for the NN interaction [34].

Of course straightforward extrapolation of such a form factor in the time-like region is meaningless, introducing an unphysical pole, and it is in disagreement with the experimental data on the time-like proton FF[35].

A not analytical extrapolation, which avoids unphysical poles, has been attempted by Etim and Malecki[36]: $F_V=1/(1+|Q|^2/\Lambda^2)$. In this model Bloom-Gilman[37] duality is invoked to justify a form factor also for a vector meson, like for any hadron. Using standard even daughters and demanding the QCD asymptotic behaviour, a good fit is achieved for the space-like and pre-APPLE time-like proton FF data, assuming $\Lambda \approx 0.7$ GeV. Then it is predicted in the range $3.5 \leq Q^2 \leq 10$ GeV^2:

$$\sigma(e^+e^- \to n\bar{n}) / \sigma(e^+e^- \to p\bar{p}) \approx 1$$

This prediction agrees with the common opinion that ρ and its recurrences dominate e^+e^- annihilation into hadrons. In the nucleon case it is anyway very dependent on the high mass recurrencies and it will change if the aforementioned doubts about $\rho'(1.6)$ are well grounded.

Another improvement upon simple EVMD is the implementation in the FF of the cuts on the Q^2 real axis above inelastic thresholds, demanded by the unitarity.

Dubnicka [38] has conceived a formula which has almost all the required analytical properties and asymptotic behaviours. Free parameters are pole positions, coupling constants on the poles, and an effective threshold for the unitarity cuts. In this formula again the effective coupling constants vary with Q^2 but the early asymptotic behaviour may be achieved without fixing the number of poles, as it has been required in simple EVMD. Leaving as parameters the still unknown masses, widths and coupling constants, a good fit is obtained for π^+, K^+, K^o, p (still considering pre-APPLE data only) and n FF, simultaneously.

It is worthwhile to note that Dubnicka's fit reproduces in part the double structure of the so called $\rho'(1.6)$ and another resonance, a $\rho'(2.2)$, which is above the $N\bar{N}$ threshold, is expected. This fit predicts on the whole range $3.5 \le Q^2 \le 10$ GeV^2 :

$$\sigma(e^+e^- \to n\bar{n}) / \sigma(e^+e^- \to p\bar{p}) \approx 25$$

There is another orthogonal approach, which provides poles peculiar to the baryon structure near $N\bar{N}$ threshold, namely a $B\bar{B}$ potential model[39]. Such a potential should be derived by the One Boson Exchange NN potential: repulsive terms become attractive if G = -1 for the exchanged meson, an imaginary part is added, related to the $N\bar{N}$ total annihilation cross section, and in principle no further parameter must be included. The non evidence of a long sought baryonium[40] has put this appealing model in the shade. However there are evidences for narrow structures in e^+e^- annihilation into hadrons just near the $N\bar{N}$ threshold[41]. Dalkarov[42] predicted these structures years ago.

Polikarpov and Van der Velde[39] also gave a similar prediction together with large neutron time-like FF near the threshold:

$$\sigma(e^+e^- \to n\bar{n}) / \sigma(e^+e^- \to p\bar{p}) \approx 2.$$

The new APPLE results strongly suggest a recovery of this appproach to $N\bar{N}$ annihilation near the threshold.

HYBRYD MODELS AND SKKYRME MODEL OF THE NUCLEON

A very interesting model has been worked out by Gari and his collaborators[43], attempting to merge VMD and PQCD in a suitable formula to have an overall fit of the nucleon FF.

When dual descriptions are merged together the main problem is to avoid double counting. For that only ρ and ω poles have been considered and the contributions coming from their recurrences or their quark structure have been lumped in a direct coupling term. Finally, all these terms have been weighted by two universal vector meson form factors. This approach is a modern, refined, version of the old Massam-Zichichi formula, as resumed by Iachello, Jackson and Lande [32]. To simplify the formulation, ρ and ω contributions, isoscalar and isovector vector mesons form factors, have been retained similar, even if there is no fundamental reason for that.

In detail:

$$F_1^{V,S} = (c\, m_\rho^2/(m_\rho^2 - Q^2) + 1-c)\ F_1(Q^2)$$
$$k_{\rho,\omega} F_2^{V,S} = (k_{\rho,\omega}\, c\, m_\rho^2/(m_\rho^2 - Q^2) + 1-c)\ F_2(Q^2)$$

Afterwards, in the limit of low Q^2 the vector mesons form factors must become $F_1 \approx F_2 \approx \Lambda^2/(\Lambda^2-Q^2)$, whereas, in the limit of high Q^2, PQCD must be achieved, that is: $F_1 \propto 1/(Q^2\log(Q^2))$ and $F_2 \propto F_1/Q^2$. An interpolating formula between these two extreme regimes is given by:

$$F_1 = \Lambda_1^2/(\Lambda_1^2-q^2)\ \Lambda_2^2/(\Lambda_2^2-q^2)$$
$$F_2 = \Lambda_1^2/\ (\Lambda_1^2-q^2)\ (\Lambda_2^2/(\Lambda_2^2-q^2))^2$$
$$q^2 = Q^2 \log((\Lambda_2^2-Q^2)/\Lambda_0^2/\log(\Lambda_2^2/\Lambda_0^2)$$

The new cutoff Λ_2 is a peculiar ingredient of the Gari formulation and it may be considered as a phenomenological estimation of the long sought energy scale beyond which PQCD predictions are achieved.

A fit of the space-like data gives: $\Lambda_0 \approx 0.3$ GeV, $\Lambda_1 \approx 0.8$ GeV and $\Lambda_2 \approx 2.2$ GeV. The authors have not quoted the errors on the estimated parameters. Anyhow the values obtained for Λ_0 and Λ_1

are well within the expectation [4,32] and the value of Λ_2 would mean, for our purposes, that $Q^2 \approx 4$ GeV^2 is almost asymptotic if there is asymptotic symmetry between space-like and time-like regions.

Peculiar to this model is that $F_1{}^n \approx 0$, hence for the neutron Pauli ff dominates, and it is expected at high Q^2:

$\sigma(e^+e^- \rightarrow n\bar{n}) / \sigma(e^+e^- \rightarrow p\bar{p}) \propto 1/Q^2$

Unfortunately no time-like extrapolation has been done up to now for this model.

As it has been anticipated, FF varying with Q^2 are naturally embodied in the Skyrme model of the nucleon. This model of strongly interacting particles (whose basis have been conceived many years ago[44]) is the only one without quarks available on the market. In this and other related approaches there is an elementary but self-interacting pion field, instead of the quark field.

Very roughly the corresponding lagrangian may be derived considering at first massless quarks inside a bag, demanding chiral invariance and introducing another term,which contains a gauge pseudoscalar field ϕ for it: $L=L_{bag}+ f_\pi{}^2 /4\ Tr(D_\mu U D_\mu U^+)$, where $U=\exp(i\phi\,\tau)$. The added term may be regarded as a first order, S wave, in a $D_\mu U$ power expansion and the next order may be related to the introduction of vector mesons. Then, the S wave coupling constant may be identified with the structure constant in the pion weak decay f_π , and the P wave coupling constant with the $\pi\pi\rho$ coupling constant g_ρ. At this moment the quark field may be avoided and a theory without quark is achieved[33].

The equations of motion have a solitonic solution, which may be identified as a baryon since it may behave like a fermion: that is the wave function changes sign under a 2π full rotation.

This paradox can be realised taking into account that an half integer angular momentum is obtained adding up an infinite number of even and odd angular momenta.

The dipole fit of the proton FF is predicted by the Skyrme model (see Fig.13) inasmuch as it corresponds to the vector meson propagator times the source dimensions.

In the simplest version of this baryons achievement, vector meson parameters are not free[33], but $g_\rho=2\pi$ and $m_\rho=\sqrt{2}f_\pi g_\rho$.

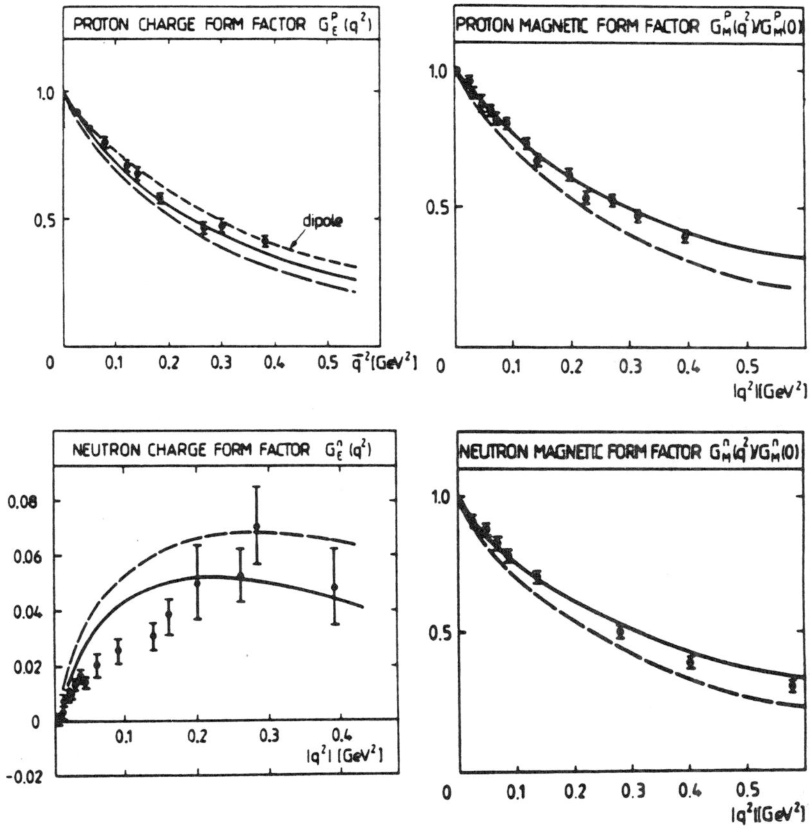

FIG. 13 - Skyrme model predictions on nucleon FF.

Remarkably, in this crude theory many mesons and baryons properties are reproduced within a 50% accuracy (see Table II).

If the Skyrme model conjectures are right, the source of the proton spin is not the spin of the quarks, which is the case according to the quoted EMC results[1]! Such a model may be not so crazy, if we take into account that it has been demonstrated how QCD becomes a local field theory of coupled mesons in the limit of an infinite number of colours[45].

TABLE II - Nucleon properties as predicted by a minimal Skyrme model.

	MODEL	EXPERIMENT
$< M_E^2 >_p$ (fm^2)	0.85	0.74±0.02
$< r_E^2 >_n$ (fm^2)	-0.22	-0.119±0.004
$< r_M^2 >_p$ (fm^2)	0.71	0.74±0.1
$< r_E^2 >_n$ (fm^2)	0.72	0.77±0.14
μ_p	3.36	2.79
μ_n	-2.57	-1.91
g_A	0.88	1.25
$< r_A^2 >_n$ (fm^2)	0.41	0.39±0.06
m_p (MeV)	826.	770.±3.
g_p	6.28	≈ 6.1±0.5

Yet the philosophical implications of a strongly interacting particles theory without quarks are so relevant that it should pursued perse.

PREDICTIONS FROM DATA ON STRANGE BARYONS

Two measurements may be employed to infer two neutron time-like measurements, making use of the SU_3 flavour symmetry and U-spin relationships. Namely:

- the only available measurement of Λ FF[8],
- the available J/Ψ baryonic branching ratios[17].

The U-spin relationship[46] between Λ and neutron magnetic ff is $G_M{}^n = 2\ G_M{}^\Lambda$, if SU_3 flavour symmetry is attained. The difference in mass $m_\phi - m_\rho$, or twice $m_\Lambda - m_n$, may be employed as a correction in Q for small symmetry violation.

In short, it is foreseen $|G_M n| = 0.24 \pm 0.05$ at $Q^2 = 4.6\ GeV^2$, to be compared to $G_M{}^p \approx 0.25$: the neutron ff is equal or greater than the proton ff, at threshold.!

Concerning the J/Ψ baryonic decays three amplitudes must be taken into account[50]: an isoscalar direct decay amplitude (see Fig.7a) and two e.m. corrections (see Fig.7b,7c), where the amplitude in Fig.7b corresponds to the FF just before the J/Ψ, amplified as the $\mu\mu$ amplitude. The direct decay is supposed to dominate and only projections on it are retained for e.m. amplitudes. This approximation should be irrelevant if PQCD holds, because e.m. and OZI amplitudes are expected to be mainly real. Furthermore, the amplitude in Fig.7c is expected to be proportional to the baryon electric charge and it does not contribute to the neutron.

The e.m. amplitude could be identified with the magnetic contribution, either because the J/Ψ is still not far from any $B\overline{B}$ threshold and the electric contribution is lowered by a $2M^2/Q^2$ factor or because the Pauli ff is small.

SU_3 flavour symmetry and U-spin relationships may be applied, once the baryon phase space $\sqrt{\beta_B}$ has been factorized. The direct decay is decomposed in a SU_3 flavour symmetric amplitude A and in SU_3 flavour symmetry breaking amplitudes B and C, related to the hypercharge as usual[47]. There are two U-spin invariant e.m. amplitudes, D and F, and U-spin violations are dealt as before. In short the relations expressed in Fig.14 hold.

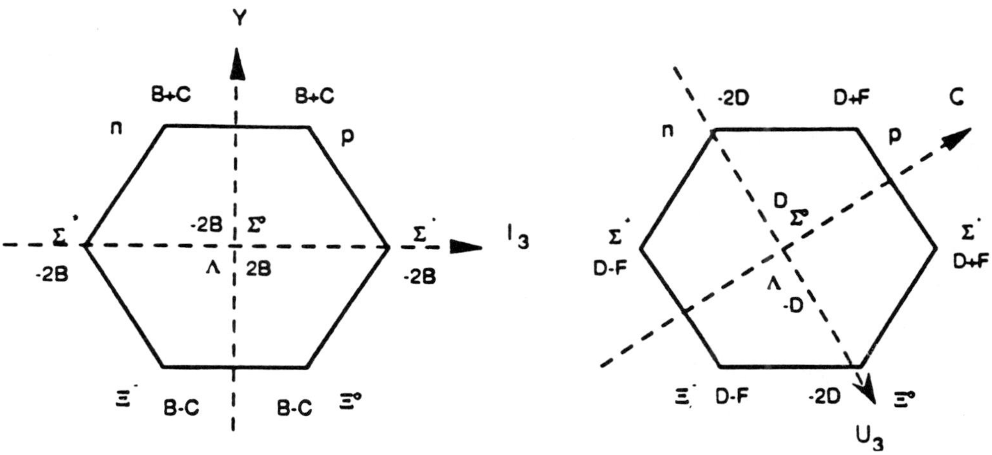

FIG. 14 - SU_3 flavour symmetry breaking and U-spin amplitudes.

For the neutron it is deduced[48]: $G_M{}^n = -0.007 \pm 0.007$ at $Q^2 = 8.1\ GeV^2$. For the proton an evaluation does not make sense because it takes a contribution from the very poorly measured branching ratio $J/\Psi \to n\overline{n}$[17]. Anyhow this neutron ff is definitively very small respect to any proton measurements extrapolation which should be about 0.04! Such a steep behaviour with Q^2 would indicate that the Pauli ff dominates the neutron time-like FF as it does in the neutron space-like FF.

The smallness of an imaginary part among direct and e.m. amplitudes, assumed in the previous reasoning, has been questioned [49]. A good measurement of $J/\Psi \rightarrow n\bar{n}, \Sigma^-\Sigma$ will allow a good check.

In conclusion neutron time-like FF equal or higher than the proton FF at threshold and a steeper neutron slope with Q^2 would be a compromise in agreement with all the expectations.

THE FENICE EXPERIMENT

A new experiment[51], FENICE, is collecting data at the renewed storage ring ADONE. In fact a new radiofrequency cavity, a new optics and a wiggler have been installed: as a consequence shorter bunches and possibly higher luminosities than in the past are available.

The detector is a 4π calorimeter made of iron, streamer tubes, scintillation counters and large area resistive plate counters (see Fig.15).

At the trigger level only the antineutron is demanded. Actually the antineutron annihilation pattern and its time of flight should also be enough to identify $e^+e^- \rightarrow n\bar{n}$, at least near threshold. However the neutron is detected in about 20% of the events.

The unknown FF of n, Λ and Σ should be measured with more than hundreds of events if the cross sections are higher than 10^{-34} cm^2. At the J/Ψ the unknown baryonic branching ratios will be measured with an overall relative error less than 10%. Furthermore the large solid angle and a second level, low threshold, trigger allow for a good measurement of the total cross section.

At present (August '90) data taking has started, 25 nb^{-1} have been collected at Q^2=4 Gev^2 and few nb^{-1} at the J/Ψ for calibration purposes. Present detector performances may be resumed by the mean time of flight resolution respect to the beam crossing ($S=\pm0.7$ ns to be compared to the intrinsic source time length $S=\pm0.35$ ns) and by the resolution in measuring Bhabha showers energy by means the numer of fired strips

($S=\pm 0.21/\sqrt{E(GeV)}$ at E=1 GeV). In Fig.16 are shown the apparatus in the two views, one orthogonal and the other parallel to the beams, with a $n\bar{n}$ candidate, selected according to time of flight requests and with the pattern of the antineutron annihilation star.

Cosmic rays background may be evaluated looking at the time distribution respect to the beam crossing. After a very preliminary scanning, taking into account the cosmic rays background, the following result has been achieved:

$$\sigma(e^+e^- \to n\bar{n}) = 1.5 \pm 0.5 \text{ nb} \quad \text{at} \quad Q^2 = 4 \text{ Gev}^2$$

that is a neutron FF equal or greater than the proton FF. Further analisys is in progress to confirm this result and give a precise measurement.

FIG. 15 - FENICE sketch, orthogonal to the beam axis.

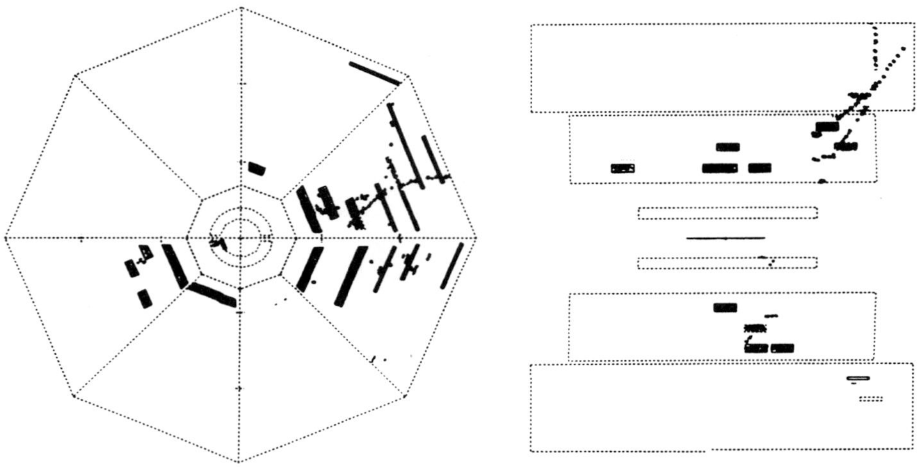

FIG. 16 - A $n\bar{n}$ candidate seen in FENICE.

REFERENCES

[1] European Muon Collaboration, Phys.Lett.B 206 364 (1988) [5]

[2] G.Bardin, G.Burgun, R.Calabrese, G.Capon, R.Carlin, P.Dal Piaz, P.F.Dal Piaz, J.Derre, U.Dosselli, J.Duclos, J.L.Faure, F. asparini, M.Huet, C.Kochowski, S.Limentani, E.Luppi, G.Marel, E.Mazzuccato, F.Petrucci, M.Posocco, M.Savrie', R.Stroili, L.Tecchio, C.Voci, N.Zekri, to be published in the Proc. of LEAP 90, Stockolm (1990)

[3] R.Baldini-Celio, Proc. of the 27th Course of the Int. Schoool of Subn. Physics on the Challenging Questions, Erice (1989)

[4] M.Gourdin, Phys.Rep. 11:29 (1974)

[5] F.Halzen, A.D.Martin, 'Quarks and Leptons', Wiley (1984)

[6] S.R.Amendolia, B.Badelek, G.Batignani, G.A.Beck, E.H.Bellamy, E.Bertolucci, D.Bettoni, H.Bilokon, A.Bizzeti, G.Bologna, L.Bosisio, C.Bradaschia, M.Budinich, M.Dell'Orso, B.D'Ettore Piazzoli, M.Enorini, F.L.Fabbri, F.Fidecaro, L.Foà, E.Focardi, S.G.F.Frank, P.Giannetti, A.Giazzotto, M.A.Giorgi, J.Harvey, G.P.Heath, M.P.J. Landon, P.Laurelli, F.Liello, G.Mannocchi, P.V.March, P.S.Marrocchesi, D.Menasci, A.Menzione, E.Meroni, E.Milotti, L.Moroni, P.Picchi, F.Ragusa, L.Ristori, L.Rolandi, C.Saltmarsh, A.Saoucha, L.Satta, A.Scribano, A.Stefanini, J.A.Strong, R.Tenchini, G.Tonelli, G.Triggiani, A.Zallo, Phys. Lett. 138B:454 (1984)

[7] S.Platchkov, A.Amroun, S.Auffret, J.M.Cavedon, P.Dreux, J.Duclos, B.Frois, D.Goutte, H.Hachemi, J.Martino, X.H.Phan, I.Sick, Proc.XII Conf. on Few Body problems in Physics Vancouver (1989)

[8] D.Bisello, G.Busetto, A.Castro, M.Nigro, L.Pescara, M.Posocco, P.Sartori, L.Stanco, Z.Ajaltouni, A.Falvard, J.Jousset, B.Michel, J.C.Montret, A.Antonelli, R.Baldini, S.Calcaterra, M.Schioppa, J.E.Augustin, G.Cosme, F.Couchot, B.Dudelzak, F.Fulda, G.Grosdidier, B.Jean-Marie, S.Jullian, D.Lalanne, V.Lepeltier, F.Mane, C.Paulot, R.Riskalla, Ph.Roy, G.Szklarz, Internal Note LAL 88-58 (1988)

[9] S.J.Brodsky, G.R.Farrar, Phys. Rev. D 11:1309 (1975)

[10] S.J.Brodsky, B.T.Chertok, Phys. Rev. D 14:3003 (1976)

[11] J.D.Stack, Phys. Rev. 164:1904 (1967)

[12] V.A.Matveev, R.M.Muradyan, A.N.Tavkhelidze, Lett. Nuovo Cimento 7:719 (1973)

[13] V.L.Chernyak, I.R.Zhitnitsky, Nucl. Phys. B 246 52 (1984)

[14] Chueng-Ryong Ji, Proc. Summer School of Computational Atomic and nuclear Physics (1989)
Chueng-Ryong Ji, A.F.Sill, R.M.Lombard-Nelsen, Internal Note SLAC-PUB-4068 (1986)
[15] M.A;Shifman, A.I.Vainshtain, V.I.Zakharov, Nucl. Phys. B 147:385 (1979)
[16] J.S.Bell, R.A.Bertlmann, Internal Note TH 2880-CERN (1980)
J.S.Bell, R.A.Bertlmann, Nucl. Phys. B 187:285 (1981)
J.S.Bell, R.A.Bertlmann, Internal Note TH 3540-CERN (1983)
[17] Particle Data Group, Rewiev of Particle Properties, Phys. Lett. B 204 (1988)
[18] S.J.Brodsky, G.P.Lepage, San Fu Tuan, Phys.Rev. Lett. 59:621 (1987)
[19] P.G.O.Freund, Y.Nambu, Phys. Rev. Lett. 34:1645 (1975)
[20] G.Martinelli, C.T.Sachrajda, Internal Note CERN TH 5042:88
[21] J.J.Sakurai, Ann. of Phys. 11:1 (1960)
[22] R.Gatto, N.Cabibbo, Phys. Rev. 124:1577 (1961)
[23] A.Bramon, E.Etim, M.Greco, Phys. Lett. B 41:609 (1972)
J.J.Sakurai, Phys. Lett. B 46:207 (1973)
[24] H.Fritzsch, Internal Note CERN TH 5569/89 (1989)
G.Hohler, E.Pietarinen, I.Sabba-Stefanescu, F.Borkowski, G.G.Simon, V.H.Walther, R.D.Wendling, H.Genz, G.Hohler, Phys. Lett. B 61:389 (1976)
R.L.Jaffe, Phys. Lett. B 229:275 (1989)
[25] G.Veneziano, Nuovo Cim. 57A:190 (1986)
G.Veneziano, Phys. Rep. 9 (1974)
[26] J.G.Korner, M.Kuroda, Phys. Rev. D 16:2165 (1977)
[27] M.Greco, Nucl. Phys.B 63:398 (1973)
[28] P.Cesselli, M.Nigro, C.Voci, Proc. of Workshop on Physics at LEAR Erice (1982)
[29] A.Donnachie, Internal Note CERN TH 5246/88 (1988)
[30] D.Bisello, G.Busetto, M.Nigro, L.Pescara, M.Posocco, P.Sartori, L.Stanco, Z.Ajaltouni, A.Falvard, J.Jousset, B.Michel, J.C.Montret, R.Baldini, S.Calcaterra, J.E.Augustin, G.Cosme, F.Couchot, B.Dudelzak, F.Fulda, B.Grelaud, G.Grosdidier, B.Jean-Marie, S.Jullian, D.Lalanne, V.Lepeltier, F.Mane, C.Paulot, R.Riskalla, Ph.Roy, G.Szklarz, Internal Note LAL 90-35 (1990)
[31] D.Bisello, G.Busetto, M.Nigro, L.Pescara, M.Posocco, P.Sartori, L.Stanco, Z.Ajaltouni, A.Falvard, J.Jousset, B.Michel, J.C.Montret,A.Antonelli, R.Baldini, S.Calcaterra, M.Schioppa,J.E.Augustin, G.Cosme, F.Couchot, B.Dudelzak,

F.Fulda, B.Grelaud, G.Grosdidier, B.Jean-Marie, S.Jullian, D.Lalanne,
V.Lepeltier, F.Mane, C.Paulot, R.Riskalla, Ph.Roy, G.Szklarz, Phys.Lett.B 212:133 (1988)
S.Fukui et al., Phys.Lett.B 202:441 (1988)
[32] T.Massam, A.Zichichi, Nuovo Cim. 43:1137 (1966)
F.Iachello, A.D.Jackson, A.Lande, Phys. Lett. 43B:191 (1973)
[33] R.K.Bhaduri, 'Model of Nucleon', Addison-Wesley (1988)
[34] R.Machleidt,K.Holinde,Ch.Elster, Phys.Rep. 149 (1987)
[35] R.Felst, Internal Note DESY 73/56 (1973)
[36] E.Etim,A.Malecki, Internal Note LNF-89-023 (1989)
[37] E.D.Bloom, F.J.Gilman, Phys. Rev. Lett. 25:1140 (1970)
[38] S.Dubnicka, E.Etim, Internal Note LNF-89/013(PT) (1989)
[39] M.Van Der Velde, M.I.Polikarpov, Jour. Nucl. Phys. 35:180 (1982)
[40] B.O.Kerbikov, I.S.Shapiro, Internal Note ITEP-159 (1978)
[41] R.Baldini Ferroli Celio, Nucleon Structure Workshop Frascati 1988)
[42] O.D.Dalkarov, V.G.Ksenzov, Pis'ma v ZhETF 30:74 (1979)
O.D.Dalkarov, K.V.Protasov, Mod. Phys. Lett.A 4:1203 (1989)
[43] M.Gari, W.Krumpelmann, Internal Note SLAC-PUB-3398 (1984)
[44] T.H.R.Skyrme, Proc. Royal Soc.A 260:127 (1961)
[45] E.Witten, Nucl. Phys.B 223:422 (1983)
[46] S.Coleman, S.L.Glashow, Phys. Rev. Lett. 6:423 (1961)
[47] M.Gell-Mann, Y.Ne'eman, 'The Eightfold Way', Benjamin (1964)
[48] M.E.Biagini, E.Pasqualucci, Internal Note LNF 90/028
[49] M.Fukugita,J.Kwiecinski, Internal Note RL-79-045 (1979)
[50] M.Claudson,S.L.Glashow,M.B.Wise, Phys.Rev.25:1345 (1982)
[51] FENICE Collaboration, Internal Note LNF-87/18(R) (1987)

Chairman: R. Baldini-Celio

Scientific Secretaries: G. Ambrosi, T. Haas

Discussion

– *Zichichi:*

Guessing from the data available, what would the ratio $\sigma(e^+e^- \to n\bar{n})/\sigma(e^+e^- \to p\bar{p})$ be?

– *Baldini-Celio:*

I guess the value would be about two.

– *Hsu:*

You noted that there are no available data for the process $e^+e^- \to n\bar{n}$. Is this due to the difficulty of detecting neutrons? For instance, how do you tell neutrons from other strongly interacting neutral particles, such as neutral $\pi's$?

– *Baldini-Celio:*

Concerning the π^0: you distinguish an antineutron star from the two photons produced by the π^0 by simply looking at the event-pattern in your detector. As for the rest: if there is multiparticle production you see the other particles in your apparatus – the solid angle covered by the detector is quite large! In particular you can have a neutron-antinuetron-star together with a π^0. We assume that in first approximation this cross-section is zero. However, you could see the photons coming from the π^0. Close to the threshold this process is negligible. Far away from the threshold that may not be true.

– *Brahm:*

In order to explain the anomalously large branching ratio $J/\psi \to \rho\pi/\psi' \to \rho\pi$ you made strong assumptions about the properties of a 1^{--} glueball: a mass of 3 GeV and a width of 150 MeV. Are there lattice calculations or any other analysis that corroborate these assumptions?

– *Baldini-Celio:*

To my knowledge, no. And this is not my hypothesis. Brodsky, for instance, has made such an hypothesis. The first one to point out the anomaly was Nambu in 1975. I do not know whether you find evidence anywhere else. But I do not think so.

– *Brahm:*

I suppose our knowledge of glueballs is very poor?

– *Baldini-Celio:*

Indeed, there is no calculation for such a heavy glueball. There is a problem with this interpretation: another branching ratio is a little bit in trouble, namely $J/\psi \rightarrow \pi\omega$, which is higher than you expect. This is strange since $\pi\omega$ has isospin 1. It can be produced from $J\psi \rightarrow \gamma \rightarrow \pi\omega$. Only in this case you have an isospin violation. So, if the interpretation is a mixing of two states, then the particle would not be a real glueball but some hybrid state. There is really something missing in the J/ψ! People have found a mystery in the decay of $\Upsilon_{4s} \rightarrow J/\psi +$ (other particles). To my knowledge there is a number of $J/\psi's$ with high momentum, much higher than you expect from Υ_{1s}. There is no explanation for this.

– *Lu:*

There is an approach to calculate the form-factor based on a diquark model, at least in the space-like region. Is the calculation valid in the time-like region? Are there predictions based on this model?

– *Baldini-Celio:*

Do you wonder whether the calculation is valid in the time-like region?

– *Lu:*

I wonder whether there is a result for the ratio of the form-factors of the neutron and the proton.

– *Baldini-Celio:*

I do not claim that these results can be extrapolated to the time-like region. But I am saying that within the qualitative picture of a leading quark the form-factor of the neutron should be lower than that of the proton. Of course, in this picture it is also essential to assume that the Dirac-form-factor dominates. That may not be true. However, it is quite strange that due to the smallness of the u and d quark masses the Pauli form-factor is so high, at least in the domain of perturbative QCD.

– *Lu:*

So is the ratio $\sigma(e^+e^- \rightarrow n\overline{n})/\sigma(e^+e^- \rightarrow p\overline{p})$ smaller or greater than one?

– *Baldini-Celio:*

In any QCD calculation I expect this ratio to come out smaller than one. Why, in any simple QCD picture, should the neutron, which has no electric charge and smaller magnetic moment than the parton, have greater electromagnetic interactions?

– *Brown:*

Why do calculations based on the Vector-Dominance-Model (VDM) give the value of $\sigma(e^+e^- \to n\bar{n})/\sigma(e^+e^- \to p\bar{p})$ to be as high as 100? Usually we expect vector mesons to couple to neutrons and protons in a similar way.

– *Baldini-Celio:*

I do not have a physical explanation why the VDM gives such high neutron-antineutron production or such high neutron form-factor. When you do the calculation in the VDM in the space-like region, in order to obtain a very small electric neutron form-factor you need a cancellation between the Dirac- and Pauli-form-factors. In front of the Pauli-form-factor there is a factor $\frac{q^2}{m^2}$. When you change the sign of q^2, to go to the time-like region, the cancellation becomes an enhancement. However, I do not know any deep reason why this happens in the VDM.

– *Kastening:*

Why has the process $e^+e^- \to n\bar{n}$ never been measured before?

– *Baldini-Celio:*

This is a very difficult experiment. Before the discovery of the streamer tube it was very hard to build an affordable big calorimeter with enough granularity. This was really a technological problem.

– *Kastening:*

Could you not have captured the neutrons in some material?

– *Baldini-Celio:*

You could perform the experiment by detecting not the antineutron but the neutron. You would need a very thick scintillation counter. For neutrons of these energies the detection efficiency is roughly 1% per centimetre of scintillator. So, if you want to get 100% detection efficiency, you need 100 cm of scintillator. This is very expensive.

– *Kastening:*

You have mentioned discrepancies in the predictions of the "Skyrme" model for baryons. Could you give us some more information on what they are?

– *Baldini-Celio:*

I am not an expert on the "Skyrme" model, but to my knowledge the biggest discrepancy is in the value of g_A for the proton. The "Skyrme" model predicts a value of 0.88 while the measured value is 1.25. Now there are a lot of theorists who are playing with the model. If you introduce enough free parameters you achieve better agreement with the data.

– *Zichichi:*

When will you have definitive data on the neutron form factor?

– *Baldini-Celio:*

I do not know. We have to increase the luminosity of the storage ring. It is now less than $10^{29}\frac{1}{cm^2 s}$. With this luminosity we will need a couple of months next year to do the measurement.

THE INTRINSIC SHORT–DISTANCE STRUCTURE OF HADRONS IN QCD*

STANLEY J. BRODSKY

Stanford Linear Accelerator Center
Stanford University, Stanford, California 94309, USA

1. INTRODUCTION

A central problem of particle physics is to determine the composition of hadrons in terms of their fundamental quark and gluon degrees of freedom. The structure of hadronic bound-states in quantum chromodynamics plays a role in virtually every aspect of high energy and weak interaction phenomenology, including jet hadronization, heavy particle production processes at colliders, general exclusive and inclusive reactions, and electro-weak decay matrix-elements. Although the QCD Lagrangian has an elegant simplicity, the structure of its bound state solutions can be extraordinarily rich and complex. In these lectures I will focus on QCD phenomena which reflect the coherence and composition of hadron wave functions as relativistic many–body systems of quark and gluon quanta.[1]

There are many different ways in which experiment can resolve the short-distance structure of hadrons. It is necessary to distinguish "intrinsic" versus "extrinsic" contributions to scattering reactions. The intrinsic contributions, which are associated with multiparticle interactions within the hadronic bound state, have lifetimes much longer than that of the time of collision; they are thus formed before the collision and lead to process–independent Feynman scaling production cross sections. Extrinsic contributions, on the other hand, are controlled by the high momentum transfer scale of the collision process itself and have short lifetimes of the same order as that of the collision time. Extrinsic contributions provide the leading twist radiative corrections associated with the renormalization of single quark or gluon lines and the QCD evolution of structure functions.

Data from many sources suggest that the intrinsic bound state structure of the nucleon has a non–negligible strange– and charm–quark content in addition to the extrinsic sources of heavy quarks created in the collision itself. In fact, QCD predicts that the hadronic wave-function has an "intrinsic hardness,"[2] which leads to a non-negligible probability for high mass and high momentum short-range fluctuations. In fact, because of asymptotic freedom, one can analyze short–distance, high momentum transfer, and heavy quark fluctuations of a hadronic wave function perturbatively. The probability that a hadronic wave functions has far–off–shell fluctuations is only power–law suppressed in QCD because of the point–like character of the quark–gluon

* Work supported by the Department of Energy, contract DE–AC03–76SF00515.

interactions. For example, the probability that a heavy quark pair exists virtually in a light hadron only decreases as $P_{Q\overline{Q}} \sim \alpha_s^2(M_Q^2)/M_Q^2$. This key property of the theory has a number of important implications for the production of heavy quark and other massive systems with large momentum fractions x in inclusive reactions and also in exclusive reactions at threshold. The intrinsic fluctuations have a Lorentz–boosted virtual lifetime of order $\tau \propto \gamma/M_Q^2$. Thus they can be materialized in high energy collisions as projectile fragments. The dependence of the hidden and open heavy quark production cross sections on the size of a nuclear target can be used as a filter to identify these intrinsic heavy quark processes. Further discussion will be given in Sections 2 and 5.

An important tool for analyzing the underlying structure of a complex system is to change its external conditions. In the case of quantum electrodynamics, one can use external Stark and Zeeman fields to perturb atomic wave functions and probe their composite structure. Analogously, in QCD, we can use a nuclear medium to modify and probe short-distance hadronic structure and dynamics. In fact, as I discuss in Section 2, we can use the nucleus as a differential "color filter"[3,4] to separate Fock components (or fluctuations) of different transverse size in the projectile's wave function and to separate perturbative short–distance subprocesses from non–perturbative mechanisms. I also will discuss "color transparency"[5] as a way to isolate strictly perturbative contributions to large angle exclusive scattering In this analysis[6] we will see how strong binding effects at the charm threshold complicates perturbative QCD predictions, explaining both the anomalous spin correlation A_{NN} observed in large angle pp scattering at $\sqrt{s}$ and the anomalous decrease of color transparency seen in quasi-elastic pp scattering in nuclei at the same energy. The strong interactions of colored particles at small relative velocity also leads to other interesting phenomena, including the production of nuclear-bound charmonium near threshold and the suppression of J/psi production coalescence of charm quarks with co-moving spectators.[7] I also will discuss in Section 6 a new approach[8] to shadowing and anti–shadowing of nuclear structure functions, and how these phenomena can provide information on the phase and magnitude of quark or gluon scattering amplitudes in the nuclear medium.

In Section 3 I will present some new results for the intrinsic polarized and unpolarized gluon distributions of the proton which are associated with hadron binding.[9] These discussions are important in regard to understanding the EMC spin crisis problem.[10] It is also important to note that the conventionally-defined "valence" distributions measured in deep inelastic lepton scattering are actually not identical to the bound state valence quark distributions because of a subtle effect due to Pauli blocking.[11] This is discussed in detail in Section 4.

The above ingredients provide the foundations for analyzing many novel features of hadronic and heavy quark processes in high energy collisions including color transparency and intrinsic charm reactions.

1.1. Relativistic Wavefunctions in Gauge Theory

How can one define a wave function of a composite system in a relativistic quantum gauge field theory? A natural description, similar physically to that of the parton model, is to utilize a Fock expansion at fixed time $\tau = t - z/c$ on the light cone. This description is particularly simple since the perturbative vacuum is an apparent eigenstate of the full theory. As discussed recently by Werner,[12] the rigorous quantization of gauge theories on the light cone allows zero mode degrees of freedom of the gauge field in the vacuum sector which corresponds to non–zero chiral charge and other topological vacuum properties. In the particle sector of the theory, where one can quantize the theory in the light cone gauge $A^+ = 0$, one obtains a Fock basis containing only physical degrees of freedom.

The hadron eigenstate state can thus be expanded on the complete set of free quark and gluon eigenstates of the free QCD Hamiltonian which have the same global quantum numbers as the hadron: *e.g.*:

$$
\begin{aligned}
|\Psi_p\rangle = \sum |n\rangle \langle n \,|\, \Psi_p\rangle \\
= \quad & |uud\rangle\, \psi_{uud}(x_i, k_{\perp i}, \lambda_i) \\
& + |uudg\rangle\, \psi_{uudg}(x_i, k_{\perp i}, \lambda_i) \\
& + |uudq\overline{q}\rangle\, \psi_{uudq\overline{q}}(x_i, k_{\perp i}, \lambda_i) \\
& + \ldots
\end{aligned}
\tag{1}
$$

The x_i are the light–cone momentum fractions $x_i = (k^0+k^z)/(P^0+P^z)$, with $\sum_{i=1}^n x_i = 1$, and $\sum k_{\perp i} = 0$. The wave functions $\psi_n(x_1, k_{\perp i}, \lambda_i)$ appearing in the Fock–state expansion contain the physics of the hadron entering scattering amplitudes. For example, the structure functions measured in deep inelastic scattering are constructed as probability distributions in x from the sum of the squares of the light–cone wave functions $\psi_n(x_i, k_{\perp i}, \lambda_i)$. (See Section 4.) Similarly, since the current is a simple diagonal local operator on the free quark basis, form factors can be computed from a simple overlap integral of the ψ_n. More generally, high momentum transfer exclusive reactions in QCD are sensitive to each hadron's distribution amplitude $\phi(x_i, Q)$, which is the valence Fock amplitude integrated over transverse momentum up to the scale Q.

The problem of solving QCD, including its bound state color-singlet spectrum and wavefunctions, is equivalent to the diagonalization of the QCD Hamiltonian. A frame-invariant Hamiltonian operator can be obtained by quantizing the theory at fixed light-cone time. This is the "light-front" formulation of Hamiltonian theory described by Dirac which produces the maximal number of interaction-free commuting invariants including the total light-cone momentum P^+ and transverse momentum $P_\perp$. The τ evolution operator $P^- = P^0 - P^z$ may be written in the general form $P^- = (H_{LC} + P_\perp^2)/P^+$ so that the eigenvalues of the operator H_{LC} are exactly the squares of invariant masses of the spectrum. The eigenvalue problem is thus $H_{LC}|\Psi >= \mathcal{M}^{\epsilon}|\Theta >$. It should be emphasized that the light-cone Hamiltonian is completely independent of the total momentum of the system P^+ and $P_\perp$ and is a Lorentz scalar. If we choose the light-cone Fock representation described above, then we obtain a covariant Heisenberg matrix representation of the theory: $< n|H_{LC}|m >< m|\Psi >= \mathcal{M}^{\epsilon}|\Theta. >$ The projections of the eigenfunctions on this basis are precisely the wavefunctions needed for phenomenology. The QCD Hamiltonian can be elegantly quantized on the light–cone in $A^+ = 0$ gauge without resort to unphysical ghost quanta, even in non-Abelian gauge theory.[1]

1.2. Discretized Light-Cone Quantization

In order to make the eigenvalue problem tractable for numerical analysis, it is convenient to choose a discrete momentum Fock basis. In the method of Discretized Light–Cone Quantization,[13] one constructs a discrete basis of discrete momentum–space color–singlet free gluon and quark light-cone Fock states satisfying periodic and anti–periodic boundary conditions, respectively. Again the analysis is frame independent. The matrix elements of the QCD Hamiltonian are extremely simple in this basis. Because of momentum and flavor conservation laws obeyed by the interaction Hamiltonian, the matrix is sparse far from the diagonal. A covariant truncation to a finite system is obtained by choosing a global, gauge invariant cut-off on the maximum invariant mass of the Fock state. A local cut-off on the change of mass across the matrix element can serve as an ultraviolet cut-off. In each case, the regulators refer to the continuum theory. The discretization is not used to regularize the theory. In DLCQ one can either numerically diagonalize the light-cone Hamiltonian on the Fock basis, or project the eigenvalue problem onto the lowest number (valence) Fock component and then numerically solve the resulting integral equation. The DLCQ formalism is defined such that each step, including ultraviolet regularization and Fock space truncation, is Lorentz–frame independent. The computer program only involves relative coordinates and is independent of the total momentum $(P^+, P_\perp.)$ The discrete formulation thus provides a finite, Lorentz invariant, and faithfully renormalizable representation of the full quantum field theory in the physical particle sector. IN DLCQ the discretization can be kept independent of the ultraviolet regulators which are set by the continuum theory. Again, unlike lattice gauge theory, there are no special difficulties with fermions.

In principle, the eigenvalues obtained by diagonalizing of the light-cone Hamiltonian in the DLCQ basis provide the entire invariant mass spectrum and eigenfunctions needed to compute the hadron matrix elements,form factors, and the structure functions and distribution amplitudes entering QCD factorization formulae. A major success of DLCQ has been its applications to gauge theories in one–space and one–time dimensions.[14] For example, the complete spectrum and the respective structure functions of mesons, baryons, and nuclei in QCD(1+1) for $SU(3)_C$ have been obtained as a function of mass and coupling constant. Results for the structure function of the lowest mass meson and baryon at weak and strong coupling are shown in Fig. 1.

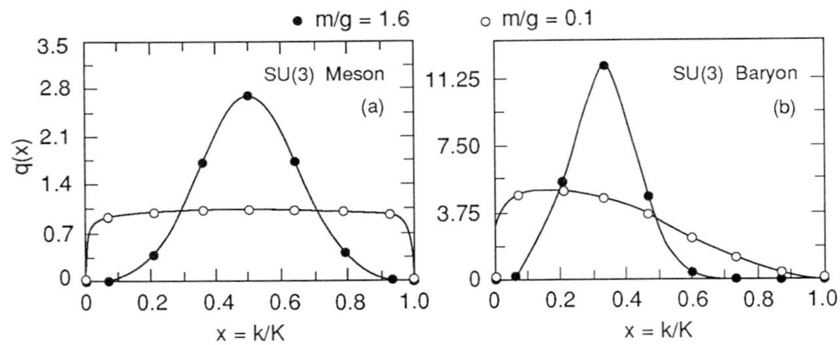

Figure 1. Valence structure functions of the baryon and meson in QCD in one–space and one–time dimension. The results are for one quark flavor and three colors.

The application of DLCQ to gauge theory in three–space and one–time dimensions is a much more challenging computational task, but significant progress has recently been made computing the fine structure and hyperfine spectrum of positronium in strong coupled QED.[15]

Important constraints and information on the non–perturbative structure of the proton in QCD models have also been obtained using bag models, quark–diquark schemes, QCD sum rules, non–relativistic quark models, and lattice gauge theory. A summary and references may be found in Ref. 1.

2. THE NUCLEUS AS A QCD FILTER

There are many important ways in which a nuclear target can probe fundamental aspects of QCD. A primary concept is that of the "color filter":[3,4] if the interactions of an incident hadron are controlled by gluon exchange, then the nucleus will be transparent to those fluctuations of the incident hadron wave function which have small transverse size. Such Fock components have a small color dipole moment and thus will interact weakly in the nucleus; conversely, Fock components of normal hadronic size will interact strongly and be absorbed during their passage through the nucleus.[3] For example, large momentum transfer quasi-exclusive reactions,[16] are controlled in perturbative QCD by small color-singlet valence-quark Fock components of transverse size $b_\perp \sim 1/Q$; thus initial-state and final-state corrections to these hard reactions are suppressed at large momentum transfer, and they can occur in a nucleus without initial or final state absorption or multiple scattering of the interacting hadrons. Thus, at large momentum transfer and energies, quasi-elastic exclusive reactions are predicted to occur uniformly in the nuclear volume. This remarkable phenomenon is called "color transparency."[5] Thus QCD predicts that the transparency ratio of quasi-elastic annihilation of the anti-proton in the $\overline{p}p \to \ell\overline{\ell}$ reaction will be additive in proton number in a nuclear target:[17, 18]

$$\frac{\frac{d\sigma}{dQ^2}(\overline{p}A \to \ell\overline{\ell}(A-1))}{\frac{d\sigma}{dQ^2}(p\overline{p} \to \ell\overline{\ell})} \to Z^1 \tag{2}$$

for large pair-mass squared Q^2. In contrast to the QCD color transparency prediction, the traditional (Glauber) theory of nuclear absorption predicts that quasi-elastic scattering occurs primarily on the front surface of the nucleus. The above ratio thus should be proportional to $Z^{2/3}$, *i.e.* the number of protons exposed on the nuclear surface.

Conditions for Color Transparency — Color transparency is a striking prediction of perturbative QCD at high momentum transfers. There are two conditions which set the kinematic scale where the effect should be evident. First, the hard scattering subprocess must occur at a sufficiently large momentum transfer so that only small transverse size wave function components $\psi(x_i, b_\perp \sim 1/Q)$ with small color dipole moments dominate the reaction. Second, the state must remain small during its transit through the nucleus. The expansion distance is controlled by the time in which the small Fock component mixes with other Fock components. By Lorentz invariance, the time scale $\tau = 2E_{\overline{p}}/\Delta\mathcal{M}^2$ grows linearly with the energy of the hadron in the nuclear rest frame, where $\Delta\mathcal{M}^2$ is the difference of invariant mass squared of the Fock components. Estimates for the expansion time are given in Refs. 4, 19, and 20.

There are a number of important tests of color transparency and color filter that can be carried out with anti-proton beams of moderate energy.[21] Since total annihilation processes such as $p\overline{p} \to \ell\overline{\ell}$ or $p\overline{p} \to \gamma\gamma$ and $p\overline{p} \to J/\psi$ automatically involve short distances, the first condition for color transparency should be satisfied. The study of the energy dependence of these processes inside nuclei (quasi-elastic reactions, integrated over Fermi-motion) can clarify the role of the expansion time scale τ. A recent analysis by Jennings and Miller[19] shows that $\tau = 2E_{\overline{p}}/\Delta\mathcal{M}^2$ is controlled by the mass difference of states which are close in mass to that of the asymptotic hadronic state. Thus color transparency may well be visible in low energy anti-proton annihilation processes, including quasi-elastic $\overline{p}p \to J/\psi$ and $\overline{p}p \to \overline{\ell}\ell$ annihilation in the nucleus.

The only existing test of color transparency is the measurement of quasi-elastic large angle pp scattering in nuclei at Brookhaven.[18] The transparency ratio is observed to increase as the momentum transfer increases, in agreement with the color transparency prediction. However, in contradiction to perturbative QCD expectations, the data suggests, surprisingly, that normal Glauber absorption seems to recur at the highest energies of the experiment $p_{lab} \sim 12\ GeV/c$. It should be noted that this is the same kinematic domain where a strong spin correlation A_{NN} is observed.[22] The probability of protons scattering with their spins parallel and normal to the scattering plane is found to be **four times** that of anti-parallel scattering, which is again in strong contradiction to QCD expectations. However, Guy De Teramond and I[6] have noted that the breakdown of color transparency and the onset of strong spin–spin correlations can both be explained by the fact that the charm threshold occurs in pp collisions at $\sqrt{s} \sim 5\ GeV$ or $p_{lab} \sim 12\ GeV/c$. At this energy the charm quarks are produced at rest in the center of mass. Since all of the eight quarks have zero relative velocity, they can resonate to give a strong threshold effect in the $J = L = S = 1$ partial wave. (The orbital angular momentum of the pp state must be odd since the charm and anti-charm quarks have opposite parity.) This partial wave predicts maximal spin correlation in A_{NN}. Most important, such a threshold or resonant effect couples to hadrons of conventional size which will have normal absorption in the nucleus. If this non-perturbative $pp \to pp$ amplitude dominates over the perturbative QCD amplitude, one can explain both the large spin correlation and the breakdown of color transparency at the charm threshold. Thus the nucleus acts as a filter, absorbing the non-perturbative contribution to elastic pp scattering, while allowing the hard scattering perturbative QCD processes to occur additively throughout the nuclear volume.[23] Similarly, one expects that the charm threshold will modify the color transparency and hard-scattering behavior of quasi-elastic $\bar{p}p$ reactions in nuclei at energies $\sqrt{s} \sim 3\ GeV$.

Diffractive Production of Jets in Anti–Proton Nuclear Reactions — In our original paper on the color filter, Bertsch, Goldhaber, Gunion, and I[3] suggested that diffractive nuclear reactions could be used as a color filter, *i.e.* fluctuations of an incident hadron with small color dipole moments and hence could emerge unscathed after transit through a nucleus without nuclear excitation. In the case of anti–proton reactions, the fluctuations of the valence Fock state where the three anti–quarks has small transverse separation and thus small color dipole moment will be produced in the form of three jets on the back side of the nucleus. The longitudinal and transverse momentum dependence of the $\bar{p}A \to A\ Jet\ Jet\ Jet$ cross section will reflect the $\overline{qqq}$ composition of the incident anti–proton wave function.

The Color Filter and Hadron Fragmentation in Nuclei — Recently, Hoyer and I[24] have shown that the color filter ansatz can explain the empirical rule that the nuclear dependence of hadronic spectra $\frac{d\sigma/dx_F(HA \to H'X)}{d\sigma/dx_F(HN \to H'X)} = A^{\alpha(x_F)}$, is nearly independent of particle type H'. The essential point is that fluctuations of the initial hadron H which have the small transverse size have the least differential energy loss in the nucleus.

Color Transparency and Intrinsic Charm — A remarkable feature of the hadronic production of the J/ψ by protons in nuclei[25,26] is the fact that the cross section persists to high x_F, but with a strongly suppressed nuclear dependence, $A^{\alpha(x_F)} \sim 0.7$. The magnitude of the cross sections for high momentum charmonium reported by the NA–3 group[25] at CERN is, in fact, far in excess of what is predicted from gluon fusion or quark anti–quark annihilation subprocesses. Both the anomalous A-dependence and the high-x_F excess can be explained by assuming the presence of intrinsic charm components of the incident hadron wave–functions.[24] The essential physics point is as follows: the intrinsic charm Fock components, *e.g.* $|uudc\bar{c}\rangle$ in the proton have maximum probability when all of the quarks have equal velocities, *i.e.* when $x_i \propto \sqrt{m^2 + k_{\perp i}^2}$. This implies that the charm and anti–charm quarks have the majority of the momentum of the proton when they are present in the hadron wave function. In a high energy proton–nucleus collision, the small transverse size, high–x intrinsic $c\bar{c}$ system can penetrate the nucleus, with minimal absorption and can coalesce to produce a charmonium state at large x_F. The remaining spectators of the nucleon tend to have more normal transverse size and interact on the front surface of the nucleus, leading to a production cross section approximately proportional to $A^{0.7}$. Since the formation of the charmonium state occurs far outside the nucleus at high energies, one predicts similar $A^{\alpha(x_F)}$–dependence of the J/ψ and ψ' cross sections, in agreement with recent results reported by the E–772 experiment at Fermilab.[26] Further discussion on the implications of intrinsic charm is given in Section 5.

Shadowing, Anti–Shadowing of Inclusive Anti–Proton Reactions — In the case of inclusive reactions, such as Drell–Yan massive lepton pair production $p\bar{p} \to \ell\bar{\ell}X$, multiple scattering of the interacting partons in the nucleus can lead to shadowing and anti–shadowing of the nuclear structure functions and a shift of the pair's transverse momentum to large transverse momentum. Hung Jung Lu and I have shown that nuclear shadowing of leading–twist QCD reactions can be related to Pomeron exchange in the multiple interactions of the quark or anti–quark in the nucleus, and that the complex phase of the quark–nucleon scattering amplitude due to non–singlet Reggeon exchange leads to anti–shadowing; *i.e.* an excess of the nuclear cross section over nucleon additivity.[8] A detailed discussion will be given in Section 6.

Formation Zone Effects in Inclusive Reactions — An essential aspect of the proofs of QCD factorization of inclusive reactions such as Drell–Yan massive lepton pair production in a nuclear target is that the entire nuclear dependence of the cross section is contained in the nuclear structure functions as measured in deep inelastic lepton–nucleus scattering. Thus the factorization theorem predicts that there is no initial state absorption or scattering that can significantly modify an incident hadron's parton distributions as it propagates through the nucleus. In particular, induced hard colinear radiation due to inelastic reactions in the nucleus before the annihilation or hard–scattering subprocess occurs must be dynamically suppressed. As shown by Bodwin, Lepage, and myself,[27] this suppression occurs automatically in the nucleus due to the destructive interference of the various multiple–scattering reactions in the nucleus. The interference occurs if the inelastic processes can occur coherently in the nucleus. This requires that the momentum transfer to target nucleons must be small compared to the inverse correlation length in the nucleus; *i.e.* $E_{\bar{q}} > \Delta\mathcal{M}^2 L_A > 1$, where E_q is the laboratory energy of the annihilating anti–quark, $\Delta\mathcal{M}^2$ is the change of mass squared of the quark in the inelastic reaction (small for hard colinear gluon emission of the anti–quark), and L_A is the length between target centers in the nucleus. This formation zone effect can be studied in detail by measuring the nuclear dependence as a function of anti–quark laboratory energy in anti–proton reactions.

Exclusive Nuclear Amplitudes — Exclusive nuclear reactions such as $\bar{p}d \to \gamma n$ or $\bar{p}d \to \pi^0 n$ can provide an important test of the reduced amplitude formalism for large momentum transfer exclusive nuclear reactions. Recent measurements at SLAC[28] are in striking agreement with the reduced amplitude predictions for photo–disintegration $\gamma d \to np$ at a surprising low momentum transfer. The corresponding anti–proton reactions will allow an important test of both the scaling behavior of exclusive nuclear reactions and their crossing behavior to the annihilation channel.

Hidden Color Nuclear Components — In QCD the six–quark deuteron is a linear superposition of five color singlet states, only one of which corresponds to the conventional $n - p$ state.[29] One can search for hidden color excitations of the deuteron in $\bar{p}He^3$ elastic scattering at large angles.

Nuclear Bound Quarkonium — The production of charmonium at threshold in a nuclear target is particularly interesting since it is possible that the attractive QCD van der Waals potential due to multi–gluon exchange could actually bind the η_c to light nuclei. Consider the reaction $\bar{p}\alpha \to (c\bar{c})H^3$ where the charmonium state is produced nearly at rest. (See Fig. 2.) At the threshold for charm production, the incident nuclei will be nearly stopped (in the center of mass frame) and will fuse into a compound nucleus because of the strong attractive nuclear force. The charmonium state will be attracted to the nucleus by the QCD gluonic van der Waals force. One thus expects strong final state interactions near threshold. In fact, Guy De Teramond, Ivan Schmidt, and I[30] have argued that the $c\bar{c}$ system will bind to the H^3 nucleus. It is thus likely that a new type of exotic nuclear bound state will be formed: charmonium bound to nuclear matter. Such a state should be observable at a distinct $\bar{p}\alpha$ center of mass energy, spread by the width of the charmonium state, and it will decay to unique signatures such as $\bar{p}\alpha \to H^3\gamma\gamma$. The binding energy in the nucleus gives a measure of the charmonium's interactions with ordinary hadrons and nuclei; its hadronic decays will measure hadron–nucleus interactions and test color transparency starting from a unique initial state condition.

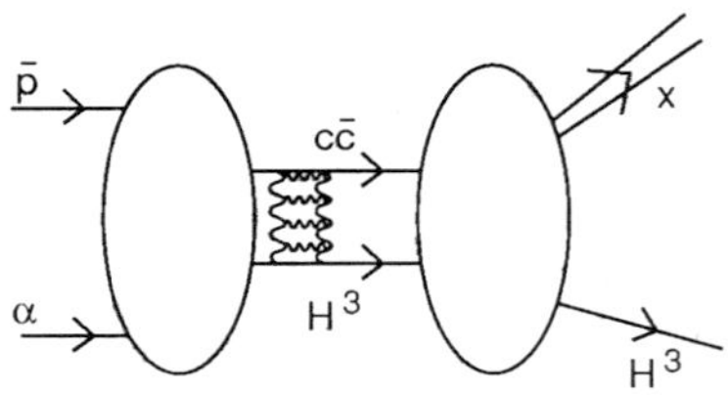

Figure 2. Formation of the $(c\bar{c}) - H^3$ bound state in the process $\bar{p}\alpha \to H^3X$.

In QCD, the nuclear forces are identified with the residual strong color interactions due to quark interchange and multiple–gluon exchange. Because of the identity of the quark constituents of nucleons, a short–range repulsive component is also present (Pauli–blocking). From this perspective, the study of heavy quarkonium interactions in nuclear matter is particularly interesting: due to the distinct flavors of the quarks involved in the quarkonium–nucleon interaction there is no quark exchange to first order in elastic processes, and thus no one–meson–exchange potential from which to build a standard nuclear potential. For the same reason, there is no Pauli–blocking and consequently no short–range nuclear repulsion. The nuclear interaction in this case is purely gluonic and thus of a different nature from the usual nuclear forces.

The production of nuclear–bound quarkonium would be the first realization of hadronic nuclei with exotic components bound by a purely gluonic potential. Furthermore, the charmonium–nucleon interaction would provide the dynamical basis for understanding the spin–spin correlation anomaly in high energy $p-p$ elastic scattering.[6] In this case, the interaction is not strong enough to produce a bound state, but it can provide a strong enough enhancement at the heavy–quark threshold characteristic of an almost–bound system.[31]

3. INTRINSIC GLUON DISTRIBUTIONS

The *intrinsic* gluon distribution $G_{g/H}(x, Q_0^2)$ describes the fractional light–cone momentum distribution of gluons associated with the bound–state dynamics of the hadron H, in distinction to the *extrinsic* distribution, which is derived from radiative processes or evolution. Given the intrinsic distribution, one can obtain the extrinsic distribution by applying the QCD evolution equations starting at the bound–state scale Q_0.

In principle, one must solve the non–perturbative bound state equation of motion to compute the intrinsic gluon distribution. In the case of positronium in quantum electrodynamics one can readily calculate the photon distribution, at least to first order in the fine structure constant α. The analysis requires coherence between amplitudes in which the electron and positron couple to the photons. In the infrared limit this coherence in the neutral atom ensures a finite photon distribution.

In the QCD case, the analysis of the intrinsic gluon distribution of a hadron is essentially non–perturbative. However, there are several theoretical constraints which limit its form:

1. In order to insure positivity of fragmentation functions, distribution functions $G_{a/b}(x)$ must behave as an odd or even power of $(1-x)$ at $x \to 1$ according to the relative statistics of a and b.[32] Thus the gluon distribution of a nucleon must have the behavior: $G_{g/N}(x) \sim (1-x)^{2k}$ at $x \to 1$ to ensure correct crossing to the fragmentation function $D_{N/g}(z)$. This result holds individually for each helicity of the gluon and the nucleon.
2. The coupling of quarks to gluons tends to match the sign of the quark helicity to the gluon helicity in the large x limit.[33] We define the helicity–aligned and anti–aligned gluon distributions: $G^+(x) = G_{g\uparrow/N\uparrow}(x)$ and $G^-(x) = G_{g\downarrow/N\uparrow}(x)$. The gauge theory couplings imply
$$\lim_{x\to 1} G^-(x)/G^+(x) \to (1-x)^2. \tag{3}$$
3. In the low x domain the quarks in the hadron radiate gluons coherently, and one must compute emission of gluons from the quark lines taking into account interference between amplitudes. We define $\Delta G(x) = G^+(x) - G^-(x)$ and $G(x) = G^+(x) + G^-(x)$. We shall show that the asymmetry ratio $\Delta G(x)/G(x)$ vanishes linearly with x; perhaps coincidentally, this is also the prediction from Reggeon exchange.[34] The coefficient at $x \to 0$ depends on the hadronic wave functions; however, for equal partition of the hadron's momentum among its constituents, we will show that
$$\lim_{x\to 0} \Delta G(x)/G(x) \to N_q\, x, \tag{4}$$
where N_q is the number of valence quarks.
4. In the $x \to 1$ limit, the stuck quark is far off–shell so that one can use perturbation theory to characterize the threshold dependence of the structure functions. We find for three–quark bound states
$$\lim_{x\to 1} G^+(x) \to C(1-x)^{2N_q-2} = C(1-x)^4, \tag{5}$$
Thus $G^-(x) \to C(1-x)^6$ at $x \sim 1$. This is equivalent to the spectator–counting rule developed in Ref. 35.

Ivan Schmidt and I[9] have proposed a simple analytic model for the intrinsic gluon distribution in the nucleon which incorporates all of the above constraints:
$$\Delta G(x) = \frac{N}{x}[5(1-x)^4 - 4(1-x)^5 - (1-x)^6] \tag{6}$$
and
$$G(x) = \frac{N}{x}[5(1-x)^4 - 4(1-x)^5 + (1-x)^6] \tag{7}$$
In this model the momentum fraction carried by intrinsic gluons in the nucleon is $< x_g >= \int_0^1 dx x G(x) = (10/21)N$, and the helicity carried by the intrinsic gluons is $\Delta G \equiv \int_0^1 dx \Delta G(x) = 7/6N$. The ratio $\Delta G/ < x_g >= 49/20$ for the intrinsic gluon distribution is independent of the normalization N. Phenomenological analyses imply that the gluons carry approximately one-half of the proton's momentum:

$<x_{g/N}\approx<0.5$. We shall assume that this is a good characterization of the intrinsic gluon distribution. The momentum sum rule then implies $N\sim 1$ and $\Delta G\sim 1.2$. A review of the present experimental and theoretical limits on gluon and quark spin in the nucleon is given in Ref. 10.

In the following sections I will discuss an analysis by Schmidt and myself[9] of both the polarized and unpolarized intrinsic gluon distribution functions. First we study the behavior of the gluon asymmetry (unpolarized over polarized distributions) in the small x region where it turns out to be approximately independent on the details of the bound–state wave function. The logarithmic ultra–violet cut–off dependence of the intrinsic distribution matches with the lower cut–off of the extrinsic distribution; the Q^2 evolution of the extrinsic distribution is studied in detail in Ref. 36.

In Section 3 we shall show that the intrinsic gluon distribution is related to the retarded part of the spin–dependent bound–state potential $-\left\langle\frac{\Delta\partial V}{\partial M_B^2}\right\rangle_{hfs}$. This allows us to derive sum rules for the difference of gluon distribution (and fragmentation) functions for hadrons with different spin in terms of the spin–dependent part of the bound–state potential.

3.1. Intrinsic Gauge Field Distributions

A general bound–state wave function can be expanded in terms of (Fock) states of definite number (n) of elementary free fields. We define the Fock expansion at equal "time" $\tau = t+z$ in the light cone gauge $A^+ = A^0 + A^3 = 0$. Labelling the corresponding renormalized amplitudes as $\psi^{(Q)}_{n/B}(x_i, \vec{k}_{\perp_i}, \lambda_i)$, the distribution function for a constituent $\underline{a}$ in the bound state $\underline{B}$ (see Ref. 37 for details and definitions) is given by:

$$G_{a/B}(x,Q^2) = \sum_{n,\lambda_i}\int\prod_i \frac{dx_i d^2k_{\perp_i}}{16\pi^3}\,|\psi^{(Q)}_{n/B}\,(\psi_i, \vec{k}_{\perp_i}, \lambda_i)|^2 \sum_b \delta(x_b - x) \quad . \tag{8}$$

We first consider positronium as an example, and calculate the intrinsic distribution function of photons $G_{\gamma/positronium}$. To leading order in the binding energy we can neglect pair annihilation, pair production, and higher particle number Fock states.

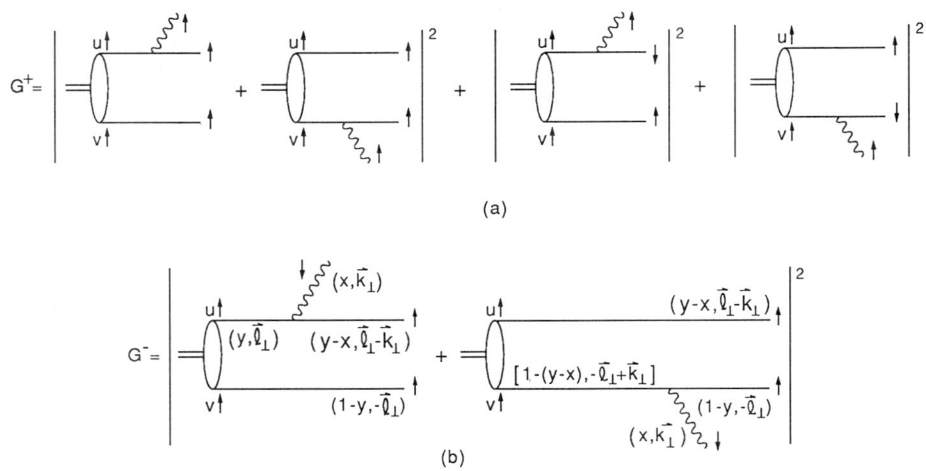

Figure 3. Diagrams that contribute to the distribution function for positive polarized photons (a), and for negative polarized photons (b), for $J_z = +1$ ortho–positronium ($u_\uparrow\overline{v}_\uparrow$).

The distribution function for positive helicity photons G^+ is calculated from the diagrams of Fig. 3(a) for the case of $J_z = +1$ ortho–positronium ($u_\uparrow \overline{v}_\uparrow$). Similarly, the corresponding diagrams for negative helicity photons are shown in Fig. 3(b), where an arrow up $\uparrow$ (down $\downarrow$) indicates positive (negative) helicity. In the diagrams, the upper fermion line corresponds to a particle (electron), and the lower to an antiparticle (positron). We have also indicated the light–cone parameterization of momenta that we will use when the photon couples to an electron or positron. With this choice, the photon is always parameterized by $(x, \vec{k}_\perp)$ and the final state has the same form in all cases. The appropriate matrix elements for the various helicity transitions are given in Ref. 38.

The calculation is now straightforward. If we denote by $\psi(y, \vec{\ell}_\perp)$ the two–body bound–state valence wave function (lowest Fock state amplitude), the results are:

$$
\begin{aligned}
G^+_{\gamma/\mathrm{ortho}\uparrow}(x, \vec{k}_\perp) &= \frac{\alpha}{2\pi^2} \int \frac{d^2\ell_\perp}{2(2\pi)^3} \int_x^1 dy \\
&\times \Bigg\{ \left[\psi(y, \vec{\ell}_\perp) \frac{y\vec{k}_\perp - x\vec{\ell}_\perp}{y - x} - \psi(y - x, \vec{\ell}_\perp - \vec{k}_\perp) \frac{(1-y)\vec{k}_\perp + x\vec{\ell}_\perp}{1 - y} \right]^2 \frac{1}{x^3} \\
&+ \left[\left| \psi(y, \vec{\ell}_\perp) \right|^2 \frac{1}{(y-x)^2\, y^2} \right. \\
&+ \left. \left| \psi(y - x, \vec{\ell}_\perp - \vec{k}_\perp) \right|^2 \frac{1}{(1-y)^2\, (1 - [y - x])^2} \right] x\, m^2 \Bigg\} \frac{1}{D^2} \ , \qquad (9)
\end{aligned}
$$

$$
\begin{aligned}
G^-_{\gamma/\mathrm{ortho}\uparrow}(x, \vec{k}_\perp) &= \frac{\alpha}{2\pi^2} \int \frac{d^2\ell_\perp}{2(2\pi)^3} \int_x^1 dy \\
&\times \left[\psi(y, \vec{\ell}_\perp) \frac{y\vec{k}_\perp - x\vec{\ell}_\perp}{y} - \psi(y - x, \vec{\ell}_\perp - \vec{k}_\perp) \frac{(1-y)\vec{k}_\perp + x\vec{\ell}_\perp}{1 - (y - x)} \right]^2 \frac{1}{x^3 D^2} \ ,
\end{aligned}
$$

where

$$
D = M_B^2 - \frac{(\vec{\ell}_\perp - \vec{k}_\perp)^2 + m^2}{y - x} - \frac{\vec{\ell}_\perp^{\,2} + m^2}{1 - y} - \frac{\vec{k}_\perp^{\,2}}{x} \ . \qquad (10)
$$

Here m and M_B are the electron and bound state masses, respectively. The intrinsic gluon distribution defined in Eq. (8) is obtained by integrating these expressions over the transverse momentum up to the cut–off Q_0^2. The same approach gives

$$
G^+_{\gamma/\mathrm{para}} = G^-_{\gamma/\mathrm{para}} \ , \qquad (11)
$$

$$
G^+_{\gamma/\mathrm{ortho}\ J_z=0} = G^-_{\gamma/\mathrm{ortho}\ J_z=0} \ .
$$

The polarized and unpolarized photon distribution functions are given by:

$$
\begin{aligned}
\Delta G(x, \vec{k}_\perp) &\equiv G^+(x, \vec{k}_\perp) - G^-(x, \vec{k}_\perp) \ , \\
G(x, \vec{k}_\perp) &\equiv G^+(x, \vec{k}_\perp) + G^-(x, \vec{k}_\perp) \ .
\end{aligned} \qquad (12)
$$

Let us now consider the small x limit for these functions. Expanding around $x = 0$, we readily obtain:

$$\Delta G(x\sim 0,\vec{k}_\perp) = \frac{\alpha}{\pi^2\vec{k}_\perp^2}\int\frac{d^2\ell_\perp}{2(2\pi)^3}\int_0^1 dy\,\Big[\,\psi(y,\vec{\ell}_\perp)-\psi(y,\vec{\ell}_\perp-\vec{k}_\perp)\,\Big]$$
$$\times\left[\frac{\psi(y,\vec{\ell}_\perp)}{y}-\frac{\psi(y,\vec{\ell}_\perp-\vec{k}_\perp)}{1-y}\right] \tag{13}$$

and

$$G(x\sim 0,\vec{k}_\perp) = \frac{\alpha}{\pi^2\vec{k}_\perp^2 x}\int\frac{d^2\ell_\perp}{2(2\pi)^3}\int_0^1 dy\,\Big[\,\psi(y,\vec{\ell}_\perp)-\psi(y,\vec{\ell}_\perp-\vec{k}_\perp)\,\Big]^2 .$$

The infrared singularity at $\vec{k}_\perp^2\to 0$ is eliminated because of the neutrality of the atom.

It should be noted that the singularity in $G(x,\vec{k}_\perp)$ at $x\to 0$ is actually an ultraviolet singularity for any non–zero value of $\vec{k}_\perp$ since $x=(k^0+k^z)/(p^0+p^z)$ can only be zero if $k_z\to-\infty$. By definition, the intrinsic distribution $G(x,Q_0^2)$ refers to Fock states with limited parton invariant mass $\mathcal{M}$: $\mathcal{M}^2=\sum_i\left[\frac{\vec{k}_\perp^2+m^2}{x}\right]_i<Q_0^2$. This restriction regularizes the $x\to 0$, $\vec{k}_\perp\neq 0$ region. On the other hand, the extrinsic contribution is derived from Fock states exceeding this cut–off, $Q_0^2<\mathcal{M}^2<Q^2$. Physical quantities are independent of the intermediate cut–off Q_0; the logarithmic dependence on Q_0 cancels in the sum of intrinsic and extrinsic structure functions.

Note that the integral of xG, the momentum fraction carried by intrinsic photons, is always well–defined. In order to proceed further, we shall assume that the wave function $\psi(y,\vec{\ell}_\perp)$ is peaked at $y\simeq 1/2$. We then obtain

$$\frac{\Delta G(x,\vec{k}_\perp)}{G(x,\vec{k}_\perp)}\simeq x\left\langle\frac{1}{y}\right\rangle\simeq 2x\qquad(x\to 0)\quad , \tag{14}$$

for the polarization asymmetry. We have found that this result is numerically accurate for a large range of positronium wave functions.

The opposite region $(x\to 1)$, where the fermions emit hard photons, can be also readily studied. After changing variables $(1-y)=(1-x)(1-\tau)$, and expanding around $(1-x)\to 0$, we obtain:

$$G^+(x,\vec{k}_\perp) = \frac{(1-x)}{2\pi^2\,2(2\pi)^3}\int_0^1 d\tau$$
$$\left\{\left[\psi(y,\vec{\ell}_\perp)\,\frac{\vec{k}_\perp-\vec{\ell}_\perp}{\tau}-\psi(y-x,\,\vec{\ell}_\perp-\vec{k}_\perp)\,\frac{\vec{k}_\perp}{1-\tau}\right]^2\right.$$
$$\left.+\,m^2\left[\frac{\left|\psi(y,\vec{\ell}_\perp)\right|^2}{\tau^2}+\frac{\left|\psi(y-x,\vec{\ell}_\perp-\vec{k}_\perp)\right|^2}{(1-\tau)^2}\right]\right\}$$
$$\times\frac{1}{\left[\frac{(\vec{\ell}_\perp-\vec{k}_\perp)^2+m^2}{\tau}+\frac{\vec{\ell}_\perp^2+m^2}{1-\tau}\right]^2}\quad , \tag{15}$$

$$G^-(x, \vec{k}_\perp) = \frac{(1-x)^3}{2\pi^2\, 2(2\pi)^3} \int_0^1 d\tau$$

$$\left[\, \psi(y, \vec{\ell}_\perp)\,(\vec{k}_\perp - \vec{\ell}_\perp) \; - \; \psi(y-x, \vec{\ell}_\perp - \vec{k}_\perp)\,\vec{\ell}_\perp \,\right]^2$$

$$\times \; \frac{1}{\left[\frac{(\vec{\ell}_\perp - \vec{k}_\perp)^2 + m^2}{\tau} + \frac{\vec{\ell}_\perp^{\,2} + m^2}{1-\tau}\right]^2} \; .$$

Thus the $x \to 1$ behavior depends on the endpoint behavior of the wave function $\psi(y, \vec{\ell}_\perp)$. Let us assume that $\psi(y, \vec{\ell}_\perp) \sim y^p$ for $y \to 0$, and $\sim (1-y)^q$ for $y \to 1$. If $p > q$, then the terms that contain $\left| \psi(y, \vec{\ell}_\perp) \right|^2$ dominate at $x \to 1$ since $y > x$. This regime corresponds to the photon taking most of the longitudinal momentum of the bound state from the electron. If $p < q$, the terms that contain $\left| \psi(y-x, \vec{\ell}_\perp - \vec{k}_\perp) \right|^2$ will dominate, which corresponds to the photon taking its large momentum from the positron. Then

$$\begin{aligned} G^+ &= \text{constant } (1-x)^{1+2h} \\ G^- &= \text{constant } (1-x)^{3+2h} \end{aligned} \qquad (x \to 1) \quad , \tag{16}$$

where $h = min(p, q)$ is the lowest endpoint power $(y \to 0, y \to 1)$ behavior of $\psi(y, \vec{\ell}_\perp)$. If $\psi(y, \vec{\ell}_\perp)$ is invariant under $y \to (1-y)$, then the two endpoint powers are the same. In any case:

$$\frac{\Delta G(x, \vec{k}_\perp)}{G(x, \vec{k}_\perp)} \to 1 \qquad (x \to 1) \quad ; \tag{17}$$

i.e. the helicity of the photon tends to be aligned with that of the bound state at large x. In the case of relativistic positronium $h = 1$.[39]

We now extend this analysis to QCD bound states. A perturbative analysis is certainly justified for heavy quark systems.[40] Since the general structure of the fermion $\to$ fermion plus gluon vertices given in Table I is dictated by Lorentz invariance and parity conservation, we will assume that this perturbative structure is also applicable to light–quark systems. We thus analyze the intrinsic gluon distribution retaining only first order corrections to the valence Fock state. The appropriate color factor is obtained by the replacement of (α) by $(C_F \alpha_s)$ where $C_F = 4/3$ for $N_C = 3$. We find similar endpoint behavior to that found in the abelian calculation. In particular, the gluon asymmetry at $x \to 0$ is $\Delta G(x)/G(x) \simeq < 1/y > x \simeq N_q x$ where N_q is the number of fermions in the valence Fock state. The $x \to 1$ behavior for the three–quark proton can also be determined[41]

$$\begin{aligned} G^+ &\sim (1-x)^4 \\ G^- &\sim (1-x)^6 \end{aligned} \qquad (x \to 1) \quad . \tag{18}$$

3.2. Connection with the Bound State Potential

On general grounds we expect a connection between the probability for emission (distribution function of photons or gluons) and the hyperfine interaction part of the bound state potential since both depend on the exchange of transverse gauge quanta. In fact, each diagram that contributes to the transverse potential has a corresponding cut–diagram in the expression for the distribution function. In the actual calculation, these quantities differ by just a denominator D. Thus

$$\int_0^1 dx \; G_{g/B}\,(x, Q_0^2) \;=\; -\left\langle \frac{\partial V}{\partial M_B^2} \right\rangle_{Q_0^2} \quad , \tag{19}$$

where $G_{g/B}$ is the unpolarized distribution function of gauge fields g in the bound state B, V is the potential due to gluon exchange and self–energy corrections, and M_B is the bound–state mass. Note that the instantaneous (non–retarded) piece does not depend on M_B, so it does not contribute. As discussed above, these quantities are regulated at $x \to 0$ by the ultraviolet cutoff Q_0^2 in the invariant mass. This singularity cancels in the hyperfine splitting:

$$\int_0^1 dx \; \left[\, G_{\gamma/\mathrm{ortho}\uparrow}\,(x) \; - G_{\gamma/\mathrm{para}}\,(x)\right] \;=\; -\left\langle \frac{\Delta \partial V}{\partial M_B^2} \right\rangle_{hfs} \quad . \tag{20}$$

where $\langle\ \rangle_{hfs}$ refers to the spin–dependent part of the bound state potential.

In the case of gluons in QCD bound states, we obtain analogous results:

$$\int_0^1 dx \; \left[\, G_{g/\rho}\,(x) \; - G_{g/\pi}\,(x)\right] \;=\; -\left\langle \frac{\Delta \partial V}{\partial M_B^2} \right\rangle_{hfs} \quad , \tag{21}$$

for mesons (ρ and π), and

$$\int_0^1 dx \; \left[\, G_{g/p}\,(x) \; - G_{g/\Delta}\,(x)\right] \;=\; -\left\langle \frac{\Delta \partial V}{\partial M_B^2} \right\rangle_{hfs} \quad , \tag{22}$$

for baryons (p and Δ).

These expressions can be analytically continued, relating the difference of fragmentation functions of gluons $D_{H/g}(z, Q^2)$ into hadrons H of different spin to the hyperfine splitting piece of the bound state potential.

3.3. Summary on the Intrinsic Gluon Distribution

The gluon distribution of a hadron is usually assumed to be generated from QCD evolution of the quark structure functions beginning at an initial scale Q_0^2.[42] In such a model there are no gluons in the hadron at a resolution scale below Q_0. The evolution is completely incoherent; *i.e.*, each quark in the hadron radiates independently.

In the approach presented here it is recognized that the bound state wave function itself generates gluons. This is clear from the relationship between the gluon distribution and the transverse part of the bound–state potential. To the extent that gluons generate the binding, they also must appear in the intrinsic gluon distribution. We emphasize that the diagrams in which gluons connect one quark to another are not present in the usual QCD evolution equations. Evolution contributions correspond in the bound–state equation to self–energy corrections to the quark lines at resolution scales $\mathcal{M}^2 > Q_0^2$.

The model forms given in Eqs. (6) and (7) provide a convenient model for the nucleon's polarized and unpolarized intrinsic gluon distributions which takes into account coherence at low x and perturbative constraints at high x. It is expected that this should be a good characterization of the gluon distribution at the resolution scale $Q_0^2 \simeq M_p^2$.

It is well–known that the leading power at $x \sim 1$ is increased when QCD evolution is taken into account. The change in power is[32]

$$\Delta p_g(Q^2) = 4C_A\, \zeta(Q^2, Q_0^2) = \frac{1}{\pi} \int_{Q_0^2}^{Q^2} \frac{d\kappa^2}{\kappa^2} \alpha_s(\kappa^2), \tag{23}$$

where $C_A = 3$ in QCD. For typical values of $Q_0 \sim 1\ GeV$, $\Lambda_{\overline{MS}} \sim 0.2\ GeV$ the change in power is moderate: $\Delta p_g(2\ GeV^2) = 0.28$, $\Delta p_g(10\ GeV^2) = 0.78$. A recent determination of the unpolarized gluon distribution of the proton at $Q^2 = 2\ GeV^2$ using direct photon and deep inelastic data has been given in Ref. 43. The best fit over the interval $0.05 \leq x \leq 0.75$ assuming the form $xG(x, Q^2 = 2\ GeV^2) = A(1-x)^{\eta_g}$ gives $\eta_g = 3.9 \pm 0.11(+0.8 - 0.6)$, where the errors in parenthesis allow for systematic uncertainties. This result is compatible with the prediction $\eta_g = 4$ for the intrinsic gluon distribution at the bound–state scale, allowing for the increase in the power due to evolution.

4. BOUND VALENCE-QUARK DISTRIBUTIONS

Deep inelastic lepton scattering and lepton–pair production experiments measure the light–cone longitudinal momentum distributions $x = (k_q^0 + k_q^z)/(p_H^0 + p_H^z)$ of quarks in hadrons through the relation

$$F_2^H(x, Q^2) = \sum_q e_q^2 x G_{q/H}(x, Q^2). \tag{24}$$

$F_2^H(x, Q^2)$ is the leading–twist structure function at the momentum transfer scale Q, Four–momentum conservation at large Q^2 then leads to the identification $x = x_{Bj} = Q^2/2p \cdot q$. In principle, the distribution functions $G_{q/H}$ could be computed from the bound state solutions of QCD.[44] For example, given the wave functions $\psi_{n/H}^{(Q)}(x_i, \vec{k}_{\perp_i}, \lambda_i)$ in the light–cone Fock expansion of the hadronic state, one can write the distribution function in the form[45]

$$G_{q/H}(x, Q^2) = \sum_{n,\lambda_i} \int \prod_i \frac{dx_i d^2 k_{\perp_i}}{16\pi^3} |\psi_{n/H}^{(Q)}(x_i, \vec{k}_{\perp_i}, \lambda_i)|^2 \sum_{b=q} \delta(x_b - x). \tag{25}$$

Here $x_i = k_i^+/p_H^+ = (k_i^0 + k_i^z)/(p_H^0 + p_H^z)$ is the light–cone momentum fraction of each constituent, where $\sum_i x_i = 1$ and $\sum_i k_{\perp_i} = 0$ in each Fock state n. The sum is over all Fock components n and helicities λ_i, integrated over the unconstrained constituent momenta.

An important concept in the description of any bound state is the definition of "valence" constituents. In atomic physics the term "valence electrons" refers to the electrons beyond the closed shells which give an atom its chemical properties. Correspondingly, the term "valence quarks" refers to the quarks which give the bound state hadron its global quantum numbers. In quantum field theory, bound states of fixed particle number do not exist; however, the light-cone expansion allows a consistent definition of the valence quarks of a hadron: the valence quarks appear in each Fock state together with any number of gluons and quark–anti–quark pairs; each component thus has the global quantum numbers of the hadron.

How can one identify the contribution of the valence quarks of the bound state with the phenomenological structure functions? Traditionally, the distribution function $G_{q/H}$ has been separated into "valence" and "sea" contributions:[46] $G_{q/H} = G_{q/H}^{\rm val} + G_{q/H}^{\rm sea}$, where, as an operational definition, one assumes

$$G_{q/H}^{\rm sea}(x, Q^2) = G_{\overline{q}/H}^{\rm sea}(x, Q^2), \qquad (0 < x < 1), \tag{26}$$

and thus $G_{q/H}^{\rm val}(x, Q^2) = G_{q/H}(x, Q^2) - G_{\overline{q}/H}(x, Q^2)$. The assumption of identical quark and anti–quark sea distributions is reasonable for the s and $\overline{s}$ quarks in the proton.

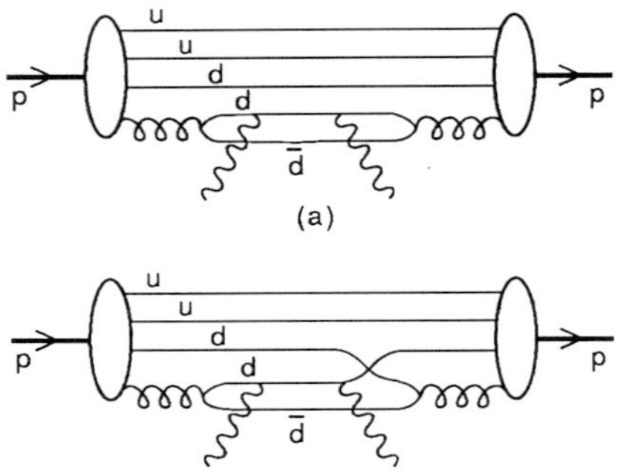

Figure 4. Structure function contributions from the three–quark plus one pair Fock state of the proton. The $d\overline{d}$ pair in diagram (a) contributes to the sea distribution, but diagram (b) due to anti–symmetrization of the d–quarks cannot be separated uniquely into "valence" versus "sea" parts.

However, in the case of the u and d quark contributions to the sea, anti–symmetrization of identical quarks in the higher Fock states implies non–identical q and $\overline{q}$ sea contributions. This is immediately apparent in the case of atomic physics, where Bethe–Heitler pair production in the field of an atom does not give symmetric electron and positron distributions since electron capture is blocked in states where an atomic electron is already present. Similarly in QCD, the $q\overline{q}$ pairs which arise from gluon splitting as in Fig. 4(a) do not have identical quark and anti–quark sea distributions; contributions from interference diagrams such as Fig. 4(b), which arise from the anti–symmetrization of the higher Fock state wave functions, must be taken into account. Although the integral of the conventional valence distribution gives correct charge sum rules, such as $\int_0^1 dx(G_{q/H}(x) - G_{\overline{q}/H}(x))$, it can give a misleading reading of the actual momentum distribution of the valence quarks. It is also interesting to notice that the Gottfried sum rule assumes the equality of anti–up and anti–down quarks in the proton. Because of the stronger Pauli blocking of up quarks, however, one would expect a relative suppression of anti–up quarks in the proton, giving an important correction to the sum rule.

The standard definition also has the difficulty that the derived valence quark distributions are apparently singular in the limit $x \to 0$. For example, standard phenomenology indicates that the valence up–quark distribution in the proton behaves as $G^{\mathrm{val}}_{u/p} \sim x^{-\alpha_R}$ for small x[47,46] where $\alpha_R \approx 0.5$.[48] This implies that quantities that depend on the $< 1/x >$ moment of the valence distribution diverge. This is the case for the "sigma term" in current algebra and the $J = 0$ fixed pole in Compton scattering.[49] Furthermore, it has been shown[50] that the change in mass of the proton when the quark mass is varied in the light–cone Hamiltonian is given by an extension of the Feynman–Hellmann theorem:

$$\frac{\partial M_p^2}{\partial m_q^2\,(Q^2)} = \int_0^1 \frac{dx}{x}\, G_{q/p}\,(x, Q^2). \tag{27}$$

In principle, this formula allows one to compute the contribution to the proton–neutron mass difference due to the difference of up and down quark masses. However, again, with the standard definition of the valence quark distribution, the integration is undefined at low x. Even more seriously, the expectation value of the light–cone kinetic energy operator

$$\int_0^1 dx \, \frac{< k_\perp^2 > + m^2}{x} \, G_{q/p}(x, Q). \tag{28}$$

is infinite for valence quarks if one uses the traditional definition. There is no apparent way of associating this divergence of the kinetic energy operator with renormalization.[51] Notice that a divergence at $x = 0$ is an ultraviolet infinity for a massive quark, since it implies $k^+ = k^0 + k^z = 0$; *i.e.* $k^z \to -\infty$. A bound state wave function would not be expected to have support for arbitrarily large momentum components.

Part of the difficulty with identifying bound state contributions to the proton structure functions is that many physical processes contribute to the deep inelastic lepton–proton cross section: From the perspective of the laboratory or center of mass frame, the virtual photon can scatter out a bound–state quark as in the atomic physics photoelectric process, or the photon can first make a $q\bar{q}$ pair, either of which can interact in the target. As we emphasize here, in such pair–production processes, one must take into account the Pauli principle which forbids creation of a quark in the same state as one already present in the bound state wave function. Thus the lepton interacts with quarks which are both *intrinsic* to the proton's bound–state structure, and with quarks which are *extrinsic; i.e.* created in the electron–proton collision itself. Note that such extrinsic processes would occur in electroproduction even if the valence quarks had no charge. Thus much of the phenomena observed in electroproduction at small values of x, such as Regge behavior, sea distributions associated with photon–gluon fusion processes, and shadowing in nuclear structure functions should be identified with the extrinsic interactions, rather than processes directly connected with the proton's bound–state structure.

In Ref. 11 Ivan Schmidt and I propose a definition of "bound valence–quark" distribution functions that correctly isolates the contribution of the valence constituents which give the hadron its flavor and other global quantum numbers. In this new separation, $G_{q/p}(x, Q^2) = G^{\rm BV}_{q/p}(x, Q^2) + G^{\rm NV}_{q/p}(x, Q^2)$, non–valence quark distributions are identified with the structure functions which would be measured if the valence quarks of the target hadron had zero electro–weak charge. We shall prove that with this new definition the bound valence–quark distributions $G^{\rm BV}_{q/p}(x, Q^2)$ vanish at $x \to 0$, as expected for a bound–state constituent.

4.1. Construction of Bound Valence–Quark Distributions

In order to construct the bound valence–quark distributions, we imagine a *gedanken* QCD where, in addition to the usual set of quarks $\{q\} = \{u, d, s, c, b, t\}$, there is another set $\{q_0\} = \{u_0, d_0, s_0, c_0, b_0, t_0\}$ with the same spin, masses, flavor, color, and other quantum numbers, except that their electromagnetic charges are zero.

Let us now consider replacing the target proton p in the lepton–proton scattering experiment by a chargeless proton p_0 which has valence quarks q_0 of zero electromagnetic charge. In this extended QCD the higher Fock wave functions of the proton p and the chargeless proton p_0 both contain $q\bar{q}$ and $q_0\overline{q_0}$ pairs. As far as the strong QCD interactions are concerned, the physical proton and the *gedanken* chargeless proton are equivalent.

We define the bound valence–structure function of the proton from the difference between scattering on the physical proton minus the scattering on the chargeless proton, in analogy to an "empty target" subtraction:

$$F_i^{\rm BV}(x, Q^2) \equiv F_i^p(x, Q^2) - F_i^{p_0}(x, Q^2). \tag{29}$$

The non–valence distribution is thus $F_i^{\rm NV}(x, Q^2) = F_i^{p_0}(x, Q^2)$. The $F_i(x, Q^2)$ ($i = 1, 2,$) are the leading–twist structure functions, with $F_{2H}^{\rm BV}(x, Q^2) = \sum_q e_q^2 x G^{\rm BV}_{q/H}(x, Q^2)$, etc. The situation just described is similar to the atomic physics case, where in order to correctly define photon scattering from a bound electron, one must subtract the cross section on the nucleus alone, without that bound electron present.[52] Physically the nucleus can scatter photons through virtual pair production, and this contribution has to be subtracted from the total cross section. In QCD we cannot construct protons without the valence quarks; thus we need to consider hadrons with chargeless valence constituents.[53]

Notice that the cross section measured in deep inelastic lepton scattering on p_0 is not zero. This is because the incident photon (or vector boson) creates virtual $q\overline{q}$ pairs which scatter strongly in the gluonic field of the chargeless proton target. In fact at small x the inelastic cross section is dominated by $J = 1$ gluon exchange contributions, and thus the structure functions of the physical and chargeless protons become equal:

$$\lim_{x \to 0} \left[F_i^p(x, Q^2) - F_i^{p_0}(x, Q^2) \right] = 0. \tag{30}$$

Remarkably, as we show below, the bound valence–quark distribution function $G_{q/H}^{\mathrm{BV}}$ vanishes at $x \to 0$; it has neither Pomeron x^{-1} nor Reggeon $x^{-\alpha_R}$ contributions.

Although the *gedanken* subtraction is impossible in the real world, we will show that, nevertheless, the bound valence–distribution can be analytically constrained at small x_{bj}. This opens up the opportunity to extend present phenomenology and relate measured distributions to true bound state wave functions.

In the following sections we will analyze both the atomic and hadronic cases, paying particular attention to the high energy regime.

4.2. Atomic Case

Since it contains the essential features relevant for our discussion, we will first analyze photon scattering from an atomic target. This problem contains an interesting paradox which was first resolved by Goldberger and Low in 1968.[52] Here we give a simple, but explicit, derivation of the main result.

The Kramers–Kronig dispersion relation relates the forward Compton amplitude to the total photo–absorptive cross section[54]

$$f(k) - f(0) = \frac{k^2}{2\pi^2} \int_0^\infty dk' \frac{\sigma(k')}{k'^2 - k^2 - i\epsilon}, \tag{31}$$

where k is the photon energy. One should be able to apply this formula to scattering on a bound electron (e_b) in an atom. However, there is an apparent contradiction. On the one hand, one can explicitly compute the high energy $\gamma e_b \to \gamma e_b$ forward amplitude: it tends to a constant value at $k \to \infty$, the electron Thomson term, $f(k) \to -e^2/m_b^e$, where m_b^e is the effective electron mass corrected for atomic binding.[55] On the other hand, the $\mathcal{O}(e^2)$ cross section for the photoelectric effect $\gamma e_b \to e'$ behaves as $\sigma_{\mathrm{photo}} \sim 1/k$ at high energies. But then the integral in the dispersion relation predicts logarithmic behavior for $f(k)$ at high energy in contradiction to the explicit calculation. Evidently other contributions to the inelastic cross–section cannot resolve this conflict.

This problem was solved[52] by carefully defining what one means by scattering on a bound state electron. For both the elastic Compton amplitude and the inelastic cross section one must subtract the contribution in which the photon scatters off the Coulomb field of the nucleus (empty target subtraction). Thus $\sigma(k)$ in the Kramers-Kronig relation is really the difference between the total atomic cross section $\sigma_{\mathrm{atom}}(k)$ and the nuclear cross section $\sigma_{\mathrm{nucleus}}(k)$, which is dominated by pair production. We will present a simple proof that the high energy behavior $\sim 1/k$ of the cross sections exactly cancels in this difference, which is a necessary condition for a consistent dispersion relation.

The total cross section for photon scattering on the atom is dominated by two main terms: the photoelectric contribution and e^+e^- pair production, with the produced electron going into a different state than the electron already present in the atom.[56] On the other hand, in the subtraction, pair production in the field of the nucleus is not restricted by the Pauli principle; this cross section contains a contribution where the produced electron goes into the same state as the bound state electron of the atom, plus other terms in which it goes into different states. These last contributions cancel in the difference $\sigma_{\text{atom}} - \sigma_{\text{nucleus}}$. Thus the bound-state electron photo-absorption cross section is the difference between the photoelectric cross section on the atom and the pair production *capture* cross section on the nucleus, where the produced electron is captured in the same state as the original bound state electron: $\sigma_{e_b} = \sigma_{\text{photoelectric}} - \sigma_{\text{capture}}$. This is depicted graphically in Fig. 5.

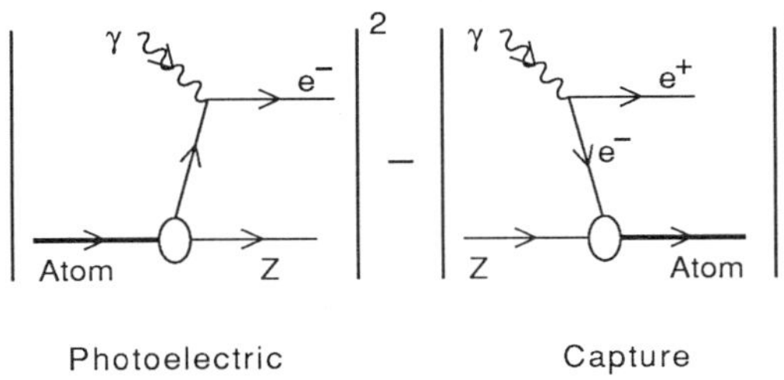

Figure 5. The bound-electron photo-absorption cross section $\sigma_{\gamma e_b}$ is defined as the difference of $\gamma - Atom$ and $\gamma - Nucleus$ cross sections. This can also be expressed as the difference between the atomic "photoelectric" cross section and the pair production "capture" cross section on the nucleus, but with the produced electron going into the same atomic state as the original bound state electron.

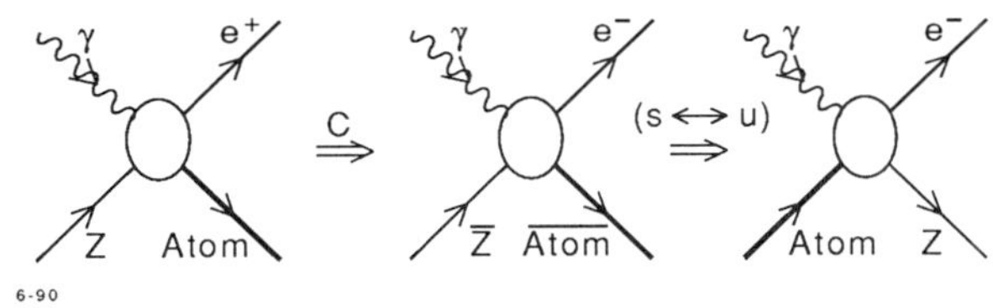

Figure 6. The helicity-summed squared amplitude for the process $\gamma Z \to e^+ Atom$ is equal, by charge conjugation, to the helicity-summed squared amplitude for $\gamma\overline{Z} \to e^-\overline{\text{Atom}}$, up to a phase. This is also equal by crossing to the helicity-summed squared amplitude for the process $\gamma Atom \to e^- Z$, but with s and u interchanged.

We next note that the squared amplitude for the capture process $\gamma Z \to e^+ Atom$ is equal, by charge conjugation, to the squared amplitude for $\gamma\overline{Z} \to e^-\overline{\text{Atom}}$. (See Fig. 6.) Furthermore, by crossing symmetry, the (helicity summed) squared amplitude for this last process is equal to the (helicity summed) squared amplitude for $\gamma Atom \to e^- Z$, with p_Z and $(-p_{\text{Atom}})$ interchanged. This is equivalent to the interchange of the Mandelstam variables $s = (p_\gamma + p_Z)^2$ and $u = (p_\gamma - p_{\text{Atom}})^2$. Thus at high photon energies (where $s \simeq -u$), the two cross sections $\sigma_{\text{photoelectric}}$ and σ_{capture} of Fig. 5 cancel, consistent with the Kramers–Kronig relation. In Regge language, the imaginary part of the $J = 0$ Compton amplitude is zero.

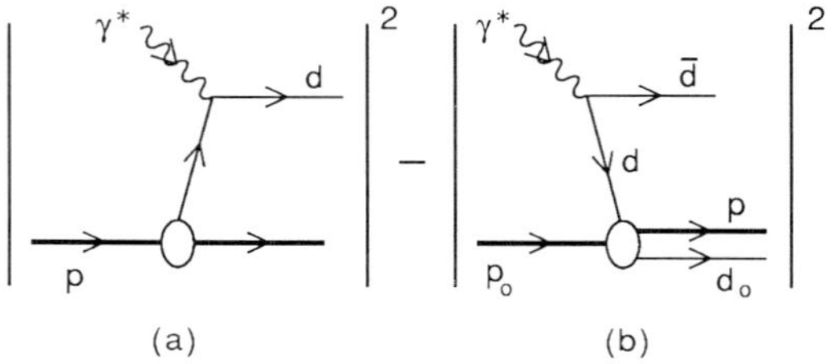

Figure 7. The bound valence–quark distribution of quark d can be calculated from the difference between (a) the cross section on the state p in which the virtual photon momentum is absorbed by the quark d, and (b) the $d\overline{d}$ pair production cross section in the field of p_0, but with the produced d quark ending in the same state as the d quark in the original proton state p.

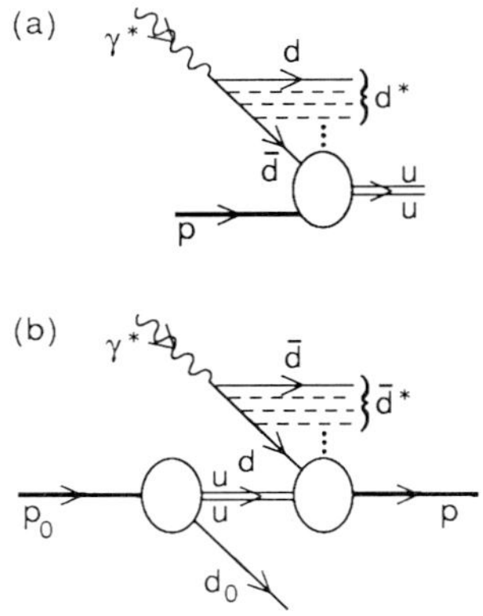

Figure 8. Amplitudes describing Reggeon behavior at small x (a) in electroproduction, and (b) in the subtraction term of Fig. 7(b).

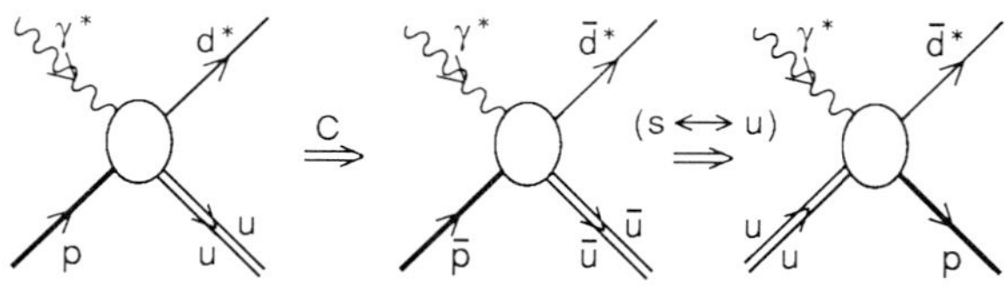

Figure 9. The helicity–summed squared amplitude for (a) $\gamma^* p \to d^*(uu)$ is equal, by charge conjugation, to the helicity–summed squared amplitude for the process (b) $\gamma^* \overline{p} \to \overline{d}^*(\overline{uu})$, up to a phase. This is also equal, by crossing symmetry, to the helicity–summed squared amplitude for (c) $\gamma^*(uu) \to \overline{d}^* p$, with s and u interchanged. Thus at high energies the Reggeon contribution from the subtraction term of Fig. 8(b) cancels the Reggeon contribution of Fig. 8(a).

The proof we have presented implicitly assumes the equality of the flux factors for the photoelectric process on the atom and the capture process on the nucleus. This is normally a good approximation since the atomic and nuclear masses are almost identical for $M_Z \gg m_e$. However, for finite mass systems such as muonic atoms, the mass of the nucleus and atom are unequal, and the cross sections do not cancel at high energy. The difficulty in this case is that the nucleus does not provide the correct "empty target" subtraction.

However, we can extend the analysis to the general atomic problem by considering hypothetical atoms A_0 consisting of null leptons ℓ_0 with normal electromagnetic and Coulomb interactions with the nucleus but with zero external charge. [In effect, we consider an extended QED with $U(1) \times U(1)$ gauge interactions, where the null lepton has charge $(-1, 0)$, and the normal lepton and nucleus have charges $(-1, -1)$ and (Z, Z), respectively.] The empty target subtraction is defined as the difference between the cross section on the normal atom $A = (Z\ell)$ and the cross section on the null atom $A_0 = (Z\ell_0)$. Since the mass and binding interactions of A and A_0 are identical, the photo–absorption flux factors are the same in both cases.

As in the earlier proof, the matrix element for the photoelectric process on the atom A becomes equal in modulus at high energies with the matrix element for the capture process on the null atom A_0. Note that in the computation of the capture process amplitude, the presence of the spectator lepton ℓ_0 is irrelevant since it remains in the original quantum state (say $1S$): The required matrix element of the current is

$$\left\langle A\ell_0(1S)\ell^+ | J^\mu | A_0 \right\rangle = \left\langle A\ell_0(1S)\ell^+ \left| \overline{\psi}_\ell \gamma^\mu \psi_\ell b^\dagger_{\ell_0}(1S) \right| Z \right\rangle = \left\langle A\ell^+ | J^\mu | Z \right\rangle .$$

By charge conjugation and crossing this is equal in modulus to

$$\left\langle Z\ell^- | J^\mu | A \right\rangle ,$$

the corresponding photoelectric matrix element with $s \to u$. Final–state interactions can only affect the phase at high energies. Thus we obtain cancellation of the photoelectric and capture cross sections at high energies, and verify the Kramers–Kronig dispersion relation for Compton scattering on leptons bound to finite mass nuclei.

4.3. Reggeon Cancellations in QCD

We now return to the analysis of the "bound valence–quark distributions" of the proton. According to the discussion of Section 4.1, the measurement of the bound valence–quark distribution requires an "empty target" subtraction:

$$\sigma(\gamma^* p \to X) - \sigma(\gamma^* p_0 \to X).$$

Both p and p_0 contain higher Fock states with arbitrary number of gluons, $q\overline{q}$, and $q_0\overline{q_0}$ pairs. It is clear that the terms associated with $J \approx 1$ Pomeron behavior due to gluon exchange cancel in the difference. In this section we shall prove that the Reggeon terms also cancel, and thus the resulting distribution of bound valence quarks $G^{\rm BV}_{q/p}(x, Q^2)$ vanishes as $x \to 0$.

As in the atomic case, we now proceed to describe the leading contributions to the scattering of a photon from both the proton p and the state p_0. For simplicity of notation, we will consider an example which isolates just the bound valence d–quark distribution of the proton $p(uud)$; in this case the subtraction term is the deep inelastic cross section on the system $p_0(uud_0)$ in which the d_0 valence quark has normal QCD interactions but does not carry electric charge. The general case, where the subtraction is on the completely neutral state $p_0(u_0u_0d_0)$, is a simple generalization. The high Q^2 virtual photo–absorption cross section on the proton (laboratory frame) contains two types of terms: contributions in which a quark in p absorbs the momentum of the virtual photon; and terms in which a $q\overline{q}$ pair is created, but the produced q is in a different quantum state than the quarks already present in the hadron. On the other hand, the cross section for scattering of the virtual photon from the state $p_0(uud_0)$ contains contributions that differ from the $p(uud)$ case in two important aspects: first the virtual photon can be absorbed only by charged quarks; and in $d\overline{d}$ pair production on the null proton p_0, the d quark can be produced in any state. Thus the difference between the cross sections off p and p_0 equals a term analogous to $\sigma_{\rm photoelectric}$, in which a d quark in p absorbs the photon momentum, minus a $d\bar{d}$ pair production contribution on p_0 analogous to σ_{capture}, in which the produced d quark ends up in the same quantum state as the d quark in the original proton state p. This is shown graphically in Fig. 7.[57]

Reggeon behavior in the electroproduction cross section can be understood as due to the appearance of a spectrum of bound $q\bar{q}$ states in the t–channel. The absorptive cross section associated with t–channel ladder diagrams is depicted in Fig. 8(a). The summation of such diagrams leads to Reggeon behavior of the deep inelastic structure functions at small x.[58] In the rest system, the virtual photon creates a $d\bar{d}$ pair at a distance proportional to $1/x$ before the target. The radiation which occurs over this distance contributes to the physics of the Reggeon behavior.

A corresponding Reggeon contribution at low x also occurs in the subtraction term indicated in Fig. 8(b). In the case of the proton target, the d–quark, after radiation, cannot appear in the quantum state already occupied by the d–quark in the proton because of the Pauli principle. However, the corresponding contribution is allowed on the p_0 target: in effect, the d–quark replaces the d_0–quark and is captured into a proton. The capture cross section is computed from the amplitude for $\gamma^* p_0 \to \bar{d}^* p\, d_0^{1S}$.[59] As in the corresponding atomic physics analysis, the spectator d_0 quark in the null target p_0 is inert and cancels out from the amplitude. Thus we only need to consider effectively the (helicity summed) squared amplitude for $\gamma^*(uu) \to \bar{d}^* p$. However, as illustrated in Fig. 9 this amplitude, after charge conjugation and crossing $s \to u$, is equal to the (helicity summed) $\gamma^* p \to d^*(uu)$ squared amplitude at small x. The flux factors for the proton and null proton target are equal.

If we write $s\sigma_{\text{photoelectric}}$ as a sum of Regge terms of the form $\beta_R |s|^{\alpha_R}$, where $\alpha_R > 0$ then the subtraction of the capture cross section on the null proton will give the net virtual photo–absorption cross section as a sum of terms $s\sigma^{\text{BV}} = \sum_R \beta_R(|s|^{\alpha_R} - |u|^{\alpha_R})$. If we ignore mass corrections in leading twist, then $s \simeq Q^2(1-x)/x$ and $u \simeq -Q^2/x$. Thus for small x every Regge term is multiplied by a factor $K_R = (-\alpha_R)x$. For example, for $\alpha_R = 1/2$ (which is the leading even charge–conjugation Reggeon contribution for non–singlet isospin structure functions), $F_2^{p(uud)} - F_2^{p_0(uud_0)} \sim x^{3/2}$. The bound valence–quark non–singlet ($I = 1$) distribution thus has leading behavior $G_{q/H}^{\text{BV}} \sim x^{1/2}$ and vanishes for $x \to 0$.

We can also understand this result from symmetry considerations. We have shown from crossing symmetry $G_{q/p}(x,Q^2) - G_{\bar{q}/p_0}(x,Q^2) \to 0$ at low x. Thus the even charge–conjugation Reggeon and Pomeron contributions decouple from the bound valence–quark distributions.

The analytic cancellation of the leading Reggeon contributions of the s–channel and u–channel contributions suggests that, given sufficiently detailed Regge fits to the data for the non–singlet structure functions, one could construct a phenomenological model for the bound valence–quark distributions. Eventually, lattice gauge theory or other non–perturbative methods for solving QCD, such as discretized light–cone quantization,[44] may provide detailed first–principle predictions for the bound valence–quark distributions which could be compared with the phenomenological forms.

4.4. Summary on Bound-Valence Quark Distributions

The observation that the deep inelastic lepton–proton cross section is non–zero, even when the quarks in the target hadron carry no charge, implies that we should distinguish two separate contributions to deep inelastic lepton scattering: *intrinsic* (bound–state) and *extrinsic* (non–bound) structure functions. The extrinsic contributions are created by the virtual strong interactions of the lepton itself, and are present even if the quark fields of the target are chargeless. The bound valence–quark distributions, defined by subtracting the distributions for a gedanken "null" hadron with chargeless valence quarks, correctly isolates the valence–quark contributions intrinsic to the bound–state structure of the target. As we have shown, both the Pomeron and leading Reggeon contributions are absent in the bound valence–quark distributions. The leading Regge contributions are thus associated with particles created by the photon–hadron scattering reaction, processes extrinsic to the bound state physics of the target hadron itself. The bound valence–quark distributions are in principle computable by solving the bound state problem in QCD. Sum rules for the proton derived from properties of the hadronic wave function thus apply to the bound valence–quark contributions. In particular, the light–cone kinetic energy of the bound valence–quarks,

$$\int_0^1 dx \; \frac{< k_\perp^2 > + m^2}{x} \; G_{q/p}^{\rm BV}(x,Q) \tag{32}$$

is finite, as expected for a bound state wave function contribution. The ultraviolet divergence of the kinetic energy obtained from the non–valence distribution is associated with the production of high mass states in the electron–proton collision, rather than the distribution of the bound–state valence quarks.

The essential reason why the new definition of the bound valence–quark distribution differs from the conventional definition of valence distributions is the Pauli principle: the anti–symmetrization of the bound state wave function for states which contain quarks of identical flavor. As we have shown, this effect plays a dynamical role at low x, eliminating leading Regge behavior in the bound valence–quark distributions. In the atomic physics case, where the leading Regge behavior corresponds to $J = \alpha_R = 0$, the analogous application of the Pauli principle leads to analytic consistency with the Kramers–Kronig dispersion relation for Compton scattering on a bound electron.

5. INTRINSIC CHARM–QUARK DISTRIBUTIONS

There are a number of striking anomalies in the data[60] for charm production which cannot be readily explained by conventional leading twist $gg \to c\bar{c}$ or $q\bar{q} \to c\bar{c}$ fusion subprocesses. The first signals for charm baryon production at large x_F were reported by the BCF and other groups at the ISR. The results are reviewed in Ref. 60. Other anomalies include:

1. The EMC data[61] for the charm structure function of the nucleon appears to be too high at large x_{Bj}.
2. The LEBC bubble chamber data[62] for charm production in pp collisions indicates an excess of D events at large x_F. The excess is not associated with $D's$ that contain the proton's valence quark.
3. The cross section measured by the WA–62 group[63] for $\Sigma^- N \to \Xi(csu)X$ is too large and flat at large x_F.
4. The NA–3 data[25] for J/ψ production in pion–nucleus and proton–nucleus collisions can be represented as two components: a normal contribution in the central region which is almost additive in nuclear number that can be accounted for by $gg \to c\bar{c}$ and $q\bar{q} \to c\bar{c}$ fusion, and a second "diffractive contribution" which dominates at large x_F and is strongly shadowed. This last contribution suggests that high momentum $c\bar{c}$ systems are being produced on the front surface of the nuclear target.

It is difficult to understand any of these anomalies, particularly the production of high x_F charmonium unless the proton itself has an intrinsic charm contribution[64] to its structure function. From the perturbative point of view, a $uudc\bar{c}$ Fock component can be generated by the $gg \to c\bar{c}$ amplitude where the gluons are emitted from two of the valence quarks. The probability for finding the heavy quark pair of mass $M_{Q\bar{Q}}$ or greater is thus of order $\alpha_s^2(M_{Q\bar{Q}}^2)/M_{Q\bar{Q}}^2$ (see the introduction). Intrinsic charm is thus a higher twist mechanism. The leading twist extrinsic charm contributions depend on the logarithm of the heavy quark mass. Since the intrinsic charm quarks are associated with the bound–state equation for the proton, then all the partons tend to have equal velocity. Unlike normal sea quarks generated by evolution, this implies that the heaviest constituents, the intrinsic charm quarks, will take a large fraction of the proton's momentum. In a hadronic collision the c and $\bar{c}$ can coalesce to produce a charmonium state with the majority of the proton's momentum.[65] The EMC charm structure function data requires a 0.3 % probability for the intrinsic charm Fock state in the nucleon.[61]

According to the hard scattering picture of QCD, production cross sections involving large momentum transfer should factorize and be approximately additive in the nucleon number, $d\sigma_A = A^{\alpha}(x_F, p_T)d\sigma_N$ with $\alpha \sim 1$, up to the small shadowing and anti–shadowing corrections seen in deep inelastic lepton–nucleus scattering. (See Section 6.) In the Drell–Yan process, large mass muon pair production, $\alpha \simeq 1$ for all x_F is indeed observed.[66] However, several experiments on open charm production show[60] that $\alpha(x_F \geq 0.2) \simeq 0.7 \ldots 0.8$. For small x_F, an indirect analysis[62] comparing different measurements of the total charm production cross section indicates $\alpha(x_F \simeq 0) \simeq 1$.

The most detailed data on the nuclear dependence of charm production is available from the hadroproduction of J/ψ. Here a decrease of α from $\alpha(x_F \simeq 0) \simeq 1$ to $\alpha(x_F \simeq 0.8) \simeq 0.8$ has been seen by several groups.[67] The analysis of Badier, *et al.*[25] is particularly interesting. They noted that the production of J/ψ at large x_F (up to $x_F \simeq 0.8$) cannot be explained by the gluon and light quark fusion mechanisms of perturbative QCD, due to the anomalous A–dependence. However, their $\pi^- A \to J/\psi + X$ data was well reproduced if, in addition to hard QCD fusion (with $\alpha = 0.97$), they included a "diffractive" component of J/ψ production at high x_F with $\alpha = 0.77$. Using their measured A–dependence to extract the "diffractive" component, they found that (for a pion beam) that the J/ψ distribution peaks at $x_F \simeq 0.5$ and dominates the hard scattering A^1 component for $x \geq 0.6$. The anomalous nuclear dependence cannot be explained by gluon shadowing since the data scale in x_F rather than the gluon momentum fraction in the nucleus x_2.[68] Final state absorption of the charmonium state would predict an increasing nuclear yield with J/ψ momentum, opposite what is seen. Furthermore, this would not explain the similar A–dependence observed by E–772[26] for J/ψ and ψ' production.

A diffractive contribution to heavy quarkonium production is consistent with QCD when one takes into account the higher twist intrinsic charm component of the projectile wave function. In high energy hadron–nucleus collisions the nucleus may be regarded as a "filter" of the hadronic wave function.[3] The argument, which relies only on general features such as time dilation, goes as follows.[69] As discussed in the introduction, one can define a Fock state expansion of a hadron in terms of its quark and gluon constituents; *e.g.* for a meson,

$$|h\rangle = |q\bar{q}\rangle + |q\bar{q}g\rangle + |q\bar{q}q\bar{q}\rangle + \ldots \tag{33}$$

The various Fock components will mix with each other during their time evolution. However, at sufficiently high hadron energies E_h, and during short times t, the mixing is negligible. Specifically, the relative phase $\exp[-i(E - E_h)t]$ of a given term in Eq. (33) is proportional to the energy difference

$$E - E_h = \left[\sum_i \frac{m_i^2 + \mathbf{p}_{Ti}^2}{x_i} - M_h^2 \right] \Big/ (2E_h) \tag{34}$$

which vanishes for $E_h \to \infty$. Hence the time evolution of the Fock expansion is, at high energies, diagonal during the time $\sim 1/R$ it takes for the hadron to cross a nucleus of radius R.

The diagonal time development means that it is possible to describe the scattering of a hadron in a nucleus in terms of the scattering of its individual Fock components. Let us explore the consequences for typical, soft collisions characterized by momentum transfers $q_T \simeq \Lambda_{QCD}$. The partons of a given Fock state will scatter independently of each other if their transverse separation is $r_T \geq 1/\Lambda_{QCD}$; *i.e.* if the state is of typical hadronic size. Conversely, the nuclear scattering will be coherent over the partons in Fock states having $r_T \ll 1/\Lambda_{QCD}$ since $e^{iq_T \cdot r_T} \simeq 1$. For color–singlet clusters, the interference between the different parton amplitudes interacting with the nuclear gluonic field is destructive. Thus the nucleus will appear nearly transparent to small, color–singlet Fock states. In an experiment detecting fast secondary hadrons the nucleus indeed serves, then, as a filter that selects the small Fock components in the incident hadrons.

Now consider the intrinsic charm state $|u\bar{d}c\bar{c}\rangle$ of a $|\pi^+\rangle$. Because of the large charm mass m_c, the energy difference in denominator of the wave function will be minimized at equal parton velocity; *i.e.* when the charm quarks carry most of the longitudinal momentum. Moreover, because m_c is large, the transverse momenta p_{Tc} of the charm quarks range up to $\mathcal{O}(m_c)$, implying that the transverse size of the $c\bar{c}$ system is $\mathcal{O}(1/m_c)$. Hence, provided only that the $c\bar{c}$ forms a color singlet, it can penetrate the nucleus with little energy loss. Thus the high momentum small transverse size $c\bar{c}$ color–singlet cluster in the incident hadron passes through the nucleus undeflected, and it can then evolve into charmonium states after transiting the nucleus.[70] In effect, the nucleus is transparent to the heavy quark pair component of the intrinsic state. The remaining cluster of light quarks in the intrinsic charm Fock state has a larger transverse size and tends to be absorbed on the front surface of the nucleus. This justifies the analysis of Badier *et al.* in which the perturbative and non–perturbative charm production mechanisms were separated on the basis of their different A–dependence ($\alpha = 0.97$ and $\alpha = 0.77$ for a pion beam, respectively). The effective x_F–dependence of α seen in charm production is explained by the different characteristics of the two production mechanisms. Hard, gluon fusion production dominates at small x_F, due to the steeply falling gluon structure function. The contribution from intrinsic charm Fock states in the beam peaks at higher x_F, due to the large momentum carried by the charm quarks. This two–component hard–scattering plus intrinsic charm model also explains why the nuclear dependence of J/ψ production depends on x_F rather than x_2, as predicted by leading twist factorization.[68]

An important consequence of this picture is that all final states produced by a penetrating intrinsic $c\bar{c}$ component will have the same A–dependence. Thus, in particular, the $\psi(2S)$ radially excited state will behave in the same way as the J/ψ, in spite of its larger size. This prediction is confirmed by the recent E–772 data.[26] The nucleus cannot influence the quark hadronization which (at high energies) takes place outside the nuclear environment.

Quarkonium production due to the intrinsic heavy quark state will fall off rapidly for p_T greater than M_Q, reflecting the fast–falling transverse momentum dependence of the higher Fock state wave function. Thus we expect the conventional fusion contributions to dominate in the large p_T region. The data are in fact consistent with a simple A^1 law for J/ψ production at large p_T. The CERN experiment of Badier *et al.*[25] finds that the ratio of nuclear cross sections is close to additive in A for all x_F

when p_T is between 2 and 3 GeV. The data of the FermiLab experiment of Katsanevas *et al.*[67] shows consistency with additivity for p_T ranging from 1.2 to 3 GeV.

As was discussed above, the probability for intrinsic heavy quark states in a light hadron wave function is expected[64,71] to scale up to logarithms inversely as the square of the heavy quark mass. This implies a production cross section proportional to $1/M_Q^4$. The total rate of heavy quark production by the intrinsic mechanism therefore decreases with quark mass relative to the leading–twist cross section which is proportional to $1/M_Q^2$. At large x the intrinsic production should still dominate, however, implying a nuclear dependence in this region characterized by $\alpha \simeq 0.7 \ldots\ 0.8$. The recent E–772 data[26] for the hadroproduction of the upsilon suggests that intrinsic beauty contributions may also be playing a role. Experimental measurements of beauty hadroproduction in nuclei over the whole range of x will be essential for unraveling the two components of the cross section.

6. SHADOWING AND ANTI–SHADOWING OF NUCLEAR STRUCTURE FUNCTIONS

The shadowing and anti–shadowing of deep inelastic nuclear structure functions refers to the depletion of the effective number of nucleons F_2^A/F_2^N at low $x \lesssim 0.1$, and the increase above nucleon additivity at $x \sim 0.15$. Results from the EMC collaboration[72] and SLAC[73] indicate that the effect is roughly Q^2–independent; *i.e.* shadowing is a leading twist in the operator product analysis. In contrast, the shadowing of the real photo–absorption cross section due to ρ–dominance[74–77] falls away as an inverse power of Q^2.

Shadowing is a destructive interference effect which causes a diminished flux and interactions in the interior and back face of the nucleus. The Glauber analysis[78] corresponds of hadron–nucleus scattering to the following: the incident hadron scatters elastically on a nucleon N_1 on the front face of the nucleus. At high energies the phase of the amplitude is imaginary. The hadron then propagates through the nucleus to nucleon N_2 where it interacts inelastically. The accumulated phase of the hadron propagator is also imaginary, so that this two–step amplitude is coherent and opposite in phase to the one–step amplitude where the beam hadron interacts directly on N_2 without initial–state interactions. Thus the target nucleon N_2 sees less incoming flux: it is shadowed by elastic interactions on the front face of the nucleus. If the hadron–nucleon cross section is large, then for large A the effective number of nucleons participating in the inelastic interactions is reduced to $\sim A^{2/3}$, the number of surface nucleons.

In the case of virtual photo–absorption, the photon converts to a $q\overline{q}$ pair at a distance before the target proportional to $\omega = x^{-1} = 2p \cdot q/Q^2$ in the laboratory frame.[79] In a physical gauge, such as the light–cone $A^+ = 0$ gauge, the final–state interactions of the quark can be neglected in the Bjorken limit, and effectively only the anti–quark interacts. The nuclear structure function F_2^A producing quark q can then be written as an integral[80,49] over the inelastic cross section $\sigma_{\overline{q}A}(s')$ where s' grows as $1/x$ for fixed space–like anti–quark mass. Similarly, the anti–quark nuclear structure function is related to inelastic quark–nucleus scattering. Thus the A–dependence of the deep inelastic nuclear structure functions cross section reflects the A–dependence of the q and $\overline{q}$ cross sections in the nucleus. Hung Jung Lu and I have recently applied the standard Glauber multi–scattering theory, to $\sigma_{\overline{q}A}$ and σ_{qA} assuming that formalism can be taken over to off–shell interactions.[8] The shadowing mechanism is illustrated in Fig. 10.

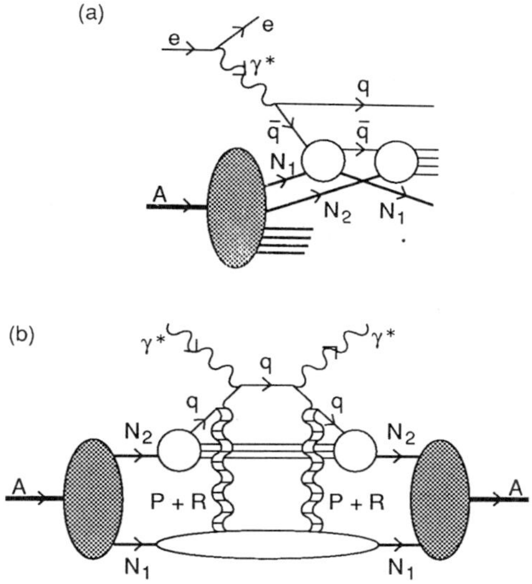

Figure 10. (a) The double–scattering amplitude that shadows the direct interaction of the anti–quark with N_2.
(b) The same mechanism as in (a), drawn in the traditional "hand–bag" form. Pomeron and Reggeon exchange between the quark line and N_1 are explicitly illustrated.

The predictions for the effective number of nucleons $A_{eff}(x)/A$ are shown in Fig. 11 for $A = 12$, 64, and 238. One observes shadowing below $x \simeq 0.1$ and an anti–shadowing peak around $x \simeq 0.15$. The shadowing effects are roughly logarithmic on the mass number A. The magnitude of shadowing predicted by the model is consistent with the data for $x > 0.01$; below this region, one expects higher–twist and vector–meson dominance shadowing to contribute. For $x > 0.2$ other nuclear effects must be taken into account. Most of the parameters used in the model are assigned typical hadronic values. The critical quantity is the effective quark–nucleon cross section σ which controls the magnitude of shadowing effect near $x = 0$: a larger value of σ implies a larger $\overline{q}^* N$ cross section and thus more shadowing. Notice that σ is the effective cross section at zero $\overline{q}$ virtuality, thus the typical value $\langle\sigma\rangle$ entering the calculation is somewhat smaller. The magnitude of anti–shadowing is determined the real–to–imaginary–part ratio of the Reggeon scattering amplitude.

Our semi–quantitative analysis shows that parton multiple–scattering process provides a mechanism for explaining the observed shadowing at low x in the EMC–SLAC data. The existence of anti–shadowing requires the presence of regions where the real part of the $\overline{q} - N$ amplitude dominates over the imaginary part. The constructive interference which gives anti–shadowing in the $x \sim 0.15$ region is due in this model to the phase of the Reggeon $\alpha = 1/2$ term. The phase follows from analyticity and is dictated by the shape of the structure functions at low x. We could utilize additional terms (at lower values of α) to parameterize other bound–state contributions which vanish as higher powers of x, but in practice their qualitative effect would be indistinguishable from the our simplified model. These results show that for reasonable values of the quark– and anti–quark–nucleon cross section, one can understand the magnitude of the shadowing effect at small x. Moreover, if one introduces an $\alpha_R \simeq 1/2$ Reggeon contribution to the $\overline{q}N$ and qN amplitudes, the real phase introduced by such a contribution automatically leads to "anti–shadowing" (effective number of nucleons $F_2^A(x,Q^2)/F_2^N(x,Q^2) > A$) at $x \simeq 0.15$ of the few percent magnitude seen by the SLAC and EMC experiments.[72,73] The analysis also provides the input or starting

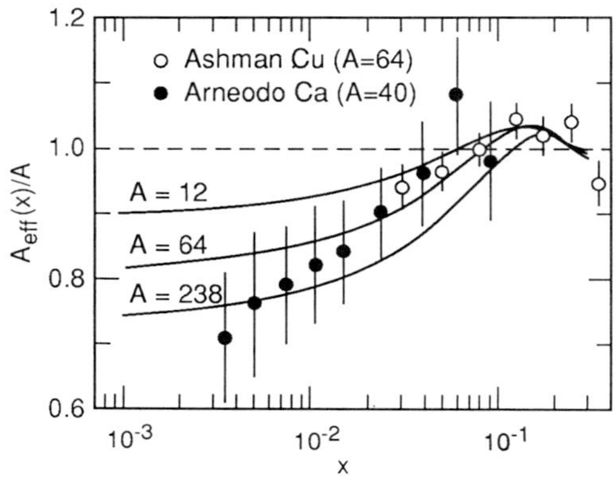

Figure 11. The predicted ratio of $A_{eff}(x)/A$ of the multi-scattering model in the low x region for different nuclear mass number. The data points are results from the EMC experiment for Cu and Ca.

point for the log Q^2 evolution of the deep inelastic structure functions, as given for example by Mueller and Qiu.[81] The parameters for the effective $\overline{q}$–nucleon cross section required to understand shadowing phenomena provide important information on the interactions of quarks in nuclear matter.

The analysis presented here correlates shadowing phenomena to microscopic quark–nucleon parameters. This approach also provides a dynamical and analytic explanation of anti–shadowing, confirming the conjecture of Nikolaev and Zakharov[82] who predicted that such an effect must exist on the basis of conservation laws. Using the perturbative QCD factorization theorem for inclusive reactions, the same analysis can be extended to Drell–Yan and other fusion processes, taking into account the separate dependence on the valence and sea quarks. Thus some shadowing and anti–shadowing should also be observable in the nuclear structure function $F_2^A(x_2, Q^2)$ extracted from massive lepton pair production on nuclear targets at low x_2. However, unlike pion excess models, the non–additive nuclear effect is not restricted to sea quarks.

This microscopic approach to shadowing and anti–shadowing analysis also has implications of the nature of particle production for virtual photo–absorption in nuclei. At high Q^2 and $x > 0.3$, hadron production should be uniform throughout the nucleus. At low x where shadowing occurs, the inelastic reaction occurs mainly at the front surface. These features can be examined in detail by studying non–additive multi–particle correlations in both the target and current fragmentation region. The same types of multi–scattering "fan" diagrams also appear in the analysis of the saturation of the gluon distribution at small x.[83]

7. CONCLUSIONS

In these lectures I have emphasized some of the remarkable features and complexities of the hadron and nuclear wave functions in QCD. By far, the simplest description of the hadron is given in terms of the light-cone Fock expansion, which provides a consistent covariant representation of the hadron wave function in terms of current quarks and gluons. This basis also allows a discrete computer simulation of the complete quantum field theory which faithfully reflects renormalizability and other properties of the continuum theory.

Nevertheless, the hadron wave-function must contains a large number of degrees of freedom in the light-cone basis. Part of the complexity is due to the lack of understanding of the QCD vacuum; it is now clear that the zero modes of the gauge field theory carry topological quantum numbers of the theory. The other complexities of the Fock basis have to do with the nature of confinement and the QCD couplings. The simulation of asymptotic freedom and chiral symmetry effects evidently require higher Fock states with multiple gluon and quark-pair components. The hardness of the QCD amplitudes implies that far-off shell fluctuations of the hadron wave function are present with a probability that only decreases as an inverse power of the virtuality. This implies relatively large probabilities for heavy flavor fluctuations, massive pair fluctuations, and high momentum components, much higher than the exponentially suppressed probabilities expected in a soft theory. The concept of "intrinsic hardness"[2] of the multi-particle amplitudes in hadron and nuclear wave functions leads to an understanding of a many diverse phenomena, such the observation of charm and charmonium at large longitudinal momentum and high-momentum "cumulative" effects in nuclear target experiments. The relatively large cross section for charm production at threshold predicted by intrinsic charm also provides an unexpected explanation for the spin correlation anomaly in elastic proton scattering and an explanation of the decrease of color transparency in quasi-elastic pp scattering at the same energy. The strong effects of coalescence particle interactions at low relative velocity also leads to an understanding of J/ψ suppression in heavy ion collisions and the prediction of nuclear bound-quarkonium.

There a number of important ways in which experiment can resolve and probe fundamental structure, particularly in electroproduction experiments where one can look at the entire final state. The traditional focus of electroproduction experiments has been tests of perturbative QCD predictions for the logarithmic evolution of deep inelastic structure functions due to gluonic radiation from the struck quark. This logarithmic effect is independent of the detailed structure of the proton itself. Tests of the leading twist PQCD evolution have been largely successful; however, at moderate values of momentum transfer there remains important questions and ambiguities concerning the magnitude and origin of higher twist corrections, the behavior of $R = \sigma_L/\sigma_T$, the origin of quark spin correlations, the shape of heavy sea components, and the gluon distribution. The non-singlet Regge behavior of structure functions has not been checked to high precision.

The interest in electroproduction has now turned to the basic problem of understanding the structure of the proton in terms of its quark and gluon degrees of freedom. Elastic and inelastic lepton scattering is still the best "microscope" for probing the fundamental structure of the nucleon. One is interested in testing the implications of non–perturbative QCD for the "intrinsic" multi–particle coherent bound–state structure of the nucleon. Information on the distribution amplitude $\phi_N(x_1, x_2, x_3, Q)$, the covariant wave function describing the correlations of valence quarks in the nucleon bound–state, can be obtained from measurements of form factors and other exclusive channels. Information on the complete multi–particle quark and gluon degrees of freedom requires detailed coincidence measurements of the final state hadrons, especially the target region. Nuclear targets play a valuable role in these measurements by modifying the hadronization and dynamics of the recoil quark jet over distances measured in fermi's, and by filtering out various components of the hadron wave function.

The following are just a few of the ways electroproduction experiments can probe basic QCD phenomena:

- Formation zone physics: study the quantum coherence and the time scales controlling the hadronization of quark jets propagating through a nucleus.
- Collision broadening: study the final state elastic interactions of the recoil quark jet as it propagates through a nuclear target.

- Anomalous spin correlations: study the spin structure functions and helicity effects in the final state, including correlations with strange particles.

- Intrinsic Charm: confirm the anomalous components of the charm structure functions of the proton seen at large x_{Bj}. Study charmonium and open charm production in the target fragmentation region.

- Color Transparency and Color Filter: study nuclear dependence of quasi-elastic reactions to separate short–range perturbative versus long–range non–perturbative phenomena. Remarkably, the cross section for large momentum transfer quasi–elastic scattering such as $ep \to ep$ in a nucleus is predicted in PQCD to be free of final state absorption corrections since the scattering is dominated by fluctuations of the valence–quark wave–functions with small transverse size.

- Photo–and electroproduction of charmonium states: provide constraints on the gluon distribution of the proton in the photon–gluon fusion model; use the nuclear dependence to test gluon shadowing and determine the ψN cross section as a measure of the basic size of quarkonium systems.

- Exclusive charm channels, e.g. $\gamma^* p \to \overline{D}\Lambda_c$: tests predictions of relatively large cross sections for the production of charm baryons near threshold as well as predictions for high momentum intrinsic charm components in the proton wave function.

- Nuclear–bound quarkonium: study electroproduction just below the threshold in $\gamma^* A \to \eta_c A$ reactions to identify nuclear-bound charmonium states such as η_c^3He, novel bound states predicted to exist due to the attractive QCD van der Waals gluonic exchange potential.

- Quark–diquark structure of the proton: study correlations of final state hadrons in the target fragmentation region.

- Exclusive nuclear amplitudes such as $\gamma^* d \to np$: test PQCD "reduced amplitude" predictions.

- Prompt photon emission: study anomalous soft–photon production as a clue to hadronization mechanisms.

- Diffractive electroproduction such as $\gamma^* p \to \rho\ p$ on proton or nuclear targets: probe Pomeron coupling to systems of variable size and measure multi–gluon exchange form factors.

- Diffractive π and η photoproduction: identify and probe the QCD "Odderon" (odd C contribution to high energy scattering from three gluon exchange, etc.) and its coupling.

- Exclusive channels, such as $ep \to eN^*$ at large Q^2: measure the fundamental distribution amplitudes (valence wave functions) of the proton and baryonic resonances; extend meson form factor measurements; test PQCD scaling laws for $\gamma^* p \to MN$ reactions.

- Exclusive Compton scattering at high momentum transfer $\gamma^* p \to \gamma p$: check perturbative QCD predictions.

- Exclusive Compton scattering on nuclei such as $\gamma^* D \to \gamma D$: search for "hidden color" multi–quark resonances predicted by QCD.

- Electron–positron asymmetry in $e^{\pm} p \to e^{\pm}\gamma X$: test fractional charges of quarks; measure new type of valence structure function.

- Cumulative Effect in $\gamma^* A \to HX$: measure the production of fast hadrons in the backward direction well beyond the kinematic limit for a proton target can identify anomalous short–range correlations predicted by PQCD.

- Intrinsic hardness: test PQCD predictions for high transverse momentum pair correlations in proton and nuclear intrinsic momentum wave functions.
- Spin–one structure functions for electroproduction on deuteron target: test PQCD predictions for high spin structure functions requiring multi-quark coherence.
- Shadowing and anti–shadowing: study deviations from uniform nuclear linear A–dependence behavior in inclusive and exclusive channels due to quark–nucleon scattering and long-range coherence.

ACKNOWLEDGEMENTS

Some of the material presented here is based on collaborations with others, particularly G. de Teramond, P. Hoyer, G. P. Lepage, H. J. Lu, A. H. Mueller, and I. Schmidt. I wish to thank Professor A. Zichichi and his staff for their outstanding hospitality and for providing such a stimulating atmosphere for physics at Erice. I also wish to thank my scientific secretaries, Boris Kastening, and Stephen Titard for their assistance in preparing the discussion session. Part of these lectures were also presented at the conference QCD90.

REFERENCES

1. For further background and references to these topics see S. J. Brodsky and G. P. Lepage, in *Quantum Chromodynamics*, edited by A. H. Mueller, (World Scientific, 1990), and S. J. Brodsky, lectures given at *The 27th International School of Subnuclear Physics: The Challenging Questions*, Erice, Italy, 1989.
2. S. J. Brodsky and P. Hoyer, to be published.
3. G. Bertsch, S. J. Brodsky, A. S. Goldhaber, and J. Gunion, Phys. Rev. Lett. 47, 297 (1981).
4. J. P. Ralston and B. Pire, Phys. Rev. Lett. 61, 1823 (1988), University of Kansas preprint 90–0548 (1990).
5. A. H. Mueller, *Proc. XVII Recontre de Moriond* (1982); S. J. Brodsky, *Proc. XIII International Symposium on Multiparticle Dynamics*, Volendam (1982); S. J. Brodsky and A. H. Mueller, Phys. Lett. 206B, 685 (1988). and references therein.
6. S. J. Brodsky and G. de Teramond, Phys. Rev. Lett. 60, 1924 (1988).
7. See S. J. Brodsky and A. H. Mueller, Ref. 5.
8. S. J. Brodsky and H. J. Lu, Phys. Rev. Lett. 64, 1342 (1990).
9. S. J. Brodsky and I. A. Schmidt, Phys. Lett. B234, 144 (1990).
10. For a review of the EMC spin crisis, see M. Karliner, 24th Rencontre de Moriond, Mar 1989; preprint TAUP 1730–89.
11. S. J. Brodsky and I. A. Schmidt, SLAC-PUB-5169, 1990, to be published in Phys. Rev. D.
12. T. Heinzl, S. Krusche, E. Werner, Regensburg University preprint TPR–90–44 (1990).
13. H. C. Pauli and S. J. Brodsky, Phys. Rev. D32, 1993, 2001 (1985); T. Eller, H. C. Pauli and S. J. Brodsky, Phys. Rev. D35, 1493, (1987). A detailed report on the development of DLCQ for gauge theories is in preparation.

14. K. Hornbostel, S. J. Brodsky, and H.-C. Pauli, Phys. Rev. D41, 3814 (1990); K. Hornbostel, SLAC-0333, Ph. D. thesis, unpublished (1988); and M. Burkardt, Nucl. Phys. A504, 762 (1989); and references therein.

15. A. Tang, S. Brodsky, H. C. Pauli, and M. Krautgartner, to be published.

16. Exclusive processes in QCD are reviewed in S. J. Brodsky and G. P. Lepage, Ref. 1.)

17. By definition, quasi-elastic processes are nearly coplanar, integrated over the Fermi motion of the protons in the nucleus. Such processes are nearly exclusive in the sense that no extra hadrons are allowed in the final state.

18. A. S. Carroll, *et al.*, Phys. Rev. Lett. 61, 1698 (1988).

19. B. K. Jennings and G. A. Miller, Phys. Lett. B236, 209 (1990). and University of Washington preprint (1990).

20. G. R. Farrar, H. Liu, L. L. Frankfurt, M. I. Strikman, Phys. Rev. Lett. 61, 686 (1988).

21. For additional discussion, see also, A. B. Kaidalov, presented at the LEAR Workshop (1990).

22. G. R. Court, *et al.*, Phys. Rev. Lett. 57, 507 (1986).

23. For an alternative explanation, see Ref. 4.

24. S. J. Brodsky and P. Hoyer, Phys. Rev. Lett. 63, 1566 (1990); S. J. Brodsky, C. Peterson. and N. Sakai, Phys. Rev. D23, 2745 (1981). E. Hoffmann and R. Moore, Z. Phys. C20, 91 (1990). For a recent estimate of intrinsic charm matrix elements from OZI-violation in baryons, see T. Hatsuda and T. Kunihiro, CERN-TH. 5386 (1990).

25. J. Badier, *et al.*, Z. Phys. C20, 101 (1983); Phys. Lett. 104B, 335 (1981). P. Charpentier, Saclay Thesis CEA-N-237 (1984).

26. D. M. Alde, *et al.*, Phys. Rev. Lett. 64, 2479 (1990), Los Alamos preprint LA-UR-90-2331 (1990), and references therein.

27. G. T. Bodwin, S. J. Brodsky, and G. P. Lepage, Phys. Rev. D39, 3287 (1989).

28. J. Napolitano *et al.*, Phys. Rev. Lett. 61, 2530 (1988).

29. For a recent review on the status of di-baryon hidden color resonances, see E. Lomon, MIT preprint (1990).

30. S. J. Brodsky, G. de Teramond, and I. Schmidt, Phys. Rev. Lett. 64, 1011 (1990).

31. The signal for the production of almost-bound nucleon (or nuclear) charmonium systems near threshold is the isotropic production of the recoil nucleon (or nucleus) at large invariant mass $M_X \simeq M_{\eta_c}$.

32. V. N. Gribov and L. N. Lipatov, Sov. J. Nucl. Phys. 15, 438 and 675 (1972).

33. J. D. Bjorken, Phys. Rev. D1, 1376 (1970).

34. See, for example, S. J. Brodsky, J. Ellis, M. Karliner, Phys. Lett. B206, 309 (1988).

35. R. Blankenbecler, S. J. Brodsky Phys. Rev. D10, 2973 (1974); J. F. Gunion, Phys. Rev. D10, 242 (1974); S. J. Brodsky and J. F. Gunion, Phys. Rev. D19, 1005 (1979).

36. M. B. Einhorn and J. Soffer, Nucl. Phys. B274, 714 (1986).

37. S. J. Brodsky and G. P. Lepage, Phys. Rev. D22, 2157 (1980).

38. J.D. Bjorken, J. Kogut, D. Soper, Phys. Rev. D3, 1382 (1971).

39. Z. F. Ezawa, Nuovo Cim. 23A, 271 (1974).

40. G. P. Lepage, Proceedings, SLAC Summer Institute (1981).

41. S. J. Brodsky and G. P. Lepage, *Proc. of the Int. Symp. on High Energy Physics with Polarized Beams and Polarized Targets*, Lausanne, Switzerland (1980).

42. See, e. g., F. Martin, Phys. Rev. D19, 1382 (1979); M. Glück, E. Reya, and W. Vogelsang, Dortmund University preprint DO–TH–89/3 (1989).

43. P. Aurenche, *et al.*, Phys. Rev. D39, 3275 (1989).

44. Structure functions for meson, baryon, and nuclear bound states in QCD in one-space and one-time dimensions have been computed numerically by diagonalizing the light-cone Hamiltonian. See Ref. 14.

45. See Ref. 37.

46. F. E. Close, *An Introduction to Quarks and Partons*, Academic Press (1979).

47. J. Kuti and V. F. Weisskopf, Phys. Rev. D4, 3418 (1971).

48. The position α_R of J-plane singularities in the forward virtual Compton amplitude are Q^2-independent, and thus the non-singlet Reggeon behavior $F_2^{NS}(x, Q^2) \sim x^{1-\alpha_R}$ at $x \to 0$ must be unaffected by QCD evolution. We thank J. Collins for conversations on this point.

49. S. J. Brodsky, F. E. Close and J. F. Gunion, Phys. Rev. D5, 1384 (1972).

50. W. Weisberger, Phys. Rev. D5, 2600 (1972); S. J. Brodsky, F. E. Close and J. F. Gunion, Phys. Rev. D8, 3678 (1973); S. J. Brodsky and I. Schmidt, unpublished. This result assumes that the valence quark mass only appears in the light-cone kinetic energy $\sum(k_\perp^2 + m^2)/x$.

51. One can identify the ultraviolet divergences at $x \to 0$ for sea quarks with the renormalization of the gluon propagator (which cancels in the proton-neutron mass difference.)

52. M. L. Goldberger and F. E. Low, Phys. Rev. 176, 1778 (1968); see also: T. Erber, Ann. Phys. 6, 319 (1959).

53. Strictly speaking, the bound-valence structure function of the proton refers to a proton in the extended QCD which contains chargeless quark pairs. Since these quarks are spectators in the deep inelastic lepton-proton cross sections, this should be a minor effect.

54. For discussion and references see M. Damashek and F. J. Gilman, Phys. Rev. D1, 319 (1970).

55. We neglect here the $\mathcal{O}(e^6)$ light-by-light scattering contributions to the forward Compton amplitude.

56. To leading order in α we can ignore higher order processes in the total cross section such as Compton scattering.

57. We neglect processes higher order in $\alpha_{\rm QED}$ such as pair production in the electromagnetic field of the quark d.

58. For a derivation of the Reggeon contribution to the leading twist structure functions, see P. V. Landshoff, J. C. Polkinghorne and R. Short, Nucl. Phys. B28, 224 (1971), and S. J. Brodsky, F. E. Close, and J. F. Gunion, Ref. 50.

59. As usual we neglect hadronization effects and other final–state interactions at high energies.

60. A. Kernan, G. VanDalen Phys. Rep. 106, 207 (1984); S. P. K. Tavernier, Rep. Prog. Phys. 50, 1439 (1987); U. Gasparini, *Proc. XXIV Int. Conf. on High Energy Physics*, (R. Kotthaus and J. H. Kühn, Eds., Springer 1989), p. 971.

61. Detailed predictions for the contribution of intrinsic charm to the nucleon charmed quark structure functions and comparisons with existing lepto–production data are given by E. Hoffmann and R. Moore, Ref. 24.

62. M. MacDermott and S. Reucroft, Phys. Lett. 184B, 108 (1987).

63. S. F. Biagi, *et al.*, Z. Phys. C28, 175 (1985). See also P. Coteus, *et al.*. Phys. Rev. Lett. 59, 5030 (1987).

64. S. J. Brodsky, P. Hoyer, C. Peterson, and N. Sakai, Phys. Lett. 93B, 451 (1980); S. J. Brodsky, C. Peterson, and N. Sakai, Ref. 24.

65. S. J. Brodsky, J. F. Gunion, and D. E. Soper, Phys. Rev. D36, 2710 (1987).

66. K. J. Anderson, *et al.*, Phys. Rev. Lett. 42, 944 (1979); A. S. Ito, *et al.*, Phys. Rev. D23, 604 (1981); P. Bordalo, *et al.*, Phys. Lett. 193B, 368 (1987).

67. Yu. M. Antipov, *et al.*, Phys. Lett. 76B, 235 (1978); M. J. Corden, *et al.*, Phys. Lett. 110B, 415 (1982); J. Badier, *et al.*, Ref. 25; S. Katsanevas, *et al.*, Phys. Rev. Lett. 60, 2121 (1988).

68. P. Hoyer, M. Vanttinen, and U. Sukhatme, HU–TFT–90–14, (1990.)

69. S. J. Brodsky and P. Hoyer, Ref. 24.

70. Alternatively, the individual charmed quarks can fragment into final state charmed hadrons either by hadronization or by coalescing with co–moving light quark spectators from the beam. This effect can explain the depletion of the ratio of J/ψ to continuum muon pairs in high transverse energy nucleus–nucleus collisions. See S J. Brodsky and A. H. Mueller, Ref 5.

71. S. J. Brodsky, H. E. Haber, and J. F. Gunion, in *Anti–pp Options for the Supercollider,* Division of Particles and Fields Workshop, Chicago, IL, 1984, edited by J. E. Pilcher and A. R. White (SSC–ANL Report No. 84/01/13, Argonne, IL, 1984), p. 100. S. J. Brodsky, J. C. Collins, S. D. Ellis, J. F. Gunion, and A. H. Mueller, published in Snowmass Summer Study 1984, p. 227.

72. J. Ashman *et al.*, Phys. Lett. 202B, 603 (1988) and CERN–EP/88–06 (1988) M. Arneodo *et al.*, Phys. Lett. 211B, 493 (1988)

73. R. G. Arnold *et al.*Phys. Rev. Lett. 52, 727 (1984) and SLAC–PUB–3257 (1983)

74. J. S. Bell, Phys. Rev. Lett. 13, 57 (1964)

75. L. Stodolsky, Phys. Rev. Lett. 18, 135 (1967)

76. S. J. Brodsky and J. Pumplin, Phys. Rev. 182, 1794 (1969)

77. J. J. Sakurai and D. Schildknecht, Phys. Lett. 40B, 121 (1972); Phys. Lett. 41B, 489 (1972); Phys. Lett. 42B, 216 (1972)

78. R. J. Glauber, in *Lectures in Theoretical Physics*, edited by W. E. Brittin *et al.* (Interscience, New York, 1959), vol. I

79. T. H. Bauer, R. D. Spital, D. R. Yennie and F. M. Pipkin, Rev. Mod. Phys. 50, 261 (1978)

80. P. V. Landshoff, J. C. Polkinghorne and R. D. Short, Nucl. Phys. B28, 225 (1971)

81. A. H. Mueller and J. Qiu, Nucl. Phys. B268, 427 (1986)
J. Qiu, Nucl. Phys. B291, 746 (1987)

82. N.N. Nikolaev and V. I. Zakharov, MPI–PAE–PTh–11/90, (1990); Physics Letters 55B, 397 (1975); Sov. J. Nucl. Phys. 21, 227 (1975).

83. A.H. Mueller, CU–TP–475, DESY Lectures given at the Small–x Workshop, DESY, May (1990).

Lecturer: S. J. Brodsky

Scientific Secretaries: Boris Kastening, Stephan Titard

Discussion

– *Grossman:*

You mentioned some results of Discretized Light Cone Quantization (DLCQ) for a (1+1)-dimensional U(1) theory where the number of degrees of freedom is rather small. Could you give us an idea of the complexity of the problem for 4-dimensional SU(3); e.g. what is the dimension of the Fock space one most probably has to consider in order to get reliable results?

– *Brodsky:*

As I discussed in my lectures, DLCQ quantization has permitted the efficient numerical diagonalization of the QCD Hamiltonian for 2, 3 and 4 colors in one space dimension at fixed light-cone time. The (1+1) theory is certainly much less complicated than the (3+1) theory since in the former there are no transverse dimensions and no dynamical gluons in the Fock space using the physical light-cone gauge. DLCQ has now been successfully applied to Abelian QED in physical space-time; the results are encouraging: in minimal computer time one obtains the fine-structure and hyperfine splitting of the low-lying Bohr spectrum to a few percent accuracy for positronium with $\alpha = 0.3$. In these initial studies one already obtains a relativistic Coulomb spectrum just by considering the Fock states $|e^+e^- >$ and $|e^+e^-\gamma >$. To see the Lamb shift and retardation effects one needs to consider the $|e^+e^-\gamma\gamma >$ Fock states. In the QCD(3+1) case, one will also need to incorporate Fock states with at least four particles to reproduce running coupling constant effects even in the analysis of the simplest case, heavy quarkonium. Studies are now proceeding to examine the convergence of the spectrum as a function of longitudinal and transverse discretization. Recent results for the spectrum of positronium suggest that for moderate α one can get reasonable results for the lowest levels allowing for a total of 100 or more points of discretization for the longitudinal momentum fraction and transverse dimensions. In the QCD case one has to allow for different color singlet representations as well as the helicity degrees of freedom, so the dimension of the matrix representation increases rapidly with the number of particles appearing in the Fock state. It is important to pre-diagonalize the commuting operators and use irreducible representations of the color group to minimize the matrix dimension. The Hamiltonian matrix is very sparse due to its conservation rules, so that one can apply Lanczos and other methods. This is being studied by Kaluza. Burkardt, Pauli, and Krautgartner have also found it advantageous to convert the light-cone Hamiltonian problem to an effective integral equation in the lowest Fock state which can then be solved numerically using an optimal discretization. Tang has also utilized variational methods. In addition, it is crucial to use a special method to allow a discrete integration over the Coulomb singularity. We are now preparing a detailed report on these methods.

– *Titard:*

In order to compute an improved valence quark distribution function F_2^{BV}, you introduced a new set of quarks and the resulting composite p_0. What is p_0?

– *Brodsky:*

The construction of the bound-valence distribution is based on a "gedanken" experiment. The idea is to isolate the proton's valence quark contributions to the lepton-proton inelastic cross section by subtracting off the cross section obtained as if the valence quarks had no electroweak charge. In this imaginary experiment one obtains the bound-valence quark structure function by first measuring the deep inelastic scattering cross section on a proton target and then subtracting the result for a target of charge-less protons p_0. The null proton p_0 contains valence null quarks u_0, u_0, d_0 having no electroweak charge, only color, plus a sea of gluons, $q\overline{q}$, and $q_0\overline{q}_0$ pairs. The null quarks are identical to the ordinary quarks except for their zero electroweak coupling. Remarkably, as Schmidt and I have shown, the difference of cross sections on p and p_0 and the resulting structure function F_2^{BV} is devoid of leading Regge behavior, as required for sensible distributions computed from the bound-state wave function. Thus the singular Kuti-Weisskopf non-singlet Reggeon contributions are associated with "extrinsic" contributions to the structure functions rather than "intrinsic" multi-particle bound-state wavefunction contributions.

– *Titard:*

What is the present status of factorization theorems?

– *Brodsky:*

There are several categories of factorization theorems for gauge theories so I will try to be brief. In the case of inclusive reactions, the general theorem for the factorization of process-independent hadron (or nuclear) structure functions from the hard subprocess cross section has been proven in QCD in the case of the Drell-Yan and related reactions by Bodwin and by Collins, Soper, and Sterman. Induced radiation and other initial– or final–state interaction corrections are the main complications that make the proof of QCD factorization so difficult for hadron-hadron collisions. As shown by Bodwin, Lepage, and myself, such corrections vanish in the leading twist cross section once the Drell-Yan cross-section is integrated over the pair transverse momentum, provided that the "formation length" condition is satisfied. To my knowledge the same rigorous analysis has not been systematically extended to other inclusive processes, but an outline of a proof has been given by Collins et al. for reactions such as the fusion process contributions to heavy quark production. Recently, Sterman and Qiu have extended the factorization theorems to allow for the first non-trivial higher twist contributions, including the $1/Q^2$ contributions to structure functions. As shown by Berger and myself, such contributions can dominate over the leading twist contributions at $x \sim 1$. An important example is the longitudinal polarization contribution to lepton pair production in meson-induced reactions. Outlines of proofs have also been given to extend QCD factorization to large transverse momentum inclusive cross sections based on $2 \rightarrow 3$ particle subprocesses. This is discussed by Catani et al. and Collins et al. There has also been extensive work proving factorization theorems for exclusive processes. For example, Lepage and I have demonstrated factorization of the universal distribution amplitude $\phi_H(x_i, Q)$ from the hard scattering quark-gluon scattering amplitude T_H to all orders in perturbation theory for exclusive large momentum transfer reactions such as $\gamma\gamma \rightarrow M\overline{M}$ and meson form factors.

– *Kastening:*

You have told us that experiments at HERA examine the structure of quarks and gluons rather than probe the structure functions of the proton. In this view what significance have the results of HERA experiments for hadron-hadron colliders?

– *Brodsky:*

At the very large momentum transfers and very small x probed by HERA, experiment is not very sensitive to intrinsic proton structure – the main effects are associated with gluon and quark evolution. Nevertheless, given QCD factorization, one needs the structure functions obtained at HERA over its huge range of x and Q^2 as input structure functions to accurately predict hard inclusive cross sections at proton-proton and proton–antiproton colliders. However, when the normalization of the gluon distribution becomes very large, then the leading twist approximation breaks down, and one inevitably needs higher-twist multi-scattering diagrams just to restore unitarity. It seems to me that there are even more complicated classes of such multi-particle processes contributing in hadron-hadron collisions involving multi-gluon exchange between partons of the two incident beams and the particles of the hard subprocesses, making it unlikely that one can obtain a true factorization theorem relating these higher-twist contributions to the HERA data. This is probably only a serious problem at small x where higher twist contributions enter to restore unitarity, but the general issues need to be examined carefully. A review of this topic can be found in my summary talk at the recent DESY workshop on low x physics.

– *Kastening:*

You told us that for the calculations you are doing, lattice calculations are worse than analytical ones. Is that a general trend or is it possible that, by using new types of computers like the connection machines, one could do better than with ordinary machines?

– *Brodsky:*

Certainly analytic results are almost always more useful than numerical results such as those obtained using lattice gauge theory or light-cone Hamiltonian methods. The DLCQ method, however, does seem to have advantages over traditional lattice methods. For example, in the case of QCD in (1+1) dimensions, the entire meson, baryon, and even nuclear spectrum and wavefunctions can be computed to high accuracy. The only high precision lattice result that has been published is that due to C. Hamer. In Hamer's Hamiltonian lattice calculations for QCD(1+1), one obtains only the first few low-lying states with reasonable accuracy. For QCD(3+1), it is too early to decide which method will ultimately work better, but both are highly computer intensive. As you suggest, specific computer architectures could be of benefit. In any case, the main problem is to identify the most effective degrees of freedom of QCD. In the case of DLCQ, dynamical fermions are actually easier to include than dynamical bosons. DLCQ provides a discrete, covariant, and renormalizable representation of the quantum field theory, yielding spectra and light-cone wave functions directly connected to phenomenology; e.g. the absolute squares of the $\psi(x_i, k_{Ti}, \lambda_i)$ yield the deep inelastic structure functions. In traditional lattice gauge theory it is difficult to extract information on the structure functions or distribution amplitudes beyond lowest moments. On the other hand there are questions involving phase structure and high temperature effects which are naturally analyzed using stochastic methods such as lattice gauge theory.

– *Lu:*

Is it correct that one cannot separate the bound-valence from the non-valence part of quark structure functions? Can we obtain an asymmetric sea distribution?

– *Brodsky:*

Schmidt and I could not think of a practical way to obtain the bound-valence distributions directly from experiment. However, our analysis shows that the bound-valence structure function differs by a factor of x times a computable constant for every term in the Regge expansion of the traditional valence structure function. This analytic method could be used to construct the bound-valence structure functions from the data. In principle one can measure any asymmetry of the sea quarks and anti-quarks. For example, in deep inelastic lepton scattering the exchanged photon does not distinguish the charm quark from the anti-charm quark. However, by studying fragmentation of the recoil jet one could identify whether it represents a c or $\overline{c}$ jet. In addition, charged current experiments can distinguish the strange and anti-strange distributions. Physically, one does expect an asymmetry between s and $\overline{s}$ distributions in the proton since they participate differently in the nucleon's lowest lying kaon-hyperon pair fluctuations. The same fluctuations can produce a spin correlation between the strange quark and proton helicities. In addition, there can be a substantial difference between the distribution of $\overline{u}$ and $\overline{d}$ quarks in the proton since they experience different Pauli-blocking in the $|uudq\overline{q}>$ Fock state.

– *Lu:*

You have indicated that nucleons have an intrinsic gluon content. Can you explain what are the problems if one assumes just three valence quarks for an initial distribution and then does evolution afterwards?

– *Brodsky:*

The idea that one can create the proton's gluon distribution entirely through perturbative QCD evolution goes back a number of years to papers by F. Martin and others. In particular, Glück and Reya constructed a detailed phenomenology based on this ansatz, but they found it was not completely compatible with the available data. These authors have recently advocated an intrinsic gluon distribution which is present at the very lowest resolution scales; however, unlike the model presented by Schmidt and myself, the GR model assumes that the starting gluon and quark distributions have identical shapes in x. Better data will be required to distinguish the various predictions. It should be emphasized that the gluons obtained by cutting the gluon-exchange potential in the proton's equation of motion are not given by QCD evolution; the logarithmic evolution corrections are associated with a single quark line.

– *Brown:*

What is the status of your method of calculating QCD effects in exclusive processes, which Isgur and Llewellyn Smith have called into question possibly because of problem with soft gluons. I am thinking particularly of the pion form factor where they have claimed that unexpected wave functions are needed to get agreement with experiments?

– *Brodsky:*

The work of Isgur and Llewellyn Smith has been helpful for understanding the region of validity of perturbative QCD predictions for large momentum transfer exclusive processes. These authors do not question the validity of the large Q^2 formulae; in fact, in their paper they re-derive the earlier results. They do question what is the minimum momentum transfer Q^2 required to test the predictions, particularly if one uses the non-perturbative hadron distribution amplitudes obtained by Chernyak et al. using QCD sum-rule constraints. In that case the integration over the hard scattering amplitude gives its largest contributions when the gluon momentum transfer is only a fraction of the overall momentum transfer; however, in this region one may need to consider non-perturbative corrections such as an effective dynamical gluon mass in the gluon propagator. Isgur and Llewellyn Smith thus advocate restricting the integration region of the PQCD expressions such that the exchanged gluon is off-shell by at least 1 GeV^2. I prefer to look at the problem from a different direction. Phenomenologically, the leading twist power-law behavior predicted by PQCD and dimensional counting laws work remarkably well, including the predicted onset of color transparency. There is no indication that the dominance of leading twist contributions is postponed to very high momentum transfer. There are a number of reason why 'precocious scaling" could be valid: First, the regime of validity of the perturbative form for the gluon propagator attaching to an off-shell quark line may well extend down to low momentum transfers, $k^2 \sim \Lambda^2_{\overline{MS}} \sim 0.04\ GeV^2$; this type of scaling also is assumed for the computation of higher loop and evolution corrections to inclusive processes. Second, since one expects significant K-factor corrections to the normalization of exclusive processes, the QCD sum-rule wavefunctions are not required to reproduce the normalization of the observed meson and baryon form-factors. It is likely that the actual non-perturbative wavefunctions fall-off rapidly in the end-point integration regions. Thus it would be much more natural to apply sum-rule moment constraints to an exponentially suppressed basis, rather than the polynomial basis used up to now. (This point has recently been raised by Adam Szczepaniak.) Thus one could well have compatibility between existing exclusive scattering phenomenology and the magnitude and scaling of leading twist perturbative QCD predictions.

– *Lu:*

In exotic atoms one can understand that the muon feels an average potential coming from the different nucleons in the nucleus. But in the case of η_c moving inside ^3He, the c and $\bar{c}$ can interact individually with the other quarks in the ^3He nucleus. Why is the correlation between the c and $\bar{c}$ supposed to be stronger than the correlation between c (or $\bar{c}$) with the other quarks?

– *Brodsky:*

The η_c is the lowest mass charmonium state so there is a natural preference for the c and $\bar{c}$ to form that long-lived state over more massive systems such as a charmed nucleus – charmed meson combination.

– *Hsu:*

Could you tell us more about the Van der Waals force in QCD and whether it could be studied in systems other than η_c^3He?

– *Brodsky:*

The QCD van der Waals potential has a short range, characteristic of the lightest inverse gluonium mass, so that it is normally difficult to separate its effect from ordinary neutral iso-singlet meson exchange. Effects due to pure gluonic exchange will thus be incorporated into the meson exchange potential of traditional low-energy nuclear physics. However, nuclear-bound quarkonium ($\eta_c A$) would provide an ideal system for studying this pure gluonic potential since the only interaction between the η_c and the rest of the nucleus is the QCD Van der Waals force. For example, the measurement of the momentum spectrum of the recoil nucleus in the formation experiment automatically yields the system's Schroindinger wavefunction in momentum space. One can then compute the η_c–nuclear potential directly from the Schroindinger equation.

– *Hsu:*

I did not quite follow your derivation of the form of the potential. It ends up looking like a massive particle being exchanged, essentially like a Yukawa interaction. Why is it also called the Van der Waals interactions?

– *Brodsky:*

The attractive inverse power-law potential due to two-photon exchange between neutral molecules or atoms in QED is called the Van der Waals potential. In QCD, the analogous interaction is the multi-gluon exchange potential which couples the color dipole moments of (color-singlet) hadrons. From the standpoint of t-channel physics and dispersion theory, the effective potential has a short range controlled by the lightest mass of the gluonium system. The Yukawa potential which de Teramond, Schmidt, and I used is just a first approximation to the actual form which would be obtained by summing over all gluonic exchanges.

– *Rizvi:*

You talked about the intrinsic contribution of s,c, and b to the structure function of the proton. What about the top quark? Can we see the effects of the top, say at HERA energies? If yes, can we set some limit on the mass of the top?

– *Brodsky:*

The idea that pairs of extremely massive objects can appear as virtual fluctuations in the wavefunction of a normal hadron requires an extrapolation of QCD to a new domain of coherent phenomena. However QCD is a renormalizable theory, so such intrinsic fluctuations are suppressed only by an inverse power of the mass. Thus the probability for finding heavy quark pairs such as $t\bar{t}$ (created by fusion processes coupled to valence quarks) in the nucleon falls off only as $1/m_Q^2$ no matter how high the quark mass. Thus, in principle, one could observe "intrinsic truth" in the proton's structure function in deep inelastic ep scattering, provided one is sufficiently above threshold, although the $1/m_t^2$ suppression makes this measurement a formidable challenge. In any high energy hadronic collisions one can search at large x_f for leading heavy hadrons such as Λ_t. The heavy quark production cross section is predicted to decrease as the intrinsic quark probability times the transverse size of the partons in the hadron wavefunction coupling to the intrinsic quarks. It is crucial to confirm the intrinsic beauty predictions. Recall that the BCF group has reported the observation of leading Λ_b baryons in pp collisions at the ISR. In addition, the anomalous nuclear dependence of the J/ψ, ψ', and Υ production in pA collisions recently reported by the E772 group at Fermilab is consistent with the existence of intrinsic charm and intrinsic beauty in the proton.

– *Rizvi:*

This is a philosophical question. You have brought out a bewildering complexity in proton structure. Is this complexity inherent in nature, or are we "making epicycles"? Do you think QCD is an adequate theory?

– *Brodsky:*

The structure of hadrons in QCD is remarkably complex, and has become a science all of its own. Part of the complexity we see in the proton is simply due to the existence of multiple quark flavors, leading to a progression of phenomena as we probe to higher energy and finer resolution. QCD has mainly been tested in the processes which are essentially extrinsic to the structure of the proton. We have much more to do to unravel hadron structure itself and at every stage challenge non-perturbative QCD. Thus far QCD has successfully accounted for most experimental phenomena, but there are challenges such as the 4:1 spin correlation in pp elastic scattering reported by Krisch's group, the breakdown of color transparency at the same energy, the EMC "spin crisis", and the anomalies observed in open and hidden charm production. All of these incompatibilities with the simplest expectations of QCD show that there is much more to understand and interpret within the conventional theory.

– *Dokshitzer:*

I would like to make a short comment on Rizvi's first question. If the top quark is heavier than the W, it will never be possible to see an intrinsic contribution of the top through toponium production. The reason is that it will decay before strongly interacting.

– *Brodsky:*

I agree; however, there are interesting gluonic effects which modify heavy quark pair production at threshold. This has been studied by Peskin and others.

– *Pittori:*

In looking for anomalous heavy quark effects, what can be done through lattice simulations?

– *Brodsky:*

Let me answer in two parts. First, it is important to compute reliable forms and parameters for the QCD Van der Waals potential in order to accurately predict the spectra of nuclear-bound quarkonium. In the case of lattice gauge theory, one should be able to consider two static heavy quark mesons and study the potential as a function of their color dipole moments, spin, and separation. Second, it is also important to compute the intrinsic heavy quark content of the proton wave function. This may be a formidable problem for lattice gauge theory due to the difficulties of including dynamical fermions and the fact that it is difficult to handle phenomena involving widely different mass scales. An analogous problem, which may be simpler to handle, would be to compute the distribution of intrinsic heavy spin-zero quarks in a gluonic or hadronic bound state. Hornbostel has studied the heavy quark content of light hadrons in QCD(1+1) using DLCQ.

– *Hsu:*

It seems there is at least one important implication of the top inside the proton. If you look at the Weinberg operator, you get a large contribution from the virtual top which runs in the loop of the three-gluon operator. At the level of effective field theory where you integrate out the top, these operators are clearly left over.

– *Brodsky:*

You are correct. It should be noted that the gluonic operators and hadronic matrix elements which appear in the computation of the heavy quark contributions to the electric dipole moment of the neutron involve the same wave functions and scaling laws that control the amplitude for intrinsic heavy quark fluctuations in the proton. I should also emphasize that detailed knowledge of hadronic wavefunctions is required in order to compute the hadronic matrix elements controlling weak decays and parity violation in hadronic amplitudes. Recently Henley, Szczepaniak, and I have used the QCD analysis of exclusive processes to calculate B decays into light mesons in terms of hadron distribution amplitudes. Thus there is an important interdependence between QCD and weak interaction phenomena.

THE MAIN ACHIEVEMENTS OF THE LAA PROJECT

A. Zichichi

CERN, Geneva

Switzerland

I. INTRODUCTION

I.1. General philosophy of the LAA Project

The LAA Project [1, 2, 3, 4, 5, 6] is an intensive programme to develop new High Energy Physics experimental techniques. The ultimate goal is to prove, on the basis of prototypes, the feasibility of essential components for a detector to operate in a future multi-TeV Hadron Collider.

The LAA Project is open to all physicists and engineers who are interested in participating.

This project represents a unique opportunity for Europe to have a leading role in the advanced technology for High Energy Physics.

I.2. The basic data

Let me remind briefly the origin of the LAA Project and its basic features. As already explained in Ref. 5 and 6, at present no one knows how to build an (e^+e^-) collider even at 1 TeV, while the conceptual design of a collider for (pp) at 100 TeV already exists (the ELN Project [7]). On the contrary, no one knows how to build detectors for a 100 TeV (pp) collider, while a detector for a 1 TeV (e^+e^-) machine is within the present-day technological possibilities. This is illustrated in Fig. I.2.1.

Physics Up to 200 TeV
Edited by A. Zichichi, Plenum Press, New York, 1991

LAA

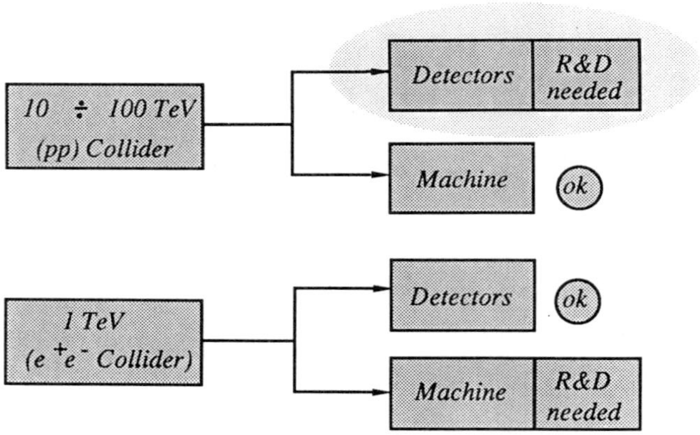

Fig. 1.2.1. The present status of Colliders and Detectors.

One of the main problems for detectors comes from the requirement that the average number of events per bunch crossing, $< n >$, must be one if the missing energy is to be used as a signature for rare events. Another limiting factor towards very high luminosities is the radiation dose the detectors can withstand.

The following machine parameters:

$$\Delta t_b \sim 100 \; ns, \text{ and}$$

$$L_{pp} \sim 10^{32} \; cm^{-2} s^{-1},$$

are well within reach from a technological point of view; on the other hand, the total (pp) cross-section is expected to be:

$$\sigma_{pp} \cong 100 \; mb = 10^{-25} cm^2.$$

These three values together produce $< n > \sim 1$. But what is wanted is $< n >= 1$ at higher luminosities. And therefore the radiation levels, the minimum bias event rate and the Δ_{tb} reach prohibitive figures:

$$\begin{array}{rccccc} L = 10^{33} & \rightarrow & 10^{34} & \rightarrow & 10^{35} cm^{-2} s^{-1}, \\ \downarrow & & \downarrow & & \downarrow \\ \Delta t_b = 10 & \rightarrow & 1 & \rightarrow & 0.1 \; ns. \end{array}$$

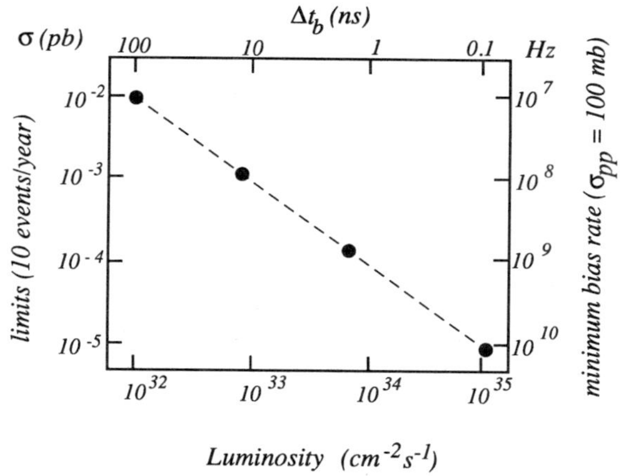

Fig. 1.2.2. Discovery limit for rare events, total rate and bunch spacing (for $< n >= 1$) as a function of the Luminosity.

In conclusion, the main items where R&D is needed in order to cope with the challenging experimental environment of the next generation of (pp) colliders are:

1. RADIATION HARDNESS,
2. HERMETICITY,
3. RATE CAPABILITY,
4. TRACK & SPACE RESOLUTION,
5. ENERGY RESOLUTION,
6. MOMENTUM RESOLUTION,
7. TIME RESOLUTION,
8. PARTICLE IDENTIFICATION,

and these are the basic parts of the LAA Project.

It is very important to notice that, up to now, the first three items were not of such a great relevance.

The potentials for new discoveries in a multi-TeV (pp) collider are summarized in Fig. I.2.2. The "magic" limit of $10^{-40}cm^2$ in the cross-section limit can be reached only if the luminosity is pushed to high levels: $L = 10^{33} - 10^{34}cm^{-2}s^{-1}$ or even higher.

Figure I.2.3 shows the correlation between Luminosity, bunch spacing, and radiation dose under the $< n >= 1$ condition.

Moreover, (pp) interactions at 100 TeV are expected to produce hundreds of particles per event. All these particles must be observed, measured (in terms of energy and/or momentum), and, possibly, identified with the best achievable precision.

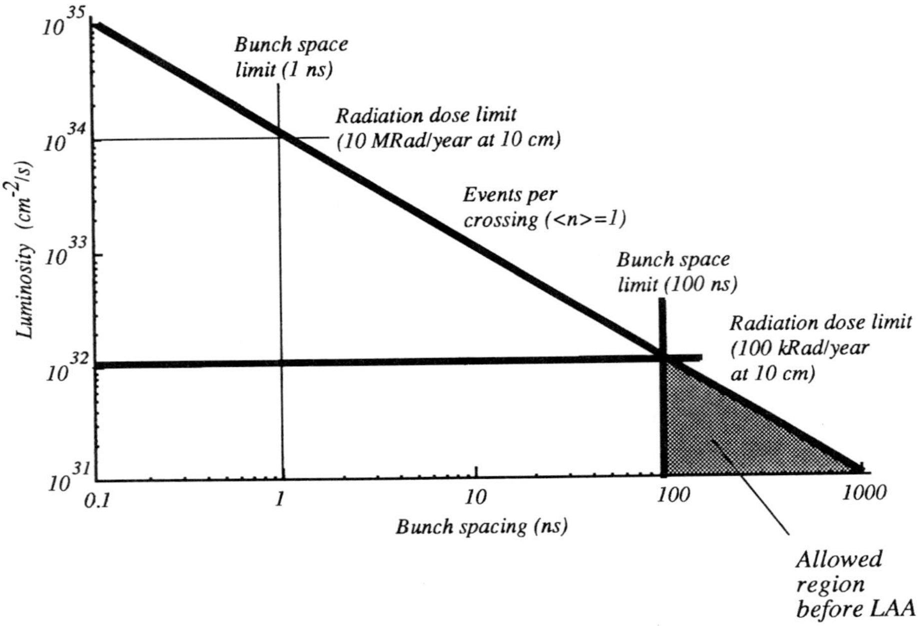

Fig. 1.2.3. Limits on Luminosity as function of bunch spacing. Also shown is the radiation dose level at different Luminosities. The grey region was allowed before LAA.

I.3. The Physics at the new generation of Hadron Colliders

What Physics are we facing with the new generation of Hadron Colliders? There are four basic theoretical problems to be solved in the multi-TeV ($10 \div 200$ TeV) energy domain:

i) The **family** problem.
ii) The **hierarchy** problem.
iii) The **proliferation** problem.
iv) The **compositeness** problem.

The questions which arise when translating these theoretical problems into physically observable phenomena are the following:

i) Do **new, heavier, quarks** and **leptons** exist?
ii) Are there **other intermediate vector bosons?**
iii) Do **Higgs bosons** exist?
iv) Do **supersymmetric partners** exist?
v) Are **quarks** and **leptons composite?**
vi) Would some **unexpected exotic process** occur?

In order to go from theory to "down-to-earth" physics, all detectors operating in the multi-TeV domain should detect and measure:

i) **electrons** and **photons**;
ii) **muons** and other **lepton-like long-lived particles**;
iii) **neutrinos** and other **non-interacting particles** (i.e. missing objects);
iv) **leading protons** (for hermeticity and new physics);
v) **hadrons** and **jets**, with and without leptons inside.

I.4. Present structure of the LAA project

The LAA project consists of eleven main parts (Fig. I.4.1):

1. HIGH PRECISION TRACKING

Three components:

a) Gaseous detectors.
b) Scintillating fibres.
c) Microstrip GaAs.

2. CALORIMETRY

Three components:

a) High precision electromagnetic.
b) Compact EM+Hadronic.
c) "Perfect" Calorimetry.

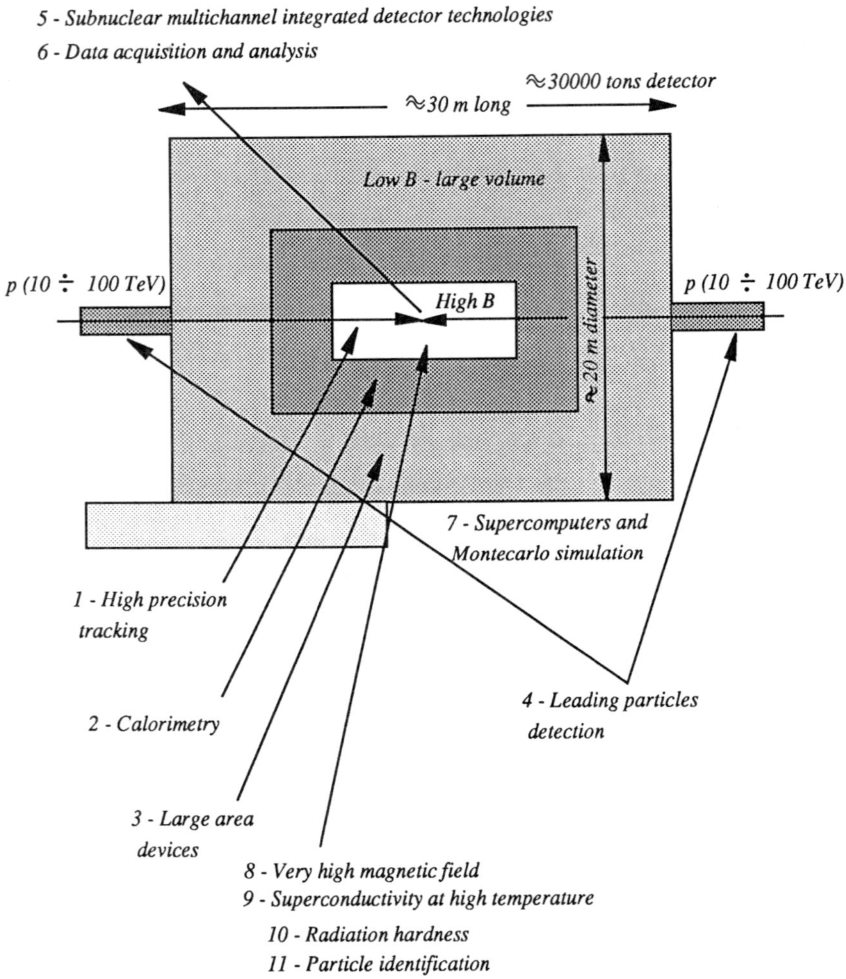

Fig. I.4.1. The eleven components of LAA.

3. LARGE AREA DEVICES

Two parts:

a) Construction .

b) Alignment.

4. LEADING PARTICLE DETECTION

5. SMIDT (Subnuclear Multichannel Integrated Detector Technologies)

Two parts:

a) Microelectronics.

b) New, Radiation-resistant Technologies.

6. DATA ACQUISITION AND ANALYSIS

Three parts:

a) Real Time Data Acquisition.

b) FASTBUS RISC computer.

c) Fine-grained Parallel Processor.

7. SUPERCOMPUTERS AND MONTECARLO SIMULATIONS

8. VERY HIGH MAGNETIC FIELDS

9. SUPERCONDUCTIVITY AT HIGH TEMPERATURE

10. RADIATION HARDNESS

11. PARTICLE IDENTIFICATION

I.5 Participants

The following Physicists, Engineers and Technicians contribute to the activity of the LAA Project:

D. Acosta[39)], J. Alberty[48)], J. Alsford[38)], C. Alvisi[2)], G. Ambrosi[37)], F. Anghinolfi[7)], F. Anselmo[7)], G. Anzivino[20)], M. Arneodo[44)], R. Arnold[12)], F. Arzarello[2)], P. Aspell[7)], R. Ayad[48)], L.E. Barberio[48)], G. Bari[2)], T. Barillari[5)], M. Basile[2,3)], R. Battiston[37)], S.P. Beaumont[21)], U. Becker[29)], L. Bellagamba[2)], M. Bénot[7)], P. Benvenuto[20)], J. Berbiers[7)], J. Berdugo[7)], F. Bergsma[7)], R. Bertin[7)], N. Bingefors[46)], D. Bisello[33)], R.K. Bock[7)], A. Bonanno[5)], D. Boscherini[2)], M. Bosteels[7)], R. Bouclier[7)], M. Bramhall[38)], G. Bruni[2)], P. Bruni[2)], S. Buontempo[7)], V. Buzuloiu[48)], L. Calôba[11)], M. Camarca[5)], M. Campbell[7)], G. Cara Romeo[2)], M. Caria[4)], R. Casaccia[20)],

H. Castro[48], S. Ceresara[17], G. Charpak[7], E. Chesi[7], M. Chiarini[48],
J. Christiansen[7], E. Christofel[12], L. Cifarelli[2,31], F. Cindolo[20],
F. Ciralli[48], E. Colavita[5], F. Coninckx[7], A. Contin[7], M. Costa[48],
F. Crea[5], I. Crotty[7], G. D'Ali[2,34], C. D'Ambrosio[7], S. D'Auria[2],
M. Dardo[45], C. Del Papa[2,3], G. Della Gatta[13], S. De Pasquale[2],
R. De Salvo[7], P. Destruel[43], J. deWitt[40], O. Di Rosa[7], D. Dorfan[40],
E. Duchovni[47], J. Dupont[7], J. Dupraz[7], J. Egger[49], T. Ekelöf[46],
C.C. Enz[42], A. Ereditato[31], J.P. Fabre[7], R. Ferrari[36], F. Fiori[2,3], P. Ford[7],
F. Frasconi[2], M. Fraternali[36], M. French[38], M. Fuchs[7], G. Fumagalli[35,36],
K. Gabathuler[49], J. Galvez[48], R. Garré[17], J. Gaudaen[7], Y. Giomataris[7],
J.P. Girod[7], P. Giusti[2], K. Goebel[7], C. Grinnel[7], H. Güsten[24],
J.L. Guyonnet[12], T. Gys[7], F.G. Hartjes[32], W. Hao[48], D. Hazifotiadu[48],
E. Heijne[7], T. Henkes[27], A.M. Henriques[7], M. Hourican[7], G. Iacobucci[2],
G. Iuvino[36], P. Jarron[7], P. Jenni[7], J.P. Jobez[9], C. Joram[6],W. Kluge[6],
W. Krisher[7], F. Krummenacher[16], A. Kuzucu[48], I. Laakso[20],J.C. Labbé[7],
G. La Commare[7], H. Larsen[7], G. Laurenti[7], T.D. Lee[10], M. Letheren[7],
H. Leutz[7], G. Levi[37], L. Levinson[47], Q. Lin[48], L. Linssen[7], B. Lisowski[2],
A. Litke[40], M. Livan[36], C. Ljuslin[7], J.G. Lynch[21], G. Maccarrone[2],
D.McIntyre[21], A. Maio[25], L. Mapelli[7], A. Marchioro[7], A. Margotti[2,3],
M. Marino[48], T. Marsico[5], T. Massam[2], T. Matsuda[48], T. Matsuura[14],
D. Mattern[7], G. Meddeler[7], K.H. Meier[7], R. Meng[14], G. Mikenberg[47],
M.R. Mondardini[23], G. Mörk[46], B. Musso[1], R. Nania[2], C. Nemoz[7],
S. Newett[38], B. Ong[39], V. O'Shea[2], N. Ozdes[48], H.P. Paar[39], F. Palmonari[3],
G. Passardi[7], F. Pastore[36], P. Pelfer[18,19], M. Pereira[25], C. Peroni[44,45],
E. Perotto[8], V. Peskov[48], D. Piedigrossi[7], R. Pilastrini[2], D. Pitzl[40],
L. Poggioli[26], G. Polesello[7], S. Qian[20], A. Racz[7], C. Raine[21], F. Rivera[48],
F. Riccardi[30,31], A. Rimoldi[35,36], T. Ruf[6], H. Sadrozinski[40], A. Sandoval[7],
G. Sannier[7], J.C. Santiard[7], G. Sartorelli[2,3], P. Sartori[33], F. Sauli[7],
C. Scheel[7,32], M. Schioppa[48], J. Schipper[7], H. Schönbacher[7], D. Scigocki[7],
M. Scioni[2,3], J. Seguinot[9], A. Seiden[40], W. Seidl[7], J.M. Seixas[48],
A. Sharma[48], P. Sharp[38], J. Schmitz[32], I.O. Skillicorn[21], A. Simon[27],
G. Simonet[7], M. Sivertz[39], K. Smith[21], P. Sonderegger[7], M.N. Souza[11],
E. Spencer[40], L. Sportelli[5], A. Staiano[44], G.C. Susinno[5,20], S. Tailhardat[7],
M. Taufer[7], N. Tartoni[18,19], M. Tavlet[7], A.E. Terraneo[10], Z.D. Thomé[11],
R. Timellini[48], J. Tischhauser[7], J. Tocqueville[9], R.M. Turnbull[21],
V. Valencic[28], B. Van Eijk[7], G. Vanstraelen[22], V. Vercesi[7], L. Votano[20],
Y. Wang[48],N. Wauquier[7], H. Wenninger[7], R. Wigmans[7], C. Williams[7],
C. Xu[48], K. You[48], T. Ypsilantis[6,8], N. Zaganidis[7] and A. Zichichi[7]

1) Ansaldo, Genoa, Italy
2) INFN, Bologna, Italy
3) University of Bologna, Italy
4) University of Cagliari, Italy
5) University of Calabria, Italy
6) University of Karlsruhe, FRG
7) CERN, Geneva, Switzerland
8) CNR, Pisa, Italy
9) Collège de France, Paris, France
10) Columbia University, New York, USA
11) COPPE, Rio de Janeiro, Brazil
12) CRN, Strasbourg, France
13) Digital Italia, Rome, Italy
14) DESY, Hamburg, FRG
15) Draper Laboratories, Cambridge, Mass., USA
16) EPFL, Lausanne, Switzerland
17) Europa Metalli - LMI, Florence, Italy
18) INFN, Firenze, Italy
19) University of Firenze, Italy
20) INFN, Laboratori Nazionali di Frascati, Italy
21) University of Glasgow, United Kingdom
22) IMEC/ASP, Leuven, Belgium
23) Cornell University, Itaca, New York, USA
24) Kern Forschungszentrum Karlsruhe, FRG
25) LIP, Lisbon, Portugal
26) LPNHE University, Paris, France
27) MAX Planck Inst., Heidelberg, FRG
28) Mead SA, Lausanne, Switzerland
29) MIT, Cambridge, Mass., USA
30) INFN, Naples, Italy
31) University of Naples, Italy
32) NIKHEF, Amsterdam, Netherlands
33) University of Padova, Italy
34) University of Palermo, Italy
35) INFN, Pavia, Italy
36) University of Pavia, Italy
37) University of Perugia, Italy
38) Rutherford Appleton Laboratory, United Kingdom

39) University of California, San Diego, USA
40) University of California, Santa Cruz, USA
41) Senter for Industriforsk., Oslo, Norway
42) Smart Silicon System, Lausanne, Switzerland
43) University of Toulouse, France
44) INFN, Torino, Italy
45) University of Torino, Italy
46) Uppsala University, Sweden
47) Weizmann Institute, Rehovoth, Israel
48) World Lab, Geneva, Switzerland
49) Paul Scherrer Institute, Zurich, Switzerland

I.6. The Main achievements of the LAA Project

The main achievements of the LAA Project, in terms of *discoveries, records, new developments* and *inventions,* are summarized in Tables I.4.I and I.4.II. They are fully described in chapter II.

Table I.6.I. Main achievements of the LAA Project in terms of: D=Discoveries, R=Records, N=New developments, I=Inventions

	D	R	N	I
1. HIGH PRECISION TRACKING				
a) Gaseous detectors		2	1	
b) Scintillating fibres	1	2	1	
c) Microstrip GaAs			1	
2. CALORIMETRY				
a) High precision EM	2		1	
b) Compact EM+Hadronic	1	2		2
c) "Perfect" Calorimetry	1	1	1	
3. LARGE AREA DEVICES				
a) Construction				2
b) Alignment		1	1	
4. LEADING PARTICLE DETECTION		1	1	
5. SMIDT				
a) Microelectronics			2	
b) New, Radiation-resistant Technologies		1		
6. DATA ACQUISITION AND ANALYSIS				
a) Real Time Data Acquisition			1	
b) FASTBUS RISC computer		1		
c) Fine-grained Parallel Processor			1	
7. SUPERCOMPUTERS AND MONTECARLO SIMULATIONS			1	
TOTAL	5	11	12	4

Table I.6.II. The main achievements of the LAA Project

1. HIGH PRECISION TRACKING	
a) Gaseous detectors	
• Records:	1. Rad-hard wire: $\gg$ 1 MRad 2. Rate resistant: $\approx 2 \times 10^6$ particles/cm^2
• New developments:	1. Mechanically reliable MWPC (MDM)
b) Scintillating fibres	
• Discoveries:	1. PMP
• Records:	1. Smallest diameter sc. fibres: 15 μm 2. Rad-hard fibres:$\approx$ 1 MRad
• New developments:	1. 1×1 mm^2 bundles with 900 fibres each
c) Microstrip GaAs	
• New developments:	1. GaAs particle detectors
2. CALORIMETRY	
a) High precision EM	
• Discoveries:	1. EF 2. New inorganic scintillators
• New developments:	1. SSAC
b) Compact EM+Hadronic	
• Discoveries:	1. Intrinsic resolution better in Pb than U
• Records:	1. Hadronic energy resolution: 27% $\sqrt{E}$ 2. Rad-hard fibres: >1 MRad
• Inventions:	1. Spaghetti calorimeter 2. e/π rejection by timing
c) "Perfect" Calorimetry	
• Discoveries:	1. CsI+TMAE adsorbed layer quantum efficiency
• Records:	1. Highest efficiency photocathode (CsI+TMAE)
• New developments:	1. Systematic studies of liquid Xe

Table I.6.II. The main achievements of the LAA Project (cont.)

3. LARGE AREA DEVICES	
a) Construction	
• Inventions:	1. Blade chambers 2. Gaseous pixel chambers
b) Alignment	
• Records:	1. Best length/angles/linearity measurements
• New developments:	1. New instruments for length/angles/linearity measurement
4. LEADING PARTICLE DETECTION	
• Records:	1. Best way to cut silicon: $< 50\ \mu m$ cracks
• New Developments:	1. Leading particle detector problems solved
5. SMIDT	
a) Microelectronics	
• New developments:	1. All components of HARP done 2. Read-out electronics for pixel detector
b) New, Radiation-resistant Tech.	
• Records:	1. Rad-hard amplifier: > 1MRad
6. DATA ACQUISITION AND ANALYSIS	
a) Real Time Data Acquisition	
• New developments:	1. Test of commercial architectures with algorithms for feature extraction
b) FASTBUS RISC computer	
• Records:	1. Highest computer power in a FASTBUS board: 50 VAX-equivalent
c) Fine-grained Parallel Processor	
• New developments:	1. ASP construction and test of chips
7. SUPERCOMPUTERS AND MONTECARLO SIMULATIONS	
• New developments:	1. Full MonteCarlo Chain

II. STATUS OF THE PROJECT

I.1. High Precision Tracking

In order to measure tracks with high precision in future Hadron Colliders, three parts are needed.

i) The *closest to the vertex* → GASEOUS DETECTORS

ii) The *surrounding one* → SCINTILLATING FIBRES

iii) *Along the beam* → GALLIUM ARSENIDE (GaAs)

This is illustrated in Fig. II.1.1, where the detectors developed within are shown in a cut view perpendicular to the beam axis.

II.1.a. Gaseous detectors

The **new development** of the MultiDrift Modules (MDM), a mechanically reliable, modular MWPC, with high redundancy and precision: $\approx 60\ \mu m$, brought us to establish two **records**:

- The most radiation resistant combination of gas (DME) and wire (stainless steel): well above 1 Mrad.
- The drift chamber most resistant to rate: $\approx 2 \times 10^6$ particles/cm^2.

Before LAA, the typical measurement precision of the standard drift chambers used for the central tracking, was in the range between 100 and 200 μm [8]. The rate capability of these chambers was, at most, about 10^5 particles/cm^2 [9] and the radiation resistance was much lower than 1 MRad [8].

The present limits of the LAA MultiDrift Modules are shown in Figs. II.1.2, II.1.3, and II.1.4: precision $= 60\ \mu m$, rate capability $= 2 \times 10^6$ particle/cm^2, and radiation resistance well above 1 MRad, respectively.

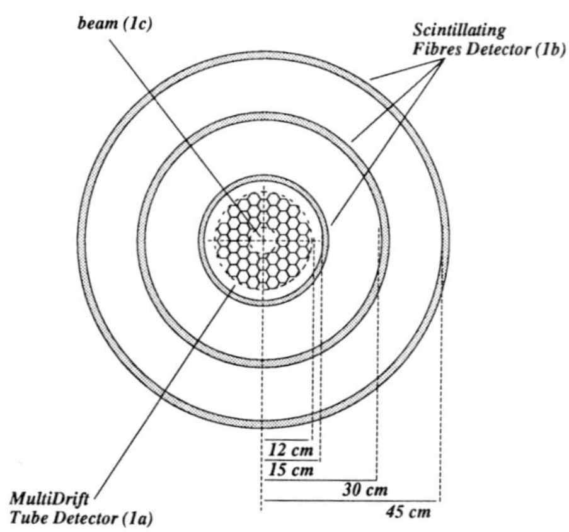

Fig. II.1.1. Vertex and tracking detectors: cut view.

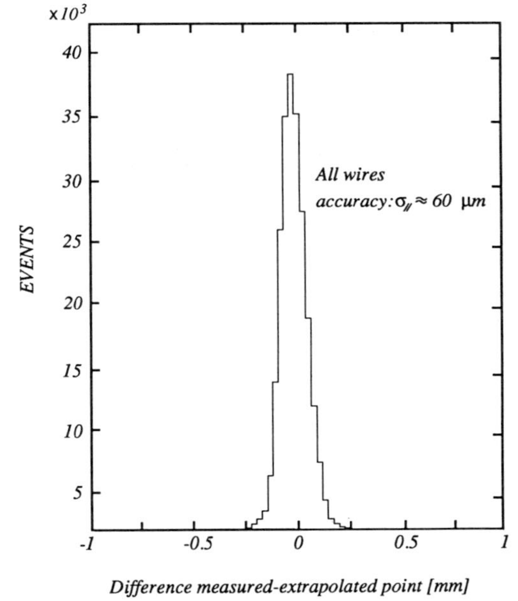

Fig. II.1.2. Position measurement perpendicular to the wires: distribution of the differences between measured and extrapolated points.

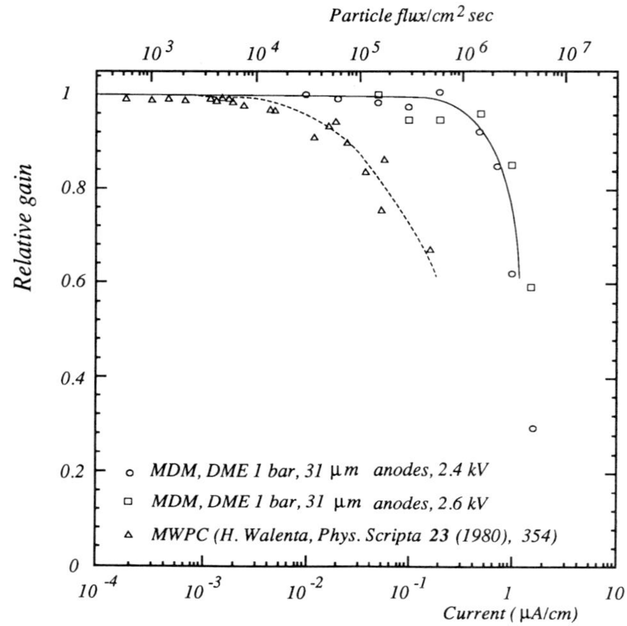

Gain drop vs wire current

Fig. II.1.3. Rate dependence of gain measured in the multidrift module with Dimethyl Ether, as a function of current and equivalent particle rate (full curve). For reference, the same measurement is shown for a conventional multiwire chamber (points on the dashed curve).

II.1.b. Scintillating fibres

Thanks to the discovery of a new scintillator with a high Stoke-shift, the PMP (see Figs. II.1.5 and II.1.6), two records were established by this component of LAA:

- The production of the smallest scintillating fibre - 15 μm diameter - with an attenuation length sufficiently large to be used in a real physics experiment. This is an improvement of nearly two orders of magnitude with respect to the situation before LAA [10].

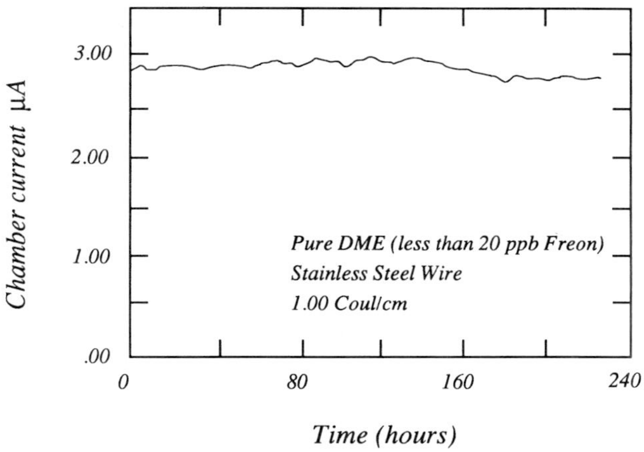

Fig. II.1.4. *Radiation resistance of the modules with Dimethyl Ether using different wires and gas purities. In the best case, no degradation is observed after an irradiation corresponding to above 1 Coul/cm (= 1 MRad).*

- The highest radiation resistance for small diameter fibre: about 1 MRad. Again, an improvement here by a factor 2 has been obtained by LAA [11].

Furthermore, a technological breakthrough came from the **development** of a new method for producing 1×1 mm^2 fibre bundles with about 900 fibres/bundle. These bundles are necessary, since the individual fibres cannot be handled any more at such small diameters. The bundles can be mechanically arranged to form the final tracker shells.

Figures II.1.7, 8, and 9 show the mechanical arrangement of the fibres in the fibre bundles.

PMP

1-phenyl-3-mesityl-4-pyrazoline

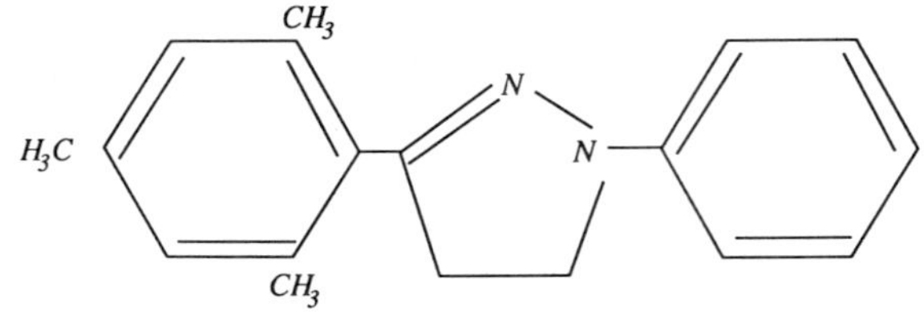

Fig. II.1.5. The PMP molecule.

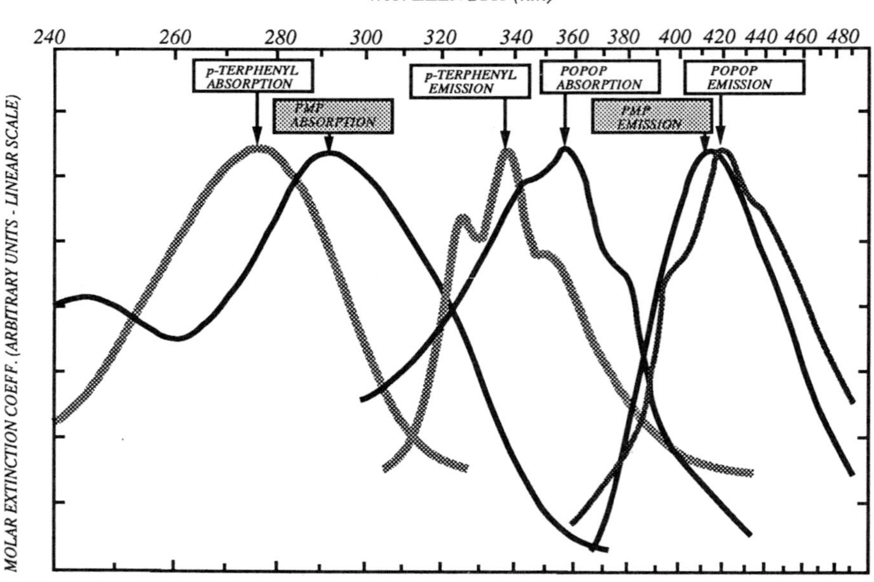

Fig. II.1.6. Absorption and emission spectra of PMP, P-TERPHENIL, and POPOP.

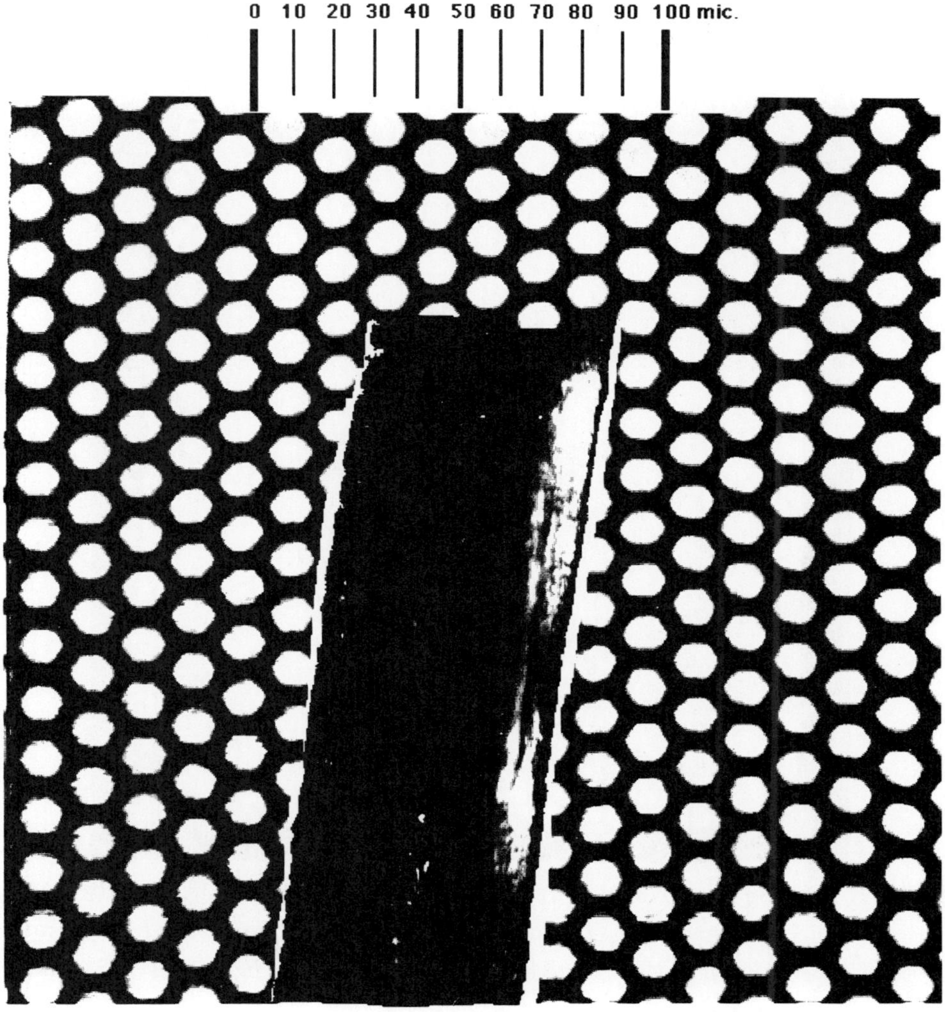

Fig. II.1.7. *Enlarged section of a fibre bundle containing 15 μm diameter fibres. For comparison, a female hair with 60 μm diameter is also shown.*

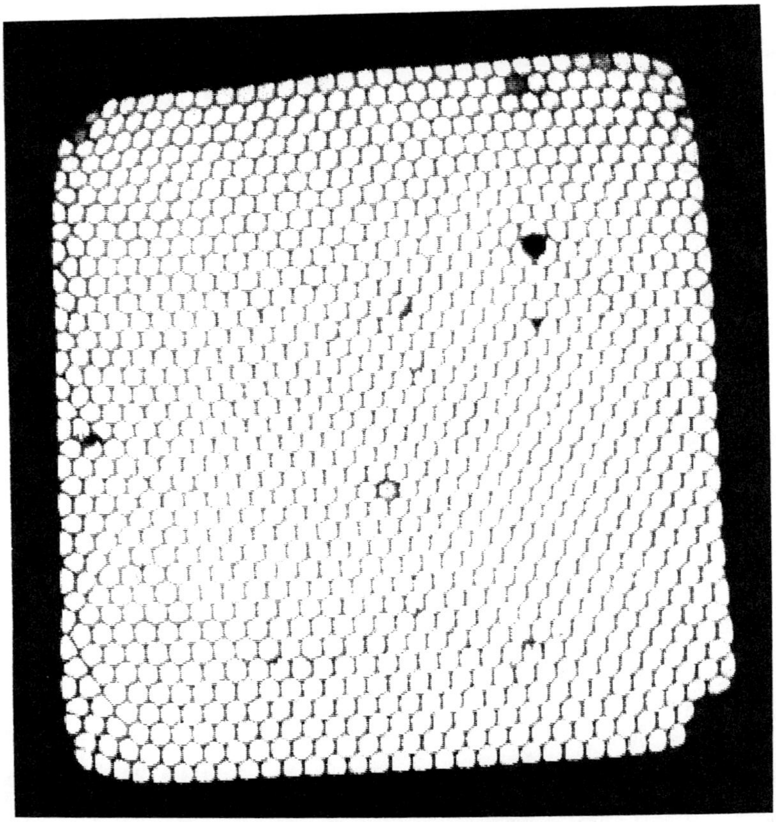

Fig. II.1.8. *Square (1 mm × 1 mm) fibre bundle containing 900 individual hexagonal fibres of 30* μm *"diameter". This picture is taken through 1 m bundle length, using white light and 175 times magnification.*

Fig. II.1.9. *Magnified picture of the same fibre bundle with only one individual fibre illuminated. Fibre diameter is 30 μm. The cladding thickness separating it from the surrounding fibres is 6 μm.*

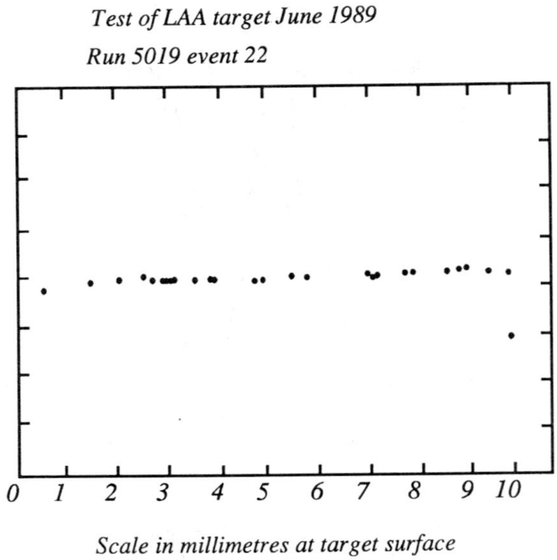

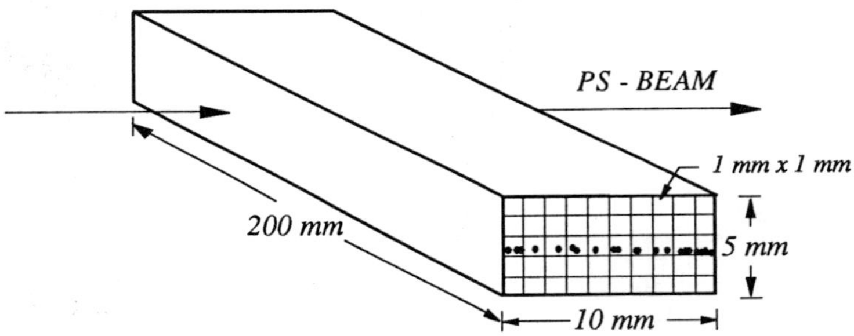

Fig.II.1.10 *Particle track seen by an assembly of multi-bundles, which represents a part of a tracker-shell. The PS-beam traverses this target like a charged particle emerging from the interaction point of a collider.*

The equivalent of one shell of the final tracker was tested in a particle beam. Figure II.1.10 shows how a single track is seen in the fibre bundle assembly.

Figure II.1.11 shows the results from 272 tracks in terms of precision and two-track resolution:

2-track resolution: 83 μm,

tracking precision: 35 μm.

Note that in classical drift chambers, the tracking precision is between 100 and 200 μm, and the two-track resolution is about 1 mm.

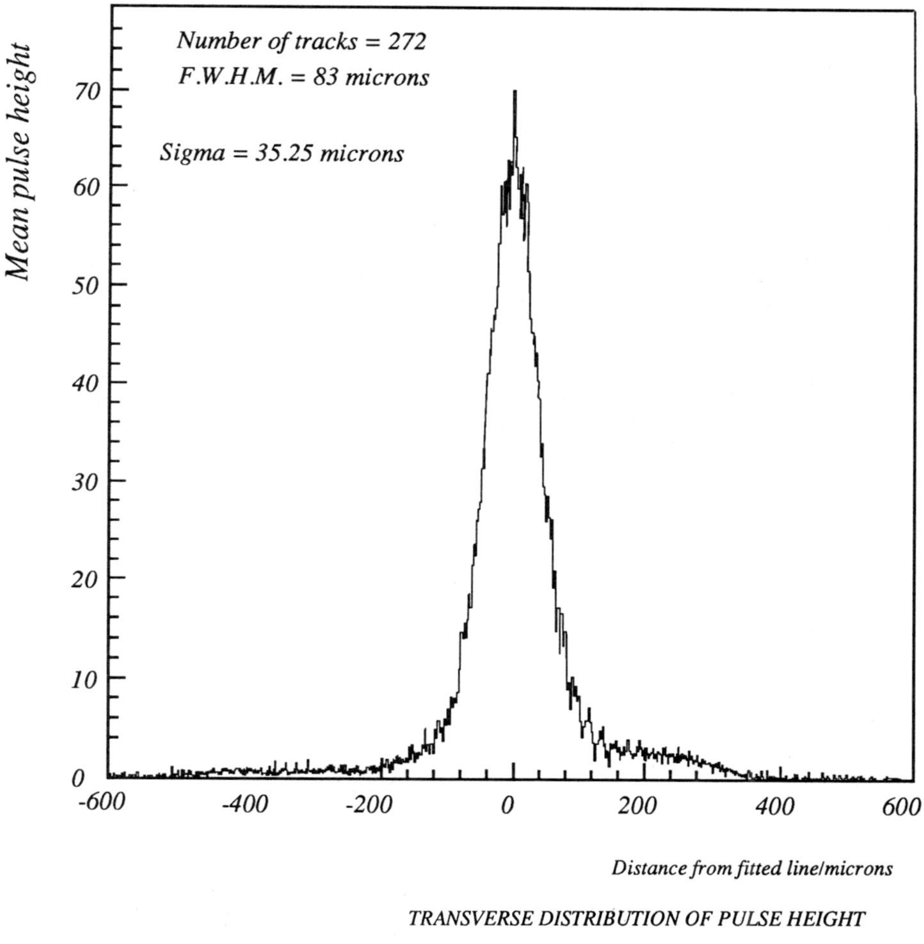

Fig. II.1.11. Histogram of 272 particle tracks showing 83 μm two-track resolution and 35 μm spatial precision.

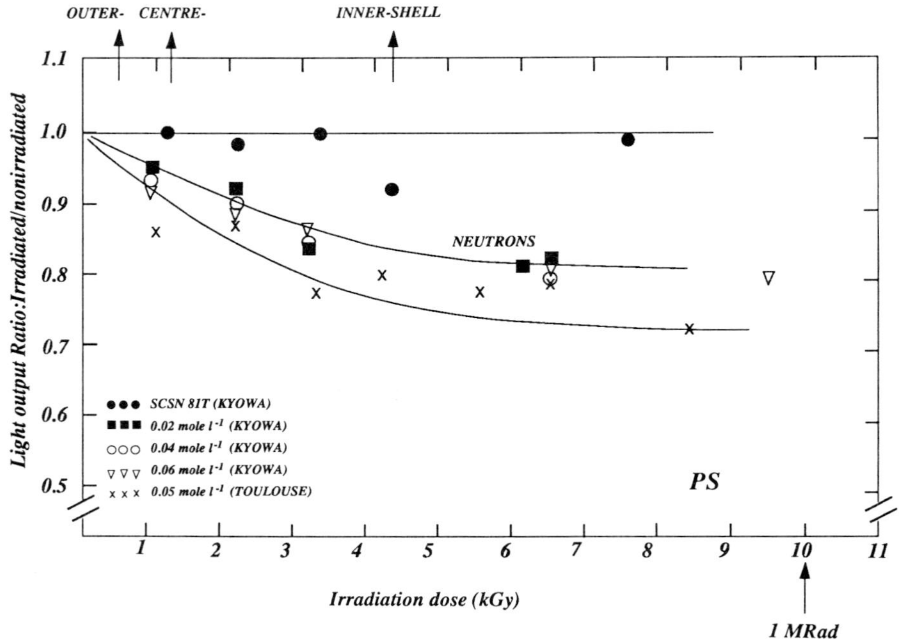

Fig. II.1.12. *Light output as a function of irradiation dose for different samples of scintillating fibres. The optimal concentration of PMP is of 0.04 model/l. The arrows show the expected level of irradiation of the three tracker shells after one year running at a luminosity of* $10^{33} cm^{-2} s^{-1}$. *The curves are free interpolation of data to guide the eye.*

The radiation resistance of the 30 μm fibres has been tested up to 1 MRad. Figure II.1.12 shows the results: at a luminosity of $10^{33} cm^{-2} s^{-1}$, and after one year running, the loss in light output due to radiation damage is of the order of 20% for the innermost tracker shell. This loss is completely acceptable for this kind of detector.

II.1.c. Microstrip GaAs

The **development** of prototype GaAs detectors by this component of LAA has shown that this material can indeed be used for building particle detectors.

The GaAs basic material for the test has been grown with the LEC (liquid-encapsulated Czochralski) method. Its relevant characteristics are shown in Table II.1.I.

Table II.1.I. Relevant properties of the GaAs as claimed by the manufacturer

DIAMETER (mm)	50.9
ORIENTATION	(100)
DOPANT	Nil
WAFER FLATNESS (μm)	< 3
WAFER THICKNESS (μm)	500 ± 25
WEIGHT (g)	5.37
RESISTIVITY ($\Omega\ cm$)	$7.5 \times 10^7 \div 1.2 \times 10^8$
MOBILITY ($cm^2\ V^{-1}s^{-1}$)	$6.8 \times 10^3 \div 6.6 \times 10^3$
CARRIER CONCENTRATION (cm^{-3})	$1.2 \times 10^7 \div 8.2 \times 10^6$
EPD (cm^{-2})	$2.1 \times 10^4 \div 3.0 \times 10^4$

We have used a 6 GeV/c hadron beam (mainly pions) to check the response of our detectors to MIPs. In Fig. II.1.13, a spectrum of energy deposited by the beam in the GaAs detector is shown. A detection efficiency curve as a function of the bias voltage is presented in Fig. II.1.14, showing that we reach $\sim$ 98% detection efficiency starting at − 40 V. It should be noticed that the spectrum presented in Fig. II.1.13 shows some edge effects of the diode. This is seen as a low number of counts that joins the pedestal to the signal spectrum. The uncertainty related to these edge effects, due to the limited acceptance of the experimental set-up, is a major contribution to the missing 2% of efficiency.

II.2. CALORIMETRY

II.2.a. High precision electro-magnetic

Through the **discoveries** of a new vapour, EF (Ethyl Ferrocene), photosensitive to the BaF_2 fast light emission (see Fig. II.2.1), and of a new family of inorganic scintillators with lower wavelength emission (ex.: $KLuF_4$, see Fig. II.2.2), the concept of SSPC (Solid State Proportional Counter) has **developed** to the SSAC (Solid State Avalanche Counter). The characteristics of these detectors are shown schematically in Figs. II.2.3 and II.2.4.

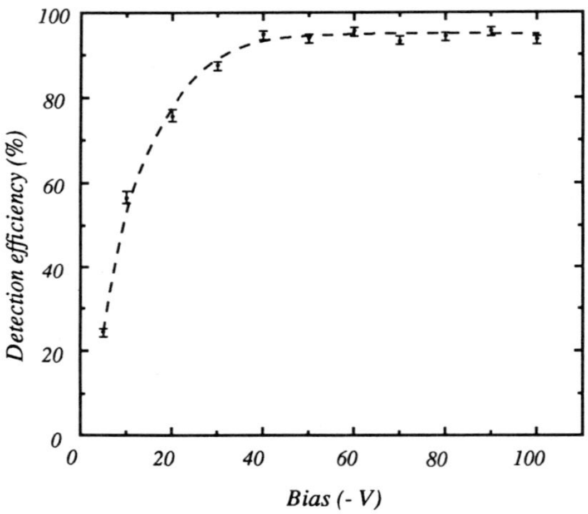

Fig. II.1.14. Efficiency of the detector as function of the bias voltage.

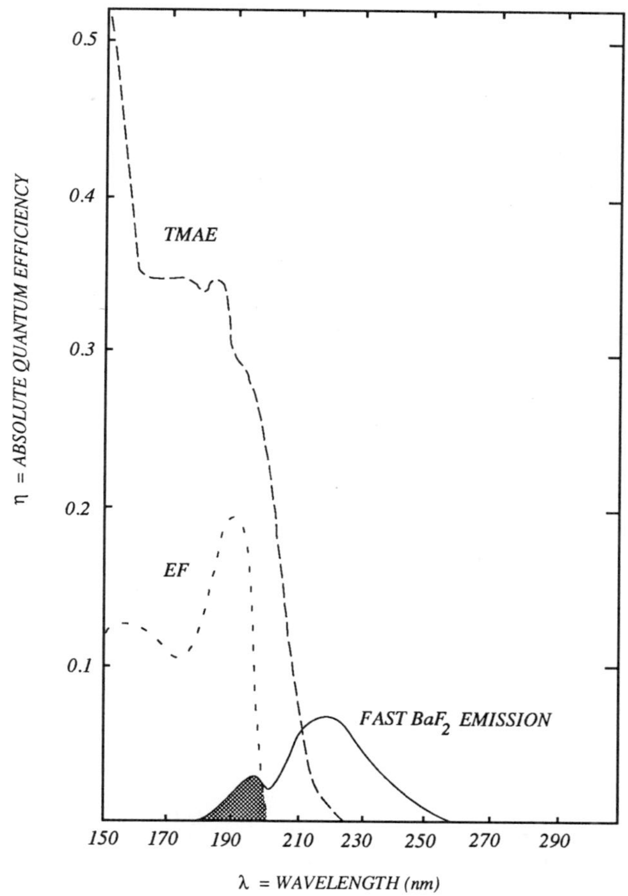

Fig. II.2.1. *Ethyl-Ferrocene and TMAE quantum efficiencies. The fast (600 ps) BaF_2 emission spectrum is also shown. The dashed area of the BaF_2 emission spectrum corresponds to the overlapping component between EF and BsF_2.*

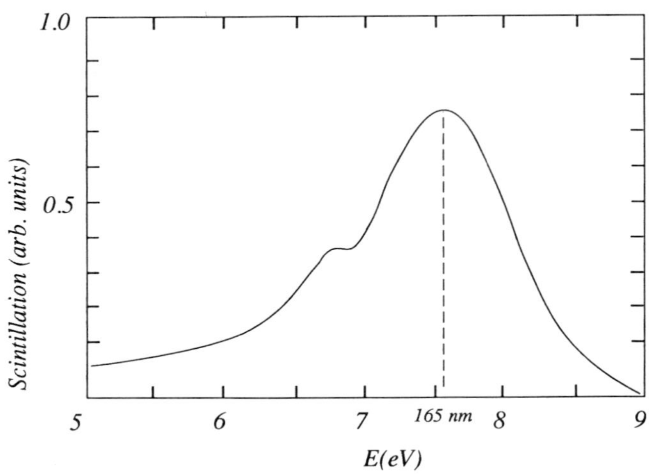

Fig. II.2.2. Luminescence emission spectra of $KLuF_4$.

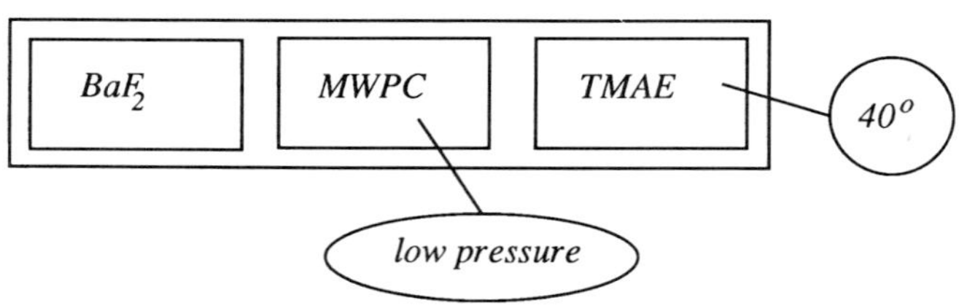

Fig. II.2.3. The essential ingredients of the "dreamt" SSPC (Solid Scintillator Proportional Counter). The aim of LAA was to get rid of low pressure and 40^0, thus achieving a counter able to work at atmospheric pressure and room temperature.

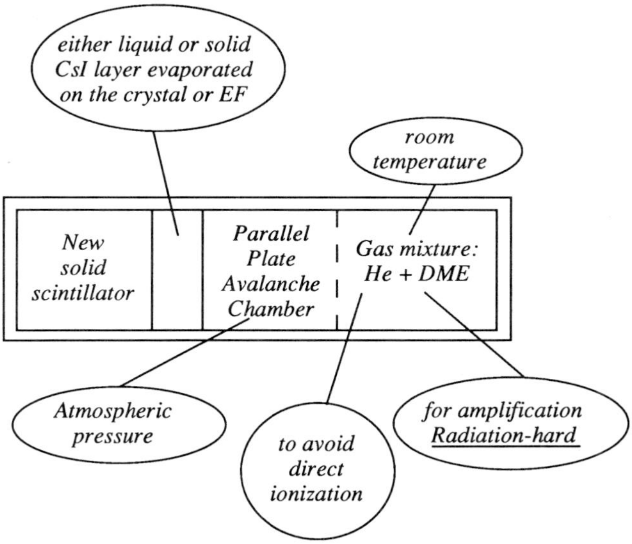

Fig. II.2.4. The Solid Scintillator Avalanche Counter.

It is interesting to know how the way was opened to the discovery of a new family of inorganic scintillators with properties similar to those of BaF_2; (they emit light in a wavelength region between 140 nm and 230 nm). This came from the complete understanding of the BaF_2 fast component light emission process (called CRL=CRoss Luminescence). The proof that the emission wavelength of the scintillators can be controlled is given by the measurement of the emission spectra of KF (potassium fluoride) when Rb^+ (Rubidium) concentration is continuously changed. Figure II.2.5 shows the CRL spectra of $K_{(1-x)}Rb_{(x)}F$ with:

$$x = 0(a), x = 10^{-3}(b), x = 10^{-2}(c), x = 10^{-1}(d), x = 1(e).$$

An example of a new scintillator is the KL_uF4 (Potassium Lutetium Fluoride), whose characteristics are better than BaF_2:

	$KLuF_4$	BaF_2
density	4.8 gr/cm^3	4.9 gr/cm^3
X_o	2 cm	2.05 cm

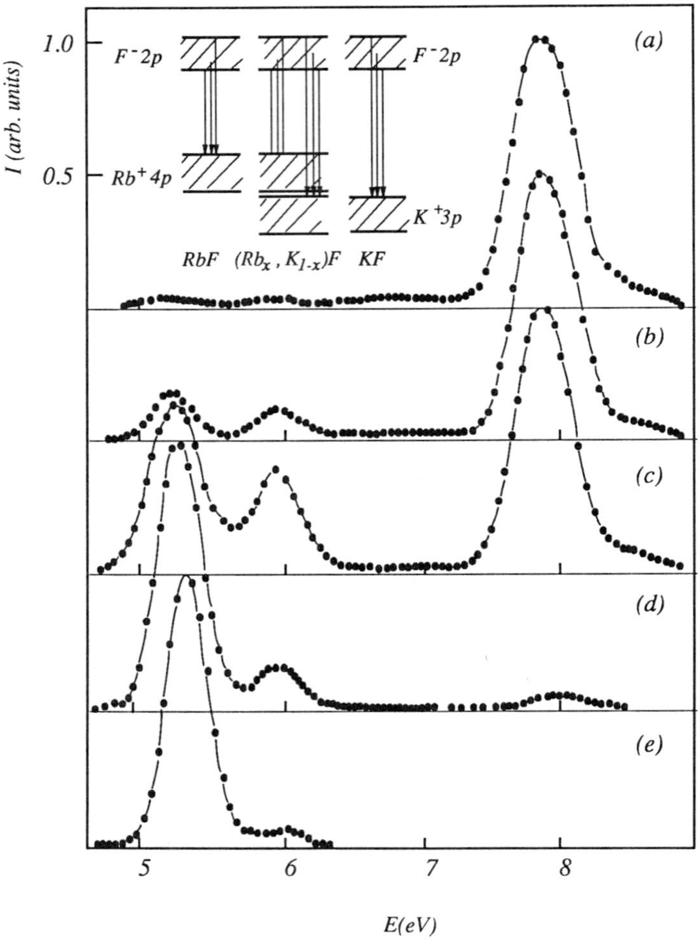

Fig. II.2.5. CRL emission spectra of $K_{1-x)}RB_{x)}F$ solid solutions, with varying x: a) 0; B 0.001; c) 0.01; d) 0.1; e) 1.

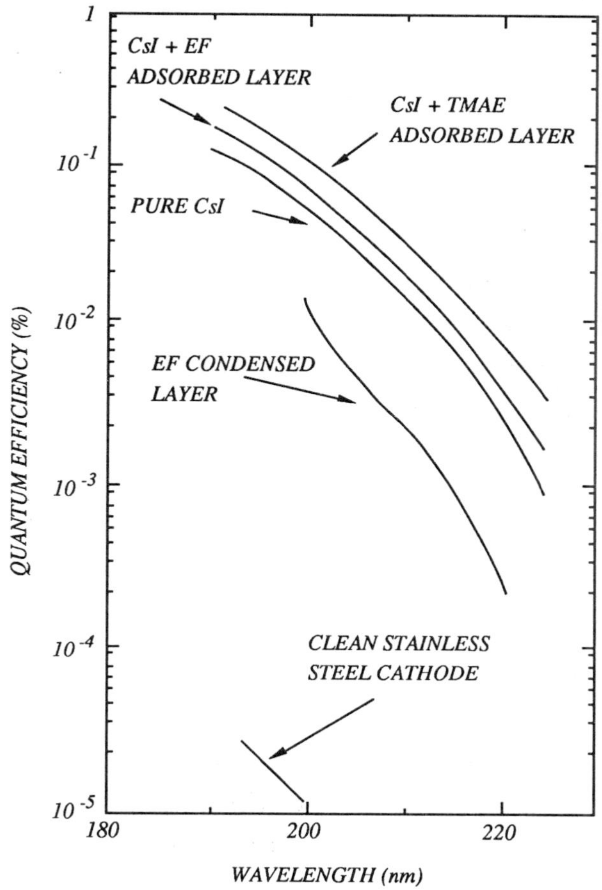

Fig. II.2.6. *Quantum efficiencies measured with a single-wire counter for CsI with TMAE adsorbed layer, CsI with EF adsorbed layer, pure CsI, EF liquid layer, and clean copper cathode.*

In addition to the points mentioned in the caption of Fig. II.2.3., the disadvantages of the "old" SSPC were:

i) the small overlap between the BaF_2 emission band and the TMAE photosensitive band;

ii) the fast ageing effects due to polymerization of TMAE in a wire chamber; and

iii) the fact that TMAE vapour is very reactive to air and corrosive for many materials. The use of liquid TMAE as a photocathode was not possible because of its low efficiency.

Another important discovery (due to V. Peskov) is the use of thin layers of photosensitive materials, either liquid (EF) or solid (CsI), deposited directly on the scintillator surface. This increases the quantum efficiency by $10^3 - 10^4$

compared to clean metallic photocathodes (Fig. II.2.6). The use of thin layers allows to eliminate the gap needed to convert the VUV light emission in the photosensitive vapour. Therefore, the newly developed Parallel Plate Avalanche Chamber (PPAC) can be used as amplification element. Conclusions: The new SSAC can operate at atmospheric pressure and room temperature.

In addition to these basic advantages the results achieved, for the new SSAC, can be summarized as follows:

- Compactness: 4 mm thick conversion gap.
  (All electrons are produced in the same superthin ($< 1\mu m$) layer of either solid (CsI) or liquid (EF), therefore the amplification signal has no jitter).
- Very fast collection time: $< 10ns$.
- Very good time resolution: $< 1ns$.
- Negligible sensitivity to direct ionization.
- Negligible sensitivity to $(\overline{B})$.
- High gain: single photoelectron detection.
- High rate capability: no space charge effect.
- No ageing due to polymerization.

The efficiency of the SSAC is equal to traditionally "dreamt" SSPC (with MWPC and TMAE) and comparable to that of a traditional PhotoMultiplier (10%). This opens up a new way towards large surface detectors.

II.2.b. Compact EM+Hadronic

A direct comparison of the results from two compensating hadron calorimeters of the lead/scintillator and uranium/scintillator sandwich type, lead to the **discovery** that the intrinsic resolution of a lead hadronic calorimeter is better than the intrinsic resolution of an uranium one.

Moreover, a deep theoretical understanding of the hadronic shower development in a variety of materials (lead, uranium, iron, silicium, etc.), has indicated that a lead/scintillator calorimeter can be compensating (i.e., it gives equal response to the electromagnetic and to the hadronic part of the shower), provided a ratio of 1:4 is kept between the two materials. This is illustrated in Fig. II.2.7, where the expected resolution of a lead/scintillator calorimeter is plotted as a function of the thickness of the lead (or of the scintillator). The experimental data obtained in the tests mentioned above are also shown in the same figure.

Both this discovery and the theoretical understanding of the shower development, have lead to the **invention** of a new kind of calorimeter: the spaghetti (lead/scintillating fibres) calorimeter.

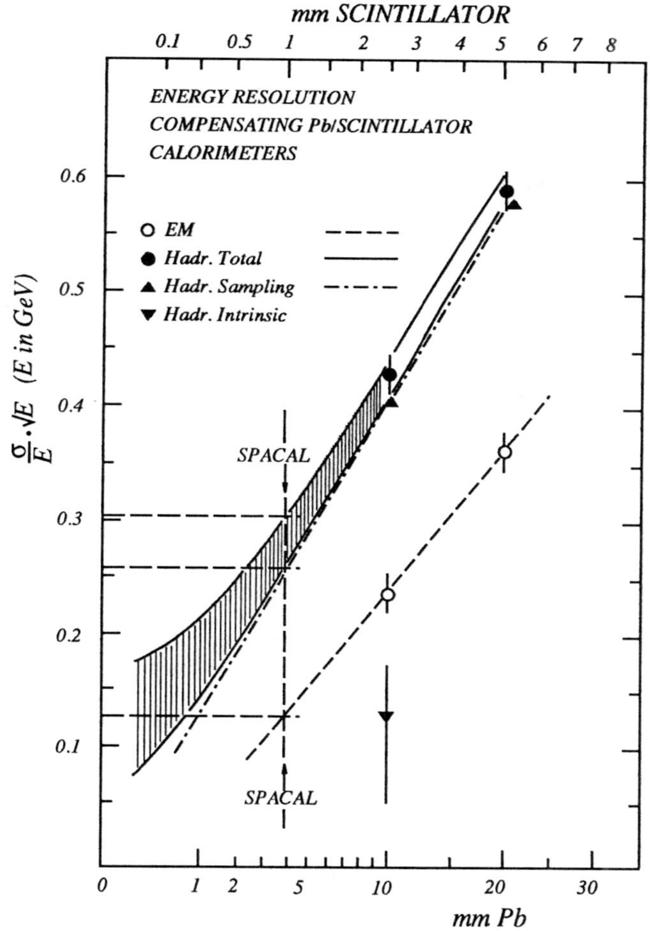

Fig. II.2.7. Energy resolution compensating lead/scintillator calorimeters.

A number of different problems have been solved during the first three years of the LAA project. Among others:

- Embedding of a very large number of fibres in lead. Figure II.2.8 shows the high uniformity of the hole dimensions and spacing obtained in prototype modules (hexagonal shape, 48.7 cm^2 cross-section, 8.6 cm apex-to-apex).
- Lateral and longitudinal uniformity of the calorimeter modules. Many details in the design of the modules have been optimized in order to obtain a "perfect" transversal (of particular importance to obtain the best energy resolution for electrons see Fig. II.2.9) and longitudinal (more important for energy resolution of hadrons see Fig. II.2.10) uniformity of light collection.
- Fibres read-out.

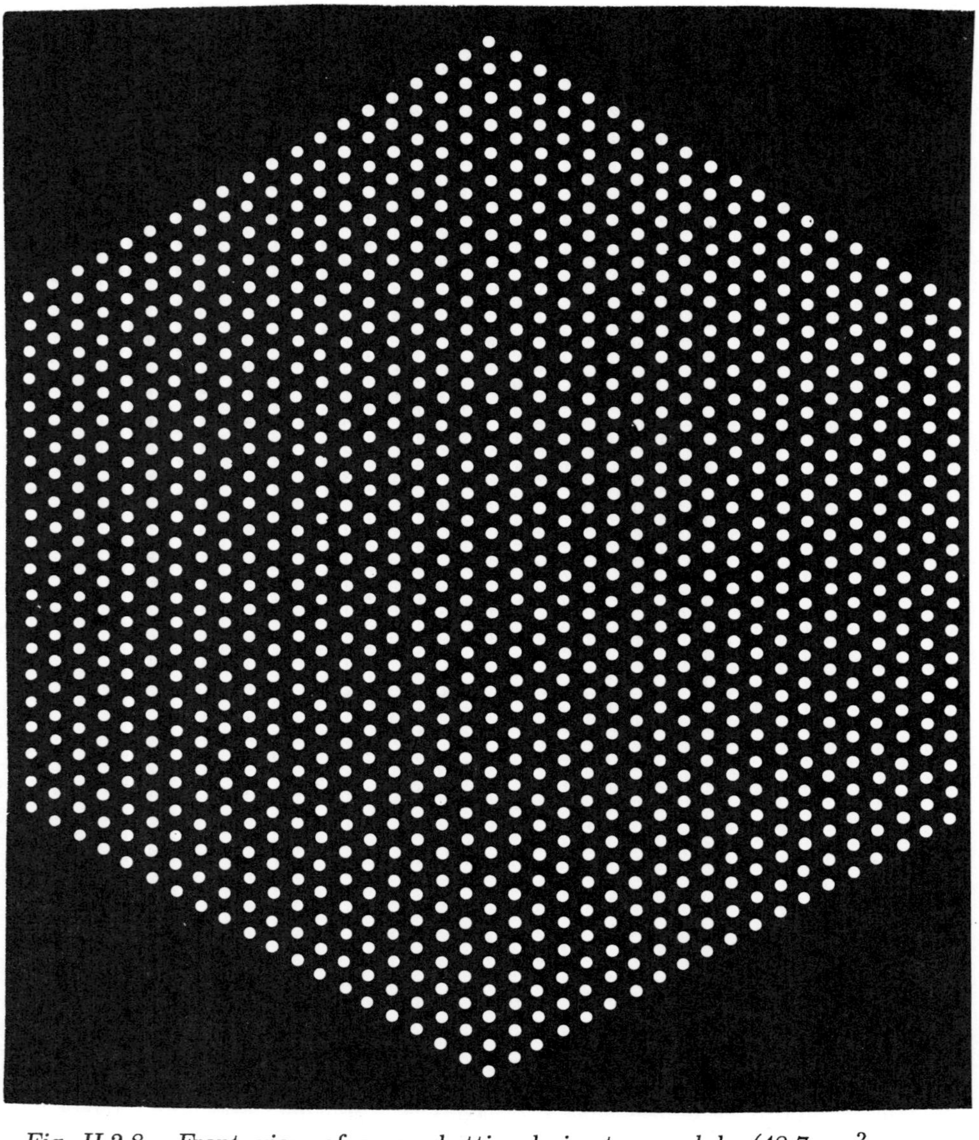

Fig. II.2.8. *Front view of a spaghetti calorimeter module (48.7 cm^2 cross-section). The fibres are inserted in the holes, running parallel to the direction of the incident particles.*

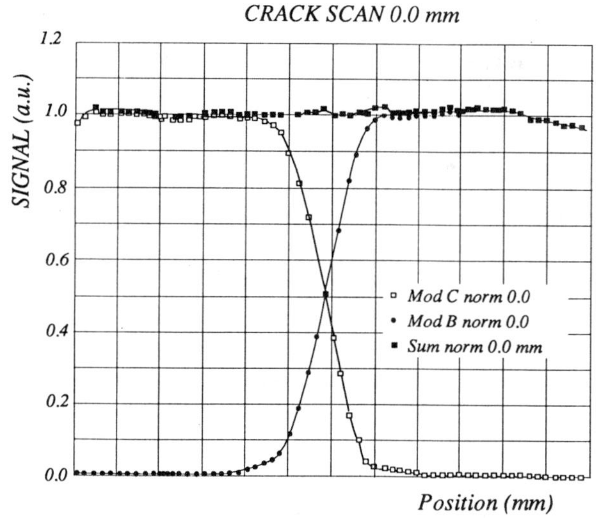

Fig. II.2.9. The calorimeter response to a collimated β-source, which is moved in small steps over the surface of two adjacent calorimeter modules. Shown are the signals in these two modules, and their sum. Results for second-generation prototype modules.

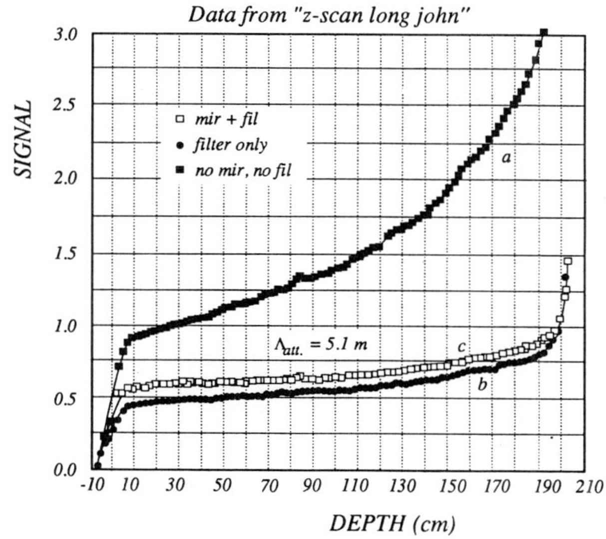

Fig. II.2.10. The calorimeter response to gamma's from a ^{60}Co ring source surrounding the module, as a function of the distance to the PM's. The measurements were done without a mirror/filter (curve a), with a yellow filter (b), and with a filter and a mirror (c).

The main motivations for using lead in place of uranium are listed in the following.

Compensating Pb/scintillator calorimeters:

- Are as compact as compensating U/scintillator, because the effective interaction length is the same for both calorimeters:

$$\sigma_{eff}^{int} \approx 20\ cm.$$

- Give better energy resolution, because the intrinsic resolution is better:

$$\left\{\frac{\sigma_{int}}{\sqrt{E}}\right\}_{Pb} = 11\%\ , \quad \left\{\frac{\sigma_{int}}{\sqrt{E}}\right\}_{U} = 19\%$$

- Give faster and better localized signals, because the neutrons from Pb have a much longer mean free path, and thus they do not contribute to the slow component (s) signal at large distances from the shower core.
- Cause less radiation damage, because the number of neutrons produced in the shower is three times smaller in Pb than in Uranium.

As to calibration, there is nothing that one can do with the U-activity that one cannot do better with removable sources inserted in a fibre slot. In particular, when such sources are used, one does not have to worry about:

- ^{235}U content fluctuations,
- plate thickness fluctuations,
- cladding thickness fluctuations ($\beta's$),
- build-up of long-lived activity in bulk material.

Moreover, lead is:

- Easy to machine.
- Non-radioactive.
- Cheap.
- Unlimitedly available.

A prototype consisting of 20 modules of the Spaghetti Calorimeter (see Fig. II.2.11), each 2 m long and with an hexagonal cross-section of 48 cm^2, was tested in high-energy particle beams from the SPS.

This prototype was used to develop the second, very important, **invention** of this component of LAA: a new method for distinguishing electrons from pions, thanks to the extremely fast (some ns) response of the lead and of the scintillating fibres. The fibres used in our calorimeter have a signal decay time of 1.5 ns.

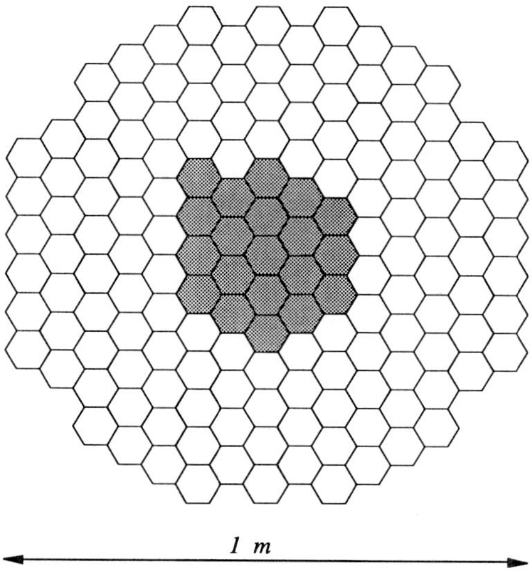

Fig. II.2.11. *Front view of the full 155-modules spaghetti calorimeter prototype. The results discussed in the text have been obtained with a smaller prototype, consisting of the grayed modules.*

This important feature paid off in an extremely interesting way. We found that the calorimeter response is so fast, that electrons and pions can be efficiently distinguished through subtle differences in the time structure of the signals, which occur on a 10 ns time scale. This is illustrated in Figures II.2.12 and II.2.13. Figure II.2.12 shows e^- and π^- signals measured with a 2 m long module. In Fig. II.2.13, the distribution of the FWFM (20%) of several hundred electron and pion signals is given. It shows a clear separation between the two types of particles, and it demonstrates that an e/π separation at the 10^{-3} level can be achieved with this completely new technique which, unlike the classical methods employing shower profiles, does not require any calorimeter segmentation whatsoever.

Another **record** of the LAA Project has also been obtained by this prototype calorimeter: the best hadron energy resolution obtained so far, i.e. $27\%/\sqrt{E}$ (test runs with a prototype for the ZEUS calorimeter gave $\sim 34\%/\sqrt{E}$ [12]). Figure II.2.14 shows the result, which, for the time being, is obtained as an extrapolation at zero shower leakage. The 20 tons full prototype, containing 155 modules, will directly measure this quantity.

A vast programme of radiation damage studies of scintillating plastic fibres has also been carried out, leading to another **record**: the development of polystyrene fibres doped with 3-hydoxyflavone (3-HF), which show only a small deterioration of the attenuation length (after recovery) up to radiation doses of 10 MRad (100 kGy), as shown in Fig. II.2.15.

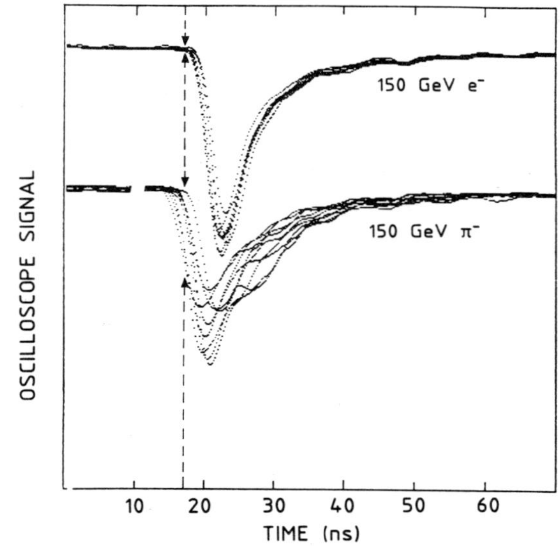

Fig. II.2.12. *Calorimeter signals from 10 different 150 GeV electron showers and from 10 different 150 GeV pion showers.*

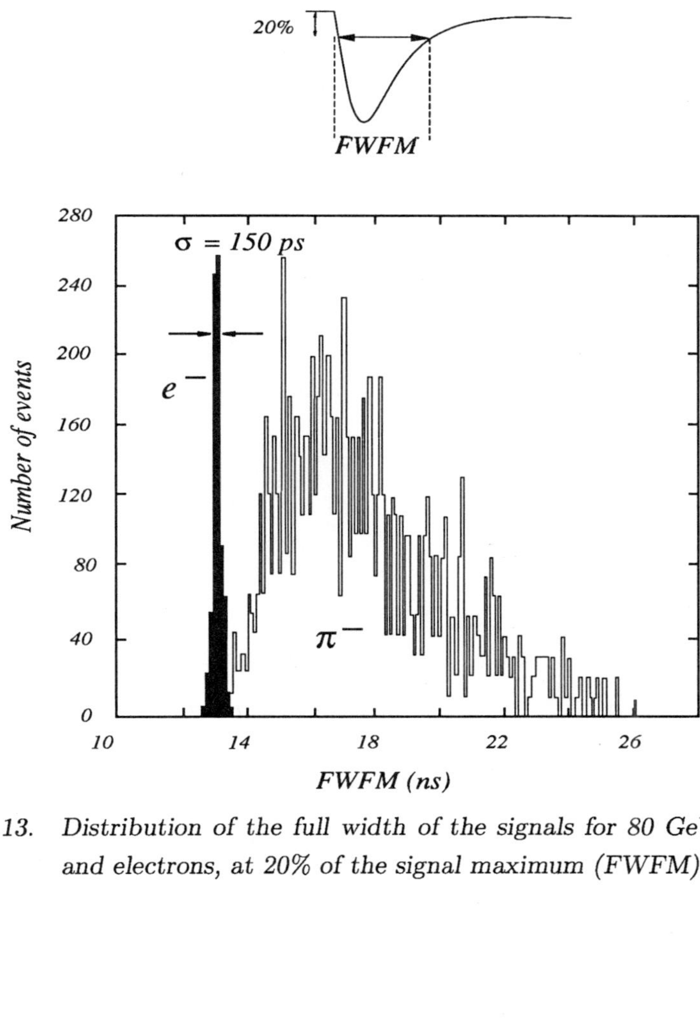

Fig. II.2.13. Distribution of the full width of the signals for 80 GeV/c pions and electrons, at 20% of the signal maximum (FWFM).

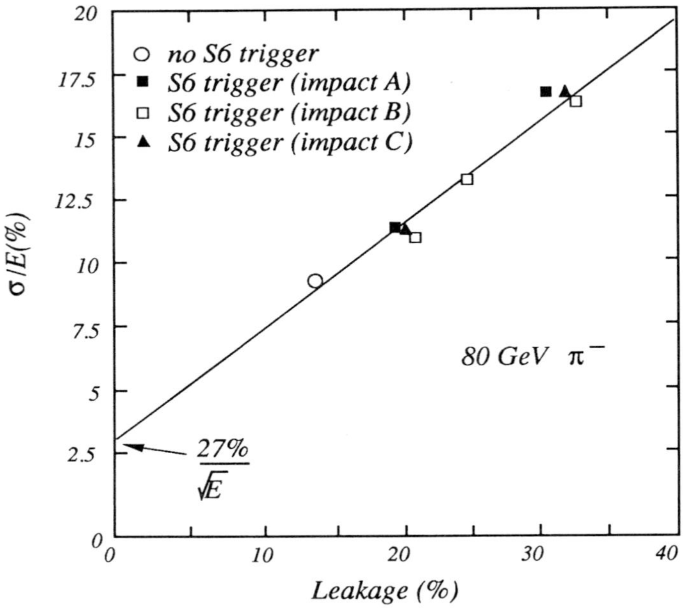

Fig. II.2.14. Energy resolution versus leakage for 80 GeV/c pions. The extrapolation at zero-leakage gives an energy resolution of $27\%\sqrt{E}$.

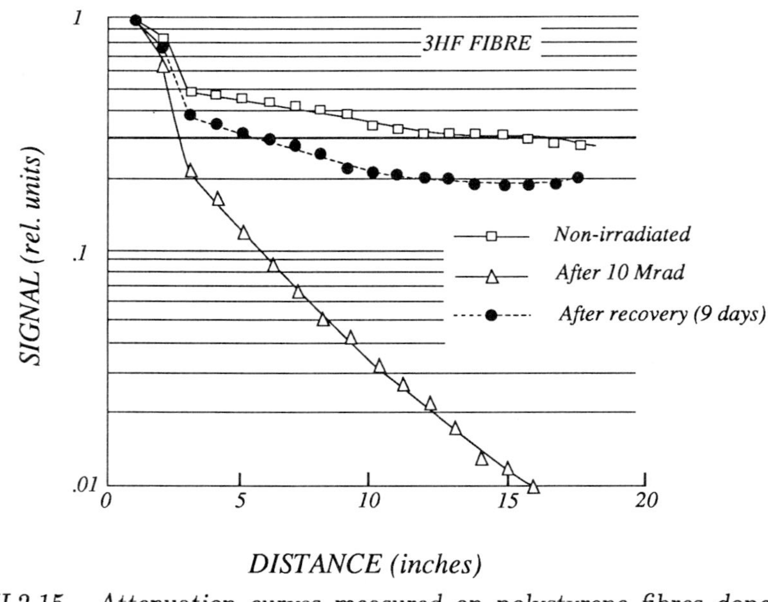

Fig. II.2.15. Attenuation curves measured on polystyrene fibres doped with 3-HF, before and after irradiation with 10 MRad.

II.2.c. "Perfect" Calorimetry

The most important **discovery** of this component of LAA is the increase in quantum efficiency of a CsI photocathode if a monomulecular adsorbed layer of TMAE is deposited on it. The experiment has been done, first by depositing a CsI reflective photocathode on a metal substrate, then by flowing (for about 1 hour) a low concentration of TMAE gas over the cathode surface. In this way a monomolecular film of TMAE is formed. This "sensitized" cathode has a higher quantum efficiency than the cathode with CsI alone. At the Xenon scintillation wavelength, this cathode has a **record** high quantum efficiency: ≈ 46%, as shown in Fig. II.2.16.

On the way towards the construction of the "perfect" electromagnetic calorimeter, the knowledge of the properties of liquid Xenon has been further **developed**, in terms of:

- scintillation speed;
- transmission of light;
- photon production in the scintillation process.

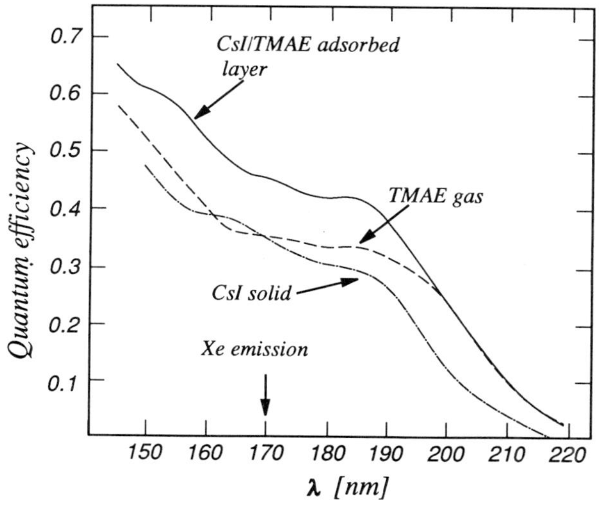

Fig. II.2.16. Quantum efficiency of different photocathodes as function of the wavelength. The liquid Xenon scintillation wavelength is $\lambda \approx 170$ nm.

A pulsed 100KV electron accelerator has been built and tested as a source for excitation of liquid xenon. A light beam from a H_2 flash lamp , with a FWHM pulse width of 15 ns, is focused onto a CsI vacuum reflective photocathode to produce the electrons which are accelerated. This device can deliver as little as 10MeV (400 electrons each with 25 KeV) or as much as 1 TeV (10^7 electrons each with 100 KeV) of excitation energy (per pulse) into the liquid Xenon, permitting the investigation of the scintillation and ionization processes over a wide range of energies.

Recently, a cleaning and condensation system was successfully inaugurated. Xenon was condensed and its scintillation lifetime in the gaseous and liquid state was measured as shown in Fig. II.2.17. The liquid lifetime is less than the observed width of 25 ns because of the lamp pulse of 15 ns. The quadrature difference gives a value of 20 ns, in agreement with other measurements of the long life triplet state of the Xe_2 molecule. The faster singlet state (3 ns) cannot be discerned with this long lamp pulse. The lifetime of the gaseous Xenon scintillation was measured to have a 60 ns component due to recombination.In conclusion, the scintillation lifetime of liquid Xenon matches very well the very short bunch spacing of multi-TeV hadron colliders.

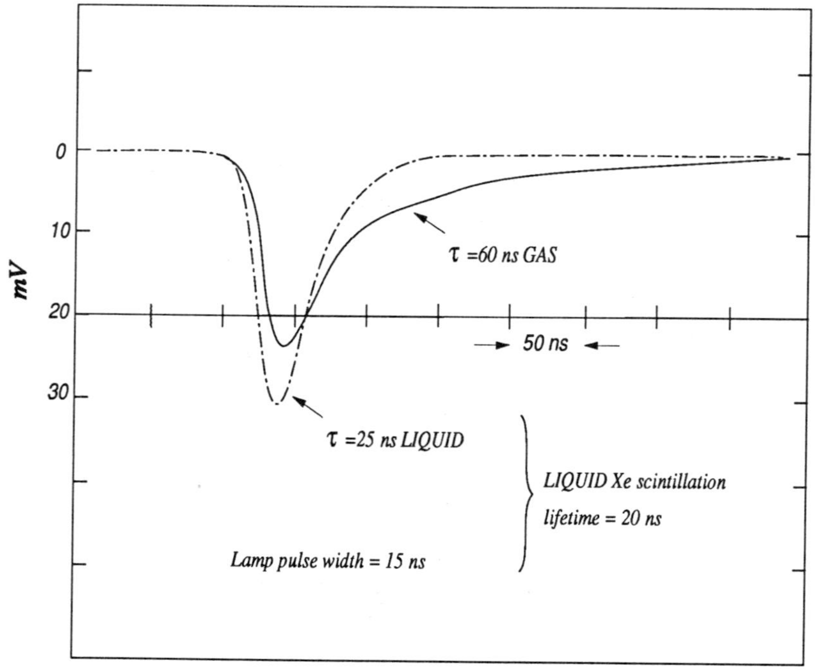

Fig. II.2.17. Liquid and gaseous Xenon scintillation light pulse duration. The scintillation was stimulated by electrons produced by a lamp with an intrinsic width of 15 ns.

The scintillation yield and transmission properties of liquid Xenon were measured by the apparatus sketched in Fig. II.2.18.

The results are shown in Fig. II.2.19. A large scintillation yield of 3×10^4 photons/MeV was measured. The quoted yield is an average with the largest pulses being two times larger (i.e. 6×10^4/MeV). The reason for this fluctuation is due to bubbling and circulation. The energy per photon is, therefore, between 33 and 16 eV, hence of the same magnitude as the ionization energy (i.e. W=15 eV). This very high photon yield promises a very good energy resolution of the final calorimeter. The transmission of a column of liquid Xenon (to its scintillation light at l=170 nm) is also shown in Fig. II.2.19. The inferred absorption length in circulating mode (i.e. during filling) is low (2.7 cm), because of bubbling. On the contrary, the photon absorption length is much greater than 20 cm in the non-circulating mode (i.e. when the liquid Xenon is left evaporating). The conclusion is that there will be no self-absorption of light in the liquid Xenon calorimeter, if this will be operated in a non-circulating mode.

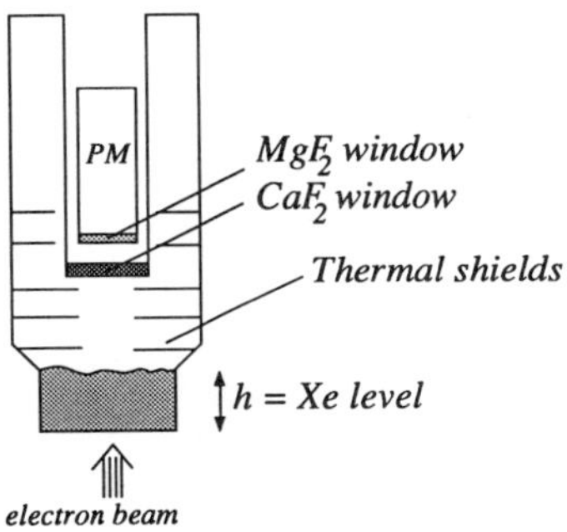

Fig. II.2.18. Principle design of the apparatus used to measure the light level and self-absorption of the liquid Xenon.

II.3. LARGE AREA DEVICES

II.3.a. Construction

The goal of this component of LAA is the development of the detectors which will cover the large area (several thousand square metres) associated with the detection of muons. The main requirements are:

- rate capability in excess of 1 Khz/cm;
- fast response time (of the order of 10 ns);
- easy construction;
- mechanical stability;
- low cost;
- easily adaptable to the geometry of the magnetic fields.

The need for a detector with unusual geometry is illustrated in Fig. II.3.1, where the principle of muon detection in the forward and backward direction is illustrated.

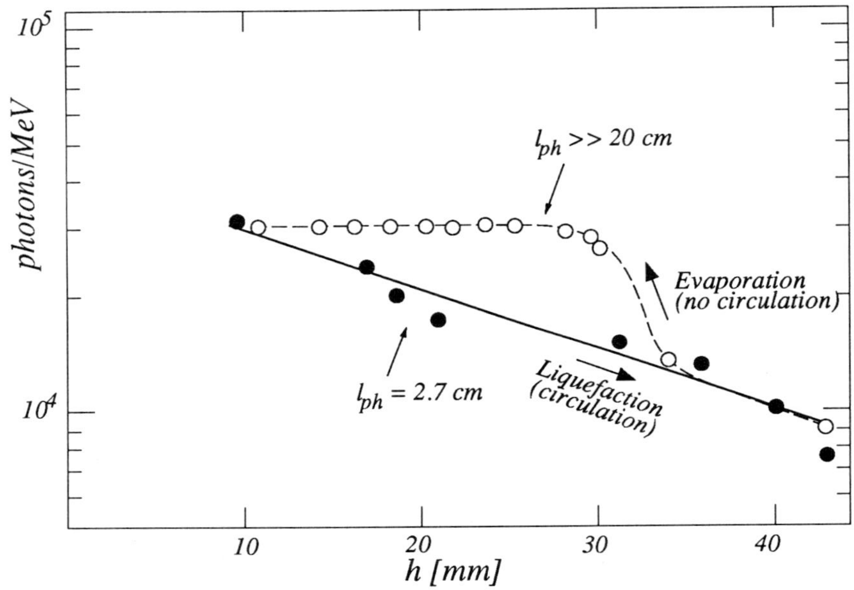

Fig. II.2.19. Absorption curves for liquid Xenon in circulating and non-circulating modes.

In these regions, a toroidal field is the only practical solution for measuring momenta. In these regions muon spectrometers require toroidal magnetic fields. In a toroidal field, a polar coordinate reading is desirable to improve trigger capability and momentum measurements. Therefore, the problem reduces to the following question: how can we build either a detector with circular cells, or a detector which can follow whatever geometry is needed? The answer to this question comes from the invention of two new types of detector:

i) A mechanically stable detector using blades instead of wires, the "blade chamber".

ii) A new wireless chamber with pixel read-out, the "gaseous pixel chambers".

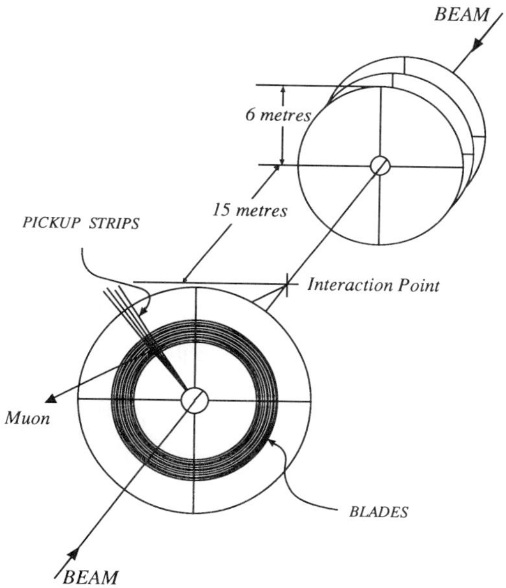

Fig. II.3.1. Geometry of the blade chambers.

Blade Chamber

The performances of a prototype "blade chamber", whose principle design is shown in Fig. II.3.2 were measured in a particle beam. The position resolution is $\sigma_{\perp} = 250\mu m$ transverse to the blades (Fig. II.3.3) and $\sigma_{\parallel} = 380\mu m$ along the blades (Fig. II.3.4). Moreover, by reading out the charge collected in the cell walls, the left-right ambiguity can be solved. This avoids the need for a double layer of chambers, as it is usually done with cell-shaped detectors. The rate limit of this new detector is 1 KHz/cm^2, matching the expected rates for muon detectors in future hadron colliders.

This new detector represents a factor 3 improvement in the precision, and a factor 10 improvement in rate capability, with respect to the classical Limited Streamer Tubes, used so far as muon detectors in big collider experiments (e.g., UA1 and ZEUS).

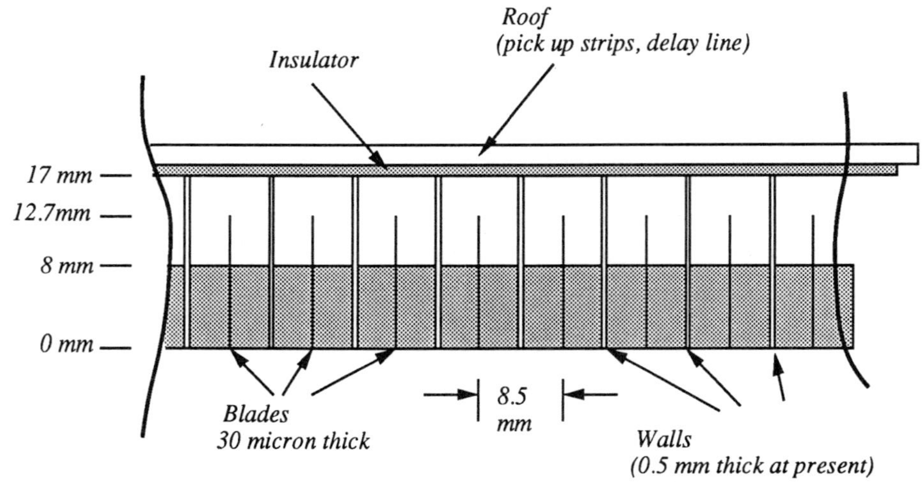

Fig. II.3.2. Cross-section of the blade chamber with insulating roof.

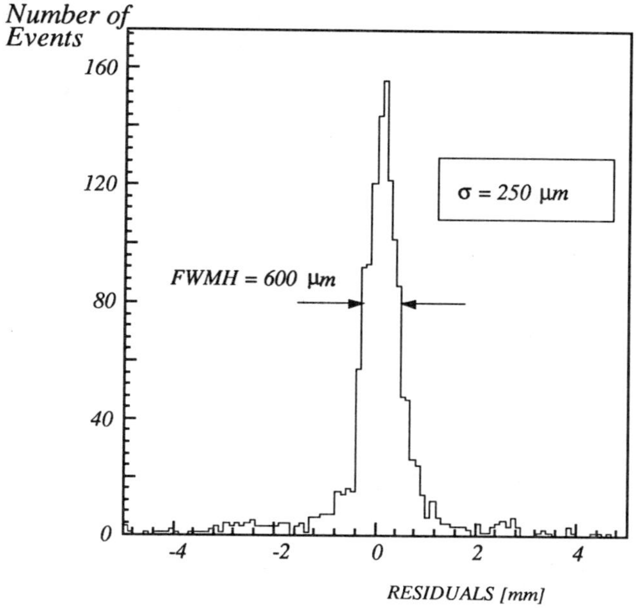

Fig. II.3.3. Spatial accuracy perpendicular to the blade from drift time measurement.

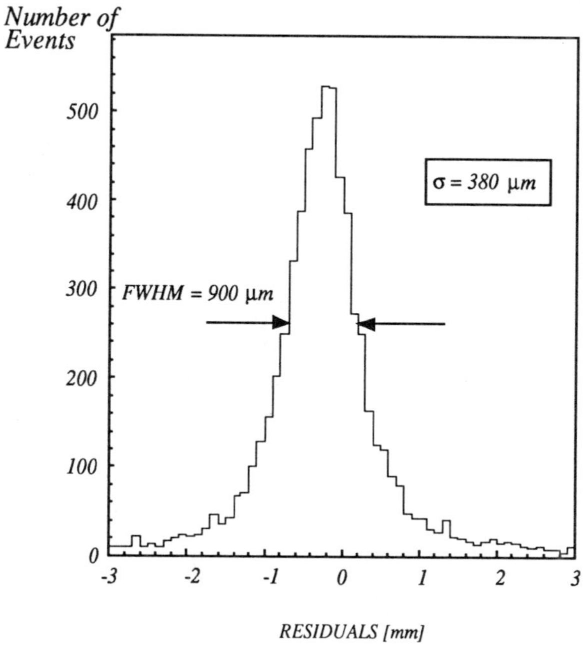

Fig. II.3.4. Position resolution along the blade using 11 mm wide pick-up strips.

Gaseous Pixel Chamber

The principle design of the gaseous pixel chamber is shown in Fig. II.3.5. It consists of a gas-filled volume delimited by a sensitive surface with pixel size of 1 cm, shown in Fig. II.3.6, and by a field-shaping roof, shown in Fig. II.3.7. The chamber works in the limited streamer mode. This device has been tested in a 6 GeV/c pion beam. Its efficiency has been measured to be 98% at an operating voltage of about 7 KV (see Fig. II.3.8). The chamber can withstand a rate of 5000 KHz/cm^2 without losses in efficiency. In conclusion, we have built a device with a very flexible geometry which can be used for triggering purposes in the whole detector.

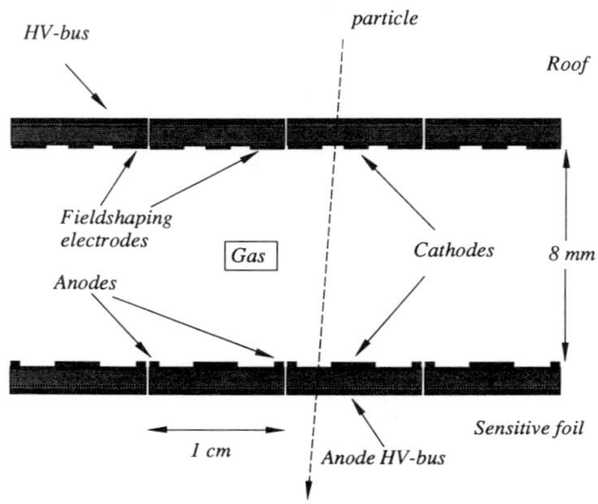

Fig. II.3.5. Principle design of the "gaseous pixel chamber".

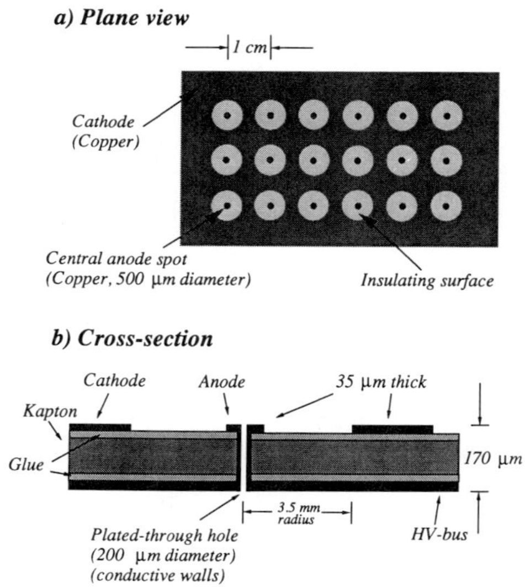

Fig. II.3.6. The "gaseous pixel chamber" sensitive surface.

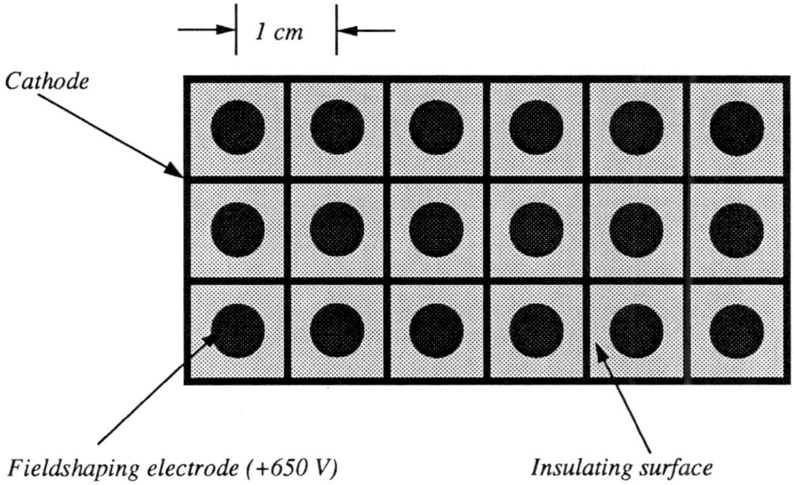

Fig. II.3.7. The "gaseous pixel chamber" roof.

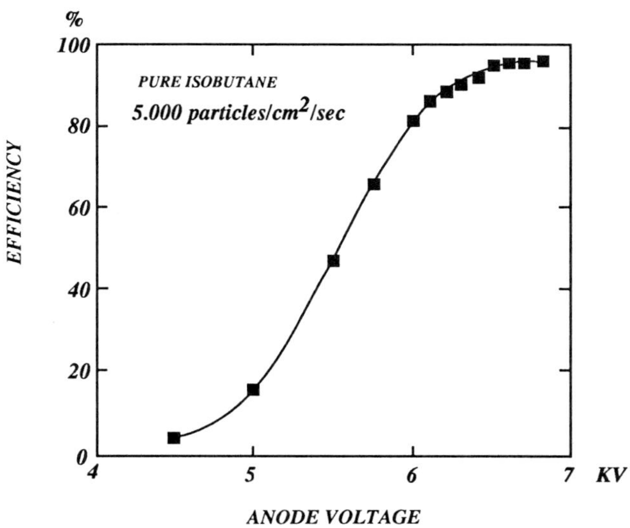

Fig. II.3.8. Efficiency versus high voltage measured in 6 GeV/c pion test beam.

II.3.b. Alignment

As already mentioned in the previous section, the muon detectors of future multi-TeV colliders will have a surface of several thousand square metres. A precision positioning of these detectors is absolutely necessary, because no detector can be better than its alignment.

Therefore, new instruments for length, angle and linearity measurements have been developed. The LAA Project has now the best set of instruments for measuring:

- lengths: 3.6 μm in 2 m;
- angles: 1 μradiant;
- linearity: 4 μm in 5.4 m.

These values, which represent a new **record** in the field of High Energy Physics, are an intermediate step towards of an overall alignment precision of 20 μm over 10 m, which is the final goal of this component of LAA. Before LAA, the most complete alignment system achieved a precision of 30 μm over 3 m [13]. In Table II.3.I, the present achievements are compared with the design goals of this component of LAA.

Table II.3.I. Design goals and achievements in aligment precision

	DESIGN GOAL	ACHIEVED
LENGTH	5 μm in 5 m	3.6 μm in 2.0 m
ANGLES	1-5 μRad	1 μRad
LINEARITY	4 - 10 μm in 10 m	4 μm in 5.4 m (needs a more powerful LASER source to increase the length)
PLANARITY	10 μm in 5×5 m^2	Still to be done (needs the instruments developed for linearity)

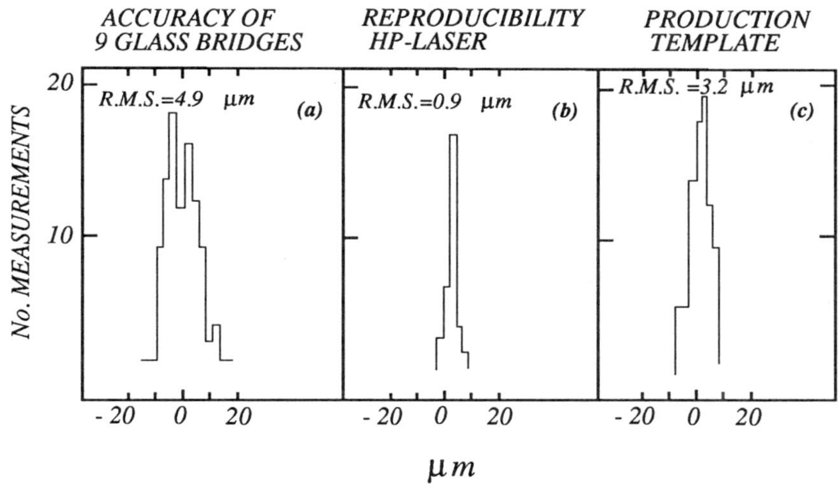

Fig. II.3.9. Error on a length standard test sample.

As an example of the measurements which are possible with the new instruments, Fig. II.3.9 shows the precision achieved with a length standard. The intrinsic error, not due to the measuring device (LASER interferometer), or to the production template, is of 3.6 μm over a length of 2 m.

II.4. LEADING PARTICLE DETECTION

The very difficult problem of designing a Leading Particle Spectrometer for a hadron collider was never tackled in the past. Therefore, all the work done by this component of the LAA Project constitutes a **new development.**

After three years of work, all instruments have been designed and built, and the various problems connected with the harsh environment of the very forward region (small space between the machine elements, radiofrequency, high background), have been solved.

In particular, one of the crucial problems was the cutting of the microstrip silicon detectors to a predefined (arbitrary) shape matching as much as possible the beam profile. This must be done because the acceptance for leading protons increases steeply for decreasing distances from the beam axis. Therefore, the detectors must be positioned as close to the beams as possible.

After several trials with different instruments, we succeeded in cutting the detectors to an elliptical shape, keeping the cracks on the surface to a **record** low level of 50 μm, i.e. 10 times better than industrial standards (see Fig. II.4.1).

Fig. II.4.1. Magnified view of the detector edge after the cutting with a diamond saw.

Some of the detectors have been received from the manufacturer (see Fig. II.4.2). Their characteristics are the following:

- Sensitive material: silicon;
- Active area: $80.96 \times 36.80\ mm^2$;
- Number of strips: 1024;
- Strip width: 25 μm;
- Strip pitch: 81.3 μm.

Fig. II.4.2. One of the silicon microstrip detectors to be used for leading particle detection.

Fig. II.4.3. The prototype pot, with the mechanics for positioning and for the fast extraction of the detector.

A collaboration has been established, which has allowed the LAA Leading Particle Detection group to make some research and development based on real life at the HERA (ep) collider. Part of this collaboration consisted in designing on the one hand the inner radiofrequency shielding and on the other hand the details of the interface between pot mechanics and the machine. Figure II.4.3 shows the complete prototype pot, with the mechanics for the fast extraction and very precise positioning of the detector.

Moreover, a direct measurement of the background conditions near the beams was done at the CERN SPS. The results are presented in Fig. II.4.4, where the extrapolations to a multi-TeV hadron collider are also shown. Notice that the background caused by beam tests is three orders of magnitude bigger than the stable beam conditions. Therefore, a continuous monitor of the beam conditions is absolutely necessary to preserve the leading particle detector from disastrous radiation damage.

In particular, the following circuits have been designed, produced and tested:

- *Front-end CMOS amplifiers.*

  Three different circuits have been successfully tested:
  - *A fast (100 ns) preamplifier in conventional CMOS technology.*
  - *A preamplifier with 1% tuneable gain.*
  - *A faster (15 ns) preamplifier "bipoltest" using lateral bipolar effect in SACMOS process.*

The output signals from the above amplifiers are shown in Figs. II.5.2 (a), (b), and (c), respectively.

- *Discriminator;*
- *Analog pipeline;*
- *12-bits A/D converter.*

A partial architecture for HARP using the functional building blocks has also been **developed.** This is a 4-channel AMPLEX upgrade in SACMOS process. It has not only preamplifier building blocks, but also track/hold and multiplexing circuits, so that it can be used in conjunction with a realistic detector module.

Another new development is the read-out electronics for a silicon micro-pattern detector (size: $200 \times 200\ \mu m^2$), working at 10 MHz. A layout of this circuit is shown in Fig. II.5.3.

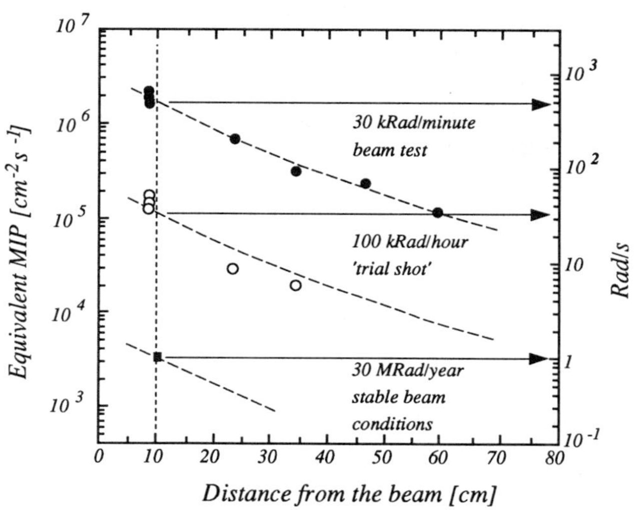

Fig. II.4.4. Left scale - results of the test at the CERN SPS ($L = 10^{29} cm^{-2} s^{-1}$) : dose rate as a function of the distance from the beam.
Right scale - extrapolation to a Super Collider ($L = 10^{33} cm^{-2} s^{-1}$).
(1 Rad corresponds to $3.6 \times 10^7\ MIPs/cm^2$).

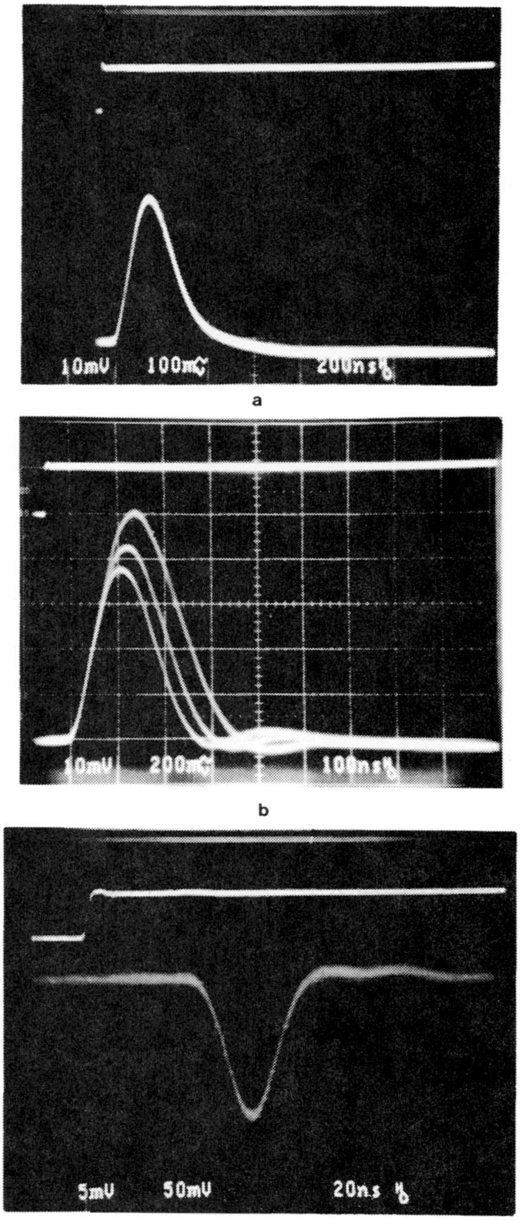

Fig. II.5.2. *(a) Output signal of the fast preamplifier (100 ns peaking time);*
(b) The effect of the pilot current on the gain (top) is obtained for $I_{pilot} = 0.2\ \mu A$*, the lowest (bottom) for* $I_{pilot} = 0.6\ \mu A$*);*
(c) Output signal of the experimental lateral bipolar amplifier in CMOS (15 ns peaking time).

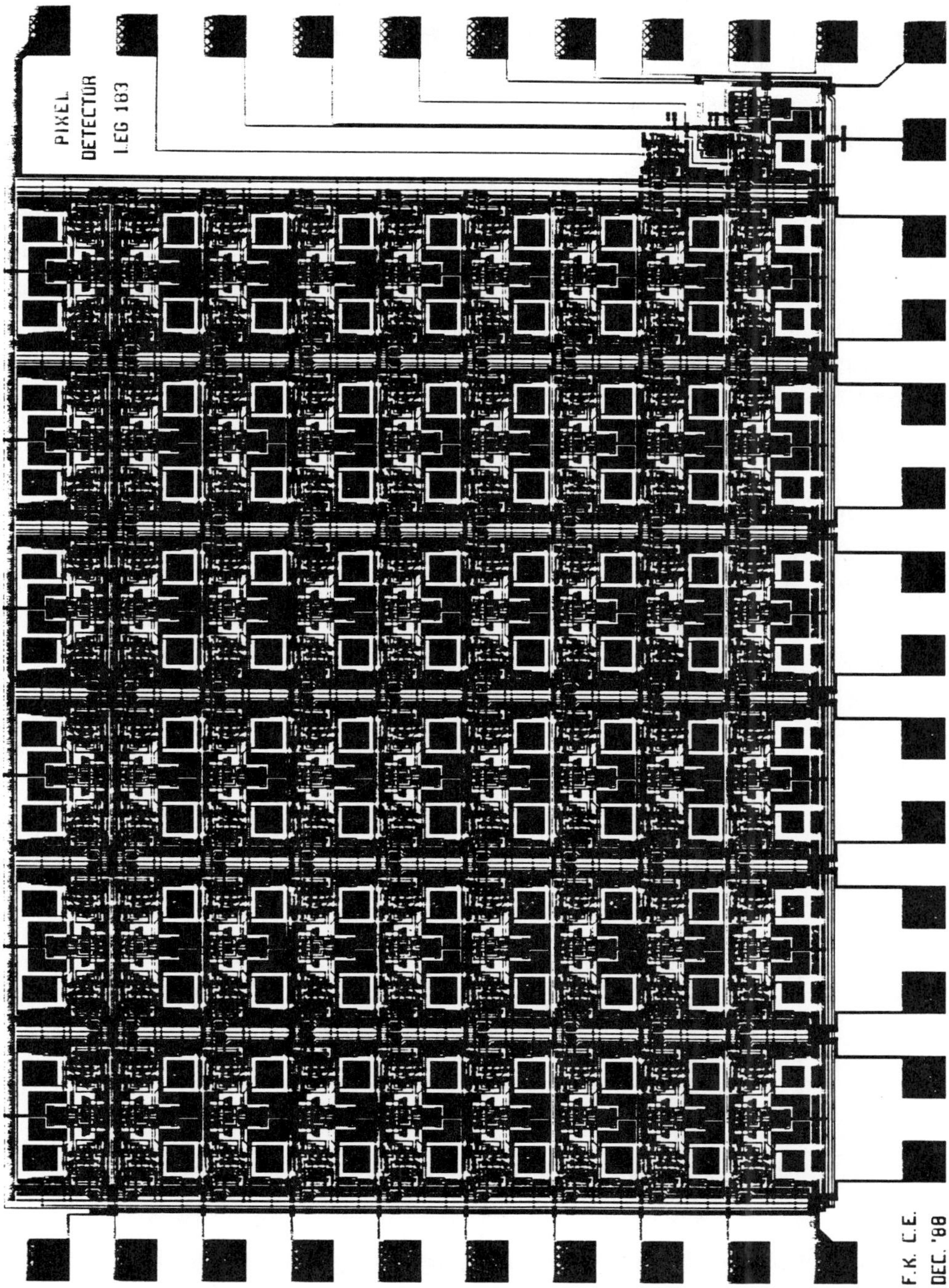

Fig. II.5.3. *The read-out electronics for the pixel detector. The size of each amplifier, matching the pixel size, is* $200 \times 200\ \mu m^2$

II.5.b. Radiation-hard technologies

The radiation-hard technology chosen by this component of the LAA Project is Silicon-on-Sapphire (SOS). The SOS technology is part of a larger category called Silicon-on-Insulator (SOI), meaning that the substrate onto which the CMOS integrated circuitry-structures are built is not made of bulk silicon, as in ordinary CMOS, but of sapphire. The intrinsic advantage of this arrangement is that, since sapphire is an insulator and not a semiconductor like silicon, there are less parasitic electrical couplings through the substrate plate. In particular, there can be no thyristor-type short-circuits so-called latch-ups appearing in the substrate. Such short-circuits may be provoked in bulk CMOS circuits by very strongly ionizing nuclear collisions in the substrate, caused, for example, by incident heavy ions or very high energy protons.

A first test circuit consisting of a cascode amplifier stage with bias points stabilized by current mirrors, and which is foreseen as a front-end amplifier on a microstrip detector readout chip, has been designed and fabricated.

The fabricated circuit has been irradiated with a ^{60}Co source. Fig. II.5.4 shows the measured noise, normalized to 1 before irradiation, as function of radiation dose up to 400 krad. No significant variation can be seen up to 400 krad. Further tests have confirmed that, even if the chip is irradiated up to 1.4 Mrad, the noise level remains acceptable, provided that some days of recovery are allowed at a temperature of 120^0C. This constitutes a new **record** for circuits to be used in High Energy Physics.

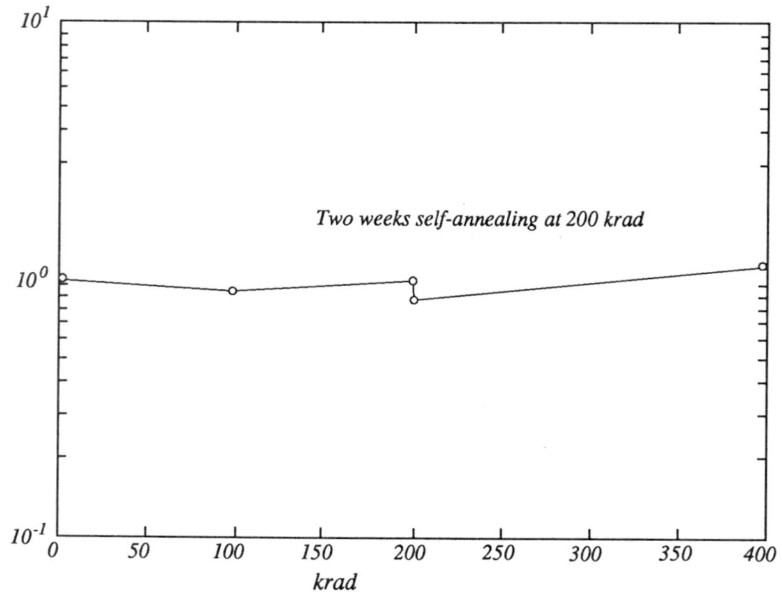

Fig. II.5.4. Measured noise as a function of radiation dose. The noise has been normalized to 1 at 0 krad.

II.6. DATA ACQUISITION AND ANALYSIS

In a future multi-TeV collider, the amount of data which has to be managed by the data acquisition system is enormous. For a luminosity ranging between 10^{33} and $10^{34} cm^{-2}s^{-1}$, the collision rate is expected to be between 10^8 and 10^9 events/s. With at least 10^6 channels per event containing significant data, coded in one byte (8 bits), we expect a total data volume of

$$10^{14} \div 10^{15} \; bytes/s.$$

Therefore, it is of paramount importance to reduce the data volume as early as possible. Ideally, this can be achieved by embedding the data reduction systems in each detector part, using simple computer architectures.

We have exposed **three selected architectures** commercially available (a Data-Cube image processing system with Sun 3 host station, a single-string 16384-processor ASP system from Brunel University, and a 48 × 48-processor GAPP system from Laben, Milano) to a set of **seven typical feature-extraction algorithms** (benchmarks): peak finder, pattern finder, calorimeter cells clustering with fixed neighbourhood, calorimeter cells clustering with variable neighbourhood, missing energy, Houghtransform for track finding, and gray level histogramming. For comparison, the same algorithms have also been run on a VAX8300. The results of the benchmarking, run either through proprietary simulators or on real hardware, are shown in Fig. II.6.1, where more or less trivial extrapolations for SIMD array sizes and technology are taken into account.

The use of commercial architectures constitutes a **new development** for the High Energy Physics community.

The conclusion is that the strong and weak points of the architectures mostly depend on the algorithms. Image processing systems are overall the easiest to cope with, and predictably the most efficient solutions: the ultimate computer performance on pixel information is obtained with one processing element per pixel.

A second path towards "intelligent detectors" consists in *embedding commercial architectures* in our environment of detectors with extremely fast response.

A second path towards intelligent detectors consists in embedding commercial architectures in our environment of detectors with extremely fast response.

The important pilot project of integrating a processor in experimental readout pipelines (typically VLSI design) has led the LAA group to defining an extension to the design of the SPARC chip. The SPARC (Scalable Processor ARChitecture, from Sun Microsystems) is an open RISC (Reduced Instruction Set Computer) processor architecture. The LAA project ultimately will intercept the SPARC at the architectural level, extending the instruction set to allow direct addressing of the Fastbus. In order to produce a usable device on a relatively short timescale,

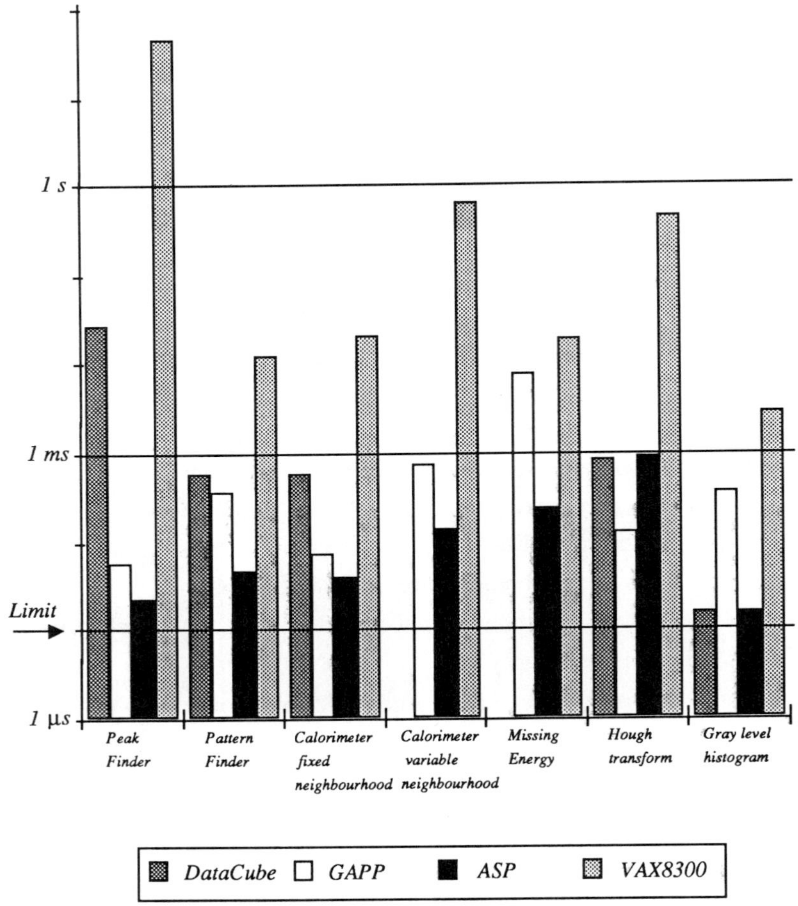

Fig. II.6.1. Performance of different computer architectures on various feature-extraction algorithms. *The 10 μs limit corresponds to the maximum time allowed to the second level trigger.*

timescale, an intermediate approach has now been chosen, interfacing standard commercial SPARC chips via a coprocessor to Fastbus.

This pilot project has now a finished design, for a 4-processor Fastbus board, as shown in Fig. II.6.2. The design has been fully simulated. Final negotiations with the chip manufacturer have started.

This LAA project will result in a commercial-like product allowing to put together experiment-dependent Fastbus boards with a record high computing power of 50 Vax equivalents with complete software support.

A third part of the "Data Acquisition and Analysis" component of LAA has developed an image processing workstation for physics analysis of very complex events (CCD images or data from future detectors with extremely high granularity): the TRAX-I. It consists of a VME-based workstation as a host processor, a series of hard wired image processing modules for data acquisition, background subtraction, gain equalization and image display, an Associative String Processor (ASP) with 16000 processing elements, and an array of Transputers interconnected through VME and a fast data channel.

The novelty in this architecture is the Associative String Processor (ASP) developed by R.M. Lea and colleagues at the University of Brunel. The ASP is a parallel processing computational structure comprising a string of identical Associative Processing Elements (APEs).

The TRAX-I architecture has now been completely specified, the first batch of VLSI chips for TRAX-I has been produced (64 processing elements per chip made in 2 μm CMOS, as shown in Fig. II.6.3) and tested. The LAA benchmarks described above have been implemented on the ASP simulator: at present the TRAX-I corresponds, for specific tasks, to 200 to 3000 VAX8300. And this is a **record.**

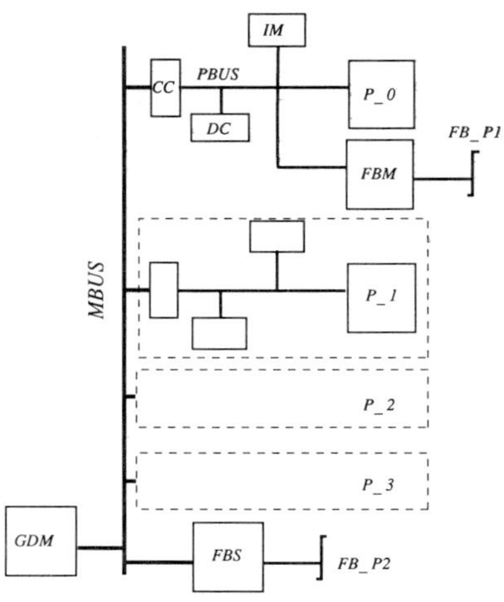

Fig. II.6.2. P_0, P_1, P_2, P_3: SPARC Processors; IM: instruction memory; FBM: Fastbus master interface; DC data cache; CC: cache controller; FBS: Fastbus slave interface, GDM: global shared memory; FB _P1,2: Fastbus backplane.

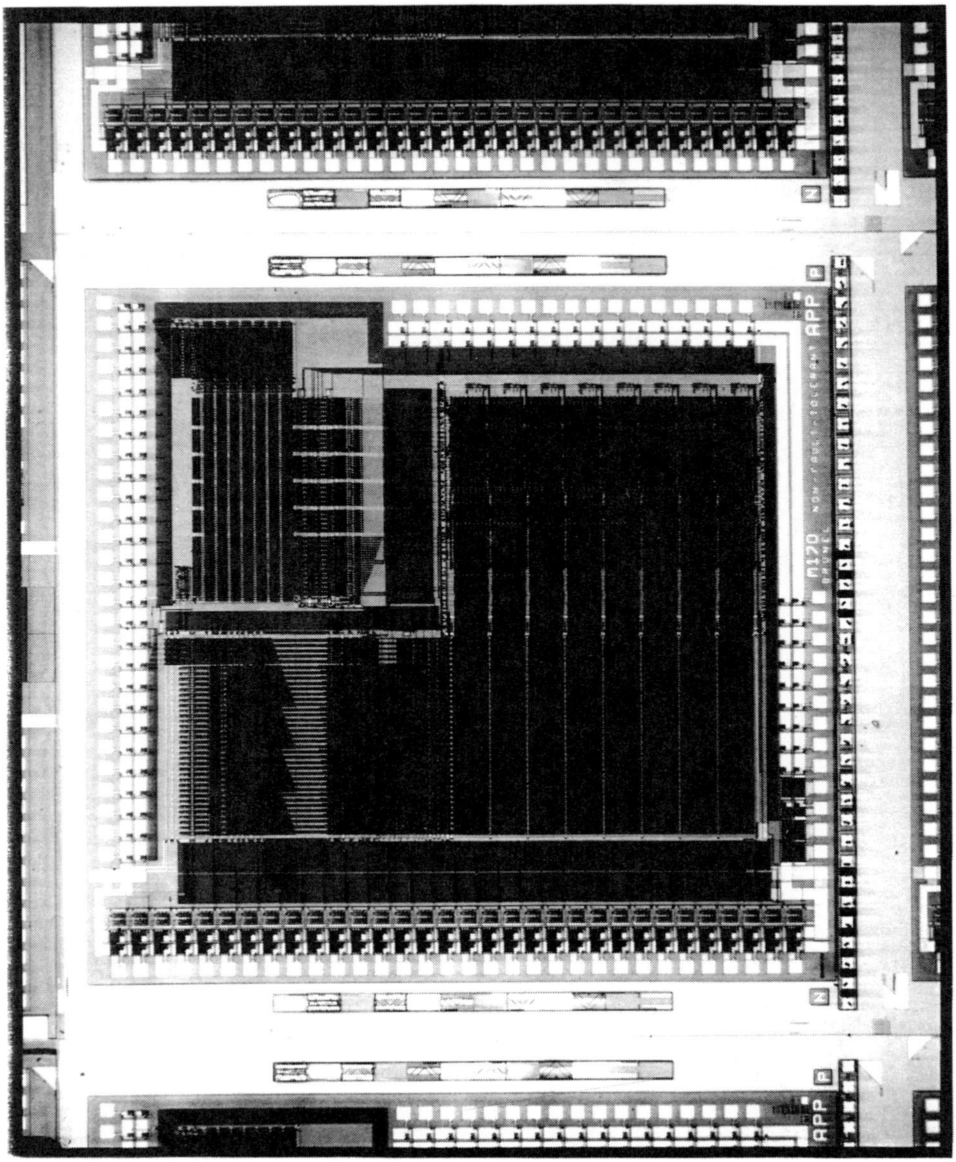

Fig. II.6.3. Picture of one VLSI chip for TRAX-I.

II.7. SUPER COMPUTERS AND MONTE CARLO SIMULATIONS

Monte Carlo Simulations for future Super Colliders are of great relevance. They are in fact needed for the prediction and/or extrapolation of the Physics scenario to very high energies, for the detector design, and for the interpretation of the experimental data.

The MonteCarlo Simulations component of LAA has proceeded along two lines:

i) *Program implementation and software tools development.*
The software tools for utilizing a full Monte Carlo simulation chain have been developed. The hardware equipment to be used by this component of LAA has been acquired and installed and the training of a group of young physicists on various event generator programs has been completed.

ii) *Monte Carlo studies and simulations.*
Studies and calculations on different event generators and on different effects (multihadronic production and leading effect, heavy flavours) have been carried out.

The main **development** of this component of LAA is a Full MonteCarlo Chain (FMC), illustrated in Fig. II.7.1, integrating different physics and detector simulation programs in a single, user-friendly package.

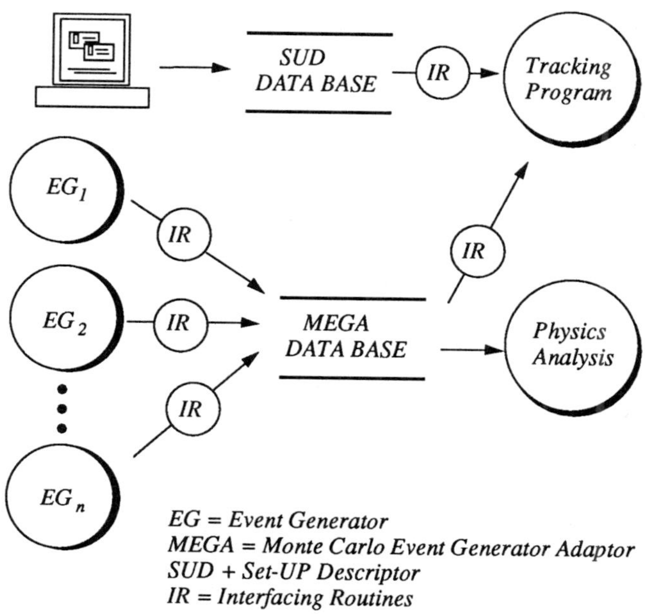

Fig. II.7.1. *Block diagram of the Full Monte Carlo Chain (FMC).*

For the first time, a modern relational data structure system, ADAMO, has been used as a framework for connecting the data from various Monte Carlo event generators existing on the market (LUND, HERWIG, ISAJET, EUROJET), to different tracking programs. Interfacing routines have been written which take into account the differences in event generator outputs and tracking program inputs. These routines, collectively named MEGA (Monte Carlo Event Generator Adaptor), use data structured into tables which are handled in a simple and flexible way. The main advantages are transparency to the user and modularity, which is suitable for the comparison of different Monte Carlo event generator outputs.

The FMC is, at present, implemented in the following computers: VAX/VMS (released), IBM/VM (released), CRAY (in preparation).

Another important software tool has been developed: the SUD (Set-Up Descriptor). SUD is a batch/interactive program to define the detector elements for the tracking program.

A new program system, which represents a natural development of the FMC, is in preparation: COSMOS (COmprehensive Super MOnteCarlo System). It will be completely modular, and it will allow the user to interactively select:

- the type of interaction;
- the centre of mass energy;
- the Feynman diagrams to be used in the interaction;
- the structure functions of the partons;
- the value of Λ_{QCD};
- the hadronization model (cluster, string, etc.);
- the level of analysis (after generation, evolution, hadronization, decay, tracking).

The logic structure of COSMOS is shown in Fig. II.7.2. The grey part has been implemented.

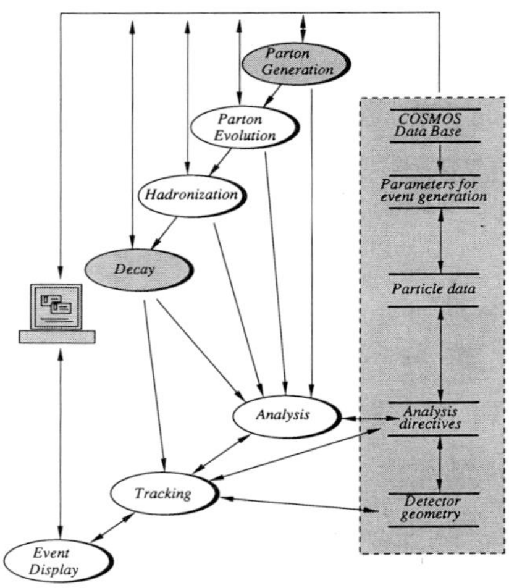

Fig. II.7.2. Schematic diagram of COSMOS.

A first detector configuration has been introduced in the FMC to simulate very high energy events. It consists in a simple tracking section (with and without a magnetic field), a Pb/scintillator calorimeter, and a muon tracker.

The complexity of the events, which are expected in a multi-TeV collider is illustrated in Figs. II.7.3 and II.7.4. These figures show different views of a $pp \to t\bar{t}X$ event at $\sqrt{s} = 16\ TeV$. The total multiplicity of the event is 483 (notice that all tracks with less than 300 MeV/c are not displayed). The top quarks (mass $= 90\ GeV/c^2$) decays into a $\mu + \nu_\mu b$ producing a well separated muon, with a momentum of 343 GeV/c, going in the forward direction.

A theoretical study of possible effects of the baryon and lepton number violation at very high energies (order of 100 TeV), has been started by the 200 TeV Club. The picture which is under development foresees that the effects of the non-conservation of the baryon number (<u>but</u> conserving the difference between baryon and lepton numbers), are too small at low energies, but rapidly increase with increasing energy. These phenomena are non-perturbative results from tunneling between different vacua of the Standard Model. Figure II.7.5 illustrates the typical feature of events, very high multiplicity of particles (of the order of $1/\alpha_{\text{weak}}$), expected.

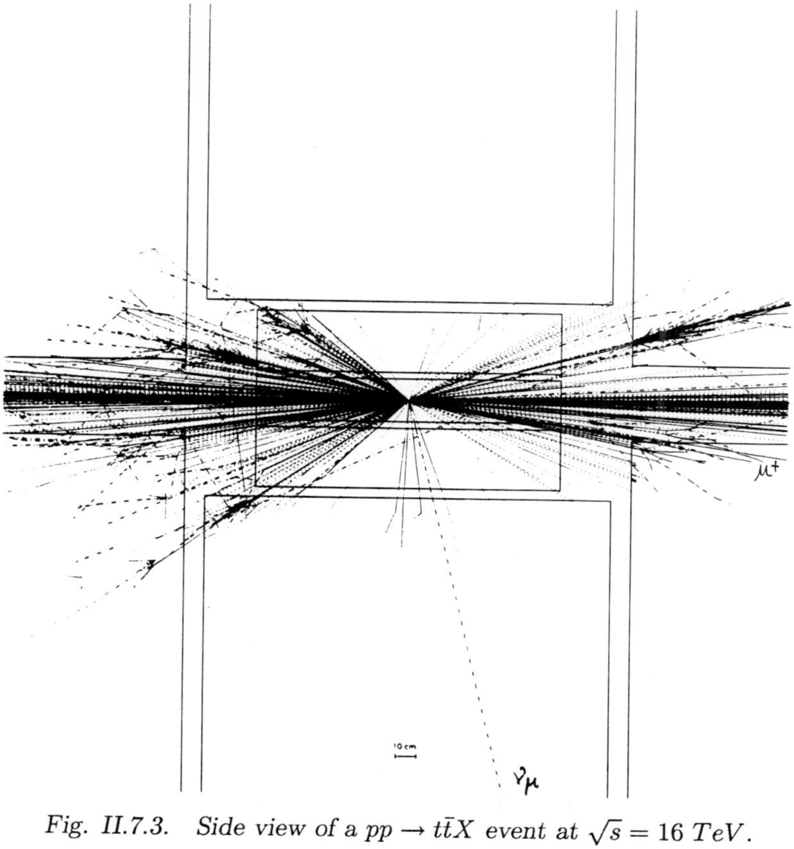

Fig. II.7.3. Side view of a $pp \to t\bar{t}X$ event at $\sqrt{s} = 16\ TeV$.

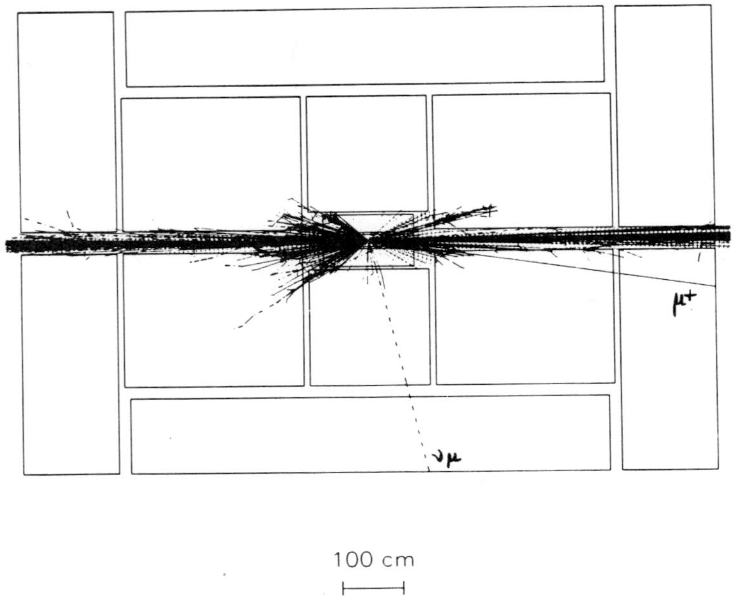

Fig. II.7.4. Same as Fig. II.7.3., but with a 4 T magnetic field in the central region.

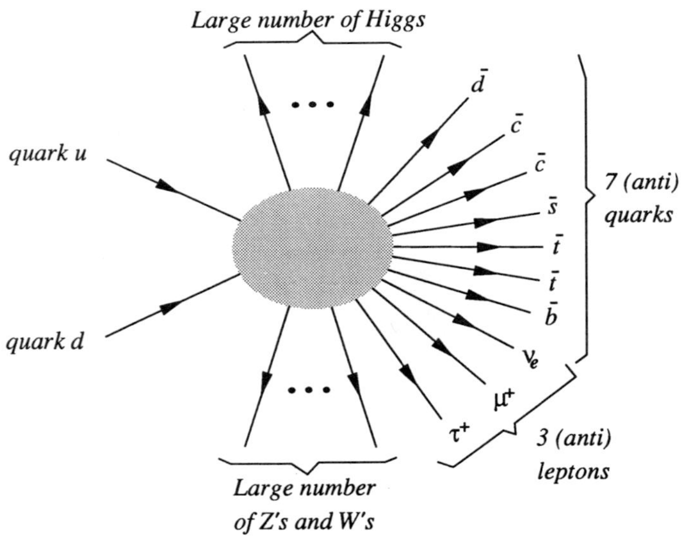

Fig. II.7.5. Typical final state produced in a baryon-number-violating event.

III. CONCLUSIONS

The results presented, i.e.: **five Discoveries, eleven Records, twelve New Developments and four Inventions,** are the best proof of the validity of the LAA Project.

References

[1] A.Zichichi, "Report on the LAA Project", Volume **1**, 15 December 1986.

[2] A.Zichichi, "Report on the LAA Project", Volume **2**, 25 June 1987.

[3] A.Zichichi, "Report on the LAA Project", Volume **3**, 19 November 1987.

[4] A.Zichichi, "Report on the LAA Project", Volume **4**, CERN-LAA/88-1, 25 July 1988.

[5] A.Zichichi, "The LAA Project", Volume **5**, CERN-LAA/88-2, 19 September 1988.

[6] A.Zichichi, "Report on the LAA Project", Volume **6**, CERN-LAA/89-1, 15 September 1989.

[7] A. Zichichi, "The ELOISATRON Project", August 1989.
A. Zichichi, "The ELOISATRON Project: Eurasiatic Long Intersecting Storage Accelerator", in *New Aspects of High-Energy Proton-Proton Collisions*, Plenum Press, 1988 (ed. A. Ali), page 1.

[8] see, for example: G.G. Hanson and M.G. Gundy, preprint SLAC-PUB-5041, July 1989, and references therein.

[9] H. Walenta, Physica Scripta, **23** (1980) 354.

[10] R.E. Ansorge et al., UA2 Collaboration, Nucl. Instr. Meth. **A265** (1988) 33.

[11] H. Schönbacker, preprint CERN-TIS/RP/201.

[12] A. Adresen et al., ZEUS Calorimeter Group, preprint DESY 89-149.

[13] L3 Collaboration, Nucl. Instr. Meth., **A277** (1989) 187.

Chairman: A. Zichichi

Secretaries: D. Brahm, S. Hsu

Discussion

– *Borden:*

Scintillation fiber tracking chambers should achieve good resolution in two dimensions (r, θ), but what about the third (z)?

– *Zichichi:*

The fiber bundles are not all parallel, they are arranged to provide high resolution in three dimensions.

– *Haas:*

Could you comment on the organizational problems involved in a collaboration of thousands of physicists? Is this environment desirable?

– *Zichichi:*

When I started to do physics, there were 10 physicists in the great Blackett group in Manchester (U.K.). Experiments were office-sized. The trend ever since has been to larger instruments and larger collaborations. This trend is inevitable if you want to do new physics. ISR made the mistake of avoiding large detectors, so this powerful collider was not fully used, and many discoveries were missed: e.g. the Upsilon.

The new particles predicted by supersymmetry are invisible, so we can only find them through missing-energy signatures. For this we require large, fast, hermetic, radiation-hard, detectors. The large dimensions in terms of people and detectors are imperative if you want to discover new phenomena in High Energy Physics. Remember, though, that there will always be the need for a single fellow with a brilliant new idea. No progress can be achieved without new ideas. The large dimensions of our enterprises should not let anybody think that new original ideas are not needed.

– *Haas:*

Are colliders the only way to find new physics?

– *Zichichi:*

Colliders were invented and implemented, first with (e^+e^-) (B. Touschek), then with (pp) (K. Johnsen). I cannot think of a better instrument. We know now that our "heavy particles" are really very light; the "truth" is at the Planck scale. Considering the consequences of our ignorance on the energy level where new physics will show up, we have to start now not only to implement LHC and SSC, but also to prepare the ground for the highest achievable energy: 200 TeV.

– *Kastening:*

The SSC has about the largest realistically imaginable ring, but is much lower than 200 TeV. How can 200 TeV be achieved? Do you anticipate stronger magnetic fields, for example, or will we use linear accelerators?

– *Zichichi:*

The Eloisatron conceptual design calls for a 300-km ring, and should reach 200 TeV, without any problem of scientific or technological nature.

– *Brahm:*

You have described a very innovative detector, probably useful for many experiments, but it is named after a fairly ordinary experiment ("Lepton Asymmetry Analyzer"). Is this a historical accident, or is this experiment particularly demanding?

– *Zichichi:*

The LAA experiment was originally planned when a 4th generation of quarks seemed likely. It was known that a very heavy quark was needed in SUSY to raise the mass of the gluino, and people did not imagine that the top was as heavy as it turns out to be. It was important, then, to determine whether a new quark was up-like (u_3) or down-like (d_4). Fortunately, scheduling problems prevented us from building the LAA then, so we continued research into detector improvements. I have since become convinced that physics must reach much higher energies, 200 TeV. The LAA project has thus become the development of a fast, hermetic, radiation-hard, detector for the next generation of colliders.

– *Ciuchini:*

We know that no decoupling theorem holds for the "low-energy" Standard Model. Do you think that high-precision measurements can replace or at least complement high-energy experiments?

– *Zichichi:*

30 years ago I worked on a high-precision experiment, the $(g-2)$ of the muon. We found that the muon is a lepton. I also did the most accurate measurement of τ_μ, i.e. of the Fermi coupling. I will never work on a high-precision experiment again! I don't think you can disentangle vacuum corrections well enough to learn much about the Standard Model from high-precision measurements. The best way to find new physics is not to worry about a high precision experiment first, but to

do a qualitatively good one in some new energy regime. Then follow it up later with high-precision measurements.

– *Kastening:*

High-precision measurements of CP violation gave ϵ'/ϵ.

– *Zichichi:*

That's true, that experiment is worth doing. But I have more interesting experiments in mind.

– *Ypsilantis:*

Would another high-precision CP-violation measurement be worthwhile, e.g. B physics?

– *Zichichi:*

It could be, but in our field we need to make choices. You must decide what priority you want to give to the many things we would like to know. My highest priority would be the search for the Higgs and the SUSY particles.

A NEW APPROACH FOR CONSTRUCTING SENSITIVE SURFACES: THE GASEOUS PIXEL CHAMBER*

D. Mattern, M.C.S. Williams and A. Zichichi

LAA group, CERN
CH - 1211 Geneva 23
Switzerland

ABSTRACT

Future Hadron colliders are expected to have an event rate so high that it will make the operation of gaseous wire chambers very difficult, except maybe for muon detection. The electrostatic attraction between the anode and cathode wires limit the density of anode wires. Many attempts have been made to construct a gaseous chamber with the anodes attached to an insulating film. Recently there have been very successful chambers built with an anode pitch of 200 mm. These are known as gaseous microstrip chambers. We also have built chambers on insulating foils, but of a completely different design. This design is more pixel in geometry and thus is known as the gaseous pixel chamber. Moreover this surface operates in limited streamer mode and is capable of working at high rates. We present here the current status of this device.

INTRODUCTION

The design luminosity at future hadron colliders is in the order of 10^{33} to 10^{34} cm^2/sec. It appears that at this luminosity gaseous wire detectors can not be used at a radius less than 50 cm.[1] However gaseous wire chambers are somewhat radiation hard, and have been in use for the last 30 years or so, thus could be labelled as 'known' technology. One of the main problems with the wire chamber is the electrostatic attraction between the anode and cathode wires. If they are too close, then either the wire tension has to be high (but the wires break at this tension) or the wire length has to be short (which introduces many boundaries and material in the endplates). There have been two recent developments which may solve this problem. The first is the gaseous microstrip chamber and the second is the gaseous pixel chamber. There are many groups[2-5] working on the gaseous microstrip chamber. This device will only be described briefly here. The gaseous pixel chamber has been developed within the LAA project[6] at CERN. Even though we are still in the early stages of the development of this device, it appears to have characteristics that will make it an important detecting device, especially at a 200 TeV collider.

* The work presented here is supported totally by the LAA project

Physics Up to 200 TeV
Edited by A. Zichichi, Plenum Press, New York, 1991

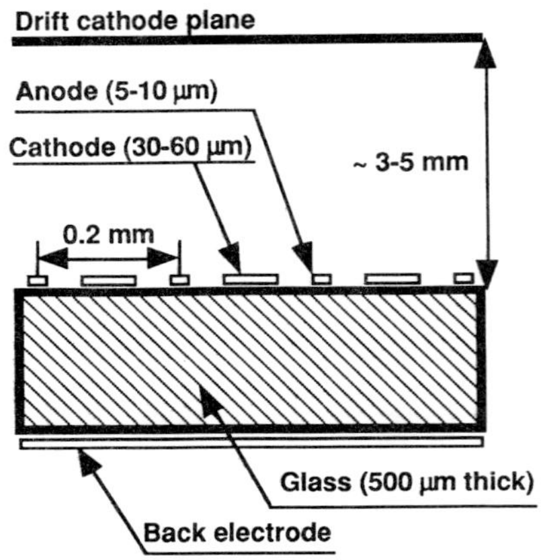

Fig. 1. Cross section of Gaseous Microstrip Chamber

THE GASEOUS MICROSTRIP CHAMBER

Many people have had the desire to build a chamber by 'printing' the lines on an insulator, using the techniques of printed circuit boards or some similar technology. In fact there have been successes. Neumann[7] et al. built a chamber where they glued wires onto a resistive foil and obtained efficiencies of 85%. Recently Oed[2] followed by others [3-5] have built a device with 10 μm wide lines and 200 μm pitch evaporated onto a glass or ceramic plate. They have obtained gas gains of 10^3 and 10^4. A cross section of a typical device is shown in figure 1. Due to the close spacing between anode and cathode, there is only limited 'gas poisoning' caused by the positive ions produced in the avalanche process. Hopefully this will allow the chamber to operate at extremely high incident particle rates. Certainly the high counting rates obtained with irradiation of the device with strong ^{55}Fe source indicate that this may be true. Another aspect of this chamber is that it has surprisingly good energy resolution. The width of the 6 keV ^{55}Fe line has σ = 12.8%. This is comparable to the best results obtained with a wire chamber. Tests of these devices in particle beams have started. These results indicate that a spatial resolution of 35 μm (or better) can be obtained. Hopefully results will soon be presented verifying the rate behaviour within a particle beam.

THE GASEOUS PIXEL CHAMBER

Rather naively we had a wish to build a large area gaseous microstrip detector. We scaled up their device and had lines of 100 μm etched onto standard Kapton and epoxy sheets with a pitch of 1 cm. After initially applying positive high voltage to the anode lines we observed large signals correlated to a ^{57}Co gamma source irradiating the surface. However these signals quickly disappeared (probably due to the surface getting charged) until only a low rate of signals remained. On investigation we found that the signals were produced just around the tips of the lines.[8] From this observation we have developed a surface of 'tips'. This is a small pad connected by a small diameter plated through hole to a high voltage bus on

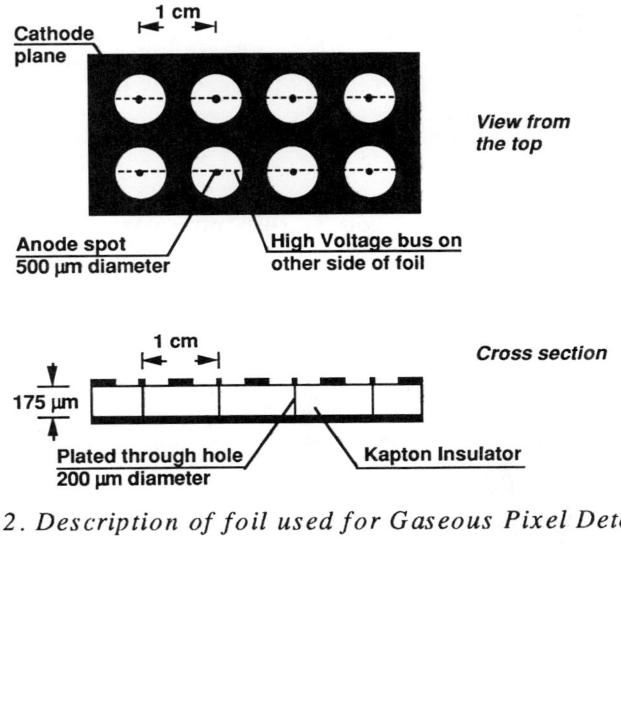

Fig.2. Description of foil used for Gaseous Pixel Detector

the other side of the insulating foil. An example of this foil geometry is shown in figure 2. We will discuss this type of surface for the remainder of this report.

DESCRIPTION OF FOIL

We have tested various foils where changes to the geometry have been made. The design of a typical foil is shown in fig. 2. The sheet of insulator is Kapton of 125 μm thickness. This is available commercially with a layer of copper (18 μm thick) glued onto each surface. The glue is a polyurethane glue and is 25 μm thick. This is treated by standard printed circuit board techniques to produce plated through holes and the desired design of copper conductor. The thickness of the copper is increased to 35 μm after this treatment. The final surface of the conductor is protected by a chemical deposit of gold. The layer of glue is left on the surface of the Kapton.

The basic cell of this surface consists of an anode surrounded at some distance by a cathode area. A typical radius for this clearance is 3.5 mm although we have tested other radii. We have also constructed surfaces where the cathode is a grid structure. These work in a similar manner. The anode is a small pad. We have tested various sizes with diameters between 200 μm and 1 mm. These pads are connected to a high voltage bus line on the other side of the insulating foil with a plated through hole of diameter 200 μm.

TESTS

We have fabricated foils of 160 x 160 mm^2. We have installed these in

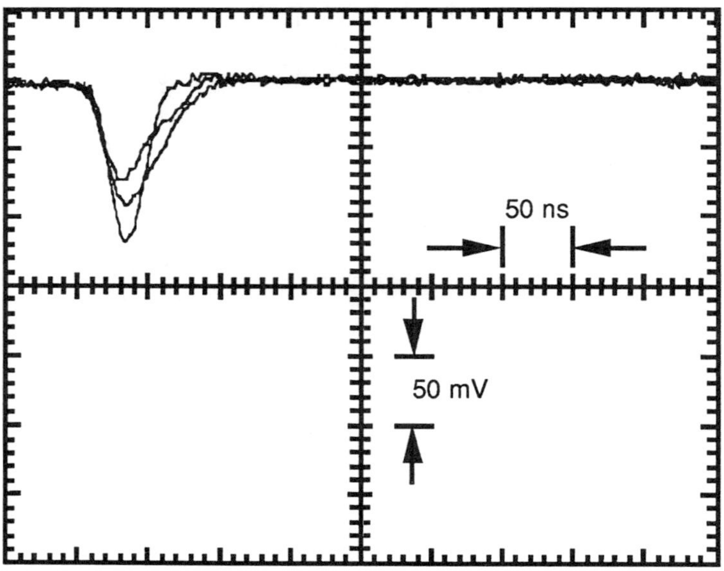

Fig. 3. Oscilloscope display of typical pulses from foil being irradiated with ^{55}Fe source (x 5 Amplifier used between foil and oscilloscope)

a gas tight box. The volume of gas above the foil was 5 cm thick. The cathode was connected to ground. The anodes were connected to a positive high voltage unit via a 10 MΩ resistor. The anode signal was decoupled with a 1 nF capacitor, then connected to a x5 LRS 335L amplifier. In case we had electrical breakdown on the foil we provided extra protection to the input of the amplifier with diodes and further decoupling with a transformer. We then studied this signal with a variety of instruments; a Tektronix 2440 digital oscilloscope, a LRS 3001 qVt or standard CAMAC ADC's and TDC's for the beam tests. In fig. 3 we show the oscilloscope traces of three successive pulses when a typical foil is irradiated with a ^{55}Fe source.

We flowed a mixture of isobutane and argon (most tests were done with pure isobutane) through the gas tight box. The box can be irradiated with a ^{55}Fe or ^{57}Co source. We built a simple cosmic ray telescope of three scintillators to study the response of the foils to minimum ionizing particles. For a more complete study we have tested some foils in a 6 GeV/c pion beam in the CERN East hall.

ELECTRIC FIELD

We have calculated the electric field produced by this surface in three dimensions. This has been done under the assumption that no field line crosses an insulating surface. We assume that initially crossing field line will transport free electrons and/or ions onto these surfaces and modify thereafter locally the field. With an ideal insulator we reach a situation where all equipotential surfaces end perpendicular on the insulators. In fig. 4 we show the potential on the surface between the anode and cathode. We superimpose two curves that fall as 1/r and as –ln(r/R). We expect the field to vary as $1/r^2$ around the anode point.

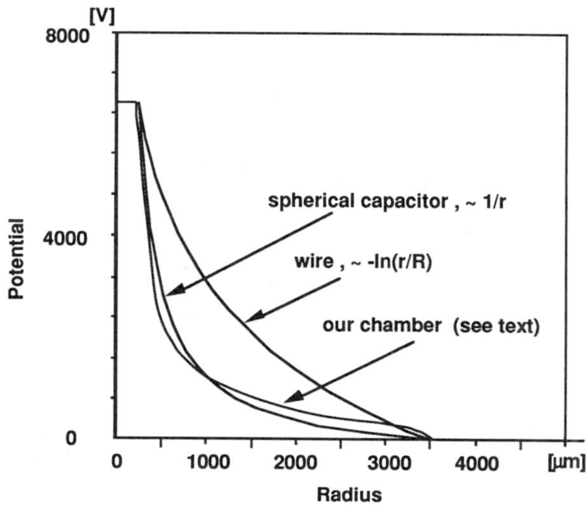

Fig. 4. Potential on surface between the anode and cathode

In reality there is a bus line on the backside of the foil that supplies the voltage to the anode spot. Immediately after turn on, the surface above this line will not be charged and field lines will pass through the Kapton and end on the high voltage bus. These lines will transport electrons to the surface of the Kapton and charge this surface. Ideally this process continues until the surface is 'fully' charged (that is, no field line enters the Kapton). However, when the surface is fully charged, then just above the surface there is a strong electric field parallel to the surface that transports the electrons to the anode. Thus, the fully charged state is unstable and the surface should not be regarded as a perfect insulator, especially above this high voltage bus line.

We have tested two identical foils except that one has the anodes strung together in a chain while the other has a single line running to each anode. This is shown in fig. 5. We installed the surface in a gas box with 5 cm of gas above the surface. We then irradiated each surface with an ^{55}Fe source. We measured the gain and counting rate as a function of high voltage. The gain of the two surfaces is identical but the counting rate was significantly higher for the surface with the single anode line, presumably this is due to primary ionization being transported to the surface above the high voltage bus rather than to the high field around the anode.

Another observation related to the high voltage bus line is that sparks are generally produced from the cathode directly above the high voltage bus. This region has a very high field produced by having the high voltage bus close to the cathode, this field is reduced by charging the surface of the Kapton insulator with electrons. In order to reduce this field to require less electrons on the surface, we mounted a ground plane below the high voltage bus (insulated by 100 µm thick Kapton tape). With this ground plane we found that the potential where sparks are produced is several hundred volts higher.

RESULTS

We have investigated the variation of gain with the size of the anode spot diameter. Again this was a foil in a gas box with 5 cm gas above the surface. Figure 6 shows our measurement of gain for a variety of anode spot diameters and also for various cathode radii. For a cathode radius of 3.5 mm there is an increase in gain by reducing the anode spot diameter from 1 mm to 600 µm. However, reducing the anode diameter further until we reach the 200 µm (which is the minimum as it is the diameter of the plated through hole) does not affect the gain. With a 500 µm anode diameter we reduced the cathode radius from 3.5 mm to 2.5 mm and finally to 1.25 mm. Reducing the radius from 3.5 to 2.5 mm does not affect the gain, however there is an increase for the 1.25 mm radius. With this small radius of 1.25 mm we reduced the anode spot diameter to 200 µm. This produced a further increase in gain. It is clear that there are a wide range of anode spot diameter sizes and cathode radii that produce pixels with the same gain. It will make it easy to fabricate surfaces with uniform operational characteristics. Another important characteristic of the foil is the voltage at which we produce sparks. The end points of the various curves in fig. 6 show the onset of sparking (this is not a very precise measurement as some surfaces would hold a voltage for 10 minutes before producing a spark). However, it is clear that a geometry with a smaller cathode radius produces sparks at a lower potential.

We have investigated the variation of the average streamer charge versus high voltage for various mixtures of argon and isobutane. This is for a foil

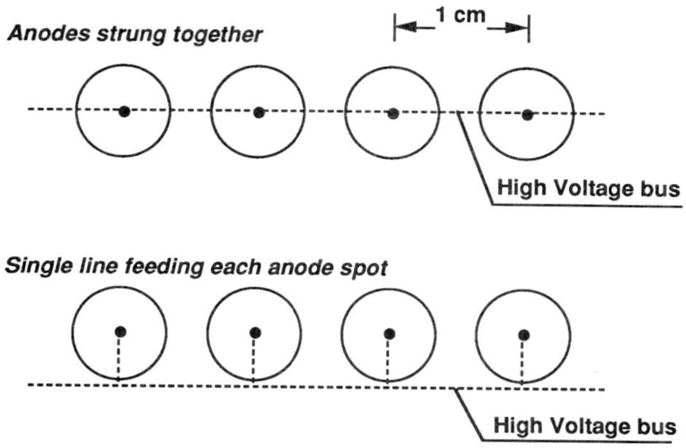

Fig. 5. The designs of the two different anode bus lines that were tested

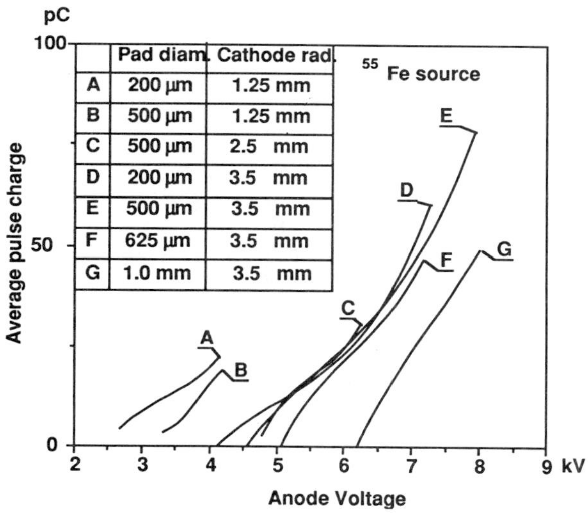

Fig. 6. Average pulse charge on various foils being irradiated with ^{55}Fe source

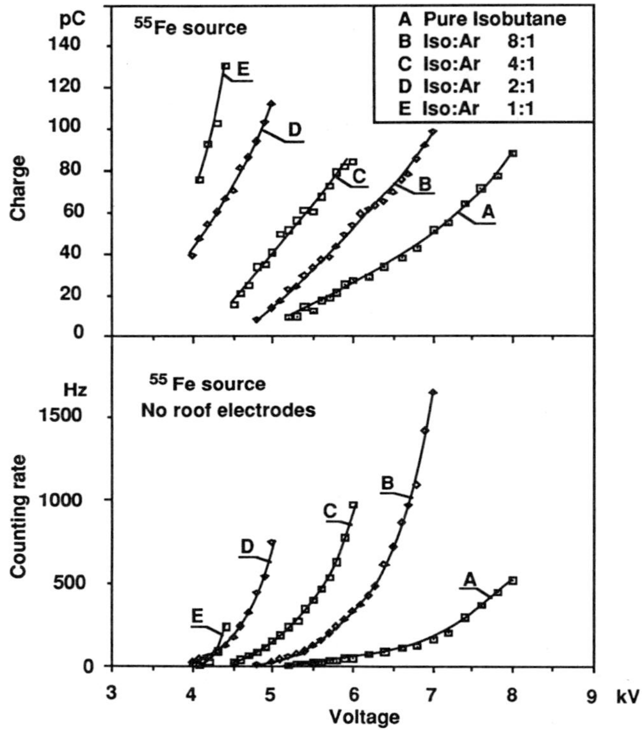

Fig. 7. Test of Gaseous Pixel Chamber with different gas mixtures

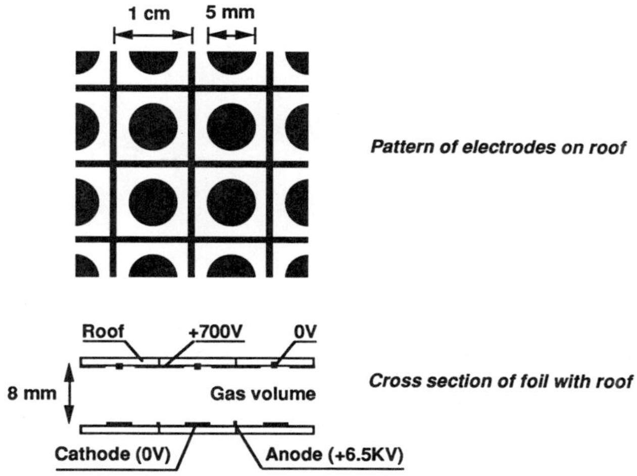

Fig. 8. Electrode design of 'roof'

with 500 µm diameter anodes with a 3.5 mm cathode radius. The results are shown in fig. 7. We also show the counting rate variation. This is for the experimental setup with a single foil within a gas-tight box. The gain is higher for an increased fraction of argon within the mixture. However, with this rather elementary setup we have a coupling between the drift field and the voltage that we apply to the anode. For example with a high fraction of argon the pixel operates at the highest gain, but the counting rate is low. Thus, the source counting rate is not too meaningful in an absolute sense except that it suggests that we will need other electrodes to shape the field in order to build an optimal detector for minimum ionizing particles.

Our first attempt for these field shaping electrodes is described in fig. 8. These electrodes are printed on a 200 µm thick epoxy-glass board. This board is mounted 8 mm above the Kapton foil. We show the efficiency versus high voltage with pure isobutane for cosmic rays for two cases in fig. 9. In the first case it is the board without the shaping electrodes and the other case is with the shaping electrodes. The effect of the shaping electrodes is clear, it allows the chamber to become efficient at a much lower voltage.

We mounted this chamber with the shaping electrodes in a test beam in the East hall at CERN. We could vary the intensity of the core of the beam spot from 5000 particles/cm^2/s to 100 000 particles/cm^2/s. The efficiency versus the voltage for various fluxes is shown in fig. 10. It is clear that at the highest flux we are becoming dominated by space charge effects. Also at this high flux we have rate induced sparking above 6.4 kV. However, at 50 000 particles/cm^2/s the onset of rate induced sparking is 6.8 kV although we are seeing a drop in efficiency due to space charge effects. We expect that the rate capability to be affected by the geometry of the cell so we will start to build chambers of different geometries to investigate and hopefully further improve the response at high rates. However, it should be noted that this chamber is already operating at a relatively high rate in a very stable manner. We have been working in a pulsed environment (one 0.4 s spill every 30 s) so it appears that worries regarding the stability of having an insulator close to the anode are unfounded.

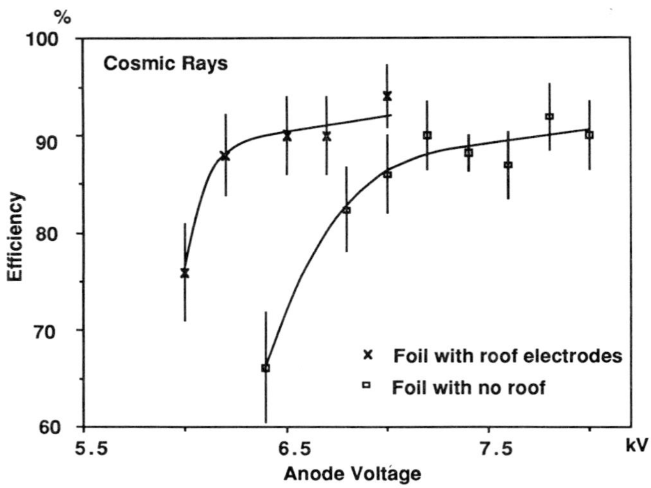

Fig. 9. Response of Gaseous Pixel Chamber to Cosmic Rays with and without a roof foil

CONCLUSIONS AND FUTURE

We have shown that we have discovered a way to construct surfaces that can detect particles in a very similar fashion to a plane of wires. However, our foils have the following advantages:

(a) They are flexible and can be bent or cut to fit any geometry.

(b) The gain is defined by the shape of the electrodes on the foil, not by the distance to some cathode plane. This makes the mechanical stability and mounting easy.

(c) Our present design is pixel-like in character, it would be easy to place these on some curve to satisfy some geometrical requirement of an experiment.

(d) The foil is easy to build using standard printed circuit board technology. Mass production can easily be contemplated.

The drift time is associated to the distance the track is from the anode. We will start on measuring the space resolution of this device once we have optimized the field shaping electrodes.

We have not worked on aging studies of the chamber yet, but do not expect it to behave badly in this respect as: (a) we can smear other materials onto the surface of the Kapton/glue insulator (such as silicon rubber) and not change the operation of the surface at all; (b) no special care is taken of the foil (sometime we wipe it with alcohol) but all the foils operate in a similar manner; (c) we can allow the surface to enter a continuous spark mode but once we reduce the voltage to the normal operation point, the foil works as before. However, more quantitative aging tests are essential and we will start on such a program in the near future.

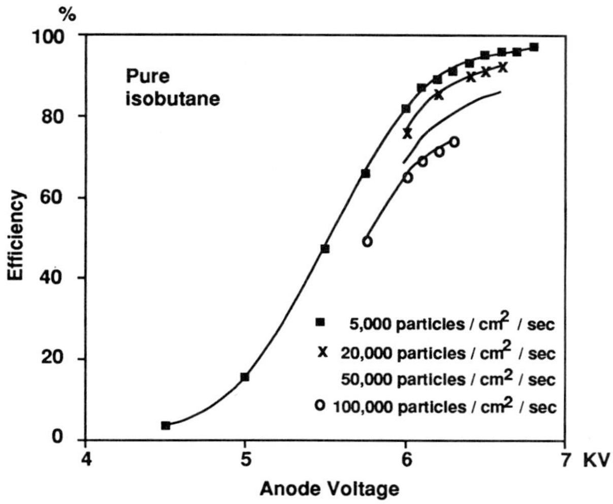

Fig. 10. Response of Gaseous Pixel Chamber to various fluxes of 6 GeV/c pions

So far we have only observed operation within the limited streamer mode of gas amplification. Obviously it would be very interesting if we can operate in a proportional mode. For this it may be necessary to work with a smaller anode diameter (and corresponding smaller anode–cathode spacing). We have built a similar but scaled down geometry on ceramic using standard thick film printing techniques. We do have ceramic surfaces that operate in a 'limited streamer' gas amplification mode. A lot more work is needed to investigate how to build this device on a small scale.

ACKNOWLEDGEMENTS

Although the foils described in this paper can be manufactured with standard printed circuit techniques, we are very fortunate to have the excellent services of the CERN printed circuit workshop run by A. Gandi. It is a pleasure to work with such a friendly and competent group of people. Also we would like to thank F. Sauli [9] for having lent to us a set of very reliable drift chambers including special preamplifiers/ discriminators. We used a Macintosh computer for data taking and the data aquisition software given to us by F. Cindolo [10] was a very convenient tool for people like us. Special thanks to D. Troester [11,12], his fast CAMAC processor and buffer, 'Pink Panther', considerably speeded up our measurements in the pulsed beam environment of the CERN–PS.

REFERENCES

[1] J.B. Dainton, Green Report CERN 89-10, EFCA 89-124 (1989) 394-412

[2] A. Oed, Nucl. Instr. and Meth. A263, 351-359 (1988)

[3] F. Angelini et al., INFN-PI/AE 89/2

[4] F. Hartjes et al. CERN/EF/4330 R/FU/ed and CERN 89-10/EFCA 89-124 455-463 (1989)

[5] H. Stahl et al., Uni Siegen preprint, SI 90-03

[6] D. Mattern, M.C.S. Williams and A. Zichichi, CERN/ECP 90-4, CERN/LAA/MC 90-06, submitted to Nuclear Instruments and Methods

[7] M.J. Neumann and T.A. Nunamaker IEEE Trans. on Nucl. Sci. 17(3) (1970) 43-49

[8] D. Mattern and M.C.S. Williams, Green Report CERN 89-10, EFCA 89-124 (1989) 435-439.

[9] A. Breskin et al., Nucl. Instr. and Meth. 124 (1975) 189.

[10] F. Cindolo and G. Prisco, BCFL Collaboration, Geneva, August 7, 1987.

[11] D.A. Troester, Yellow Report CERN 84–03.

[12] Ch. Buess and D. A. Troester, CERN CAMAC Note 75–00, April 1984.

SUPER MONTE CARLO SIMULATIONS AT 16, 40, 200 TeV

Luisa Cifarelli

University of Naples, Italy and
CERN/LAA Project, Geneva, Switzerland

1. INTRODUCTION

Future experiments at hadron Supercolliders, such as LHC (16 TeV), SSC (40 TeV) or ELOISATRON (200 TeV), will require intensive theoretical and phenomenological studies for extrapolations and predictions of possible Physics scenarios, as well as detailed detector design studies to define the desired construction parameters. These studies have become one of the most important and challenging fields of today's High Energy Physics.

The main problem in simulating hadronic interactions is to deal with the non-Abelian nature of Quantum Chromo Dynamics (QCD)[1], i.e. the most reliable theory existing so far to describe the hadronic interaction itself. There are essentially two ways of using QCD to make predictions: the "lattice" approach and the "perturbative" one. The "Supercomputers and Monte Carlo Simulations" component of the LAA Project works in both these directions. At Columbia University, Professor T.D. Lee is developing a Supercomputer for lattice dynamical QCD calculations, whereas at CERN, the so-called "Monte Carlo Simulation Laboratory" (MSL) is in charge of perturbative QCD studies based on different Monte Carlo Event Generators, and of the subsequent experimental aspects of the simulations.

During the year 1990, the activities of the MSL group of LAA have developed along the following three lines:

a) Studies of QCD Event Generators
   (section 2)
b) Programme implementation, software tool development and software engineering
   (section 3)
c) Monte Carlo simulations
   (section 4).

2. STUDIES OF QCD EVENT GENERATORS

The Event Generators are based on the Monte Carlo method, i.e. the generation of events in random sequence like numbers in a "roulette". They consist of two independent parts: one, where perturbative QCD is applicable (hard scattering plus parton evolution, if any) and one where it is not and some models are instead used (hadronisation).

Physics Up to 200 TeV
Edited by A. Zichichi, Plenum Press, New York, 1991

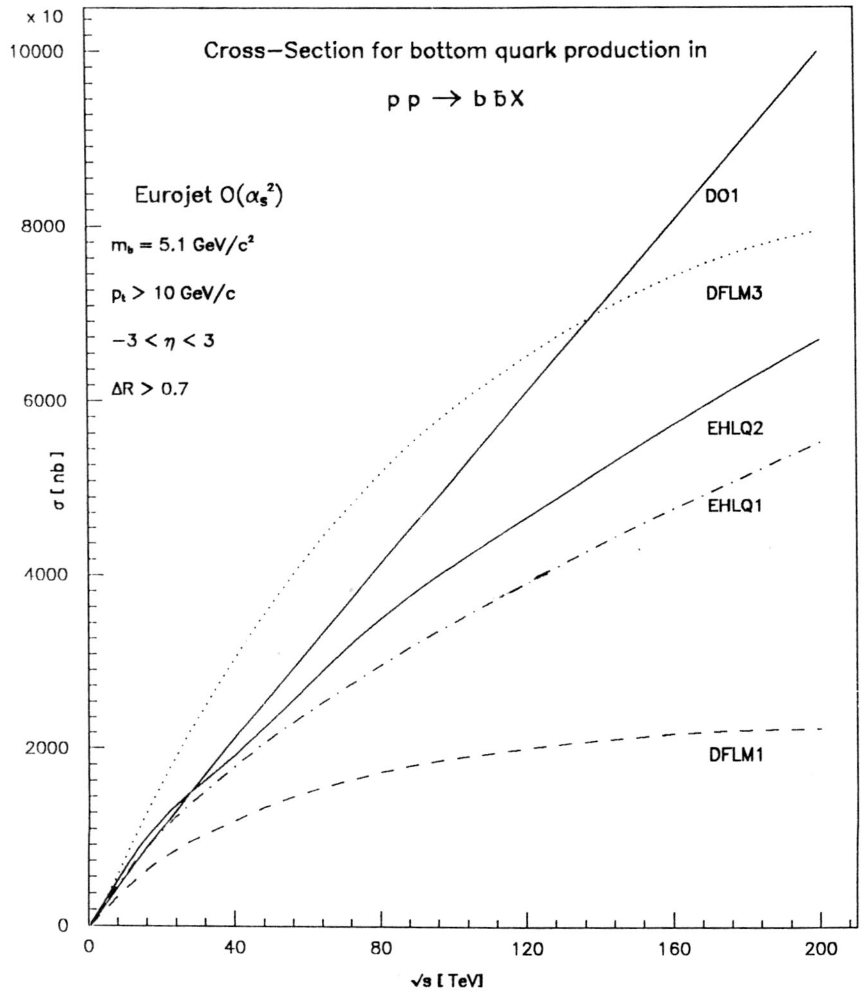
x 10
10000
8000
6000
4000
2000
0
Cross–Section for bottom quark production in
p p → b b̄ X
Eurojet O(α_s^2)
m_b = 5.1 GeV/c^2
p_t > 10 GeV/c
$-3 < \eta < 3$
ΔR > 0.7
DO1
DFLM3
EHLQ2
EHLQ1
DFLM1
σ [nb]
0
40
80
120
160
200
√s [TeV]

Figure 1

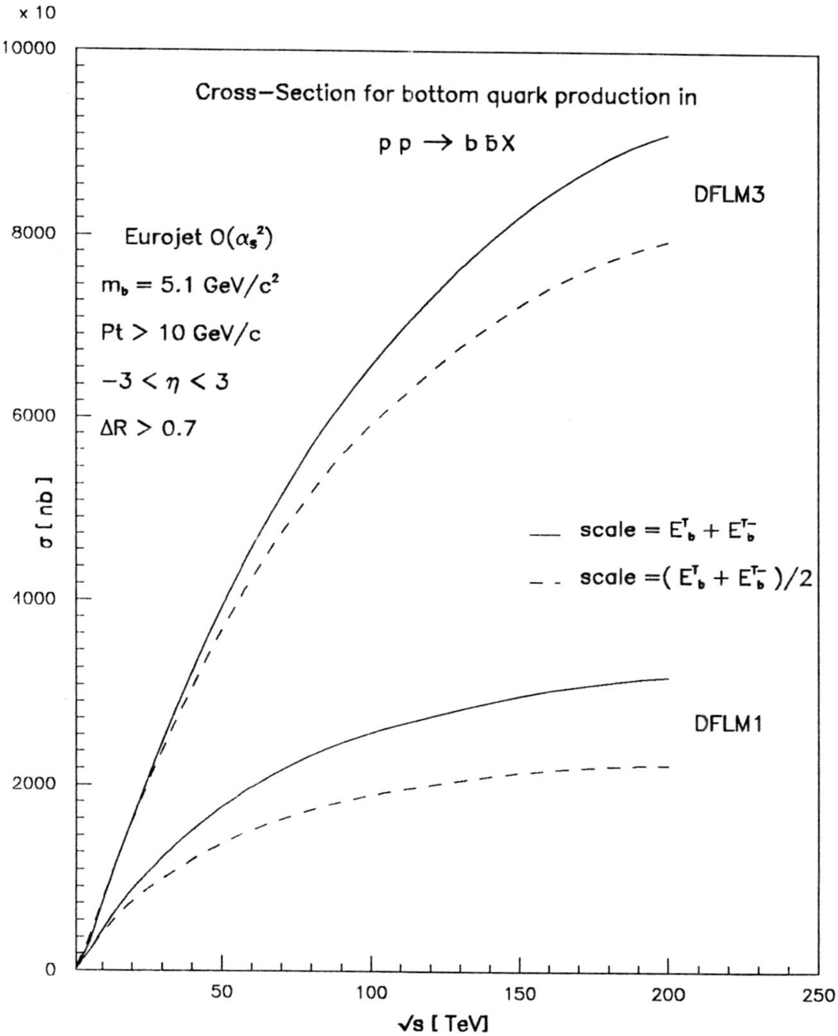
x 10
10000
8000
6000
4000
2000
0
Cross–Section for bottom quark production in
p p → b b̄ X
DFLM3
Eurojet $O(\alpha_s^2)$
$m_b = 5.1$ GeV/c^2
Pt > 10 GeV/c
$-3 < \eta < 3$
$\Delta R > 0.7$
σ [nb]
scale = $E^T_b + E^T_{\bar{b}}$
scale =($E^T_b + E^T_{\bar{b}}$)/2
DFLM1
50
100
150
200
250
√s [TeV]

Figure 2

Since the goal of the whole LAA Project and of the MSL group in particular, is to prepare the theoretical and technical support for experiments at future Supercolliders in the range of hundreds of TeV, the problem of reliable simulations is even more critical. In fact, as the collision energy increases, higher and higher α_S corrections have to be included in the cross-section computations. This is not a trivial problem at all. Present Event Generators try to solve the problem of high order corrections in two ways: either with the "Matrix Elements" (ME) approach or with the "Parton Shower" (PS) one. The former would be the correct way, that is, whenever the exact high order subprocesses matrix elements are calculated by theorists they are included in the Event Generator frame. The latter just takes into account the exact matrix element for the hard $2 \to 1$ or $2 \to 2$ subprocess and higher order corrections are included using the Leading Log Approximation Altarelli-Parisi[2] equations. The Monte Carlo implementation of the Parton Showering is relatively simple, because any higher order correction is seen as a gluon emission from gluons or quarks or as a gluon splitting in quark-antiquark pairs. Both methods have advantages and disadvantages: the exact hard scattering cross-section calculation and kinematics determination of the ME approach is an overwhelming problem as soon as the number of outgoing partons becomes large. The PS method would in principle be able to reproduce any correction, but it is anyway a Leading Log Approximation.

Among the many technical problems and theoretical uncertainties concerning event generation at hadron colliders, the one related to the structure functions cannot be ruled out very easily. Structure functions are derived from experimental data and are therefore valid within finite intervals of x (the parton fraction of hadron momentum) and Q^2 (the square of the four-momentum transfer): roughly $0.05<x<0.8$ and $0.2<Q^2<200$ GeV^2. There is no solution to the problem of extrapolation outside these intervals unless we wait for HERA data. Figure 1 shows how a cross-section computation can vary using different structure functions sets, as a function of $\sqrt{s}$. The simulation refers to $b\bar{b}$ production with the following conditions: $p_t>10$ GeV and $|\eta|<3$ for both b and $\bar{b}$, and $\Delta R>0.7$ (ΔR being the b and $\bar{b}$ separation in the η-ϕ plane). In figure 1, DO1 stands for Duke and Owens[3] set 1, EHLQ1 for Eichten, Hinchliffe, Lane and Quigg[4] set 1, DFLM for Diemoz, Ferroni, Longo and Martinelli[5]. As the energy increases, the uncertainty due to the choice of a given parametrization also increases, becoming critical at 200 TeV.

Figure 2 shows the influence of another source of uncertainty in the simulations: the arbitrary choice of the QCD scale in the cross-section calculation. Here again the uncertainty is enhanced in the 200 TeV range. Figures 1 and 2 were derived using EUROJET version 3.1[6] as Event Generator.

Nobody knows exactly how to handle the interactions between coloured partons at a scale of the order of Λ_{QCD}. Essentially three models exist with many possible variants: Independent Fragmentation, String Fragmentation and Cluster Fragmentation (for a good review, see ref. 7).

Independent Fragmentation is not really a Physics model but a simple parametrization; moreover, energy-momentum conservation is to be imposed. Therefore, it is considered as not true even if it is able to reproduce experimental data when opportunely tuned. The way hadrons are formed from partons is via quark-antiquark formation from the vacuum polarization and subsequent arrangement of quarks and antiquarks. Several schemes have been proposed to take gluons into account[7].

According to the String Fragmentation model the tubelike colour field flux lines can break, by tunnelling effect, into production of quark-antiquark pairs. Gluons are kinks in the string, giving transverse momentum.

Both Independent and String Fragmentation models do need a fragmentation function to give longitudinal momentum to hadrons and a distribution function for transverse momenta, typically a gaussian one. Many fragmentation functions exist; examples are: the Field-Feynman[8] one, the Peterson[9] one and the Lund[10] one.

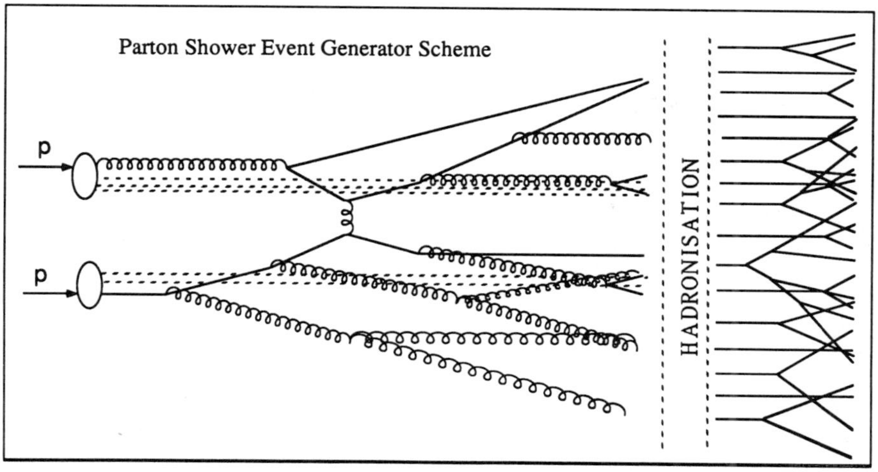

Figure 3

Table 1

QCD Event Generators				
	HARD SCATTERING	PARTON SHOWER	HADRONIZ.	DECAY
EUROJET	Multipartons*	NO	Independent fragmentation	YES
HERWIG	2 (1) **	YES	Cluster fragmentation	YES
PYTHIA	2 (1)	YES	String fragmentation	YES
ISAJET	2 (1)	YES	Independent fragmentation	YES
COJETS	2 (1)	YES	Independent fragmentation	YES

* Hard scattering subprocess up to $O(\alpha_s^3)$ already included (higher order subprocesses up to $O(\alpha_s^5)$ foreseen).

** Only 2->1 or to 2->2 hard scattering subprocesses considered.

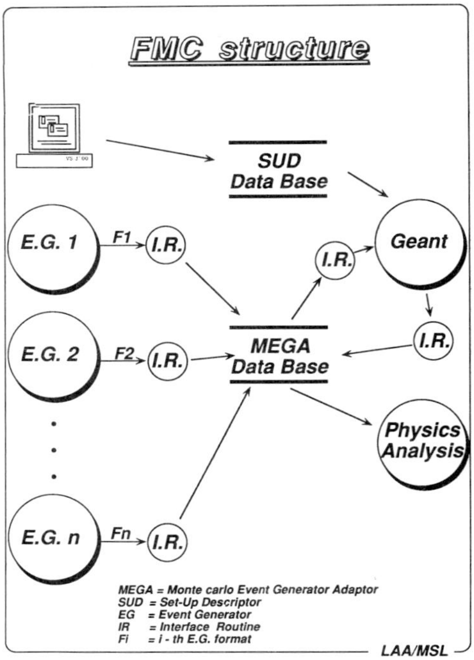

Figure 4

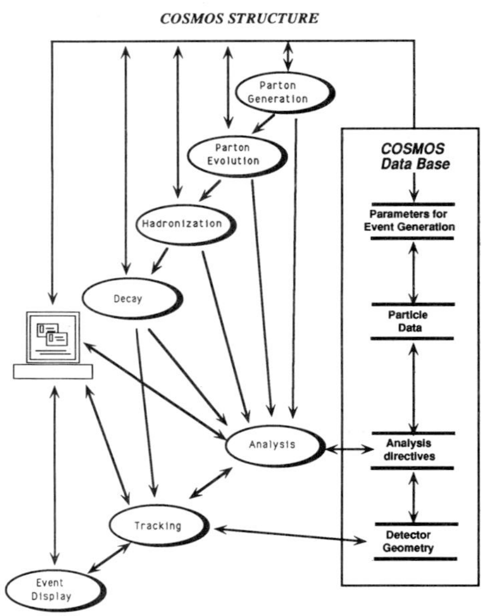

Figure 5

Multiplicity Distributions of Minimum Bias Events

at $\sqrt{s}$ = 16 TeV . Event Generator : PYTHIA 5.4

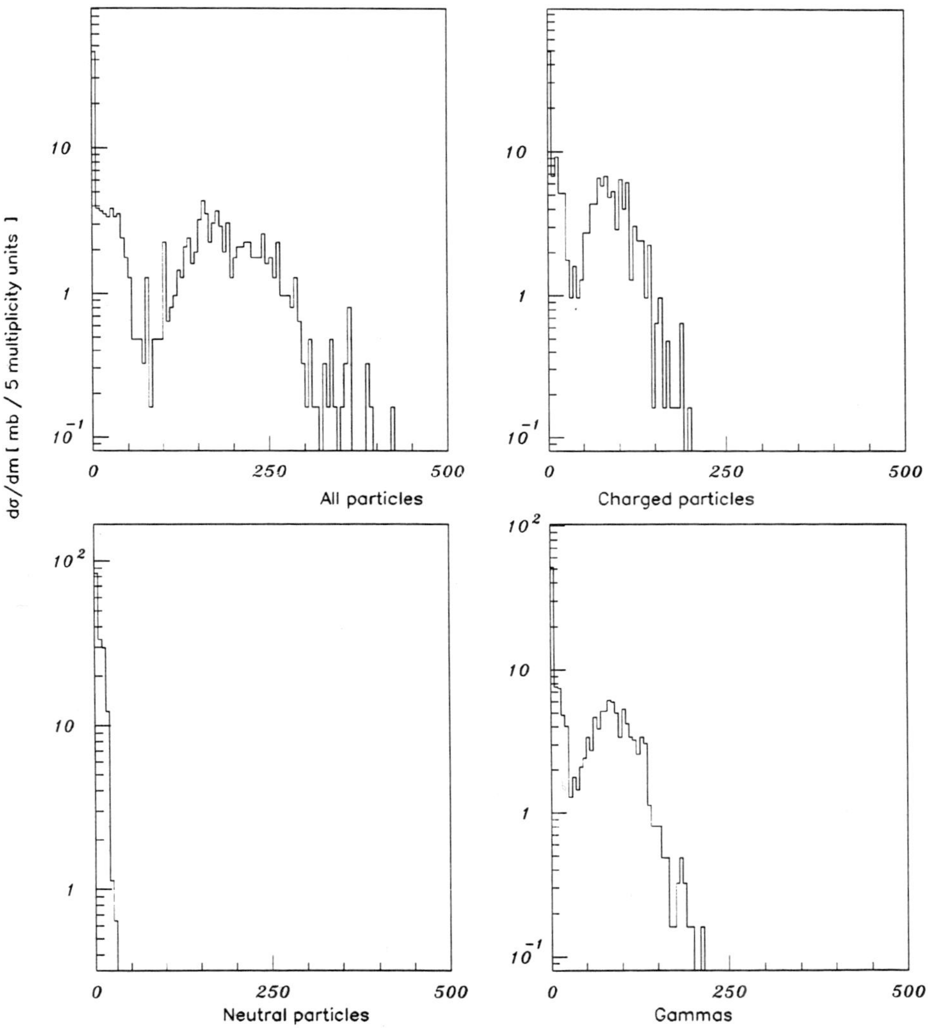

Figure 6

Cluster fragmentation has less free parameters. In the HERWIG version it is based on some preconfinement studies[11]. After the parton shower all gluons, if any, are forced to decay into quark-antiquark pairs. The quarks that are closest in the phase space arrange themselves into "clusters" of more or less the same invariant mass. Clusters are then made to decay in hadrons.

Figure 3 summarizes the various steps of an Event Generator based on the PS approach: the 2 $\rightarrow$ 2 hard scattering, the parton showering, the "black box" of the hadronisation and the decay of the produced hadrons. The dashed lines are the hadron remnants, valence quarks in the best case, whose correct treatment in a Monte Carlo Event Generator is still an open problem. In this respect, the well-known leading hadron effect (which belongs to the low p_t domain, where perturbative QCD is no more applicable) is only roughly reproduced in most of the Event Generators by means of some "recipes", appropriately tuned with existing experimental data.

Table 1 is a summary of the main features of the most known QCD Event Generators. Various tests and comparisons of different Event Generators at different energies are on the go, the goal being to choose, optimize and/or upgrade (if needed) the most promising one to produce the best extrapolations up to energies as high as a few hundred TeV.

3. DEVELOPMENT OF SOFTWARE TOOLS

Apart from understanding Physics ingredients, ad hoc models and "tricks" of QCD Event Generators, also technical problems, related to the software needed for Monte Carlo simulations, have been treated. The reason being that, as the energy and multiplicity of events increase, the detector size and complexity also increase. And all this translates into huge software packages, whose organization, handling and mutual linking become very difficult. The problem critically shows up when different Event Generators have to be interfaced to different detector simulation and/or analysis programmes. Only recently Monte Carlo authors agreed on the introduction of some standards.

Monte Carlo Event Generator Adaptor (MEGA)

A solution to this problem has been worked out, which is by far more complete than a simple output commonblock. It consists of a package of routines, collectively called MEGA (Monte Carlo Event Generator Adaptor)[12], which provides a unique standard database for all the different Event Generators. MEGA is internally organized via ADAMO[13], a modern data structure system based on the Entity-Relationship model. Data are treated as "entities", with their "attributes", and mapped onto tables which can be comfortably handled for any further application.

Full Monte Carlo Chain (FMC)

An immediate application of MEGA is in fact the FMC (Full Monte Carlo Chain)[14] package, where MEGA is interfaced with the well-known GEANT[15] Monte Carlo programme, for particle tracking inside a detector. The MEGA tables are used in input, then updated in output with the results from the tracking for further analysis. Figure 4 illustrates the FMC structure. A more detailed description can be found in the most recent LAA Report[16]. The current FMC version 1.3 uses ADAMO version 3.0 and is implemented on both IBM/VM and VAX/VMS.

Set Up Descriptor (SUD)

In order to set up the geometry of the detector for FMC, instead of making multiple calls to the GEANT routines, a package has been written which is part of FMC and which reads the detector data from ADAMO tables. The data to fill these tables reside on an ADAMO Direct Access File (DAF), generated by another software tool, SUD (Set Up Descriptor)[17], also based on ADAMO. SUD exists in both batch and interactive versions. In particular, the

Multiplicity Distributions of Minimum Bias Events

at $\sqrt{s}$ = 40 TeV . Event Generator : PYTHIA 5.4

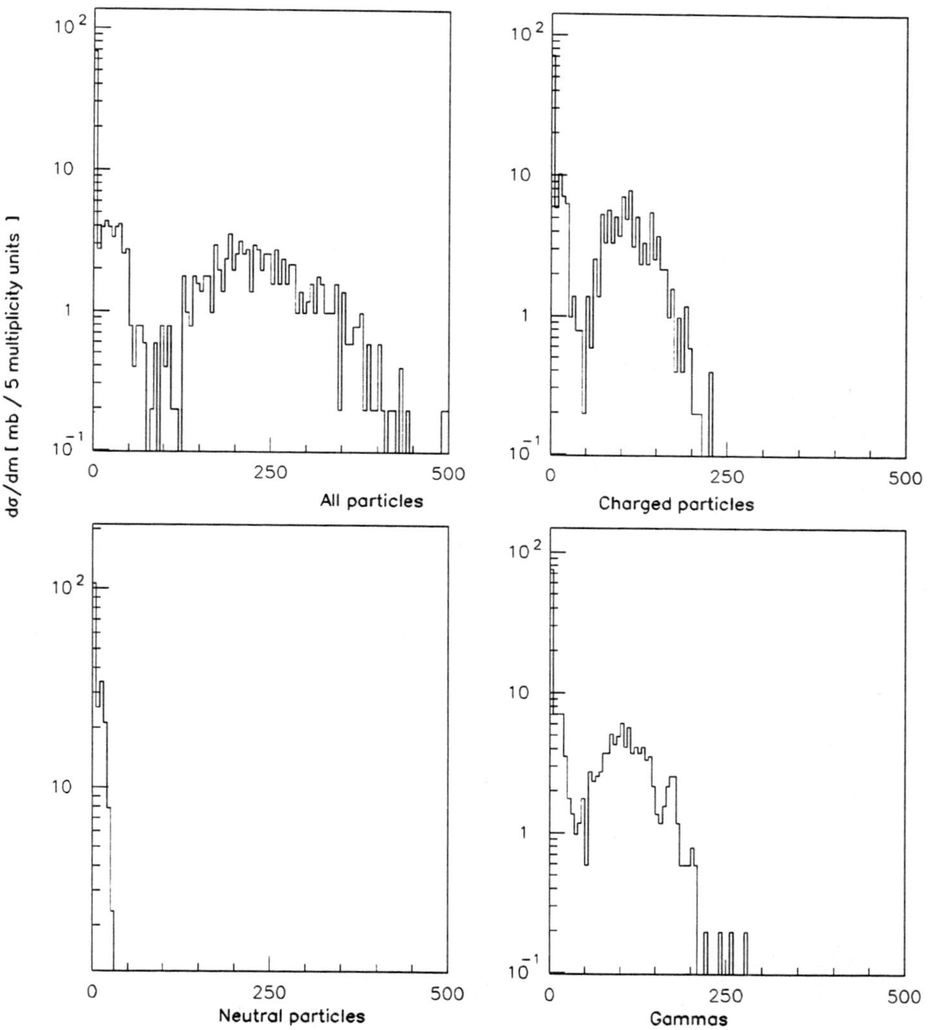

Figure 7

Multiplicity Distributions of Minimum Bias Events

at $\sqrt{s}$ = 200 TeV . Event Generator : PYTHIA 5.4

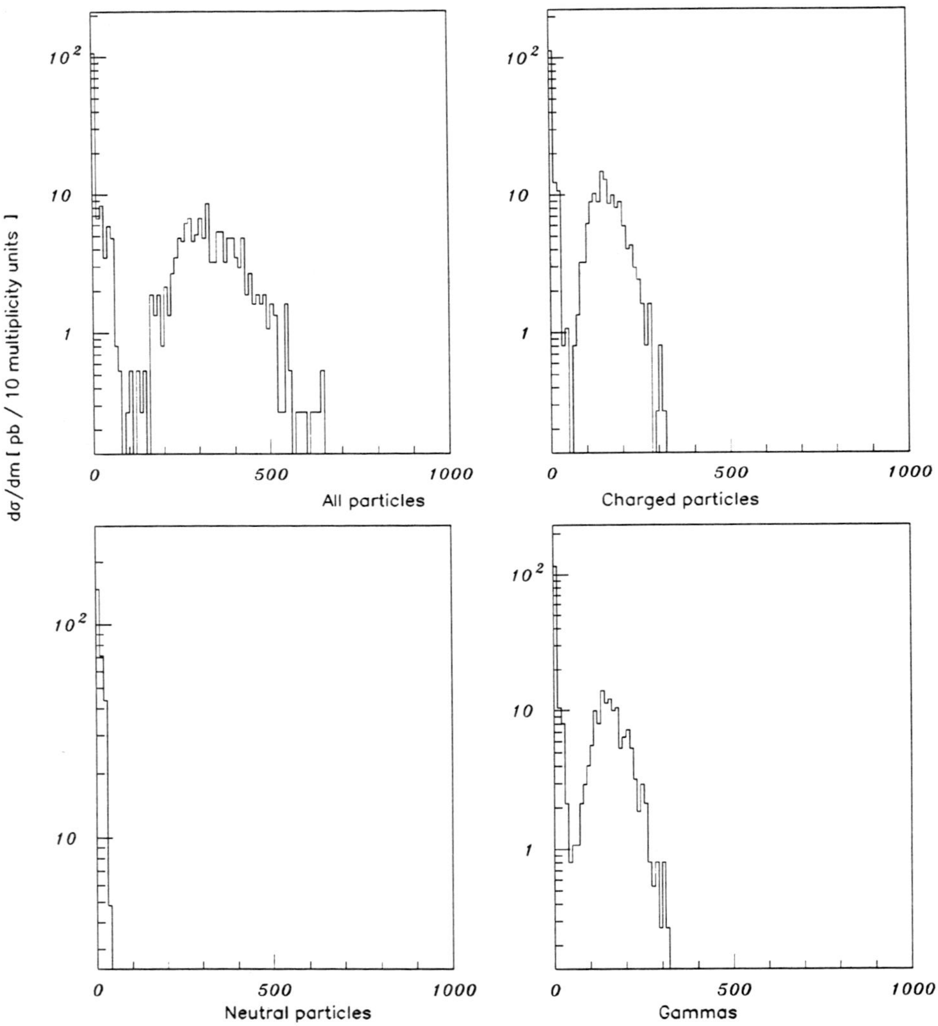

Figure 8

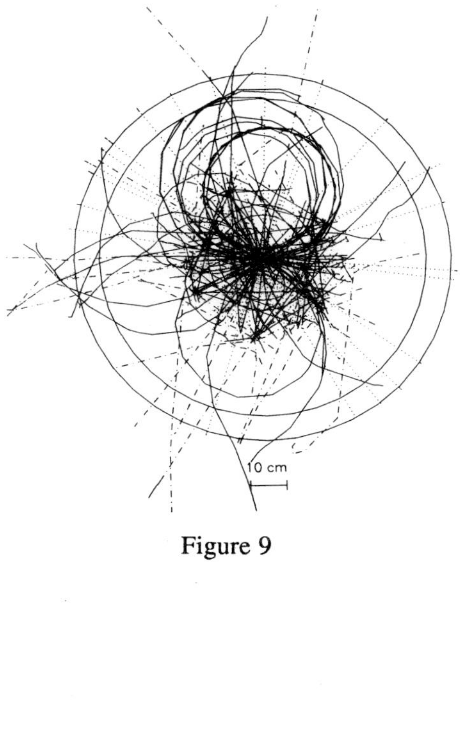

Figure 9

interactive version constitutes a very user-friendly tool. It is operated via menus based on KUIP[18] and gives various possibilities: to describe detector volumes, to define materials and tracking media, to position the volumes, to divide them into sections, etc.. SUD runs on VAX/VMS, but ADAMO tools exist for "exporting" the generated DAF to a machine-independent file and "importing" it into another computer (i.e. IBM/VM).

Comprehensive Super Monte Carlo System (COSMOS)

As a further generalization to FMC, a new project has started for the implementation of a COmprehensive Super MOnte Carlo System (COSMOS). In such a system, different Monte Carlo simulation packages are assembled, after complete modularization, organized and interfaced in a unique framework: from the early generation of the parton-parton interaction, to the ultimate analysis of the simulated detector response. The need for such a system comes from the intrinsic approximations and uncertainties of QCD, which translate into different event generation approaches (as already pointed out in section 2), leading sometimes to different qualitative and quantitative results. Therefore the basic idea behind this project is to provide an exhaustive compilation of all existing options and tools for simulation. The complete event simulation chain is seen as decomposed in independent blocks (see figure 5). In COSMOS the user can easily combine modules from different Event Generators, attach his analysis programme to any stage of the event generation, track the events, implement the detector geometry, etc.. The COSMOS system is entirely designed and configured. Among its software engineering features, it is worth mentioning the organization of data in databases: for the parameters regulating the event generation, for the particle data (mass, charge, lifetime, etc.), for the directives of the analysis and for the detector geometry. Many existing/new packages have already been included, as well as databases such as SUD and PDKDB (see below), and work is in progress.

Standard Dynamical Commonblocks

It is obvious that combining within the same package modules from different programmes imposes a need for standardisation of input and output data. It is also true that since the various Event Generators on the market have been written by different authors, there is a large variety of formats for input/output data and different mechanisms for accessing them. Towards the direction of an overall standardization a proposal has been made[19] for a number of dynamical commonblocks, which will contain and share between programme modules the standardized information on particle properties (COMMON/HEPPPR) and decays (COMMON/HEPPDE), as well as the event history (COMMON/HEPEVE).

Particle Decay Data Base (PDKDB)

A Particle and Decay Channels Data Base (PDKDB)[20] has been developed, to manage the information on particle properties and their decays, which again uses the ADAMO Entity-Relationship model. The data are obtained from the Particle Data Group (PDG), with whom a collaboration has been established. The aim is to produce a scheme for an automatic update of some universally accessible file, with the latest information on particle properties.

4. MONTE CARLO SIMULATIONS

Two kinds of simulation studies have been carried out: simulations as input to detector studies and simulations for theoretical investigation purposes.

People working on detector design for future Supercolliders do need an input of Physics event simulation to enhance the detector capabilities and performances. On the other hand, people working on Physics simulations and looking for possible signatures of new particles or new phenomena have to interact with the detector experts in order to understand which limitations and constraints might be faced. This is the reason why the MSL group has undertaken a series of simulations of various High Energy Physics phenomena at various future Supercollider

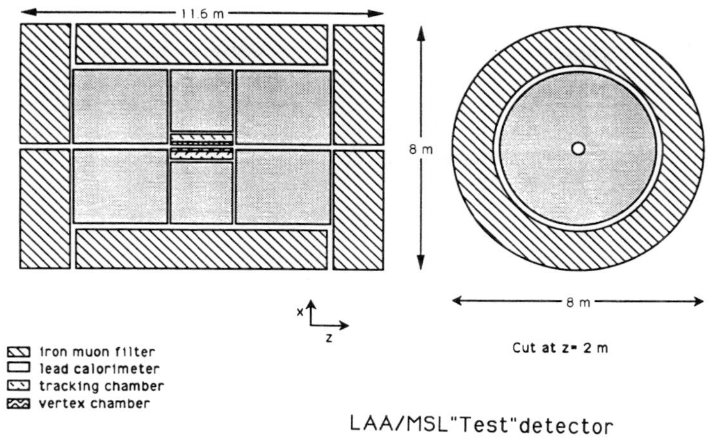

Figure 10a

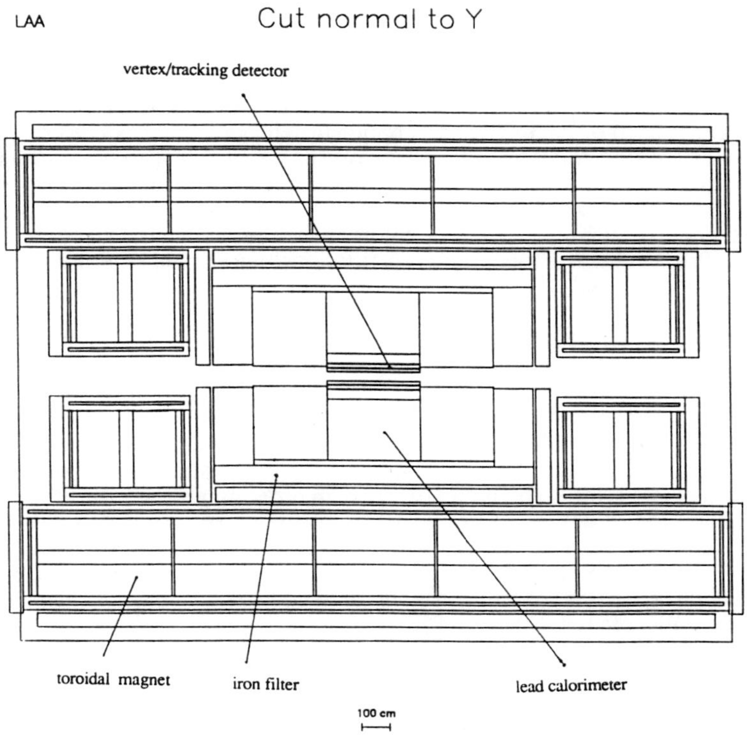

Figure 10b

energies, namely at LHC (16 TeV), SSC (40 TeV) and ELOISATRON (200 TeV) energies. The aim being also to point out if and why the few hundreds of TeV range appears to be more interesting than the few tens of TeV one.

Event Generation

Different samples of events have been generated, at various energies: "minimum bias" events (i.e. low p_t QCD, elastic and diffractive processes), high p_t QCD processes, top and Higgs production with different mass hypotheses. The Event Generator used was PYTHIA[21] version 5.4 (with default parameters). For each sample (of 1000 events), a catalogue of event features (multiplicity, transverse momentum, Feynman-x distributions for different kinds of particles) has been produced. A summary, in terms of cross-section and average multiplicity, is given in table 2.

Among the various simulations, the minimum bias ones are of particular importance for detector simulation studies since they represent the bulk of the production in hadron-hadron interactions. Some results on the multiplicity distributions of all particles, charged particles, neutral particles (i.e. neutrinos plus neutral particles other than gammas) and gammas, at $\sqrt{s}$=16 TeV (LHC), $\sqrt{s}$=40 TeV (SSC) and $\sqrt{s}$=200 TeV (ELOISATRON), are shown in figures 6, 7 and 8.

A typical $\sqrt{s}$=16 TeV minimum bias event is shown in figure 9, as an indication of the lower level of complication one should expect in the detector. The view is transverse to the beam. A 5 Tesla solenoidal magnetic field (parallel to the beam) is present in a region of 50 cm radius. All charged (plain line), gamma (dotted line) and neutral (dashed/dotted line) particles having p>100 MeV have been tracked.

Detector Simulations

Concerning studies involving a detector, first a LAA/MSL test-detector has been defined (using the above mentioned SUD package), as illustrated in figure 10a. It consists of a vertex chamber plus a tracking chamber (in the barrel region), a lead calorimeter and an iron muon filter (both in the barrel and forward regions). The presence of a magnetic field close to the beam and/or at the level of the iron filter is optional. This detector has been designed mainly for the purpose of testing software tools. However it is evolving with time from its very first stage of "embryo" to a more realistic configuration, such as, for instance, the one shown in figure 10b. In this case, without going into details, there is a unique vertex/tracking device made of layers of scintillating fibres. The calorimeter is of the "spaghetti" type, i.e. lead plus scintillating fibres. There is again an iron filter for muons, and the muon momenta are measured via air-core toroids.

Many detector design studies are presently going on in the various LAA components. Just to give an example, using the test-detector of figure 10a, the FMC package and various events samples at $\sqrt{s}$=16 TeV, some studies have been done on the problem of the overlap between charged particles and gammas[22] in the calorimeter, with or without a magnetic field. The aim was the definition of the granularity of a preradiator to be inserted before the main calorimeter, at a radius of 50 cm in the barrel region. The use of preshower devices with fine granularity before the main calorimeter is common nowadays in order to improve space resolution and to enhance electron/pion separation. The resulting probability for pion/gamma overlap is shown in figure 11, for different event samples (see table 2). Glancing at the problem, for a distance between tracks of about 5 cm (which corresponds to a typical calorimeter granularity), this probability increases by many orders of magnitude as soon as high p_t jets or heavy particles are produced (for details, see ref. 22).

The Higgs case at 200 TeV

Pioneering studies on Higgs particle detection at 200 TeV have been performed. The existence or not of the Higgs boson is a crucial check of the Standard Model. Its mass is undetermined

Table 2

Generated Event Samples

Event Generator: PYTHIA 5.4

	$\sqrt{s}$ = 16 TeV	$\sqrt{s}$ = 40 TeV	$\sqrt{s}$ = 200 TeV
1. Minimum Bias low p_t	σ =161 mb n = 113	σ = 196 mb n = 130	σ = 268 mb n = 171
2. QCD Production $p_t^* > 30$ GeV	σ = 230 nb n = 340	σ = 677 nb n = 397	
3. Top Production m(top) = 90 GeV	σ = 22 nb n = 403	σ = 91 nb n = 463	
4. Top Production $p_t^* > 30$ GeV	σ = 18 nb n = 402	σ = 75 nb n = 467	
5. Higgs Production m(Higgs) = 200 GeV m(top) = 90 GeV $p_t^* > 30$ GeV	σ = 40 pb n = 384	σ = 160 pb n = 447	
6. Higgs Production m(Higgs) = 1 TeV m(top) = 90 GeV $p_t^* > 30$ GeV			σ = 27 pb n = 584

p_t^* : transverse momentum of outgoing partons in the 2➔2 reference frame
σ : cross-section
n : mean multiplicity

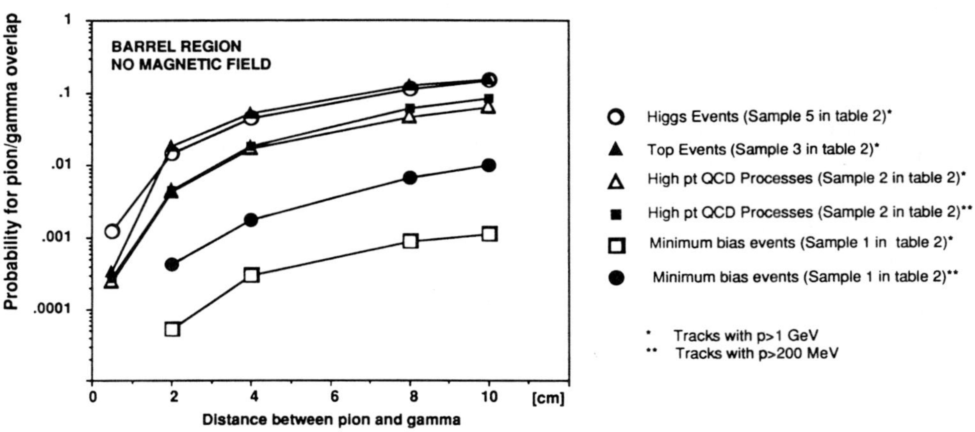

Figure 11

by the model and could be as heavy as 1 TeV. Therefore, we have to foresee the possibility that neither LHC nor SSC will discover it or demonstrate its non-existence.

If the Higgs is heavy, i.e. more than twice the Z^0 mass, the relevant production channels are the gluon-gluon fusion and the vector-boson fusion, the former being the dominant one if the Higgs mass stays below the ~800 GeV level[23]. The main decay modes are: $H^0 \to W^+W^-$, $H^0 \to Z^0Z^0$, $H^0 \to t\bar{t}$. These can be assumed to have approximately equal branching ratios. The cleanest signal would be $H^0 \to Z^0Z^0 \to l^+l^-l^+l^-$ ($l=e,\mu,\tau$) for which the main background comes from $q\bar{q} \to Z^0Z^0$ and $gg \to Z^0Z^0$, i.e. the QCD continuum production of Z^0 pairs. As the collision energy increases, the gluon-gluon background channel becomes more and more important due to the increasing gluon flux.

Table 3 shows a comparison, in terms of cross-section, of the LHC, SSC and ELOISATRON theoretical capabilities to discover a Higgs of 750 GeV. One can immediately notice a sizable increase of the signal versus background ratio when jumping from LHC or SSC energies, up to the ELOISATRON energy. In order to improve this ratio, some characteristics of the signal and background have to be underlined: the Z^0s produced by the signal have mainly high transverse momentum and are mainly longitudinally polarized whereas the background ones have lower transverse momentum and mainly transverse polarization. The Z^0 polarization is recognizable from the lepton angular distribution in the Z^0 rest frame.

Three sets of cuts were used in the simulation,[23] and their influence on the four-lepton invariant mass spectrum was studied in detail:

Set 1: $p_t(l)>20$ GeV and $|y(l)|<2.5$, where p_t is the transverse momentum and y the rapidity (in the pp c.m.s.), for each lepton.

Set 2: $p_t(Z^0)>\frac{1}{4}\,m(llll)$, where m(llll) is the four-lepton invariant mass, for each Z^0.

Set 3: $|\cos\theta^*(l)|<0.8$, where θ^* is the lepton polar angle in the Z^0 rest frame, for each lepton pair.

Figures 12,13 and 14 show the four-lepton invariant mass distribution in $gg \to H^0 \to Z^0Z^0 \to l^+l^-l^+l^-$, plus the background due to $gg \to Z^0Z^0$ and $q\bar{q} \to Z^0Z^0$, at LHC, SSC and ELOISATRON. The Higgs mass is 750 GeV, the top mass is 120 GeV and the $Z^0 \to l^+l^-$ branching ratio is 3%. In each figure there are four histograms obtained as follows:

(a) : without cuts
(b) : with cuts of set 1
(c) : with cuts of sets 1+2
(d) : with cuts of sets 1+2+3

The message is that if clever cuts are imposed on the outgoing decay particles, the Higgs boson will presumably be observable at future Supercolliders. However the confidence level of observation turns out to be much higher at ELOISATRON than at LHC or SSC.

Higgs detection in real life, i.e. in a Supercollider with the utmost luminosity, will be by far more complicated than in the present exercise. The four leptons will have to be unambiguously identified in a very high rate, high multiplicity environment. The experimental resolutions on the lepton energies and production angles will have to be folded in. Moreover, despite a clear four-lepton signature, the drawback due to the small $Z^0 \to l^+l^-$ branching ratio will have to be seriously considered and other possibilities, such as $H^0 \to Z^0Z^0 \to l^+l^-$ jet jet, for instance, investigated.

In order to get the flavour of what real life would look like, Higgs production at 200 TeV has been simulated using again PYTHIA version 5.4 as Event Generator. A Higgs mass of 1 TeV, a top mass of 90 GeV and the condition $p_t(H^0)>80$ GeV (in the parton-parton frame) were

Table 3

Higgs Production Cross-Sections

Signal versus Background

Signal: $gg \rightarrow H^0 \rightarrow Z^0 Z^0$

Background: $gg \rightarrow Z^0 Z^0$ and $q\bar{q} \rightarrow Z^0 Z^0$

Cross-sections in pb

Top mass: 120 GeV

	LHC	SSC	ELOISATRON
Signal (M_{H^0}= 500 GeV)	0.8	5	69
Signal (M_{H^0}= 750 GeV)	0.1	0.8	14
$q\bar{q}$ Background	11	32	172
gg Background	3	14	174

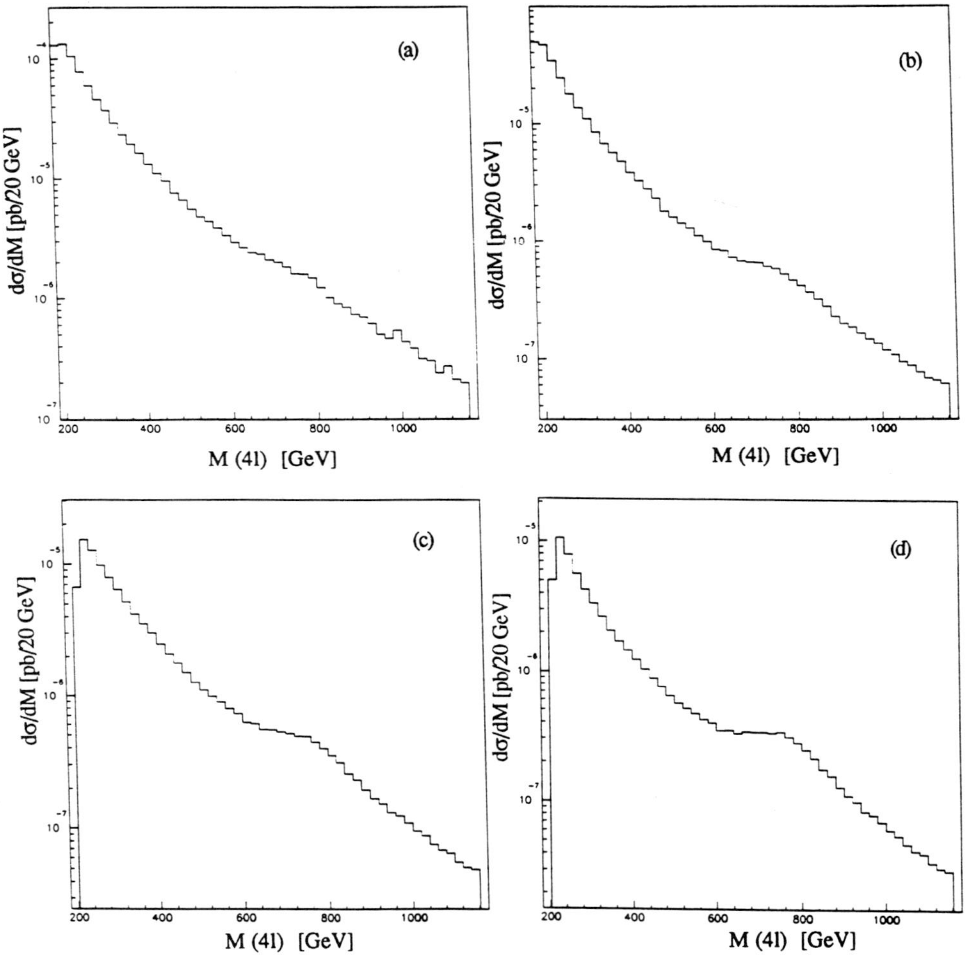

Figure 12

Higgs production at LHC (Signal plus Background)

Signal : $gg \rightarrow H^\circ \rightarrow Z^\circ Z^\circ \rightarrow l^+l^-l^+l^-$

Background : $gg \rightarrow Z^\circ Z^\circ \rightarrow l^+l^-l^+l^-$ and $q\bar{q} \rightarrow Z^\circ Z^\circ \rightarrow l^+l^-l^+l^-$

Higgs mass : 750 GeV
Top mass : 120 GeV

(a) no cuts
(b) cuts of set 1 (see text)
(c) cuts of sets 1 + 2
(d) cuts of sets 1 + 2 + 3

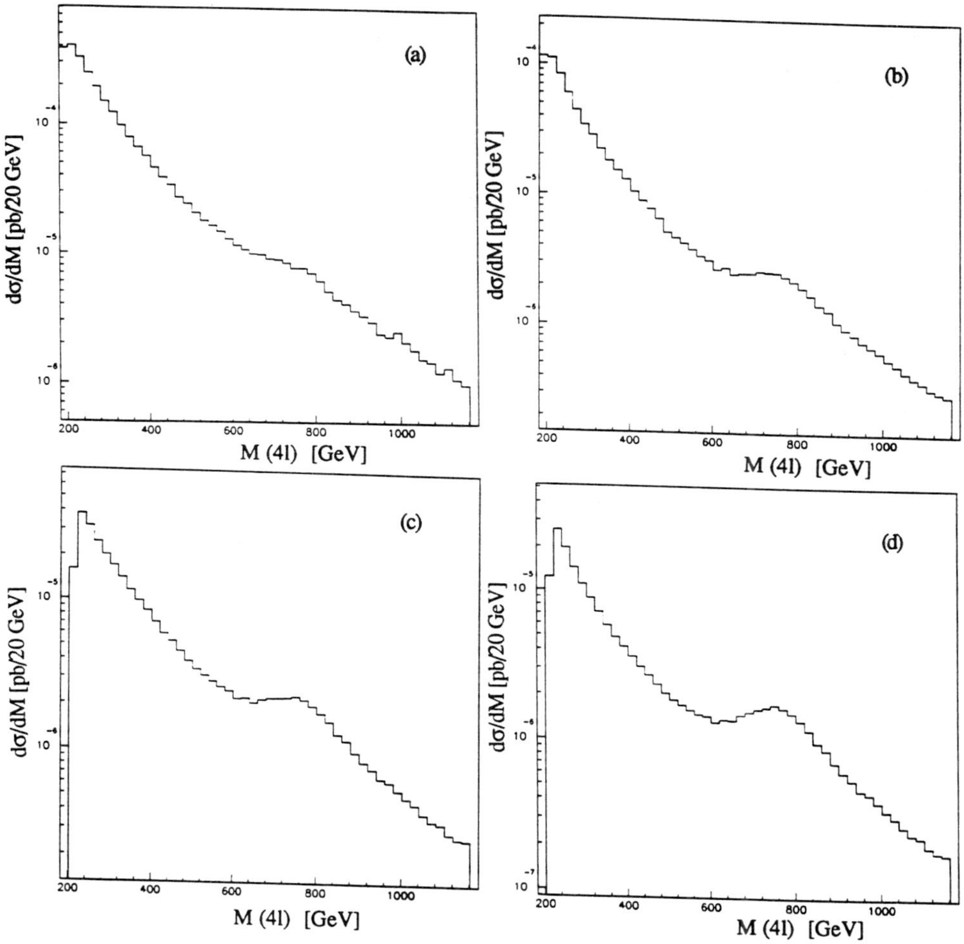

Figure 13

Higgs production at SSC (Signal plus Background)

Signal : $gg \rightarrow H^{\circ} \rightarrow Z^{\circ}Z^{\circ} \rightarrow l^{+}l^{-}l^{+}l^{-}$

Background : $gg \rightarrow Z^{\circ}Z^{\circ} \rightarrow l^{+}l^{-}l^{+}l^{-}$ and $q\bar{q} \rightarrow Z^{\circ}Z^{\circ} \rightarrow l^{+}l^{-}l^{+}l^{-}$

Higgs mass : 750 GeV
Top mass : 120 GeV

(a) no cuts
(b) cuts of set 1 (see text)
(c) cuts of sets 1 + 2
(d) cuts of sets 1 + 2 + 3

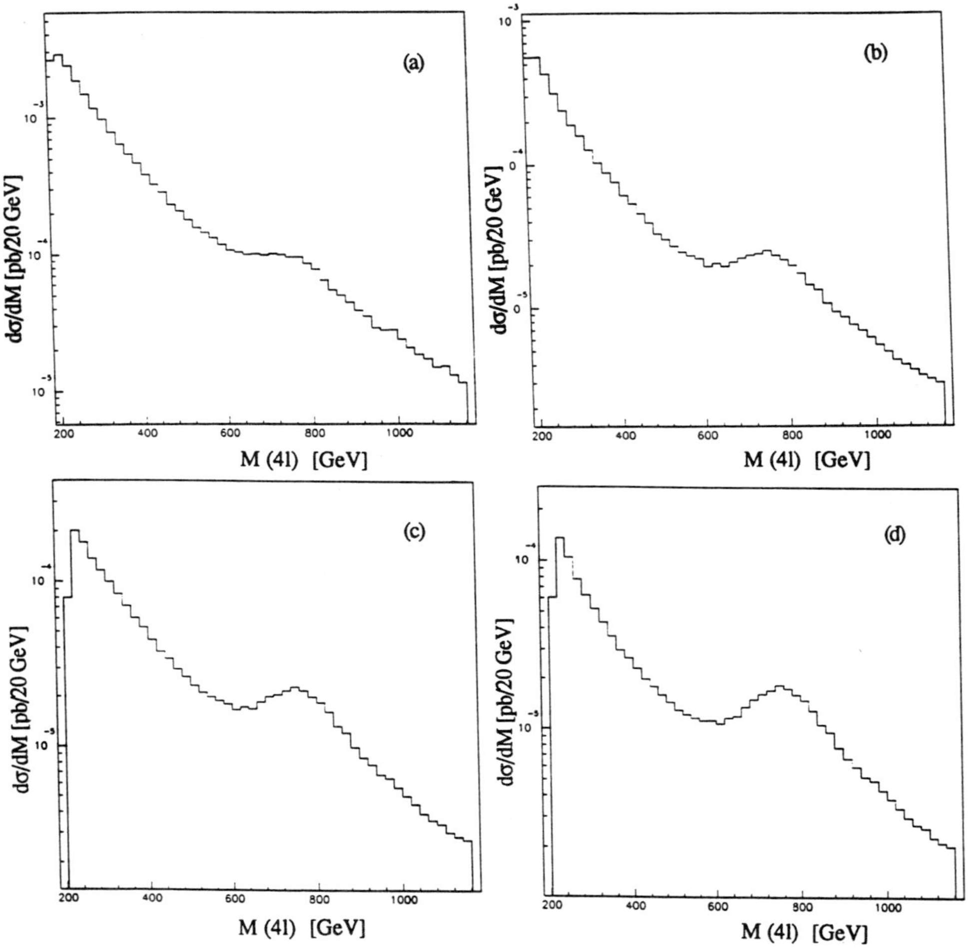

Figure 14

Higgs production at ELOISATRON (Signal plus Background)

Signal : $gg \rightarrow H^{\circ} \rightarrow Z^{\circ}Z^{\circ} \rightarrow l^{+}l^{-}l^{+}l^{-}$

Background : $gg \rightarrow Z^{\circ}Z^{\circ} \rightarrow l^{+}l^{-}l^{+}l^{-}$ and $q\bar{q} \rightarrow Z^{\circ}Z^{\circ} \rightarrow l^{+}l^{-}l^{+}l^{-}$

Higgs mass : 750 GeV

Top mass : 120 GeV

(a) no cuts

(b) cuts of set 1 (see text)

(c) cuts of sets 1 + 2

(d) cuts of sets 1 + 2 + 3

Multiplicity Distributions of Higgs Events

at $\sqrt{s}$ = 200 TeV . Event Generator : PYTHIA 5.4

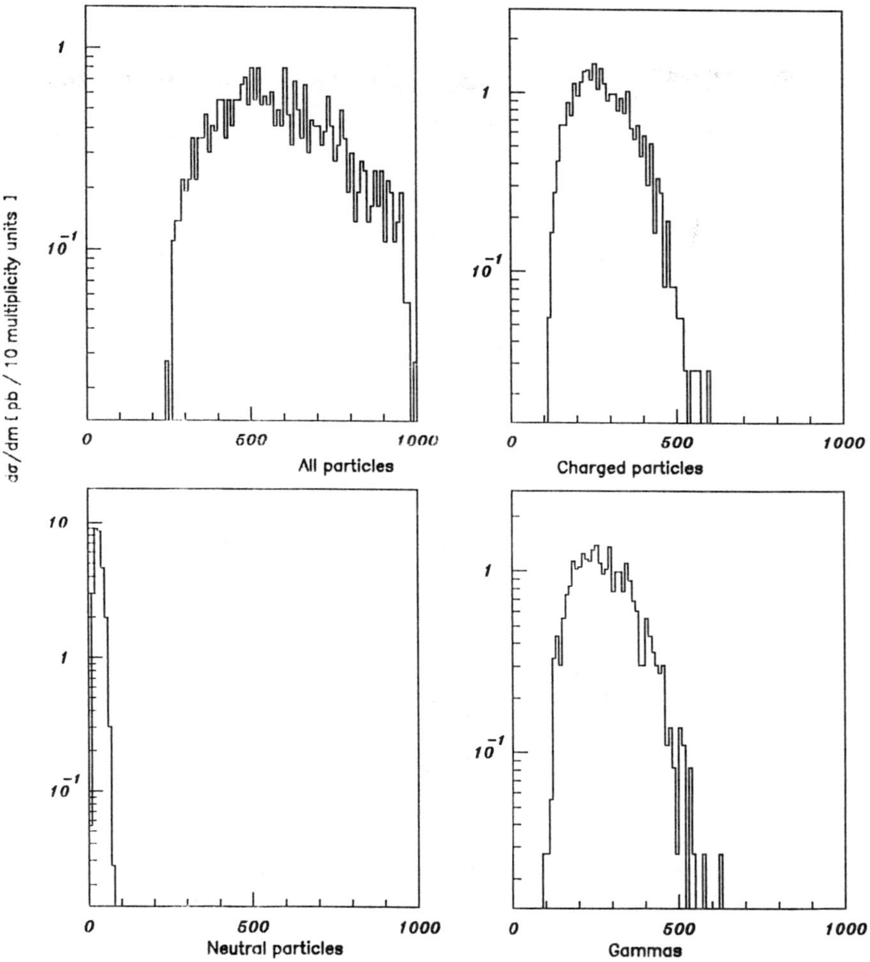

Figure 15

Higgs $\sqrt{s}$ = 200 TeV (expanded view)

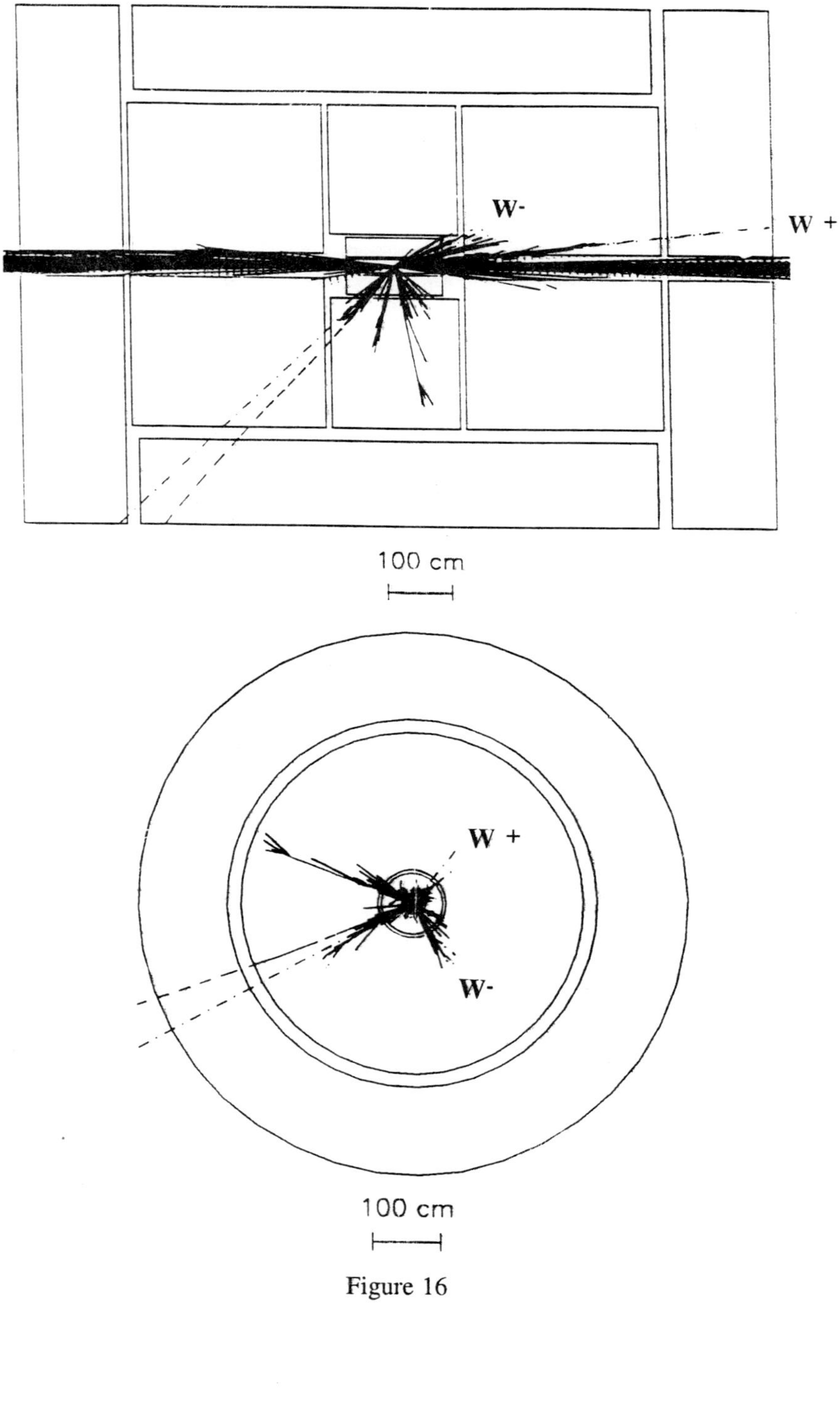

Figure 16

Higgs $\sqrt{s}$ = 200 TeV

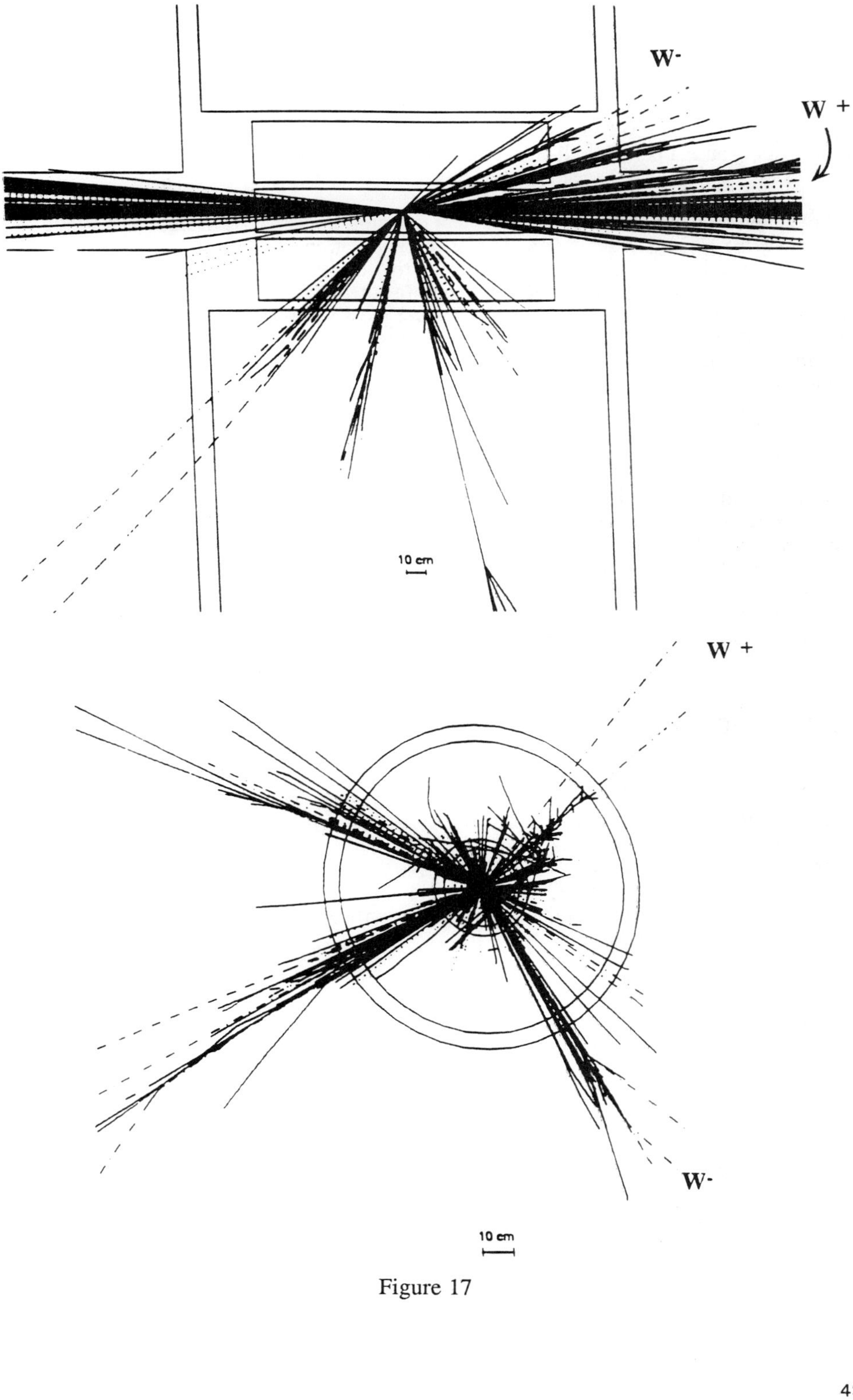

Figure 17

imposed in the generation. With these hypotheses, the total cross-section was at the level of 27 pb, the individual production mechanism contributions being:

$q\bar{q}$:	5 pb
gg :	9 pb
Z^0Z^0:	4 pb
W^+W^-:	9 pb.

Figure 15 shows the resulting multiplicity distributions, for all, charged, neutral (other than gamma) and gamma particles. The total multiplicity distribution has a mean value of about 500 particles/event, with a tail exceeding the 1000 level.

A typical event, as observed in the LAA test-detector, is displayed in figures 16 (longitudinal and transverse views), and in figure 17 after magnification. In this event, which gives an idea about the complexity of a Higgs search in a "real experiment", 670 tracks are present: 310 charged particles (plain lines), 36 neutrinos (dashed/dotted lines), and the remaining are gammas (dotted lines). Only particles with p>2 GeV have been tracked. The Higgs decays into a pair of W bosons: one of them produces jets, the other a chain of high energy (~1 TeV) neutrinos via leptonic decays. In addition to the jets in the forward region, jets are also visible in the central region, which balance the H^0 production.

5. CONCLUSIONS

The ideas and results presented herein show that Monte Carlo simulations are indeed a multitask activity, where the diversity of problems to be experienced, and hopefully solved, is practically unlimited. The efforts should concentrate on two directions. First of all, on new Physics inputs: to provide the most reliable predictions up to the highest energy one can realistically foresee for a hadron Supercollider within ten years from now. Second, on the development of software tools: to cope with the ever growing complexity of the Physics phenomena one will have to analyze in the future.

ACKNOWLEDGEMENTS

The MSL group consists of the following physicists and computer scientists of the CERN/LAA and World Laboratory/HED Projects, whose contribution to the present work I would like to acknowledge: J. Alberty, F. Anselmo, G. Bruni, O. Di Rosa, B. van Eijk, E. Eskut, D. Hatzifotiadou, G. La Commare, C. Maidantchik, M. Marino, T. Matsuura, R. Meng, R. Nania and G. Xexeo. I would also like to thank T. Sjöstrand for his suggestions on how to run the PYTHIA Event Generator at the unprecedented limit of 200 TeV. Finally my gratitude goes to Professor A. Zichichi and to the staff of the Ettore Majorana Centre for their as always warm and kind hospitality.

REFERENCES

1. W. Marciano and H. Pagels, Phys. Reports, 36C (1978) 137.

2. G. Altarelli and G. Parisi, Nucl. Phys. B126 (1977) 298.

3. D. Duke and J.F. Owens, Phys. Rev. D30 (1984) 49.

4. E. Eichten, I. Hinchliffe, K. Lane and C. Quigg, Rev. Mod. Phys. 56 (1984) 579 and 58 (1986) 1065.

5. M. Diemoz, F. Ferroni, E. Longo, G. Martinelli, Z. Physik C39 (1988) 21.

6. F. Anselmo and B. van Eijk, EUROJET, a Monte Carlo for the Simulation of Hard Hadronic Interactions, preprint in preparation.

7. T.D. Gottschalk, preprint CERN-TH 3810, CALT-68-1075.

8. R.D. Field, R.P. Feynman, Nucl. Phys. B136 (1978) 1.

9. C. Peterson, D. Schlatter, I. Schmitt, P.Zerwas, Phys. Rev. D27 (1983) 105.

10. B. Andersson, G. Gustafson, B. Söderberg, Z. Physik C20 (1983) 317.

11. G. Marchesini, L. Trentadue, G. Veneziano, Nucl. Phys. B181 (1981) 335.

12 F. Anselmo, E. Barberio, G. Bruni, P. Bruni, L. Cifarelli, O. Di Rosa, G. Iacobucci, G. La Commare, M. Marino, R. Nania and A. Zichichi, MEGA : Monte Carlo Event Generator Adaptor, preprint CERN/LAA-MSL/90-02 (submitted to Particle World Communications in Subatomic Physics)

13. The ALEPH Collaboration, ADAMO Notes, March 28, 1988.

M.G. Green, RHBNC 89-01 and CERN-DD/US/131, March 15, 1989.

14. J. Alberty, F.Anselmo, A. Bassi, G. Bruni, P. Bruni, L. Cifarelli, O. Di Rosa, D. Hatzifotiadou, G. Iacobucci, G. La Commare, M. Marino and R. Nania, FMC User's Guide, preprint CERN/LAA-MSL/90-06.

15. R. Brun, F. Bruyant, M. Maire, A.C. McPherson, P. Zanarini, GEANT 3, CERN DD/EE/84-1.

16. "The LAA Project", CERN-LAA/89-1, p.77 and p.295.

17. G. Anzivino, F. Arzarello, G. Bari, M. Basile, A. Bassi, L. Bellagamba, D. Boscherini, G. Bruni, G. Cara Romeo, M. Chiarini, L. Cifarelli, F. Ciralli, F. Cindolo, A. Contin, M. Costa, S. D'Auria, C. Del Papa, S. De Pasquale, F. Fiori, F. Frasconi, P. Giusti, G. Iacobucci, G. La Commare, A. Margotti, M. Marino, G. Maccarrone, T. Massam, N. Mc Cubbin, R. Nania, S. Quian, V. O'Shea, F. Palmonari, G. Sartorelli, M. Schioppa, G.C. Susinno, R. Timellini and A. Zichichi, "SUD: a Set-Up Descriptor for Modern High Energy Physics Experiments", submitted to Particle World Communications in Subatomic Physics.

18. R. Brun, P. Zanarini, KUIP Users' Guide, CERN-DD I202, February 26 1988.

19. F.Carminati, O. Di Rosa, B. van Eijk, D. Hatzifotiadou, I. Zacharov, Standard Interfaces between Modules of Event Generators using dynamical Common Structures, in: "Proceedings of Large Hadron Collider Workshop", G. Jarlskog, D. Rein, ed., CERN 90-10 and ECFA 90-133,vol. III, p.52.

20. G.La Commare, B. van Eijk, M. Marino, PDKDB : A Particle and Decay Data Base, to be presented at the Workshop on Detector and Event Simulation in High Energy Physics, Amsterdam, April 91.

21. H.U. Bengtsson, T. Sjöstrand, Computer Phys. Comm. 46 (1987) 43.

22. D.Hatzifotiadou, Charged particle - gamma overlap studies at $\sqrt{s}$ = 16 TeV, MSL Internal Report/90-01.

23. T. Matsuura, A Case Study on Higgs Production at 16, 40 and 200 TeV, MSL Internal Report/90-02.

Discussion

Chairman: L. Cifarelli
Scientific Secretaries : Eda Eskut, Despina Hatzifotiadou, Mark Wexler

- *Ciuchini:*

It seems that there are problems with the structure functions just by changing the input numbers for the subtraction point. This is quite surprising because of the logarithmic dependence of α_s on μ. Is it due to the presence of very different energy scales or what else?

- *Cifarelli:*

In our Monte Carlo, we have simulated $b\bar{b}$ production at different energies, using $2\to2$ diagrams. Depending on the structure function parametrization, one can get quite different results in terms of cross-sections, especially at 200 TeV. Moreover, for a given parametrization, the choice of the μ-parameter (i.e. the energy scale) in the running coupling constant $\alpha_s(\mu)$ has its own influence on the cross-section calculation. Typically one can choose for this parameter either the sum or the average of the b and $\bar{b}$ transverse energies. This corresponds to a factor of 2 difference in the μ value, namely to a difference of at least 10 GeV, since the condition $p_t>10$ GeV for both b and $\bar{b}$ was used in the simulation.

- *Ringwald:*

That difference in the predictions for $b\bar{b}$ cross-section if one uses different structure function parametrizations is probably due to the uncertainty in the gluon structure function at small x. If one goes to higher and higher $\sqrt{s}$, one is entering more and more the yet unexplored small x region.

- *Cifarelli:*

The $b\bar{b}$ case is just an example used as a starting point and many different phenomena should be studied in detail. However, what is instructive here is that in the 200 TeV range, even for standard $b\bar{b}$ production, the degree of uncertainty is very large, which is not the case at 16 or 40 TeV, although the energy is already very high. Extrapolations from LHC or SSC to ELOISATRON are far from being straightforward. As already pointed out, more information about structure functions will presumably come from deep inelastic scattering experiments at HERA, where new domains of x and Q^2 will be explored. This would be extremely useful for Monte Carlo simulation purposes.

- *Zichichi:*

This is just the beginning of a new enterprise. It is only to show how poor is our present knowledge in the field of QCD phenomenological models, even at low energy.

- *Brodsky:*

One of the most important questions in the theoretical analysis of hard inclusive reactions is the setting of the scale in the running coupling constant, as well as in the two input structure functions which, in general, have different scales. This is surprisingly a controversial question in the literature. As noted by Lepage, Mackenzie and myself, there is no ambiguity in the case of abelian theory (QED), and this can be used as a guide to the QCD case. Clearly further work is needed in this area.

- *Haas:*

You have presented Higgs production at the 16 TeV energy level. I am curious to know how much computing time is required for such an event generation and how this would increase for the production of a 1 TeV Higgs at a 200 TeV Collider?

- *Cifarelli:*

Using the IBM 3090 at CERN, the CPU time needed for 1 TeV Higgs production at 200 TeV is 10 minutes for 1000 events, with an average multiplicity of ~500 particles/event. This time increases by roughly a factor of 10 on a VAX 8350.

- *Haas:*

You have mentioned at the beginning of your talk that you think new methods were needed for developing software for Supercolliders. However you have only mentioned one tool, namely ADAMO. What other new methods do you have in mind?

- *Cifarelli:*

ADAMO is a very useful data modelling language based on the well-known ZEBRA database managing system. It drastically makes the user's life easier when dealing with huge and complex data sets, although it introduces some computing time overhead, which is the price to pay. In addition to ZEBRA, other systems exist on the market, such as BOS, CHEETA or JAZELLE, but they are in principle rather equivalent. New methods are needed, which, by tracing the ADAMO philosophy, should provide even simpler and moreover faster data handling systems for High Energy Physics.

- *Maley:*

What is the event tracking time when you put in a detailed detector configuration in the Monte Carlo?

- *Cifarelli:*

Using the GEANT programme with a full detector description for events containing many hundreds of tracks (as it is the case for Higgs production at 200 TeV) the CPU time needed for tracking can be as long as a few minutes/event on the IBM, or a few tens of minutes on the VAX.

- *Maley:*

Is anyone working on developing/testing pattern recognition/track-finding algorithms, as this seems to be incredibly hard in the 200 TeV regime?

- *Cifarelli:*

Not only track-finding, but also jet-finding algorithms. Actually our study on distances between pairs of tracks is not only related to the pion-gamma overlap problem, but also to the definition of an efficient jet-finding algorithm. On the other hand, we have also started trying to apply in our simulations the jet algorithms used at LEP. The difficulty here is due to the different event topology of hadron-hadron interactions, with a limited p_t production in the forward region which is not present in e^+e^- .

-*Borden:*

What language is the Monte Carlo source code written in ?

- *Cifarelli:*

FORTRAN 77.

- *Borden:*

Why not a more powerful language like C?

- *Cifarelli:*

C is clearly becoming very popular but you must realize that for at least 30 years all physicists have been using FORTRAN and it is hard to make them all switch to another language.

- *Civinini:*

Did you study the possibility of looking for the Higgs particle in the mass range between 70 and 160 GeV using the Higgs decay in two gammas?

- *Cifarelli:*

No, not yet.

- *Ambrosi:*

Is it possible, with the Super Monte Carlo, to simulate the actual rate of particles produced in prohibitive conditions, like $L = 10^{34}\ cm^{-2}s^{-1}$ and bunch crossing time of the order of 1 ns?

- *Cifarelli:*

Absolutely. We have generated events of the type known as "minimum bias" which represent most of the beam-beam interactions. In addition, we have generated events where some rare phenomenon occurs, such as top or Higgs production. We can then artificially pile-up these events, depending of course on the machine conditions we have in mind. The question is how to tag these overlapped events in reality. In minimum bias events, most of the products go forward, even very forward. One possibility would be to think of a detector with forward calorimeters and forward leading particle detectors, to recognize whether the event carries an overall unphysical energy.

- *Syed:*

What do you think about the GEANT programme that you are using, would it suffice the needs at 200 TeV? What about the computers required?

- *Cifarelli:*

As already pointed out, the GEANT programme is much too slow (I would say by a factor of 10), even for LHC needs. But the GEANT authors are working on the problem. About the computers, we should use either parallel or vector machines. Of course some effort will be required at the beginning to adapt our programmes to these machines. But the time needed for Monte Carlo simulations is such that one has to go in this direction.

- *Dokshitzer:*

I would like to propose a new "5-BRAIN manifold" for the benefit of the Super Monte Carlo activity. First is certainly Professor Brodsky, whose ideas should be implemented in the Monte Carlo since this would be of great interest for those who are planning to study CP-violation effects at future pp collidres (LHC, UNK...). The second part of the manifold includes LNPI theorists, E.Levin, M.Ryskin, Yu Shabelski and A.Shuvaev, who have started already (with the help of the Ettore Majorana Centre) to work on building in the event generators the QCD properties of the "Pomeron". This activity should clarify the problem of the drastic discrepancy between different structure function parametrizations as was shown by Professor Cifarelli. Implementation of the QCD knowledge to the small-x phenomena at very large s must exhibit the slow $\langle \ln s \rangle^{1/2}$ increase, the gluon shadowing, etc.

- *Zichichi:*

For the audience, I think it is interesting to know that there is a resident MSL subgroup here, in Erice, with some contribution from LNPI fellows, and that we want to enlarge it in a very near future.

- *Brodsky:*

One additional note concerning structure functions. At the very low x domain at fixed Q^2, structure functions saturate due to unitary considerations. These higher twist effects appear to be especially difficult to compute in hadron-hadron collisions since they are unlikely to factorize. It is clearly important to understand these effects theoretically and to take them into account in high energy projections.

SPONTANEOUS SYMMETRY BREAKING TWENTY-FIVE YEARS AGO

Peter Higgs

Department of Physics
University of Edinburgh
Scotland

The full title given to this lecture in the programme, "The Origin of Spontaneous Symmetry Breaking Twenty-Five Years Ago", was not my choice; it seems to imply that anything involving spontaneous symmetry breaking started only twenty-five years ago. In fact the story that I shall be telling is about how spontaneous symmetry breaking has been around in other parts of physics for quite a time before that, and how people like myself working in particle theory and field theory gradually came to understand what people in other branches of physics, particularly condensed matter physics, already knew.

So what happened twenty-five years ago that provides the excuse for the title of the lecture? To be precise, twenty-five years ago I myself was involved in trying to understand in detail the breaking of local U(1) symmetry in the simple field theoretic model about which I had written briefly the previous year and to which people subsequently attached my name. The importance of this model, which was described in detail in my 1966 Physical Review paper, was that it brought together spontaneous symmetry breaking and a relativistic field theory with a local gauge symmetry in the simplest possible way. Let me remind you of the historical background of these two ingredients.

First, gauge theories. The term gauge-invariance originates in Weyl's 1918 paper, via a mistranslation from German into English, but its current use occurs first in quantum electrodynamics. Non-Abelian gauge theory began with the attempts by Yang and Mills and by Shaw in 1954 to formulate gauge theories of SU(2) in particle physics. The problem there was that gauge theories appeared to be associated with massless spin-one bosons; to get a realistic theory with mostly massive bosons one had to put in mass terms by hand, which spoilt the gauge symmetry and the renormalisability of the theory. I think that the crucial paper for the development of all the viable models which appeared later was that by Schwinger in 1957, "Theory of the Fundamental Interactions", which contained many of the ingredients of later unified theories, but was somewhat too general and did not put the ingredients together in quite the right way. I was very much inspired by Schwinger's work. Shelly Glashow was one of Schwinger's students, and one of the applications of this type of theory was Shelly's SU(2)×U(1) model of leptonic electroweak interactions, based on a symmetry of the known leptons which Lee and Yang (1960) were the first to describe. The Glashow model, published in 1961, was already in existence thirty years ago in 1960, another crucial date in this story.

The other side of the story concerns spontaneous symmetry breaking. This arrived in theoretical physics, in the form in which we would recognise it today, in 1928, over sixty years ago, in the Heisenberg model of a ferromagnet, in which the interaction between spins results in spontaneous breaking of rotational symmetry. That is the crucial date in the context of condensed matter physics. The next development came in 1947 when Bogoliubov, discussing superfluidity in terms of Bose condensation, described a model which is the first example in which an internal symmetry, the multiplication of spinless boson wave-functions by a phase factor, is spontaneously broken. The broken symmetry here was a global U(1) symmetry. Then in 1955 came the BCS theory of superconductivity, which is the true precursor of the theories that we now use in particle physics.

That is the prehistory of spontaneous symmetry breaking from a particle theorist's point of view and it brings us to the period in which these ideas moved from condensed matter physics into particle physics, the late fifties and early sixties. In 1960 Nambu, who had learnt BCS theory from Schrieffer, published a formulation of this theory in quantum field theory language, which made it clear that Bose condensation of Cooper pairs involved spontaneous breaking of the local U(1) symmetry associated with the arbitrary phase of the electron field. This is similar to the situation in superfluidity, but there are differences. The crucial difference is that the broken symmetry is a local one, associated with the gauge transformations of electrodynamics. The other difference, which is not so relevant to this story, is that the condensate is formed from composite bosons.

Unfortunately, Schrieffer stayed at Chicago only one year, so Nambu did not learn as much about superconductivity as he would have liked. (This I learnt from Nambu himself six years ago.) What Nambu did next was to take the idea of spontaneous symmetry breaking and apply it to flavour symmetries, such as SU(2), in particle physics. The models which he published from 1960 onward followed the BCS pattern by using condensates of fermion pairs to break the symmetry. Unfortunately, it was a global SU(2) symmetry that he was breaking, and he discovered (as did other people who tried models of this type) that, when you do this, you always predict in the spectrum some massless spin-zero bosons. This result was very soon turned into a theorem, first in an informal way by Goldstone (1961) and then, more formally, by Goldstone, Salam and Weinberg (1962). The Goldstone theorem said that, if you spontaneously break an internal symmetry in a manifestly Lorentz-invariant quantum field theory, inevitably you predict massless spin-zero particles, which we now call Goldstone bosons. That theorem persuaded most particle theorists that Nambu's ideas were of no use, because one didn't know any strongly interacting massless spin-zero particles, which would have been easily detected.

Significantly, the only encouragement at the time for Nambu's ideas in particle physics came from a condensed matter theorist, Philip Anderson. He wrote a paper, which appeared in Physical Review in 1963, entitled "Plasmons, Gauge Invariance and Mass", in which he pointed out that there are no Goldstone excitations in a superconductor., The conclusion that he drew from this for particle theorists was that the Goldstone zero-mass difficulty was not a serious one because it could probably be cancelled off against an equal gauge theory zero-mass problem. This is the first formulation of what Anderson recently referred to as the so-called "Higgs" mechanism. But particle theorists didn't take any notice of him because he didn't provide an explicit relativistic example: he just said that it would be so on the basis of the dynamics of superconductivity. Because the Goldstone theorem appeared to have been proved, for relativistic theories, particle theorists tended to believe that he was mistaken.

At that time I was interested in this problem and, although I couldn't see how Anderson could be right, it seemed to me very disappointing that the Goldstone theorem ruled this out. During the following year a number of papers in Physical Review Letters debated whether the Goldstone theorem was completely foolproof. One, by Abraham Klein and Ben Lee, pointed out that the theorem could be evaded in a theory containing a special direction in space-time, such as the inertial frame of reference of a condensed matter system. They speculated that this could occur also in a relativistic theory, but in the Spring of 1964 a reply was published by Walter Gilbert, who said that there is no such thing as a preferred frame of reference in relativistic quantum theory. It was, I think, his last contribution to quantum field theory and it turned out to be wrong. I'm glad to say that he made good in another scientific field shortly afterwards.

I happened to have the good luck, just twenty-six years ago, in July 1964, to realise what was wrong with Gilbert's argument. The basis for my luck was the circumstance that, in parallel with my interest in spontaneous symmetry breaking, I had followed Schwinger's papers on whether gauge invariance forced the photon to be massless. Schwinger was a devotee of Coulomb gauge electrodynamics, in which you do have a preferred frame of reference; you impose the gauge condition in your own rest frame but, in spite of the apparent lack of Lorentz invariance, the physics remains Lorentz-invariant. I knew from Schwinger's papers what spectral representations of vacuum expectation values of commutators, such as those which were used in the proof of the Goldstone theorem, could look like in a Coulomb gauge. So it suddenly occurred to me that this was the loophole in Gilbert's argument; if you have a gauge theory in Coulomb gauge (one in which there are no unphysical states in the spectrum, unlike covariant gauges), the Goldstone theorem is no longer relevant because Lorentz-covariance is no longer manifest.

Immediately I wrote a short paper, which showed how the Goldstone theorem could be evaded in a gauge theory, for

Physics Letters and they accepted it. But all I had done so far was to show that evasion was a possibility: did it really happen? So I wrote down the simplest possible relativistic model, scalar electrodynamics with a potential for the scalar fields added so as to make the vacuum with zero scalar field expectation value unstable and have stable vacua away from the symmetry point. I studied it as a classical field theory, simply linearising the field equations about a minumum of the potential, and, lo and behold, it worked in just the way that Anderson had said it should! The gauge field and the Goldstone field combined to propagate in a way corresponding to a massive spin-one quantum, just as in a non-relativistic superconductor. I call this the relativistic Anderson mechanism because Anderson described it first: it was his misfortune not to do so explicitly enough.

I wrote a brief paper on this model and sent it off to Physics Letters: it was rejected. This was a blow, because I thought it was important. A colleague of mine who was at CERN at the time, which was where I sent it to the nearest editor, told me later that they didn't see its relevance to physics. Perhaps, at a time when quantum field theory was rather out of fashion in particle physics, it was a tactical mistake to use classical field theory and to hand-wave the quantisation by invoking the de Broglie relations for the linearised field equations, as I had done.

Being convinced that this really ought to be published, I rewrote the paper with more speculation about possible applications. In the new version I drew attention to a characteristic feature of this type of model, that it would predict incomplete particle multiplets of the broken symmetry; in particular, there would be some left-over scalar particles, those which are now called Higgs bosons. I sent it across the Atlantic to Physical Review Letters and they accepted it. This was perhaps because, as I discovered six years ago, the referee was Nambu, who was well qualified to understand the point of it.

During the following year I was busy with other things in Edinburgh and didn't develop these ideas very much. But I had been invited by Bryce DeWitt to spend a sabbatical year at Chapel Hill, and when I arrived there in September 1965 I worked out more detail of the Abelian model. This went into the paper which appeared in Physical Review in 1966.

The preprints of this paper were sent out from Chapel Hill just before Christmas 1965. In the New Year I got a very nice letter from Freeman Dyson, who was on the mailing list. He said that my paper had cleared up a number of things which had puzzled him for a long time, and invited me to the Institute at Princeton to give a seminar. I accepted the invitation for March and, in order to combine two things in the same trip, got in touch with Stanley Deser, who had the previous year invited me to give a seminar in the Boston area if an opportunity arose while I was at Chapel Hill. As a result I gave one seminar at the Institute on the 15th March and another at Harvard on the 16th.

The Princeton seminar was a rather gruelling experience. I had to follow a very high-powered talk by Dyson himself and, just to put me at my ease, Klaus Hepp, an axiomatic field theorist, assured me at tea time that what I was going to say must be nonsense because axiomatic field theorists had proved the Goldstone theorem rigorously using the methods of C*-algebras. However, I survived questions from Arthur Wightman and others, so I concluded that perhaps the C*-algebraists should look again.

The next day at Harvard was an entirely different experience. By that time, having survived Princeton, I had gained a little confidence. The seminar at Harvard, which Shelly may remember because he was there, was more like a dialogue between me and the audience - which was fun. Years later, in 1979, I ran into Sidney Coleman again, He remembered the occasion, and told me that he and his colleagues "had been looking forward to some fun tearing to pieces this idiot who thought he could get round the Goldstone theorem". So, in the event, we all had fun!

At the Harvard seminar I think I convinced people that gauge theory evaded the Goldstone theorem, but in a very negative way. As far as model building to make contact with experiment was concerned, I was very much tied, as Nambu had been, to thinking about hadronic flavour symmetries, and I didn't see how to use gauge theories in this context. So the feeling that (I think) people came away with from my seminar was that this was maybe fun as a model field theory but it wasn't any use. Afterwards, I continued trying to find a phenomenologically relevant model, but still in terms of hadronic symmetries — SU(2), SU(3), SU(3)×SU(3), and so on — and that was the wrong game. So it wasn't until 1967 that Steven Weinberg and Abdus Salam used the idea of spontaneously broken gauge symmetries in the right context, which was, in the first instance, Shelly's SU(2)×U(1) theory of leptons.

You may well ask why that didn't happen earlier, and Shelly has already speculated about this in his Nobel lecture. I hope he won't object to my quoting from it. He said this about Goldstone, Kibble and myself: "These workers never thought to apply their work on formal field theory to a phenomenologically relevant model. I had had many conversations with Goldstone and Higgs in 1960. Did I neglect to tell them about my SU(2)×U(1) model, or did they simply forget?" I'm going to answer that. I'll tell you my side of the story, why I didn't know as much as I should have known about Shelly's model.

I should confess right away that, in my 1966 Physical Review paper, among the many models of gauge theory type which I listed in the references was Shelly's paper. But, as some of you probably know, the fact that you list various papers as references in the paper which you are writing doesn't always mean that you've read them very carefully. They may be there to show that you have some knowledge of the background to your own paper.

My contention is that Shelly never told me about his model in 1960. My version of the story about why it didn't happen goes back to thirty years ago, to the very first summer school that I attended, which was the first Scottish Universities Summer School in Physics at Newbattle Abbey near

Edinburgh. I had just been appointed to the faculty of the University of Edinburgh from 1st October 1960, and I had been told that since I was coming to Edinburgh I had better do some work at the summer school. They had a job for me. The school was being funded by grants from NATO and the Scottish Universities, and there was some spare money as a consequence of one of the speakers having his fare paid from another source. They decided that they could use this money — NATO's rules were not so strict in those days! — on providing wine to go with dinner each night. I was given the job of steward, and the main task of the steward was to buy supplies of wine, to distribute it at dinner, and to conserve it otherwise.

Now among the students at that first Scottish summer school were Dr. Nicola Cabibbo, a Mr.Martinus Veltman and, last but not least, a Dr. Sheldon Glashow, who by then had already formulated his SU(2)×U(1) theory. This group, along with some others, discussed such things as weak interaction theory late into the night and they had, I think, an interest in non-conservation of wine. It wasn't until Cabibbo confessed in 1979 that I discovered that a condensate of surplus wine from dinner had been formed in the grandfather clock in the old crypt of Newbattle Abbey, where these discussions took place. You can perhaps appreciate that for social reasons — reasons of differing interests in wine conservation or non-conservation — I didn't get involved in these discussions very much myself. So unfortunately I never got to learn about Shelly's model in 1960, and by the time of his next visit to Edinburgh in 1962 or 1963 he was very much involved in hadronic symmetries and electromagnetic mass differences. I don't think we ever talked about electroweak SU(2)×U(1). So that is why, unfortunately, Shelly had to share the 1979 prize with two other guys.

THE NEW ROLE OF SCIENCE

V.F. Weisskopf

Massachusetts Institute of Technology (M.I.T.)

Cambridge, MA 02139 USA

And I gave my heart to seek and search out by wisdom concerning all things that are found under the heaven. That sore task hath Good given to the sons (and daughters) of man to be exercised herewith.

For in much wisdom is much grief and he that increaseth knowledge increaseth sorrow.

Eclesiastes

Since the beginning of culture, man has been curious about the world in which he lives; he has continually sought explanations for his own existence and for the existence of the world – how it was created, how it developed and brought forth life and humankind, and how one day it will end. Early ideas on the subject were developed in mythological, religious, or philosophical frameworks.All these ideas have a common characteristic: they are directed to the totality of the phenomena; they want to account for everything that is. They intend to present the absolute truth by attempting to give immediate answers to the fundamental questions of existence: "Why is the world the way we find it? What is life? What is the beginning and the end of the universe?

Several hundred years ago, human curiosity took a different turn: instead of reaching for the whole truth, people began to examine definable and clearly separable phenomena. They asked: "What is matter?" and "What is life?" but "What are the properties of matter?" and "How does blood flow in the blood vessels?"; not "How was the world created?" but "How do the planets move in the

Physics Up to 200 TeV
Edited by A. Zichichi, Plenum Press, New York, 1991

sky?" In other words, general questions were shunned in favour of limited ones for which it seemed easier to get direct and unambiguous answers.

By means of this detailed questioning, a framework was created for understanding the natural world. Thus, something like a scientific world view arose in the twentieth century, a synthesis of scientific insights gained over the previous five hundred years.

The world view of natural science differs in two important aspects from the religious, mythological, and philosophical ones. First, it does not directly contain concepts that are connected with the "human soul", such as faith, awe, desolation, happiness, good and evil, and so on. But is does contain these concepts in an indirect way. They appear as manifestations of certain neurophysical processes in the brain. I am convinced that some important human aspects will never be touched by these scientific investigations. They distinguish themselves from scientific treatable phenomena because they are unique and non-recurrent. We are all different and in most cases the differences are probably outside the scientific range of knowledge.

Second, the insights are "tentative"; they are considered as incomplete perceptions, as part of a greater truth hidden in the plenitude of phenomena. The insights are not based upon dogmatic principles, revealed to us by divine inspiration, or by some internal sparks of full recognition. What is perceived as "scientific truth" is steadily revealed in partial steps, sometimes big ones, sometimes small ones, and sometimes even steps backward. Some recent knowledge will turn out to be mistaken. It was rare in the past that insights turned out to be outright wrong, but some of them became and will become too limited, not general enough, misconceived, or awkwardly formulated; some will appear irrelevant in view of future deeper insights.

The applications of scientific insights proceed in two ways: first by using results of one branch of science in other branches, such as chemistry in biology; physics has led to the construction of many apparatus employed in other sciences. Second, by using scientific insights for technical purposes. Modern technology is almost exclusively developed from scientific insights. The older the scientific insights the more extended are the technical applications.

I tried to present this development of a triangle (fig. 1) with its tip on top. Time moves vertically. The present fundamental sciences that had not yet been applied are on top, such as particle physics, cosmology, the science of chaos. Somewhat lower are those sciences that are beginning to be applied. This is shown by the extensions to the right and left, the former containing scientific applications, the left ones technical ones. For example, in nuclear physics, although still fundamental, already is applied e.g. as tracer technique in biology and material science and as power plants in technology. Going further back in time, (down in the figure) we find atomic physics, condensed matter physics with more extensive applications of both kinds, finally electrodynamics and mechanics whose applications are numerous and varied.

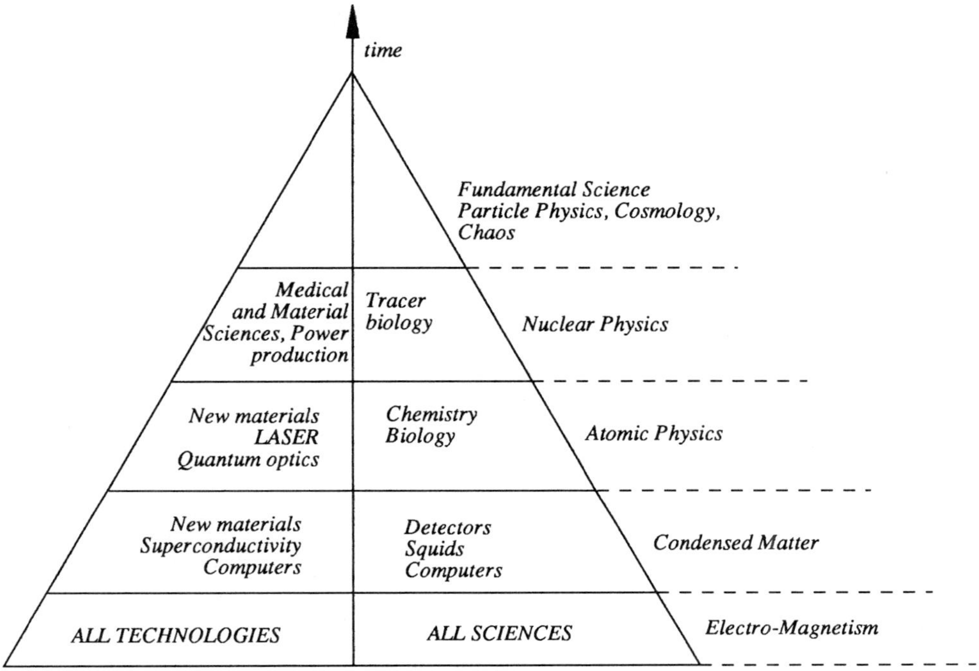

Today we observe an increasing tendency towards applications. For example, the physics of condensed matter directs its efforts more and more towards applied problems.

I observe other tendencies in fundamental physics. First a tendency of building larger accelerators. It comes from the fact that modern research deals with conditions very different from those on earth, conditions that prevail in the universe at special locations in time and space, far away in exploding supernovas, early in time shortly after the big bang. It is the "leap" into the cosmos, to situations of very high temperature and pressure, far from our earthly environment. Such instruments require large teams of several hundred collaborators, an organization of collaborators unknown a few decades ago.

Theoretical physics has also changed character. Although the fundamental approach remains similar – looking for symmetries, quantum field theory, renormalization –, the character of theory became much more mathematically sophisticated and formal, with a lot of computer use; a tendency I do regret, since it makes it much harder to recognize the essential features.

There was great progress in theoretical physics: the generalisation of electrmagnetic quantum field theory to Quantum-chromo-dynamics, which describes well the forces between quarks; the unification of the weak and electromagnetic fields are the most striking examples.

In spite of this, as we all know, many fundamental problems remain unsolved in particular the problem of the origin of masses. There are still two famous constants in nature that determine all processes in our neighborhood: the proton mass divided by the electron mass and the ratio 137 of the quantum of action times the light velocity to the square of the electric charge. These are very important numbers of which we do not know the provenance.

I was told that in the "cabala", (a sort of Jewish mysticism), every word has a symbolic number with a deep sense; the word "cabala" has the number 137; is the only explanation I can give for this number.

The fact that there is no explanation for the masses except that the Higgs field couples with every particle with an arbitrary coupling constant and gives them a mass, is certainly not very satisfactory. The Higgs field itself is a problematic concept. A field that has a constant value all over the universe is some what perverse. This is why I believe there is no Higgs particle. We may have misunderstood the Higgs situation. I believe that in some years we will express things in a different way. If you people do not find the Higgs particle, you will make me happy.

I would like now to make a remark about the foundation of quantum mechanics. There was a conference here last year on this subject. People like John Bell and others discussed what there is still to be understood; admittedly a difficult subject. I personally think that quantum mechanics is pretty good as it is.

However, one fundamental question worries me very much: in the Einstein, Podosky, Rosen discussion, as you know, there is an effect that goes faster than light. It is true that you cannot use this super-relativistic effect to transmit messages. I am deeply worried about this; the fundamental law about light velocity be a limit, is somehow not quite right.

Is it really an aim of theoretical physics to get the world formula? The greatest physicists have always thought that there was one, and that everything else could be derived from it. Einstein believed it, Heisenberg believed it, I am not such a great physicist, I do not believe it. I have the feeling that when we will exploit the SSC, LHC and Eloisatron we will always be confronted with new things which will not fit in any world formula; this, I think, is because nature is inexhaustible.

My next point is important from the practicle point of view. When one speaks about the aims of fundamental physics, let us say particle physics, I would maintain that the aims of particle physics is threefold and all three are of equal importance:

The first is to find new laws of nature, we found some of them;

the second is to find new ways of behaviour of nature, and there we found a lot: three types of electrons and neutrinos, quarks, and many other phenomena, e.g. the polarization of the vacuum;

the third one, which is usually too much neglected, is the instrumental challenge that faces us when we want to observe phenomena under very unusual conditions. It forces us to find new ways of looking at nature; these new instruments are important, not only for high energy physics, but also for other sciences and technologies.

Michael Polanyi expressed the third point very succinctly:

> *The scientific method was devised precisely for the purpose of elucidating the nature of things under more carefully controlled conditions and by more rigorous criteria than are present in situations created by practical problems. These conditions and criteria can be discovered only by taking a purely scientific interest in the matter which again exist only in the minds educated in the appreciation of scientific value. Such sensibility cannot be switched on at will for purposes alien to its inherent passion.*

Fundamental science requires new ways of looking at Nature. This is of a great practical importance. It is equally important, if not more, than the fundamental laws and the new phenomena.

The application of science plays an increasing role in Science. After all, biology is essentially applied chemistry, chemistry, like other sciences, becomes more biological within general trend to more applications. These applications have special effects on our living conditions and an important role on our civilization; medicine (the technology of biology) led to a reduction of epidemics, it introduces something we may call "death control", at least in the developed countries. These applications changed our way of life through industry; they have eliminated strenuous labour, fantastic means communications have been installed. If you compare the carriage by which Emperor Hadrian traveled from Rome to Vienna with the carriage with which, 700 years later, Mozart travaled from Vienna to Rome there is little difference.

At the end of the 19 century something happened which I would like to call the humanization of technology. When industry began, the workers were exploited, children were working and the social services were non-existent. At the beginning of this century something changed: social services were installed, the working hours reduced, children protected, and assistance for sick people was set.

But there are also negative things. Pollution: material pollution and spiritual pollution. The material pollution changed our world and probably not for the better. We influence the whole surface of the earth with the greenhouse effect, we ruin the ozone layer, we destroy forests, we make unsafe nuclear reactors; we have a population explosion because medicine has introduced death control, but we have not introduced birth control, at least in many places. Previously the application of science was directed towards improving our living conditions to make our life more pleasant. Now we have the different task to avoid effects which I call the material pollution. The application of science has to be diverted from inventions to make our life pleasant, to the avoidance of environmental catastrophies. Thomas Mann expressed it in the following way:

> *Astronony – a great science – teaches us to consider the earth as a comparison of an insignificant star in the giant cosmic turnoil, roving about at the periphery of our galaxy. This is, no doubt, correct. But I doubt that such a correctness reveals the whole truth. In the depth of my soul I believe – and consider this belief to be natural to any human soul – that this earth has a central significance in the universe. In the depth of my soul I entertain the presumption that the act of creation which called forth the inorganic world from nothingness, and the procreation of life from the inorganic world, was aimed at humanity. A great experiment was initiated, whose failure by human irresponsibility would mean the failure of the act of creation itself, its very refutation. May be it is so, may be it is not. It would be good if humanity behaved as if it were so.*

This statement gives us a warning. Avoiding environmental catastrophies is expensive and difficult, but this is the task of the New Role of Science which I would like to call the second humanization of technology.

We come now to spiritual pollution. The application of nuclear science for bomb making is a spiritual pollution. In a few decades, I hope so, the nuclear arms race will be considered as a collective mental disease of mankind. To have created 50 thousand nuclear warheads is crazy. Fortunatly, during the last few years people and their leaders began to see the craziness of the situation; there has been a little improvement, but not enough yet.

What about other spiritual pollutions? As I mentioned before, technology has freed the humanity from burdens, at least in the developed countries, people are not longer forced to fight for mere existence. The scientific industrial age has brought us enormous gains which have not been well enough exploited. It has given us the means to remove hunger, to remove want, to remove illness, to remove oppressive manual work. Humanity can be freed of its burden. Freed for what? What does one do if one doesn't need any more to fight for one's existence? Then one is thrown back upon himself, one must find something, one must find a meaning of life. Work today, apart from a selected few, is very mechanical for most of the people, mechanical or clerical. The result of the work is not the of the working person, but one of the engineers who have developed the machines. The worker has very little influence on their work or on the enterprise of which is part.

What has become then of human dignity, of individual sense and purpose? In the earlier centuries this individual sense and purpose was given by religion, but the role of religion has grown weaker nowadays.

There are, however, groups of persons for which a kind of religion furnishes sense and purpose to their lives if religion is defined to be a deep commitment to a great cause beyond one's personal interest, whose values are never questioned. There are people like social workers who work for a better life of the poor among us, by urging birth control, drug control; these activities make a lot of sense; the same goes for people who fight pollution by technical means. It is part of a fight for a second humanization of civilization and against the passivity that has entered our culture.

The scientific community also belongs to those who have "religion" in the previously defined sense. It is derived from the greatness of the scientific ideas that have today evolved, such as the origin of the universe, the development of stars, the formation of elements, the fundamental, and molecular basis of life. We begin to understand what makes the universe tick. This is an inspiration for a true scientist, it fans his enthusiasm and his commitment. For the non-scientists, however, it does not have much significance. He only knows that there are laws of nature, and therefore the bible stories can no longer be true; the significance of science for him is only derived from the practical application of physics, or chemistry, or biology. The scientists are guilty for this because they do not try sufficiently hard to tell the public about the greatness and the wonder of nature as discovered by science.

Obviously, artists belong also to this category. They devote their efforts to a purpose greater than their personal interest.

For most people, however, neither of these "idealistic" efforts mean very much. When the most important material needs are provided for, as they are in the developed countries, what is then the content of life? There is a rampant disease of senselessness, which expresses itself in an increase of criminality, of drug culture, of drop outs, of cynicism, in spite of the great improvement of our well-being in the last thirty years. There is a hole in the belly that is left by the weakening of religion, which has not been filled yet by anything else. We are psychologically, if not economically, deprived. What is sorely lacking is a fulfilled, creative content of life for the population at large.

How can we do this? It is a hard task. Among many things we need something that has to do with concept of complementarity. Bohr himself generalized this concept by saying that it is not only valid within physics when we consider the particle and wave nature of an electron, but actually all over. For instance, when looking at a sunset one can think of the scattering light or of the beautiful play of colors, or of the symbolic meaning of the end of a day; the same is true for a starry sky, one has similarly contradictory sides of the same experience, beauty, science and the awe inspiring universe. A Beethoven sonata can be described as vibration of the air and, in a complementary way, as an emotional experience.

Heisenberg and Bloch were walking one day on a beach in a beautiful day; Bloch tried to explain Heisenberg some new ideas about the geometry of space, when Heisenberg, switching into another avenue of complementarity said *"Space is blue and birds fly in it"*. This is also a true statement.

John Keats said:

> *"Do not all charms fly at the touch of cold philosophy. There was an awful rainbow once in heaven.*
> *We know her woof, her texture.*
> *She is given in the dull catalogue of common things"*

He accuses science to take poetry out of nature. I disagree. The rainbow is much more interesting if one looks not only at the esthetic beauty of it, but also enjoys the scientific explanation.

Complementarity, which started with quantum mechanics, gives us a more complete view of our life experiences. There are many pairs of complementarity views, such as the scientifc approach and poetic approach, or justice and compassion.

Bohr liked to show us a cubistic painting in his house, the kind where one can see a person at the same time from all views of all sides; this painting gave many different seemingly contradictory aspects of the person which a photography could not have shown; a typical example of complementary aspects.

History has shown that whenever one approach is developed in great force, other approaches are usually unduly neglected. For example, the religious approach dominated the Middle Ages, the religious way of looking at the world (at least in Europe) was so dominant that nobody noticed the appearance, on July 4th 1054, of a supernova ten times brighter than Venus; there is not a single reference about this new star. It depends on what one considers important. In the framework of the religious ideas of that time, a new star was just one little miracle like many others.

In China, where there was a different way of understanding things, a detailed record of this phenomenum was made; from it the exponential decay curve could be derived.

The scientific approach will only succeed if the civilization considers it as relevant.

The emphasis on religion of the Middles Ages and our present obsession with scientific and technological developments have released enourmously powerful creative forces; in the Middle Age architecture, philosophy, gothic cathedrals are wonderful examples of the creativity of that period, but there were serious abuses such as the crusades and a very low value given to individual lives.

When my students in the late 60's were excited about Vietnam and described the cruelties committed during the Vietnam war, I told them about the conquest of the city of Beziers in 1205 by the French army. The general asked the Papal legate what he should do with the population. The answer was: *"Kill them all. God will select the ones to send to heaven from the ones to be damed to hell"*. This shows the abuse of indirectional ideas. The students correctly pointed out that there can be abuses also coming from a complete emphasis on technology and science. These abuses stem from our thoughtless exploitation of nature, from our over-emphasis on material values, what I call the greed of the Western world, and the irrational production of weapons for destruction. It is clear to me that we have to face our tasks, and this is maybe the New Role of Science, to find an approach that seeks complementary solutions. In order to deal with the totality of human experience we need more than the answer provided by science, but to apply also concepts that are not in science, like the soul and the moral values.

In many cases, for moral and political decisions, the scientific insights can point out the consequences of certain actions, but the decision which actions should be taken always rests on non-scientific arguments, and this to include openness, tolerance and full understanding of the situation.

Human experience encompasses much more than any given system of thought can express within its own framework of concepts. We must be receptive to the varied, different, and apparently contradictory ways of the mind when we are faced with the reality of nature, of our imaginations, and of human relations. There are many modes of thinking and feeling: each of them contains a part of what we may consider the truth. Science and technology comprise some of the most powerful tools for deeper insight and for solving the problems we face. Some of these problems, indeed, were created by the thoughtless applications of those very tools, such as the pollution of our environment and the danger of nuclear war. But science and technology are only one of the avenues toward reality; others are equally needed to comprehend the full significance of our existence. Indeed, those other avenues are necessary for the prevention of thoughtless and inhuman abuses of the results of science. We will need all approaches to deal with the predicaments of humanity that prevent so many of our fellow beings from having a life worth living. It is the precondition for the survival of our civilisation which is severely threatened by the lack of complementary thinking.

CLOSING CEREMONY

The closing ceremony took place on Monday 23rd July 1990. The Director of the School presented the Prizes and Scholarships as specified below.

PRIZES AND SCHOLARSHIPS

Prize for Best Student

awarded to: *Stephen HSU*, Lawrence Berkeley Laboratory, Berkelcy, USA.

Eleven Scholarships were open for competition among the participants. They were awarded as follows:

Patrick M.S. Blackett Scholarship
Mark WEXLER, Princeton University, Princeton, USA.

James Chadwick Scholarship
David E. BRAHM, Lawrence Berkeley Laboratory, Berkeley, USA.

Amos De-Shalit Scholarship
Hung Jung LU, SLAC, Stanford, USA.

Paul A.M. Dirac Scholarship
Stephen HSU, Lawrence Berkeley Laboratory Berkeley, USA.

Isidor I. Rabi Scholarship
Bernd GROSSMANN, Universität Heidelberg, FRG.

Gunnar Källen Scholarship
Boris KASTENING, University of California, Los Angeles, USA.

André Lagarrigue Scholarship
Tobias HAAS, DESY, Hamburg, FRG.

Ettore Majorana Scholarship

Mark BODNER, University of California, Los Angeles, USA.

Giulio Racah Scholarship

Nicholas BROWN, Rutherford Appleton Laboratory, Chilton, UK.

Jun John Sakurai Scholarship

Ujjwal VIKAS, University of Delhi, India.

Antonio Stanghellini Scholarship

Carlotta PITTORI, Università "La Sapienza", Rome, Italy.

Prize for **Best Scientific Secretary** awarded ex-equo to:

Boris KASTENING, University of California, Los Angeles, USA. and
Stephan TITARD, University of Michigan, Ann Arbor, USA.

The following students received **Honorary Mentions** for their contributions to the activity of the School:

Marco CIUCHINI, INFN, Rome, Italy.
Finn LARSEN, Niels Bohr Institute, Copenhagen, Denmark.
Robert LEIGH, University of Texas, Austin, USA.
Janusz ROSIEK, Warsaw University, Poland.
Stephan TITARD, University of Michigan, Ann Arbor, USA.

The following participants gave their collaboration in the Scientific Secretarial work:

Kassa ADEL
Giovanni AMBROSI
Rachid AYAD
Putukuchi V.K.S. BABA
Mark BODNER
David E. BRAHM
Nicholas BROWN
Marco CIUCHINI
Eda ESKUT
Josè GALVEZ
Bernd GROSSMANN
Thomas HAAS
Despina HATZIFOTIADOU
Hafeez R. HOORANI
Stephen HSU
Boris KESTENING
Robert LEIGH
Jiangfeng LI
Hung Jung LU
Paul D. MALEY
Giampiero MANGANO
Hasan Akhtar RIZVI
Aly Amer SYED
Stephan TITARD
Patribha VIKAS
Ujjwal VIKAS
Maneesh WADHWA
Mark WEXLER
Jianming YOU
Ziping ZHANG

The following students received the **EPS** Scholarships:

Janusz ROSIEK, Warsaw University, Poland.
Martin MOJZIS, Comenius University, Mlynska Dolina, Bratislava, Czechoslovakia.

PARTICIPANTS

Kassa ADEL	Randall Laboratory of Physics University of Michigan ANN ARBOR, MI 48109, USA
Giovanni AMBROSI	Dipartimento di Fisica Università di Perugia Via G. Pascoli 06100 PERUGIA, Italy
Giuseppina ANZIVINO	INFN-Laboratori Nazionali di Frascati C.P. 13 Via Enrico Fermi, 40 00044 FRASCATI, Italy
Rachid AYAD	CERN PPE Division - HED 1211 GENEVA 23, Switzerland
Putukuchi V.K.S. BABA	CERN PPE Division - FBLJA 1211 GENEVA 23, Switzerland
Rinaldo BALDINI-CELIO	INFN-Laboratori Nazionali di Frascati C.P. 13 Via Enrico Fermi, 40 00044 FRASCATI, Italy
Teresa BARILLARI	CERN PPE Division 1211 GENEVA 23, Switzerland

Giuseppe BIMONTE
Dipartimento di Scienze Fisiche
Università di Napoli
Mostra d'Oltremare - Pad. 19
80125 NAPOLI, Italy

Mark BODNER
Department of Physics
University of California
405 Hilgard Avenue
LOS ANGELES, CA 90024, USA

Douglas BORDEN
Department of Physics
University of California
SANTA BARBARA, CA 93106, USA

David BRAHM
Lawrence Berkeley Laboratory
Building 50A
BERKELEY, CA 94720, USA

Stanley J. BRODSKY
SLAC
P.O. Box 4349
STANFORD, CA 94305, USA

Chiara BROFFERIO
Dipartimento di Fisica
Università di Milano
Via Celoria, 16
20133 MILANO, Italy

Nicholas BROWN
Rutherford Appleton Laboratory
CHILTON, Didcot, OX11 0QX, UK

Paolo BRUNI
CERN
PPE Division
1211 GENEVA 23, Switzerland

Stephen BURKE
Rutherford Appleton Laboratory
CHILTON, Didcot, OX11 0QX, UK

Francesco CARAVAGLIOS
Dipartimento di Fisica
Università di Pisa
Piazza Torricelli, 2
56100 PISA, Italy

Maria Chiara CARROZZA
INFN-Sezione di Pisa
Via Livornese, 582/a
56010 S. PIERO A GRADO, Italy

Luisa CIFARELLI	Dipartimento di Scienze Fisiche Università di Napoli Mostra d'Oltremare - Pad. 19 80125 NAPOLI, Italy
Fabio CIRALLI	CERN PPE Division 1211 GENEVA 23, Switzerland
Marco CIUCHINI	INFN- Sezione Sanità Viale Regina Elena, 299 00161 ROMA, Italy
Carlo CIVININI	Dipartimento di Fisica Università di Firenze Largo E. Fermi, 2 50125 FIRENZE, Italy
Yuri DOKSHITZER	Leningrad Inst. of Nuclear Physics Department of Theoretical Physics GATCHINA 188350, USSR
Michael DUFF	Department of Physics College of Science Texas A&M University COLLEGE STATION, TX 77843, USA
James DUNLEA	FERMILAB MS 221 - E706 P.O. Box 500 BATAVIA, IL 60510, USA
Eda ESKUT	CERN PPE Division - HED 1211 GENEVA 23, Switzerland
Sergio FERRARA	Department of Physics University of California 405 Hilgard Avenue LOS ANGELES, CA 90024, USA and CERN TH Division 1211 GENEVA 23, Switzerland

Jose GALVEZ	CERN PPE Division - HED 1211 GENEVA 23, Switzerland
Gerardo GANIS	Lab. de l'Accelerateur Linéaire Centre d'Orsay 91405 ORSAY Cedex, France
Sheldon L. GLASHOW	Department of Physics Harvard University CAMBRIDGE, MA 02138, USA
Benigno GOBBO	Istituto di Fisica Università di Trieste Via Valerio, 2 34127 TRIESTE, Italy
Luca GRIGUOLO	Dipartimento di Fisica Università di Padova Via Marzolo, 8 35100 PADOVA, ITALY
Bernd GROSSMANN	Institute für Theoretische Physik Universität Heidelberg Philosophenweg, 10 6900 HEIDELBERG, FRG
Tobias HAAS	Deutsches Elektronen-Synchrotron (DESY) Notkestrasse 85 2000 HAMBURG 52, FRG
Lawrence HALL	Lawrence Berkeley Laboratory 1 Cyclotron Road BERKELEY, CA 94720, USA
Despina HATZIFOTIADOU	CERN PPE Division 1211 GENEVA 23, Switzerland
Rudolf HEINZ	Institut für Physik Universität Mainz Postfach 3980 6500 MAINZ, FRG

Peter W. HIGGS	Department of Physics University of Edinburgh Kings Building EDINBURGH, EH9 EJZ, UK
Hafeez R. HOORANI	CERN PPE Division - FBLJA 1211 GENEVA 23, Switzerland
Stephen HSU	Lawrence Berkeley Laboratory Theoretical Physics - 50A-3115 1 Cyclotron Road BERKELEY, CA 94720, USA
Boris KASTENING	Department of Physics University of California 405 Hilgard Avenue LOS ANGELES, CA 90024-1547, USA
Ronald KLEISS	CERN TH Division 1211 GENEVA 23, Switzerland
Hanz Alfred KRAUS	Department of Physics E 15 Technical University of Munich 8046 GARCHING, FRG
Jonathan LABS	SLAC Bin 65 - P.O. Box 4349 STANFORD, CA 94309, USA
Jean François LAPORTE	CEN - SACLAY DphPE/DSM 91191 GIF-SUR-YVETTE, France
Finn LARSEN	Niels Bohr Institute Blegdamsvej 17 2100 COPENHAGEN, Denmark
Tsung Dao LEE	Department of Physics Columbia University NEW YORK, NY 10027, USA

Robert LEIGH

Department of Physics
Theory Group
University of Texas
AUSTIN, TX 78712, USA

Jianfeng LI

CERN
PPE Division - FBLJA
1211 GENEVA 23, Switzerland

Qun LI

CERN
PPE Division - FBLJA
1211 GENEVA 23, Switzerland

Hung Jung LU

SLAC
Bin 81 - P.O. Box 4349
STANFORD, CA 94309, USA

Paul D. MALEY

CERN
PPE Division
1211 GENEVA 23, Switzerland

Gerd MANDELBAUM

Institut für Theoretische Physik
Universität Bern
Sidlerstrasse 5
3012 BERN, Switzerland

Giampiero MANGANO

Dipartimento di Scienze Fisiche
Università di Napoli
Mostra d'Oltremare - Pad. 19
80125 NAPOLI, Italy

Samjid MANNAN

Physics Department
The University of Southampton
SOUTHAMPTON, SO9 5NH, UK

Enrico MEGGIOLARO

Scuola Normale Superiore
Piazza dei Cavalieri, 7
56100 PISA, Italy

Dean MILLER

Department of Physics and Astronomy
University of Oklahoma
440 West Brooks, Room 131
NORMAN, OK 73012, USA

Martin MOJZIS	Department of Theoretical Physics Comenius University Mlynska Dolina 84215 BRATISLAVA, Czechoslovakia
Holger MÖNCH	I Physikalisches Institut der Rheinisch-Westfalischen Technischen Hochschule Sommerfeldstrasse 5100 AACHEN, FRG
Marco MONTENO	Istituto di Fisica Università di Torino Via P. Giuria, 1 10125 TORINO, Italy
Enrico PASQUALUCCI	Dipartimento di Fisica II Università di Roma Via O. Raimondo 00175 ROMA, Italy
Peter PELDAN	Institute of Theoretical Physics Chalmers University 41296 GOTEBORG, Sweden
Carlotta PITTORI	Dipartimento di Fisica Università "La Sapienza" Piazza Aldo Moro, 2 00185 ROMA, Italy
Andreas RINGWALD	Deutsches Elektronen-Synchrotron (DESY) Notkestrasse 85 2000 HAMBURG 52, FRG
Hasan Akhar RIZVI	CERN PPE Division - FBLJA 1211 GENEVA 23, Switzerland
Janusz ROSIEK	Physics Department Institute of Theoretical Physics Warsaw University ul. Hoza 69 00681 WARSAW, Poland

Ada SOLANO	Istituto di Fisica Università di Torino Via P. Giuria, 1 10125 TORINO, Italy
Carlos Javier SOLANO	CERN PPE Division 1211, GENEVA 23, Switzerland
Jack STEINBERGER	CERN PPE Division 1211 GENEVA 23, Switzerland
Aly Amer SYED	CERN PPE Division 1211 GENEVA 23, Switzerland
Stephan TITARD	Randall Laboratory of Physics University of Michigan ANN ARBOR, MI 48109-1120, USA
Steen VARSTED	Niels Bohr Institutet Blegdamsvej, 17 2100 KOBENHAVN, Denmark
Patribha VIKAS	CERN PPE Division 1211 GENEVA 23, Switzerland
Ujjwal VIKAS	CERN PPE Division 1211 GENEVA 23, Switzerland
Maneesh WADHWA	CERN PPE Division - FBLJA 1211 GENEVA 23, Switzerland
Victor F. WEISSKOPF	Center for Theoretical Physics Massachusetts Inst. of Technology CAMBRIDGE, MA 02139, USA

Mark WEXLER	Department of Physics Jadwin Hall Princeton University PRINCETON, NJ 08544, USA
Crispin M.C.S. WILLIAMS	CERN PPE Division 1211 GENEVA 23, Switzerland
Qinghao YE	CERN PPE Division - FBLJA 1211 GENEVA 23, Switzerland
Thomas YPSILANTIS	CERN PPE Division 1211 GENEVA 23, Switzerland
Jianming YOU	CERN PPE Division - FBLJA 1211 GENEVA 23, Switzerland
Ziping ZHANG	CERN PPE Division - FBLJA 1211 GENEVA 23, Switzerland

INDEX